Green Alternatives to Antibiotics in Animal Health

Rifat Ullah Khan • Shabana Naz
Editors

Green Alternatives to Antibiotics in Animal Health

Springer

Editors
Rifat Ullah Khan
Physiology Lab
College of Veterinary Sciences
The University of Agriculture
Peshawar, Pakistan

Shabana Naz
Department of Zoology
Government College University
Faisalabad, Pakistan

ISBN 978-981-92-0308-6 ISBN 978-981-92-0309-3 (eBook)
https://doi.org/10.1007/978-981-92-0309-3

This Springer imprint is published by the registered company Springer Nature Singapore Pte Ltd.
The registered company address is: 152 Beach Road, #21-01/04 Gateway East, Singapore 189721, Singapore

Preface

The widespread and often indiscriminate use of antibiotics in animal production has contributed significantly to the emergence of antimicrobial resistance (AMR), posing a serious threat to animal health, public health, and global food security. As concerns over drug residues in animal-derived foods, environmental contamination, and the declining efficacy of conventional antibiotics continue to grow, the need for sustainable and safer alternatives has become more urgent than ever. This book, *Green Alternatives to Antibiotics in Animal Health*, is conceived in response to this global challenge.

In recent years, increasing attention has been directed toward natural, eco-friendly, and biologically based strategies that can maintain animal health and productivity while minimizing reliance on antibiotics. These "green alternatives" include phytogenic feed additives, probiotics, prebiotics, synbiotics, organic acids, enzymes, essential oils, bacteriophages, immunomodulators, and nutraceuticals, as well as improved management and biosecurity practices. Collectively, these approaches offer promising tools to enhance immunity, improve gut health, control pathogens, and promote sustainable livestock and poultry production.

This book aims to provide a comprehensive and up-to-date overview of green alternatives to antibiotics, integrating fundamental concepts with practical applications. Each chapter is designed to bridge the gap between research and practice, highlighting mechanisms of action, efficacy, safety, and limitations of various alternatives across different animal species. Emphasis is also placed on emerging technologies, regulatory perspectives, and future directions in antibiotic-free animal production systems.

This book is intended for researchers, veterinarians, animal nutritionists, students, policymakers, and professionals involved in livestock and poultry production. By bringing together contributions from experts in diverse disciplines, this volume seeks to foster informed decision-making and encourage the adoption of sustainable strategies that safeguard animal welfare, human health, and the environment.

It is our hope that this book will serve as a valuable reference and inspire continued research and innovation toward reducing antibiotic dependence and advancing a greener, more responsible approach to animal health management.

Peshawar, Pakistan Rifat Ullah Khan
Faisalabad, Pakistan Shabana Naz

Contents

Probiotics and Prebiotics

Other Non-antibiotic Approaches

Introduction

The Need for Green Alternatives in Animal Health

Muhammad Umar Ijaz, Rabia Azmat, Tahreem Fatima, Khalid Abbas, Tanzeela Hassan, and Muhammad Faisal Hayat

Abstract There is an urgent need for revolutionizing conventional veterinary practices to overcome the growing risk of antibiotic resistance, ethical and environmental issues caused by conventional practices. Green alternatives, including plant-based therapies, probiotics/prebiotics, and organic feed provide sustainable solutions to improve animal health and alleviate environmental damage. Therefore, this chapter focuses on the challenges of conventional veterinary practices, the role of green alternatives, and the economic and policy aspects influencing their implementation. The implementation of green alternatives establishes a sustainable framework, providing advantages for ensuring food security, animal health, and a sustainable environment. All stakeholders need to cooperate through policymaking towards achieving sustainable veterinary practices by enforcing regulations that advocate green alternatives over antibiotic misuse. For making green alternatives, cost-effective, accessible, and effective, researchers should focus on the development of green alternatives. For the successful implementation of green alternatives, veterinarians, and farmers must adopt sustainable practices and leverage advanced training and education.

Keywords Green alternatives · Animal · Health · Conventional practices · Antibiotic resistance

1 Introduction

Food security is one of the important components of the national security of a country, which helps preserve its statehood and sovereignty, and it is also a key element of demographic strategy. Food security, as a strategic goal, is intended to ensure

M. U. Ijaz (✉) · R. Azmat · T. Fatima · K. Abbas · T. Hassan · M. F. Hayat
Department of Zoology, Wildlife and Fisheries, University of Agriculture, Faisalabad, Pakistan
e-mail: umar.ijaz@uaf.edu.pk

R. U. Khan, S. Naz (eds.), *Green Alternatives to Antibiotics in Animal Health*,
https://doi.org/10.1007/978-981-92-0309-3_1

whether the population has access to high-quality and safe food for human health. Animal-based food products, including meat, milk, eggs, and fish, play a crucial role in human nutrition by providing essential proteins, vitamins, and minerals necessary for growth, development, and overall health. Additionally, livestock farming supports rural economies, generates employment, and contributes significantly to global food systems. However, meeting the increasing demand for animal-based products presents challenges, particularly regarding sustainability, environmental impact, and disease management. Whilst highlighting the need for a holistic approach for the protection of animal health and welfare, conventional livestock production regulations impose considerable restrictions on the use of many animal health inputs that are commonly administered in traditional production systems. The incorporation of antibiotics in animal feeds significantly enhances feed efficiency, promotes animal growth, and improves the quality of animal products (Vaarst et al. 2005). While these practices have significantly contributed to the expansion of the livestock industry, they have also raised serious concerns about antibiotic resistance, environmental contamination, and long-term sustainability.

According to statistics, consumption of antibiotics in animal husbandry is significantly higher than that of human healthcare (Polianciuc et al. 2020). A global analysis estimated that livestock consumed 63,151 tons of antibiotics in 2010, which is anticipated to rise by 67% by 2030, which will nearly double in some BRICS nations (Van Boeckel et al. 2015). Antibiotics play a crucial role in intensive, large-scale farming; however, their uncontrolled use raises concerns about bacterial resistance (Aarestrup et al. 1998), potentially enabling the transmission of resistant bacteria and resistance factors from animals to humans (Stanton 2013). This increased multidrug resistance (MDR) is associated with non-therapeutic antimicrobial use (Marshall and Levy 2011). However, increased antibiotic consumption has led to widespread exposure of bacteria in communities and ecosystems to antibiotic residues. In response to this concern, Sweden banned certain antibiotic growth promoters in animal feed in 1986 (Castanon 2007), followed by an EU-wide ban on all such promoters in 2006 under European Parliament and Council Regulation EC No. 1831/2003. However, the EU-wide ban on in-feed antibiotics had unintended consequences.

for the animal production industry, leading to increased infections and reduced productivity. Additionally, overall antibiotic use in animals increased as the ban led to a higher demand for therapeutic antibiotics and disinfectants due to the rise in infections (Rafiq et al. 2022).

The persisting effects of these significant changes are becoming alarming; therefore, it is imperative to develop innovative scientific approaches, focusing on environmental health and animal welfare because food safety and public health are interrelated matters (Carresi et al. 2024). In this case, green alternatives in animal health are a complementary and sustainable strategy to propose reducing chemical use, decreasing drug resistance, and persistence of residues in the environment.

2 Challenges in Animal Production Practices

Animal production faces multiple biological, environmental, and economic challenges that limit productivity, sustainability, and profitability, affecting livestock health, market stability, and ecological balance. One of the most pressing concerns in modern animal husbandry is antibiotic resistance, which reduces treatment effectiveness and poses serious public health risks (Wegener 2012). Additionally, zoonotic diseases (such as bovine tuberculosis, avian influenza, and foot-and- mouth disease), intensive selective breeding, and inadequate nutrition have weakened immunity and increased the risk of genetic disorders, disrupting supply chains in the competitive livestock, poultry, swine, and aquaculture industries (Seal et al. 2013). Furthermore, economic challenges such as high antibiotic costs, market price fluctuations, and trade barriers have burdened small- scale farmers, who have limited access to technology and markets (Laxminarayan et al. 2015). Adopting more sustainable animal health strategies is essential to improving resilience and ensuring long-term productivity.

The use of antibiotics in conventional veterinary treatments has significantly increased due to their benefits in animal health, welfare, and food safety. These include disease prevention and treatment, protection against zoonoses, improved production efficiency in limited spaces, and economic advantages for farmers and consumers (Carmo et al. 2017). The production and consumption of antibiotics are influenced by several factors, such as biosecurity level, farming methods, veterinarians' and farmers' education, as well as government and industrial regulatory frameworks (Manyi-Loh et al. 2018).

Extensive antibiotic use in animal husbandry contributes to agro-ecosystem contamination through fertilizers, wastewater irrigation, and the spread of contaminated manure on agricultural fields. Veterinary antibiotics used as growth promoters in animal feed supplements are released through the feces and urine of grazing animals on livestock farms, pasture areas, and aquaculture facilities. The use of excreted organic byproducts as alternative fertilizers (Kaczala and Blum 2016), drifting from manuring, intensive animal sewage, and aquaculture system effluents, farmers also contribute to indirect release (Kaczala and Blum 2016). The global practices of land recycling organic manure are increasing, leading to potentially significant ecological effects from these fecal matters (Motoyama et al. 2011). Veterinary antibiotics enter the environment primarily through agricultural crop fields where animal manure from the treated animals accumulates, eventually contaminating groundwater systems and surface waters through natural drainage and runoff (García-Galán et al. 2010).

Antibiotic resistance poses a significant risk to public health as antibiotic-resistant bacteria linked to animals can be readily transferred to humans through food chains and extensive environmental spread from animal waste (Manyi-Loh et al. 2018). Horizontal gene transfer within the bacterial population causes serious health risks for both humans and animals. Antibiotic-resistant genes as well as bacteria reduce the effectiveness of antibacterial drugs against diseases in humans and animals (Wegener 2012). The presence of antibiotic residues in surface water has the ability to disrupt crucial bacterial activities essential for maintaining aquatic balance, agricultural stability, and animal production. Other potential sources of antibiotic release into the environment include wastewater treatment effluents, agricultural runoff, and sewage leaks, as these chemicals don't decompose completely. The persistent public health crisis globally is attributed to drug residues, antibiotic-resistant microorganisms, and antibiotic resistance genes, which are considered environmental contaminants (Manyi-Loh et al. 2018). The major health issues associated with antibiotic-resistant microorganisms affect developing nations, where limited treatment access leads to more deadly outbreaks of infectious diseases (Samie et al. 2009). Soil and water environments serve as fundamental sources and reservoirs for antibiotic resistance, and they are affected by contamination from agricultural practices (Polianciuc et al. 2020). Antibiotic residues (including the parent compounds or their metabolites, or both) may be found in animal-derived products (muscles, kidneys, liver, fat, milk, and egg) that are suitable for human consumption, in addition to having the potential to promote antibiotic resistance in food-producing animals (Getahun et al. 2023). Studies have shown that antibiotic residues adversely affect public health and food safety, including drug toxicity, immunopathological diseases, carcinogenicity, allergic reactions, and drug sensitization.

3 Understanding Green Alternatives

Green alternatives in veterinary practices are a paradigm shift based on the principles of sustainability, safety, and welfare of animals. While addressing the drawbacks of traditional methods, these innovative solutions, such as plant-based treatments, probiotics and prebiotics, organic feed, and biological pest control (Fig. 1), provide transformative methods to manage animal health (Manyi-Loh et al. 2018). Green alternatives are characterized by using natural, environmentally friendly solutions that support animal health, environmental protection, and food security, in contrast to conventional methods, which are based on antibiotics and synthetic chemicals. The rationale behind this swift paradigm shift is to counteract globally increasing antibiotic resistance, mitigate environmental contamination, and meet the rising demand for animal products produced ethically (Van Boeckel et al. 2015).

Plant-based treatments, including plant extracts, phytochemicals, and essential oils, have recently gained popularity due to their therapeutic properties, such as antioxidants, antifungal, antimicrobial, and immunoregulatory. For example, neem, turmeric, and garlic are some of the herbs used to boost immunity, improve digestion, and prevent infection in livestock (Cheng et al. 2014; Vondruskova et al. 2010). The use of these natural remedies decreases the requirement of synthetic drugs and mitigates the risk of drug residues in animal products, providing safer food for consumers.

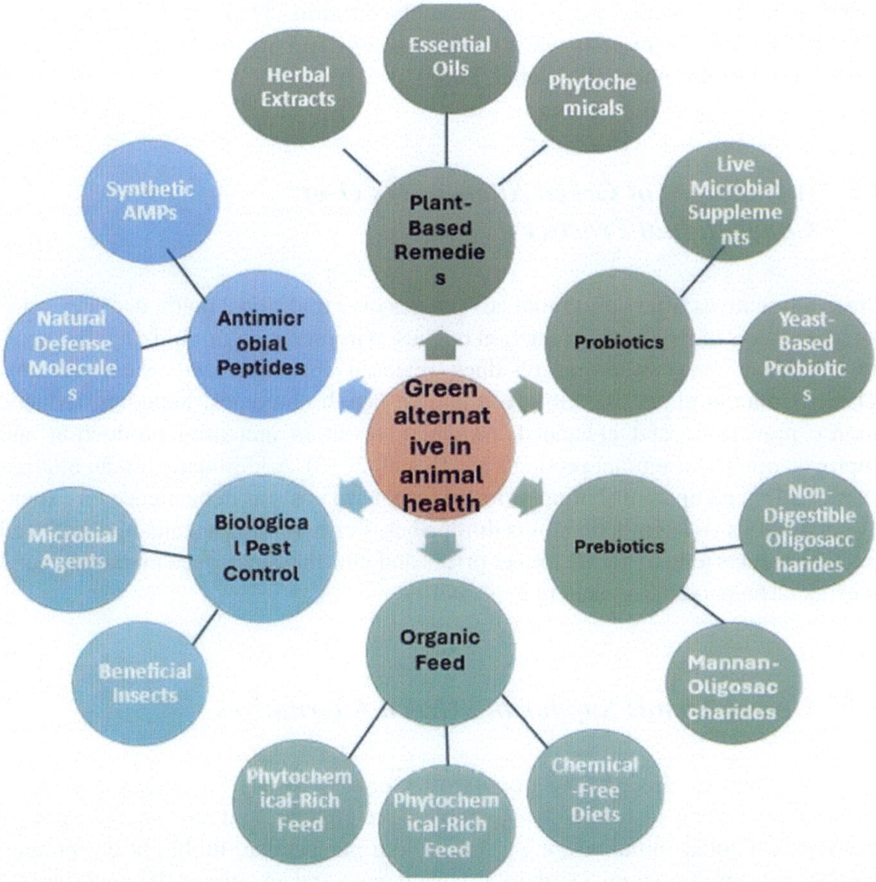

Fig. 1 Green alternatives in animal health: sustainable approaches for disease prevention and nutrition

Probiotics and prebiotics are also essential for preserving gut health and increasing immunity. Lactobacillus and Bifidobacterium are probiotics that help to balance the gut microbiota by inhibiting the noxious pathogens colonization (Gaggìa et al. 2010). Fructo-oligosaccharides and inulin are considered prebiotics that provide food for gut bacteria, promoting their proliferation and activity. Probiotics and prebiotics synergistically enhance the absorption of nutrients, reduce the prevalence of diseases, and increase animal productivity (Millet and Maertens 2011).

Organic Feed, a key part of green alternatives, is equally important as it ensures that animals obtain a nutritious, chemical-free diet. Organic feed enhances faster development and reduces the possibility of unsafe residues in animal-based products by substituting chemical additives with natural components (Sundrum 2001).

Biological pest management including the use of beneficial insects or microbiological agents, provides sustainable alternative control methods to the use of chemical pesticides that are environmentally harmful. These techniques protect crops and animals from pests while conserving biodiversity and reducing contamination of the environment (Bale et al. 2008).

3.1 Advantages of Green Alternatives Over Conventional Practices

Green alternatives offer substantial advantages over traditional methods. They mitigate the effects of livestock production on the environment by minimizing the harmful chemicals and antibiotic residues released into soil and water systems (García-Galán et al. 2010). Improved animal health outcomes, including reduced disease prevalence and enhanced immunity, result in increased production and improved quality of animal products (Seal et al. 2013). Additionally, green alternatives usually end up contributing to lower long-term costs as dependence on expensive antibiotics and synthetic inputs diminishes. Furthermore, sustainably produced animal products tend to garner higher prices and thus their increased market demand benefits farmers (van Wagenberg et al. 2017).

3.2 Global Trends Supporting Green Alternatives

Green alternatives are gaining popularity globally, owing to increased concern about antimicrobial resistance, sustainable development, and sustainable farming techniques. Global initiatives, e.g., One Health perspective, highlight the human, animal, and environmental health interdependences and encourage the implementation of sustainable veterinary practices (Robinson et al. 2016). Research and innovation in this area are growing, and the development of biotechnology, nanotechnology, and natural product research have led to the emergence of more efficient and accessible green alternatives (Plotniece et al. 2023). However, governments, researchers, and industry stakeholders are working together to put forth supportive policies, funding opportunities, and educational programs to promote the implementation of green alternatives (Manyi Loh et al., 2018).

4 Applications of Green Alternatives

The promotion of "clean, green, and ethical" (CGE) approach in animal farming has restricted the use of drugs, chemicals, growth promoters, and hormones without compromising food production or animal welfare. The excessive consumption of antibiotics and chemicals in livestock production has raised concerns related to antibiotic resistance, ecological damage, and the ethical welfare of animals (Van et al. 2020). There has been an increasing demand for the development of strategic antibiotic-free management plans and alternative feed additives to promote animal growth, gut health, and livestock productivity to combat the global rise in antibiotic resistance. To address these challenges, green alternatives have been proposed, focusing on natural disease prevention, sustainable feed solutions, eco-friendly pest control, ethical farming practices, and waste management. Effective alternatives include natural disease prevention through probiotics, prebiotics, phytogenic, antimicrobial peptides, minerals, and phytogenic feed additives like oils, herbal extracts, and organic acids, which enhance gut health and immunity. Sustainable feed alternatives have shown potential in improving animal performance and disease resistance while reducing environmental impact.

4.1 Plant-Based Therapeutics

Humans have been using plant-based therapies as a part of traditional and ethnoveterinary practices from antiquity dated back to 4000 BC in Egypt, China, and the medieval periods in Europe. Phytotherapy is a viable option due to its effectiveness, safety profile, and non-toxic effects, reducing reliance on laborious and expensive synthetic medicine (Heinrich et al. 2006). Following the ban on antibiotic usage in animal husbandry, driven by concerns over the side effects of conventional medicine, drug and microbial resistance, high production costs, and toxic residues in food, phytotherapy offers a sustainable and natural solution. Herbs are used as feed additives, zootechnical additives, growth promoters, immunostimulants, and increase fecundity and reproductive viability. About 70% of ungulates are treated with herbal remedies in the absence of veterinary services in tribal areas (Rehman et al. 2022).

Various herbs, including *Ocimum basilicum* (basil), *Citrus medica* (citron), *Allium sativum* (garlic), and *Duranta erecta* (golden dewdrop), show diverse pharmacological effects, including antimicrobial, antifungal, antiviral, anthelmintic, and acaricidal, helping in the prevention of severe infections. *Asparagus racemosus* (Shatavari) improves fertility in livestock. *Foeniculum vulgare* (Fennel) and *Trifolium pratense* (red clover) silage-based diets improve lactation performance and nitrogen (N) efficiency (Broderick et al. 2001), particularly in improving protein utilization and reducing nitrogen waste (Lee et al. 2009). Some androgenic and adaptogenic herbs, including *Tribulus terrestris*, *Withania somnifera* (Ashwagandha), *Mucuna pruriens* (Velvet Bean), interfere with the hormones directed by hypothalamic–pituitary–gonadal (HPG) axis, including LH, FSH increase, and release of testosterone, and progesterone (Sirotkin and Kolesarova 2021). These hormones integrate male reproductive function, including sexual differentiation, pubertal maturation, and spermatogenesis, ultimately boosting male fertility (Sirotkin and Kolesarova 2021).

4.1.1 Role of Medicinal Plants in Animal Health

Medicinal plants and herbal remedies have been a part of veterinary medicine for more than 3000 years, as a potential substitute for synthetic medication. Phytobiotics and phytochemicals from herbs and plants are used in animal husbandry as feed additives, growth promoters, appetite stimulants, and due to their anti-hypertensive, aphrodisiac, antiviral, anthelmintic, diuretic, digestion-promoting, and health-protective properties. (Cheng et al. 2014; Vondruskova et al. 2010). These phytogenic elements are leveraged due to their biologically active constituents and chemically and structurally diverse secondary metabolites, including terpenoids, phenolics, alkaloids, flavonoids, saponins, tannins, and glycosides (Cheng et al. 2014). Therefore, plants have conventionally been used in ethnoveterinary as complementary or alternative antibiotics to mediate nutritional metabolism and modulate immune response. Plant constituents have been a key source for 75% of approved drugs between 1981 and 2019, either directly from natural products, derivatives, or compounds mimicking natural products (Plotniece et al. 2023).

4.1.2 Common Herbs and Their Applications

Some herbs have been traditionally used in both traditional and modern medicine for animal care, serving as immunomodulators, gut health enhancers, wound healers, and antibacterial agents. The most commonly used herbs in Ayurvedic, Chinese, and Arabic ethnoveterinary medicine include neem, turmeric, garlic, aloe vera, ginger, cloves, and slippery elm.

- *Neem (Azadirachta indica)*

  Neem is a tropical tree native to South Asia, renowned for its durable wood. The non-woody parts of neem (*Azadirachta indica*) exhibit promising phytogenic properties, including antibacterial, antimicrobial, antifungal, insecticidal, and bacteriostatic effects. Neem leaves have been commonly used to enhance protein digestibility and metabolism in ruminants. Its leaves exhibit promising bioactive compounds like tannins, flavonoids, and polyphenols. The tannins in neem leaf supplementation in ruminant diet prevent rumen protein degradation, thereby enhancing amino acid absorption in the small intestine of the animal (Wanapat et al. 2009).

  The tannins have also been characterized as regulators of protein metabolism, enhancing nitrogen retention and promoting efficient nutrient utilization in livestock. Rumen fermentation has been a key focus for animal nutritionists aiming to enhance animal growth and performance (Taethaisong et al. 2023). Bioactive compounds such as tannins and saponin derivatives of neem are proposed to reduce methane emissions, enhance nitrogen utilization, and improve microbial protein synthesis, thereby promoting growth performance (Lo et al. 1985). This enhances fiber digestibility and nutrient absorption, making it a promising phytogenic feed additive for sustainable livestock production. Other parts of neem, such as its seeds and oil extracts, possess antiseptic, antifungal, and antibacterial properties and have also demonstrated anti-hyperglycemic effects. Neem flowers also exhibit potent lethal effects against the cattle filarial parasite *Setaria cervi* (Taethaisong et al. 2023). Traditionally, neem has been used to control livestock insects such as maggots, horn flies, and biting flies. Due to its diverse bioactive properties, neem is a valuable feed additive, offering sustainable solutions for livestock health, nutrition, and disease management.

- *Turmeric (Curcuma longa):*

  Turmeric is a member of the Zingiberaceae (ginger) family, commonly used as a taste-enhancing spice in South Asian cuisine. It has been extensively used in traditional medicinal systems, such as Ayurveda, due to its diverse and potent pharmacological properties, including anti- inflammatory, antioxidant, nematocidal, antiprotozoal, and antivenom effects in poultry, ruminants, and humans. The dietary inclusion of dried turmeric promotes pig growth performance (Stevenson et al. 2023). Curcumin is the primary curcuminoid in turmeric, known for its diverse biological properties. In broiler chickens, curcumin administration has been shown to decrease abdominal fat deposition and regulate lipid metabolism (Xie et al. 2019). Turmeric essential oil contains bioactive compounds, notably ar-turmerone, which exhibits significant insecticidal properties. Research indicates that ar-turmerone exhibits larvicidal activity against mosquito species such as *Aedes aegypti* and *Anopheles quadrimaculatus*, which are vectors for diseases like dengue, malaria, and chikungunya in humans. Additionally, ar-turmerone has been identified as an effective biting deterrent, contributing to its potential as a vector control agent. Dietary supplementation of turmeric has demonstrated antimicrobial properties that prevent harmful bacterial colonization in the intestine, thereby modulating the gut microbiota balance (Odo et al. 2023) (Table 1).

4.2 Role of Probiotics and Prebiotics in Animal Health

The World Health Organization defines probiotics as “live microorganisms that, when administered in adequate amounts, confer health advantages to the host.” The ability of probiotics to counter pathogenic microorganisms through bacteriocin and organic acid production contributes to beneficial gastrointestinal microbial conditions through adherence to intestinal mucosa (Fig. 2), thereby obstructing pathogen attachment and competing for nutrients, stimulating intestinal immune responses, and augmenting the absorption as well as digestion of nutrients (Silva et al. 2020). The most frequently used probiotics include *Bacillus* alongside *Lactobacillus, Lactococcus, Streptococcus, Enterococcus, Pediococcus, Bifidobacterium, Bacteroides,* and *Pseudomonas* bacteria, along with yeast and fungus species, including *Aspergillus* and *Trichoderma* (Reid 2015).

The concept of prebiotics is a relatively recent one that includes nondigestible dietary elements (such as nondigestible oligosaccharides) that are primarily fermented by bacteria that promote gut health (Swennen et al. 2006). Many studies

Table 1 Role of medicinal plants as green alternatives in animal health

Categor y	Herb/Plan t name	Method of applicatio n	Active constituents	Therapeuti c effects	Target disease	References
Rumina nts	Neem (*Azadirach ta indica*)	Leaves in feed, oil as topical	Azadirachtin, Nimbin, Nimbidin	Antimicrobial, antiparasitic, wound healing	Mastitis, skin infections, internal parasites	Gail et al. (2015), Taethaisong et al. (2023)
	Turmeric (*Curcuma longa*)	Powder in feed, paste for wounds	Curcumin, Turmerones	Anti-inflammatory, antioxidant, hepatoprotective	Mastitis, wound healing, Digestive issues	Singh et al. (2024), So-In and Sunthamala (2021)
	Garlic (*Allium sativum*)	Fresh/powdered in feed, juice extract	Allicin, sulfur compounds	Antibacterial, antifungal, immune booster	Respirator y infections, digestive disorders	Bhatwalkar et al. (2021), Kale et al. (2021)
	Aloe Vera (*aloe barbadensi s*)	Gel for topical application, juice in feed	Aloin, polysaccharides	Wound healing, anti-inflammatory, digestive aid	Wound healing, skin infections	Buttar et al. (2022), Ferrer et al. (2022)
	Fenugreek (*Trigonella foenum-graecum*)	Seeds in feed	Saponins, flavonoids	Milk production enhancer, antifungal	Mastitis, lactation issues	Posan et al. (2023)
	Holy basil (*Ocimum sanctum*)	Leaves in feed, decoction	Eugenol, ursolic acid	Antioxidant, immune modulator	Respirator y infections, stress	Dhama et al. (2023)
	Ginger (*Zingiber officinale*)	Rhizome extract in feed	Gingerols, shogaols	Anti-inflammatory, digestive aid	Indigestion, bloating	Abidi et al. (2022), and Singh et al. (2022)
	Ashwagan dha (*Withania somnifera*)	Powder in feed	Withanolides, alkaloids	Adaptogenic, stress reliever, fertility booster	Stress, reproductive issues	Rodriguez-Sánchez et al. (2024)
	Moringa (*Moringa oleifera*)	Leaves in feed, seed powder	Quercetin, Chlorogenic acid	Nutritional enhancer, immune booster	Malnutrition, general immunity	Navarro et al. (2022)
	Clove (*Syzygium aromaticu m*)	Powder in feed, oil for topical use	Eugenol, flavonoids	Antimicrobial, antifungal, analgesic	Respiratory infections, parasitic infestations	Saeed et al. (2021, 2022)

Poultry	Black cumin (*Nigella sativa*)	Seed powder in feed	Thymoquinone, alkaloids	Antioxidant, immune modulator	Viral infections, gut health	Elbaz et al. (2025), Navidshad et al. (2024)
	Mint (*Mentha* spp.)	Leaf powder in water or feed	Menthol, flavonoids	Appetizer, respiratory health	Heat stress, respiratory infections	Abbass and Abid (2023)
	Cinnamon (*Cinnamomum verum*)	Bark powder in feed	Cinnamaldehyde, eugenol	Antimicrobial, antifungal, gut health	Digestive disorders, bacterial infections	Ali et al. (2021b), Qaid et al. (2021)
	Licorice (*Glycyrrhiza glabra*)	Root powder in feed	Glycyrrhizin, flavonoids	Anti-inflammatory, stress reducer	Heat stress, respiratory infections	Abo-Samaha et al. (2022), Haleem et al. (2024) and Reda et al. (2021)
	Lemongrass (*Cymbopo gon citratus*)	Essential oil in drinking water	Citral, geraniol	Antimicrobial, appetite stimulant	Bacterial infections, heat stress	El-Sahn et al. (2024), Hasan-Al-Sharif et al. (2023) and Jegede et al. (2021)
	Bitter gourd (*Momordica charantia*)	Juice in drinking water	Charantin, Alkaloids	Anthelmintic, hypoglycemic	Intestinal parasites	Jamaluddin et al. (2023)
	Thyme (*Thymus vulgaris*)	Essential oil in feed	Thymol, Carvacrol	Antimicrobial, respiratory health	Bacterial infections, respiratory distress	Puvača et al. (2022), Taher et al. (2021)
	Papaya (*Carica papaya*)	Leaves and seeds in feed	Papain, alkaloids	Anthelmintic, digestive aid	Worm infestations, gut health	Zirintunda et al. (2024)
	Coriander (*Coriandrum sativum*)	Seeds in feed	Linalool, Coumarins	Antioxidant, appetite enhancer	Indigestion, bloating	Anwar et al. (2023)

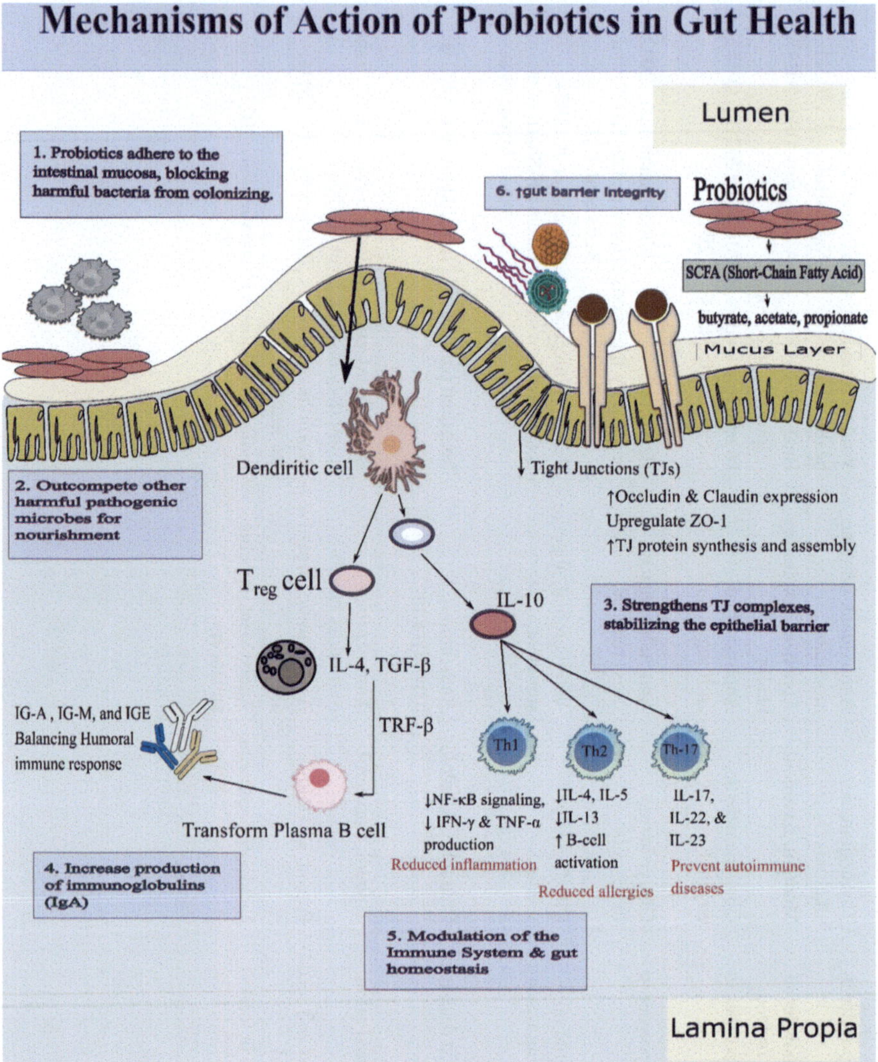

Fig. 2 Biological mechanisms of probiotics in the gut

have demonstrated that multiplication of endogenous lactic acid bacteria and Bifidobacteria in the gut improves the health of hosts. *Bifidobacteria* and *lactobacilli* may proliferate in the gut due to the activity of prebiotics, improving the host microbial stability. Prebiotics, in contrast to probiotics, promote the growth of gut bacteria that have adapted to the gastrointestinal tract's environment. The ability of prebiotics to increase the number of lactic acid bacteria (LAB) in the gut may contribute to the competitive elimination of pathogens from the bird gastrointestinal tract (Wang et al. 2020). Elevated intestinal acidity induced by prebiotics may contribute to the mitigation of infections in the gastrointestinal tract of chickens. Additionally, it has been demonstrated that chickens have an improved immunological response after prebiotics administration that facilitates faster clearance of infections (Shehata et al. 2022). For instance, prebiotics could indirectly (by means of preferred colonization of beneficial bacteria and microbial metabolites) or directly interact with immune cells of the gut (Peredo-Lovillo et al. 2020).

5 Economic and Policy Perspectives on Green Alternatives

The economic viability and sustainability of livestock production have been a persistent challenge, highlighting the need for collaborative efforts across different industries. Conventional practices have been based on using technologies that increase productivity by developing high-yielding breeds, enhanced food intake, coupled with improvements in veterinary health management. Conversely, organic techniques use cultural, biological, and mechanical methods to grow foods safe for the environment, without chemical residue, as well as with high animal welfare standards. Economic efficiency is assessed by analyzing parameters like farm revenue, expenses incurred, farm gate price premium, market risks, and employability (Fig. 3) (van Wagenberg et al. 2017).

Green dairy production offers significant economic benefits over conventional approaches, evidenced by an 84% higher farm gate price and 30% reduced variable

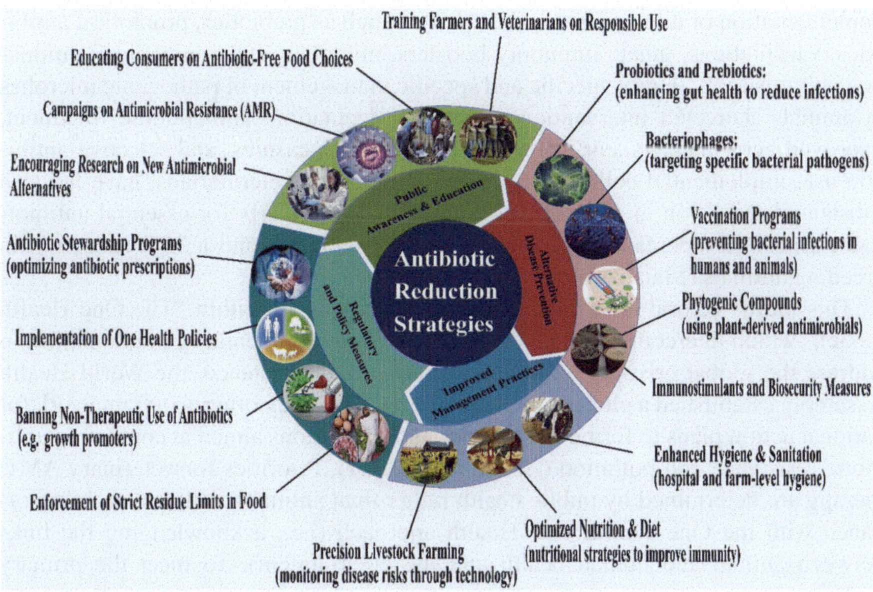

Fig. 3 Antibiotic reduction strategies in livestock production

expenses, resulting in increased farm income per cow. However, organic approaches encounter considerable cost and yield challenges. Both systems demonstrate dual but opposing effects; conventional systems necessitate less labor per unit, present diminished income risk per animal, achieve higher production and reproduction rates, exhibit higher feed efficiency, use less space, enhance animal health, and reduce levels of microbial contamination (van Wagenberg et al. 2017). On the other hand, organic systems showed higher amounts of beneficial fatty acids in cow milk, a lesser impact on biodiversity per unit product, a higher income per animal or full-time employee, and an equal or reduced chance of antibiotic resistance in bacteria (Gamage et al. 2023).

A practical evaluation of both environmental responsibility and economic viability is necessary when deciding between conventional and organic technologies (Gamage et al. 2023). Conventional systems provide stable and cost-effective production, whereas organic approaches prioritize long-term environmental integrity and consumer well-being. A hybrid strategy, leveraging the advantages of both, can establish a more sustainable and ecologically responsible animal industry. Infection prevention and control can be achieved by reducing or limiting the therapeutic use of antibiotics in animals by improving animal nutrition, living conditions, waste management, biosecurity measures, and animals' natural immunity. This can result in a reduction or limitation of the therapeutic application of antibiotics in animals. However, once the animal is exposed to the infection, the body's immune system can effectively combat the pathogen, leading to less severe symptoms, these tactics will lower the amount and kind of antibiotics required for therapy (Moryani et al. 2021).

Furthermore, decreases in antibiotic usage can be accomplished through the implementation of non-antibiotic alternatives, such as prebiotics, probiotics, antibiotics, vaccinations, innate immunity boosters, microbes, and competitive elimination cultures for both non-specific and specific management of pathogenic microbes in animals. Directed interventions, such as vaccination, anthelmintic treatment, improved herd management, enhanced biosecurity measures, and selective antibiotic use, implemented collaboratively by farmers and veterinarians, have led to a substantial reduction in antibiotic consumption, particularly for essential antibiotics, with a 52% decrease in pigs from birth to slaughter and a 32% reduction in breeding animals (Manyi-Loh et al. 2018).

This multi-sectoral issue is frequently contextualized within "The One Health model," which interconnects human, animal, and environmental health domains. To address the global crisis of antibiotic pollution and resistance, the World Health Assembly established a global action plan in 2015, thereby offering a framework for national action plans to formulate policies and regulations aimed at combating antibiotic resistance and pollution (Aslam et al. 2021). Priorities for veterinary AMD therapy are determined by public health rather than animal health issues, in accordance with the One World, One Health approach (i.e., acknowledging the links between animal and human health and the environment). To meet the primary requirement of the One Health policy, which mandates that drugs must not adversely affect human health and the environment, all regulatory authorities (EMA, FDA/EPA) must evaluate the potential of green antimicrobials as substitutes for currently available medications (Toutain et al. 2016). Regulatory authorities must evaluate the effects of all AMDs, both new and existing, on the environmental resistome prior to issuing marketing permission to enhance the competitive edge of green AMDs over traditional ones. The problem of AMR spread through environmental channels is currently not considered in guidelines for either pioneer or generic products (Kaczala and Blum 2016).

6 Future Prospects of Green Alternatives

The future of sustainable and environmentally friendly veterinary practices is green alternatives. Green alternatives in veterinary practices are being revolutionized by bioengineering and nanotechnology. These technologies, shaped by innovation, provide novel solutions for the improvement of natural therapies, promote animal health, and mitigate environmental impact (Ali et al. 2021a). Innovations in genetic engineering and microbial biotechnology have contributed to the development of plant-based medicines, antimicrobial peptides, and next-generation probiotics. Genetically engineered probiotics with improved antibiotic properties can specifically target pathogens while preserving gut microbiota (Reddy et al. 2023). Additionally, CRISPR-based technology can facilitate the development of disease-resistant animal breeds, thereby lessening the dependence on drugs. Potential applications of nanomaterials have been investigated in enhancing phytochemical bioavailability, targeted drug delivery, and increasing vaccine efficiency (Wani et al. 2023). For instance, essential oils or plant extracts are nano-encapsulated to protect their bioactive compounds from degradation and increase delivery efficacy by sustained release, enhancing therapeutic efficacy. Whereas nano-engineered probiotics or prebiotics can be formulated to preferentially colonize the gut and outcompete harmful bacteria to improve the gut health and immunity in animals (Bayer 2023).

To successfully implement green alternatives, adequate involvement by governments, researchers, industry players, and farmers is required. These collaborations can stimulate innovation, foster policy development, and promote the implementation of sustainable practices. The training of farmers and veterinarians in the application of green alternatives, including probiotics, prebiotics, and plant-based therapies, should be the focus of educational initiatives. Workshops, online courses, and field experiments can effectively address the knowledge gaps while improving confidence in these approaches. Awareness-raising campaigns aimed at the public can emphasize the advantages of using green alternatives, for example, in terms of animal health, environmental impact, and food safety.

7 Conclusion

Taken together, green alternatives are signified as an important step in ethical and sustainable animal production. Green alternatives emphasize natural solutions by addressing the limitations of conventional methods to develop a healthier world and ensure a safe food supply. The advancements of green alternatives in veterinary practice are rendering them indispensable for contemporary veterinary practices, while establishing a foundation for a sustainable future in agriculture and animal health.

References

Aarestrup F.M, Bager F, Jensen NE, Madsen M, Meyling A and Wegener HC (1998) Resistance to antimicrobial agents used for animal therapy in pathogenic-, zoonotic-and indicator bacteria isolated from different food animals in Denmark: a baseline study for the Danish Integrated Antimicrobial Resistance Monitoring Programme (DANMAP). Apmis 106:745-770. https://doi.org/10.1111/j.1699-0463.1998.tb00222.x

Abbass G, Abid AR (2023) Effect of peppermint, fenugreek and their mixture on production traits of heat stressed broiler chickens. SFS 10:2704–2716. https://doi.org/10.17762/sfs.v10i3S.980

Abidi C, Rtibi K, Boutahiri S, Tounsi H, Abdellaoui A, Wahabi S, Gressier B, Eto B, Sebai H (2022) Dose-dependent action of Zingiber officinale on colonic Dysmotility and ex vivo spontaneous intestinal contraction modulation. Dose-Response 20:15593258221127556. https://doi.org/10.1177/15593258221127556

Abo-Samaha MI, Alghamdi YS, El-Shobokshy SA, Albogami S, El-Maksoud EMA, Farrag F, Soliman MM, Shukry M, El-Hack MEA (2022) Licorice extract supplementation affects antioxidant activity, growth-related genes, lipid metabolism, and immune markers in broiler chickens. Life 12:914. https://doi.org/10.3390/life12060914

Ali A, Ijaz M, Khan YR, Sajid HA, Hussain K, Rabbani AH, Shahid M, Naseer O, Ghaffar A, Naeem MA (2021a) Role of nanotechnology in animal production and veterinary medicine. Trop Anim Health Prod 53:1–14. https://doi.org/10.1007/s11250-021-02951-5

Ali A, Ponnampalam EN, Pushpakumara G, Cottrell JJ, Suleria HAR, Dunshea FR (2021b) Cinnamon: a natural feed additive for poultry health and production—a review. Animals 11:2026. https://doi.org/10.3390/ani11072026

Anwar H, Hussain G, Muzaffar H, Hashmi BN, Mukhtar A, Malik L (2023) Coriander and physiological systems of the body. In: Handbook of coriander (Coriandrum sativum), pp 191–206. 9781003204626

Aslam B, Khurshid M, Arshad MI, Muzammil S, Rasool M, Yasmeen N, Shah T, Chaudhry TH, Rasool MH, Shahid A (2021) Antibiotic resistance: one health one world outlook. Front Cell Infect Microbiol 11:771510. https://doi.org/10.3389/fcimb.2021.771510

Bale JS, Van Lenteren JC, Bigler F (2008) Biological control and sustainable food production. Philos Trans R Soc B Biol Sci 363:761–776. https://doi.org/10.1098/rstb.2007.2182

Bayer IS (2023) Controlled drug release from nanoengineered polysaccharides. Int J Pharm 15:1364. https://doi.org/10.3390/pharmaceutics15051364

Bhatwalkar SB, Mondal R, Krishna SBN, Adam JK, Govender P, Anupam R (2021) Antibacterial properties of organosulfur compounds of garlic (*Allium sativum*). Front Microbiol 12:613077. https://doi.org/10.3389/fmicb.2021.613077

Broderick GA, Walgenbach RP, Maignan S (2001) Production of lactating dairy cows fed alfalfa or red clover silage at equal dry matter or crude protein contents in the diet. JDS 84:1728–1737. https://doi.org/10.3168/jds.S0022-0302(01)74608-5

Buttar HS, Kumar H, Tuli HS, Dhama K (2022) Potential health benefits of using aloe vera as a feed additive in Livestock: a mini-review. Indian Vet J 99:9–18

Carmo LP, Nielsen LR, Alban L, Müntener CR, Schüpbach-Regula G, Magouras I (2017) Comparison of antimicrobial consumption patterns in the Swiss and Danish cattle and swine production (2007–2013). Front Vet Sci 4:26. https://doi.org/10.3389/fvets.2017.00026

Carresi C, Pauletto M, Fiore E, Musolino V, Britti D (2024) green veterinary pharmacology and toxicology: a "One Health" approach milestone. Front Vet Sci 11:1476877

Castanon JIR (2007) History of the use of antibiotic as growth promoters in European poultry feeds. Poult Sci 86:2466–2471. https://doi.org/10.3382/ps.2007-00249

Cheng G, Hao H, Xie S, Wang X, Dai M, Huang L, Yuan Z (2014) Antibiotic alternatives: the substitution of antibiotics in animal husbandry? Front Microbiol 5:217. https://doi.org/10.3389/fmicb.2014.00217

Dhama K, Sharun K, Gugjoo MB, Tiwari R, Alagawany M, Iqbal Yatoo M, Thakur P, Iqbal H. MN, Chaicumpa W, Michalak I (2023) A comprehensive review on chemical profile and pharmacological activities of *Ocimum basilicum*. Food Rev Int 39:119–147. https://doi.org/10.1080/87559129.2021.1900230

Elbaz AM, Ashmawy ES, Farahat MAA, Abdel Maksoud A, Amin SA, Mohamed ZS (2025) Dietary Nigella sativa nanoparticles enhance broiler growth performance, antioxidant capacity, immunity, gene expression modulation, and cecal microbiota during high ambient temperatures. Sci Rep 15:1–13. https://doi.org/10.1038/s41598-024-82725-9

El-Sahn AA, Manaa EA, Amal M, Khalifah AM, Fayez S, Abdel-Daim ASA, Albadrani GM, Abdel-Daim MM, Abdel-Latif MA (2024) Lemongrass extract enhances productive performance, blood biomarkers, immunity, and gut health of broilers. Heliyon 10(18). https://doi.org/10.1016/j.heliyon.2024.e37783

Ferrer YZ, Flores LIP, Duarte RF, del Valle Laveaga D, Díaz MG, del Río MH, González MA, Bauta RB (2022) Aloe vera and Bixa orellana in a revitalizing ointment for cattle skin injuries. GSC GSCBPS 20:1–11. https://doi.org/10.30574/gscbps.2022.20.1.0256

Gaggìa F, Mattarelli P, Biavati B (2010) Probiotics and prebiotics in animal feeding for safe food production. Int J Food Microbiol 141:S15–S28. https://doi.org/10.1016/j.ijfoodmicro.2010.02.031

Gail H, Tarryn B, Oluwaseyi A, Denver D, Oluchi M, Diana G (2015) An ethnobotanical survey of medicinal plants used by traditional health practitioners to manage HIV and its related opportunistic infections in Mpoza, Eastern Cape Province, South Africa. J Ethnopharmacol 171:109–115. https://doi.org/10.1016/j.jep.2015.05.029

Gamage A, Gangahagedara R, Gamage J, Jayasinghe N, Kodikara N, Suraweera P, Merah O (2023) Role of organic farming for achieving sustainability in agriculture. Farming Syst 1:100005. https://doi.org/10.1016/j.farsys.2023.100005

García-Galán MJ, Garrido T, Fraile J, Ginebreda A, Díaz-Cruz MS, Barceló D (2010) Simultaneous occurrence of nitrates and sulfonamide antibiotics in two ground water bodies of Catalonia (Spain). J Hydrol 383:93–101. https://doi.org/10.1016/j.jhydrol.2009.06.042

Getahun M, Abebe RB, Sendekie AK, Woldeyohanis AE, Kasahun AE (2023) Evaluation of antibiotics residues in milk and meat using different analytical methods. Int J Anal Chem 2023:4380261. https://doi.org/10.1155/2023/4380261

Haleem MIA, Gaballa MMS, El-Far AH, Taie HAA, Elshopakey GE (2024) Mitigating impact of Glycyrrhiza glabra on virulent Newcastle disease virus challenge in chickens: clinical studies, histopathological alterations and molecular docking. J Ethnopharmacol 48:3823–3845. https://doi.org/10.1007/s11259-024-10530-w

Hasan-Al-Sharif M, Hossain MM, Asha AS, Al-Mamun M (2023) Lemongrass (Cymbopogon Citratus) essential oil improves gut health and production performance in broiler: a review. Int J Curr Sci 13:788–801

Heinrich M, Kufer J, Leonti M, Pardo-de-Santayana M (2006) Ethnobotany and ethnopharmacology—interdisciplinary links with the historical sciences. J Ethnopharmacol 107:157–160. https://doi.org/10.1016/j.jep.2006.05.035

Jamaluddin AW, Taufan AM, Ris A, Apada MS (2023) Activity test of bitter melon leaf infusion (Momordica charantia) as an anthelmintic toward Ascaridia galli worms in vitro. JRVI 6:30

Jegede OC, Olayemi DO, Meleni DO, Jegede TO (2021) Anthelmintic activities of Momordica charantia Linn (bitter melon): a review. J Sustain Vet Allied Sci 1(2). https://doi.org/10.54328/covm.josvas.2021.018

Kaczala F, Blum ES (2016) The occurrence of veterinary pharmaceuticals in the environment: a review. Curr Anal Chem 12:169–182. https://doi.org/10.2174/1573411012666151009193108

Kale RB, Gadge SS, Jayaswall K, Patole AO, Mahajan V, Singh M (2021) Ethno-veterinary medicinal uses of garlic (Allium sativum) by livestock rearer. IJTK 20:426–435. https://doi.org/10.56042/ijtk.v20i2.30417

Laxminarayan R, Van Boeckel T, Teillant A (2015) The economic costs of withdrawing antimicrobial growth promoters from the livestock sector. https://doi.org/10.1787/5js64kst5wvl-en

Lee MRF, Theobald VJ, Tweed JKS, Winters AL, Scollan ND (2009) Effect of feeding fresh or conditioned red clover on milk fatty acids and nitrogen utilization in lactating dairy cows. J Dairy Sci 92:1136–1147. https://doi.org/10.3168/jds.2008-1692

Lo KV, March AC, Liao PH (1985) Determination of optimal feed strength for the thermophylic production of methane from dairy manure filtrate. Energy Agric 4:217–226. https://doi.org/10.1016/0167-5826(85)90018-X

Manyi-Loh C, Mamphweli S, Meyer E, Okoh A (2018) Antibiotic use in agriculture and its consequential resistance in environmental sources: potential public health implications. Molecules 23:795. https://doi.org/10.3390/molecules23040795

Marshall BM, Levy SB (2011) Food animals and antimicrobials: impacts on human health. Clin Microbiol Rev 24:718–733. https://doi.org/10.1128/cmr.00002-11

Millet S, Maertens L (2011) The European ban on antibiotic growth promoters in animal feed: from challenges to opportunities. Vet J (London, England: 1997) 187:143–144. https://doi.org/10.1016/j.tvjl.2010.05.001

Moryani AA, Rajput N, Naeem M, Shah AH, Jahejo AR (2021) Screening of the herbs and evaluation of their combined effects on the health and immunity of coccidiosis challenged broiler chickens. Pak Vet J 41:2074–7764

Motoyama M, Nakagawa S, Tanoue R, Sato Y, Nomiyama K, Shinohara R (2011) Residues of pharmaceutical products in recycled organic manure produced from sewage sludge and solid waste from livestock and relationship to their fermentation level. Chemosphere 84:432–438. https://doi.org/10.1016/j.chemosphere.2011.03.048

Navarro CA, Góngora A, Flórez H (2022) Moringa oleifera: a forage resource for efficient livestock farming in tropical regions. Archivos de Zootecnia 71:204–214. ISSN 0004-0592

Navidshad B, Dadashi Orang A, Royan M (2024) The potential of black cumin (Nigella sativa) as a natural feed additive for poultry: an update. J World's Poult Sci:1–15. https://doi.org/10.1080/00439339.2024.2438352

Odo EO, Ikwuegbu JA, Obeagu EI, Chibueze SA, Ochiaka RE (2023) Analysis of the antibacterial effects of turmeric on particular bacteria. J Med 102:e36492. https://doi.org/10.1097/MD.0000000000036492

Peredo-Lovillo A, Romero-Luna HE, Jiménez-Fernández M (2020) Health promoting microbial metabolites produced by gut microbiota after prebiotics metabolism. Food Res 136:109473. https://doi.org/10.1016/j.foodres.2020.109473

Plotniece A, Sobolev A, Supuran CT, Carta F, Björkling F, Franzyk H, Yli-Kauhaluoma J, Augustyns K, Cos P, De Vooght L (2023) Selected strategies to fight pathogenic bacteria. J Enzyme Inhib Med Chem 38:2155816. https://doi.org/10.1080/14756366.2022.2155816

Polianciuc SI, Gurzău AE, Kiss B, Ştefan MG, Loghin F (2020) Antibiotics in the environment: causes and consequences. Med Pharm Rep 93:231. https://doi.org/10.15386/mpr-1742

Posan P, Suler A, Nistor L, Hodosan C, Marius M, Udroiu A (2023) Use of medicinal plants with galactogenic effect, as a food supplement, in order to increase milk production, in dairy animals: a review. Indian J Anim Res 57:677–683. https://doi.org/10.18805/IJAR.BF-1527

Puvača N, Tufarelli V, Giannenas I (2022) Essential oils in broiler chicken production, immunity and meat quality: review of Thymus vulgaris, Origanum vulgare, and Rosmarinus officinalis. Agriculture 12:874. https://doi.org/10.3390/agriculture12060874

Qaid MM, Al-Mufarrej SI, Azzam MM, Al-Garadi MA (2021) Anticoccidial effectivity of a traditional medicinal plant, *Cinnamomum verum*, in broiler chickens infected with *Eimeria tenella*. Poult Sci 100:100902. https://doi.org/10.1016/j.psj.2020.11.071

Rafiq K, Tofazzal Hossain M, Ahmed R, Hasan MM, Islam R, Hossen MI, Shaha SN, Islam MR (2022) Role of different growth enhancers as alternative to in-feed antibiotics in poultry industry. Front Vet Sci 8:794588. https://doi.org/10.3389/fvets.2021.794588

Reda FM, El-Saadony MT, El-Rayes TK, Farahat M, Attia G, Alagawany M (2021) Dietary effect of licorice (Glycyrrhiza glabra) on quail performance, carcass, blood metabolites and intestinal microbiota. Poult Sci 100:101266. https://doi.org/10.1016/j.psj.2021.101266

Reddy PRK, Yasaswini D, Reddy PPR, Kumar DS, Elghandour MMMY, Salem AZM (2023) Nanotechnology in veterinary sector: current applications, limitations and future perspective. In: Handbook of green and sustainable nanotechnology: fundamentals, developments and applications. Springer, pp 1541–1567. https://doi.org/10.1007/978-3-031-16101-8_8

Rehman S, Iqbal Z, Qureshi R, Rahman IU, Sakhi S, Khan I, Hashem A, Al-Arjani ABF, Almutairi KF, Abd_Allah EF (2022) Ethnoveterinary practices of medicinal plants among tribes of tribal district of North Waziristan, Khyber Pakhtunkhwa, Pakistan. Front Vet Sci 9:815294. https://doi.org/10.3389/fvets.2022.815294

Reid G (2015) The growth potential for dairy probiotics. Int Dairy J 49:16–22. https://doi.org/10.1016/j.idairyj.2015.04.004

Robinson TP, Bu DP, Carrique-Mas J, Fèvre EM, Gilbert M, Grace D, Hay SI, Jiwakanon J, Kakkar M, Kariuki S, Laxminarayan R (2016) Antibiotic resistance is the quintessential One Health issue. Trans R Soc Trop Med Hyg 110:377–380. https://doi.org/10.1093/trstmh/trw048

Rodriguez-Sánchez AJ, Meza-Herrera CA, De Santiago-Miramontes A, Navarrete-Molina C, Veliz-Deras FG, Ordoñez-Morales JZ, Flores-Salas JM, Marin-Tinoco RI (2024) Circular economy, dairy cow feed leftovers, and Withania somnifera supplementation: effects on black belly ram's libido, sperm quality, sexual behavior, and Hemogram values. Biology 13:656. https://doi.org/10.3390/biology13090656

Saeed M, Khan MS, Alagawany M, Farag MR, Alqaisi O, Aqib AI, Qumar M, Siddique F, Ramadan MF (2021) Clove (Syzygium aromaticum) and its phytochemicals in ruminant feed: an updated review. Rend Lincei Sci Fis Nat 32:273–285. https://doi.org/10.1007/s12210-021-00985-3

Saeed M, Khan MS, Younas U, Siddique F, Niaz K, Farooq Z, Farag MR, Elnesr SS, Alagawany M, Ramadan MF (2022) Clove (Syzygium aromaticum) and its derivatives in fish feed. In: Clove (Syzygium aromaticum). Elsevier, pp 257–266. https://doi.org/10.1016/B978-0-323-85177-0.00038-0

Samie A, Guerrant RL, Barrett L, Bessong PO, Igumbor EO, Obi CL (2009) Prevalence of intestinal parasitic and bacterial pathogens in diarrhoeal and non-diarroeal human stools from Vhembe district, South Africa. J Health Popul Nutr 27:739. https://doi.org/10.3329/jhpn.v27i6.4325

Seal BS, Lillehoj HS, Donovan DM, Gay CG (2013) Alternatives to antibiotics: a symposium on the challenges and solutions for animal production. Anim Health Res Rev 14:78–87

Shehata AA, Yalçın S, Latorre JD, Basiouni S, Attia YA, Abd El-Wahab A, Visscher C, El-Seedi HR, Huber C, Hafez HM (2022) Probiotics, prebiotics, and phytogenic substances for optimizing gut health in poultry. Microorganisms 10:395. https://doi.org/10.3390/microorganisms10020395

Silva DR, de Sardi JCO, de Souza Pitangui N, Roque SM, da Silva ACB, Rosalen PL (2020) Probiotics as an alternative antimicrobial therapy: current reality and future directions. J Funct Foods 73:104080. https://doi.org/10.1016/j.jff.2020.104080

Singh P, Mishra G, Pottoo FH, Singh B, Zeleke MM (2022) Zingiber officinale: its ethanobotanical uses, phytochemistry, and pharmacology. In: Edible plants in health and diseases: volume II: phytochemical and pharmacological properties. Springer, pp 1–42. https://doi.org/10.1007/978-981-16-4959-2_1

Singh A, Singh S, Singh JP, Kumar R, Singh NK (2024) In Vitro efficacy of Curcuma longa in combating Staphylococcus aureus subclinical Mastitis in Buffalo. IJVSBT 20:107

Sirotkin AV, Kolesarova A (2021) Puncture vine (Tribulus terrestris L.) in control of health and reproduction. Physiol Res 70:S657. https://doi.org/10.33549/physiolres.934711

So-In C, Sunthamala N (2021) Treatment efficacy of Thunbergia laurifolia, Curcuma longa, Garcinia mangostana, and Andrographis paniculata extracts in Staphylococcus aureus-induced rabbit dermatitis model. Vet. World 15:188. Efficacy of Garcinia mangostana Linn. and Achyranthes aspera Linn. Combined Extracts in the Prevention of Endometritis in Cattle C. https://doi.org/10.5398/tasj.2024.47.3.291

Stanton TB (2013) A call for antibiotic alternatives research Trends in Microbiology 21(3):111–113. https://doi.org/10.1016/j.tim.2012.11.002

Stevenson JL, Kalita G, Goswami R, Das H, Sarma K, Ali MA, Das BK, Samanta AK, Mayengbam P, Tolenkhomba TC (2023) Effect of dietary supplementation of turmeric (*Curcuma longa*) on Health Status of Young Pigs during Pre and Post Weaning Periods. IJBSM 14:279–284. https://doi.org/10.23910/1.2023.3307a

Sundrum A (2001) Organic livestock farming: a critical review. Livest Prod Sci 67:207–215. https://doi.org/10.1016/S0301-6226(00)00188-3

Swennen K, Courtin CM, Delcour JA (2006) Non-digestible oligosaccharides with prebiotic properties. Crit Rev Food Sci Nutr 46:459–471. https://doi.org/10.1080/10408390500215746

Taethaisong N, Paengkoum S, Kaewwongsa W, Onjai-Uea N, Thongpea S, Paengkoum P (2023) The effect of neem leaf supplementation on growth performance, rumen fermentation, and ruminal microbial population in goats. Animal 13:890. https://doi.org/10.3390/ani13050890

Taher MS, Salloom YF, Al-Asadi RAUH, Al-Mousswi ZJ, Alamrani HA (2021) The medicinal importance of Thyme plant (Thymus vulgaris). J Biomed 41:531–534. https://doi.org/10.51248/.v41i3.708

Toutain PL, Ferran AA, Bousquet-Melou A, Pelligand L, Lees P (2016) Veterinary medicine needs new green antimicrobial drugs. Front Microbiol 7:1196. https://doi.org/10.3389/fmicb.2016.01196

Vaarst M, Padel S, Hovi M, Younie D, Sundrum A (2005) Sustaining animal health and food safety in European organic livestock farming. Livest Prod Sci 94:61–69. https://doi.org/10.1016/j.livprodsci.2004.11.033

Van Boeckel TP, Brower C, Gilbert M, Grenfell BT, Levin SA, Robinson TP, Teillant A, Laxminarayan R (2015) Global trends in antimicrobial use in food animals. Proc Natl Acad Sci 112:5649–5654. https://doi.org/10.1073/pnas.1503141112

van Wagenberg CPA, de Haas Y, Hogeveen H, Van Krimpen MM, Meuwissen MPM, van Middelaar CE, Rodenburg TB (2017) Animal board invited review: comparing conventional and organic livestock production systems on different aspects of sustainability. Animal 11:1839–1851

Van TTH, Yidana Z, Smooker PM, Coloe PJ (2020) Antibiotic use in food animals worldwide, with a focus on Africa: pluses and minuses. J Glob Antimicrob Resist 20:170–177. https://doi.org/10.1016/j.jgar.2019.07.031

Vondruskova H, Slamova R, Trckova M, Zraly Z, Pavlik I (2010) Alternatives to antibiotic growth promoters in prevention of diarrhoea in weaned piglets: a review. Vet Med 55:199–224. https://doi.org/10.17221/2998-VETMED

Wanapat M, Pilajun R, Kongmun P (2009) Ruminal ecology of swamp buffalo as influenced by dietary sources. Anim Feed Sci Technol 151:205–214. https://doi.org/10.1016/j.anifeedsci.2009.01.017

Wang S, Xiao Y, Tian F, Zhao J, Zhang H, Zhai Q, Chen W (2020) Rational use of prebiotics for gut microbiota alterations: specific bacterial phylotypes and related mechanisms. J Funct Foods 66:103838. https://doi.org/10.1016/j.jff.2020.103838

Wani AK, Akhtar N, Singh R, Prakash A, Raza SHA, Cavalu S, Chopra C, Madkour M, Elolimy A, Hashem NM (2023) Genome centric engineering use ZFNs, TALENs and CRISPR-Cas9 systems for trait improvement and disease control in animals. Vet Res Commun 47:1–16. https://doi.org/10.1007/s11259-022-09967-8

Wegener HC (2012) Antibiotic resistance—linking human and animal health. In: Improving food safety through a one health approach: workshop summary 331

Xie Z, Shen G, Wang Y, Wu C (2019) Curcumin supplementation regulates lipid metabolism in broiler chickens. Poult Sci 98:422–429. https://doi.org/10.3382/ps/pey315

Zirintunda G, Kateregga J, Nalule S, Biryomumaisho S, Omujal F, Okwee-Acai J, Vudriko P (2024) Extracts of carica papaya L. and capsicum annuum L. showed comparable efficacy to piperazine citrate and levamisole hydrochloride in treatment of poultry helminths. https://doi.org/10.21203/rs.3.rs-5436954/v1

Antibiotics Discovery, Mode of Action, and Mechanism of Resistance

Syed Kashif Abbas and Saira Qadeer

Abstract Although antibiotics have revolutionized modern medicine, studies on their discovery, mechanisms of action, and the growing problem of resistance are still vital. Foundational work by pioneers such as Antonie van Leeuwenhoek, Joseph Lister, Louis Pasteur, and Robert Koch contributed significantly to the understanding of microorganisms and infectious diseases. The development of key antibiotics, including Penicillin, mycophenolic acid, synthetic Salvarsan, and Prontosil, marked significant medical advancements. In order to improve their clinical applications and output, researchers such as Andrew Moyer, Robert Coghill, Norman Heatley, Howard Florey, Ernst Chain, and Alexander Fleming were essential. Subsequent discoveries of antibiotics such as tyrothricin, streptomycin, chlortetracycline, and erythromycin further broadened treatment possibilities. Antibiotics primarily function by inhibiting cell wall synthesis, disrupting the membranes of bacterial cells, and interfering with vital biological functions like the synthesis of folic acid, DNA replication, transcription, and protein synthesis. However, resistance mechanisms—including modifications of penicillin-binding proteins, β-lactamase production, drug inactivation, reduced membrane permeability, efflux pump activity, and target site alterations—have complicated treatment effectiveness. Resistance can be innate or acquired, and its development is mostly influenced by genetic mutations and horizontal gene transfer. Human behaviors such as self-medication, improper prescription writing, and excessive use of antibiotics in agriculture all contribute to the development of resistance. The serious consequences of antibiotic resistance, which include increased hospital stays, increased medical expenses, increased mortality rates, and diminished efficacy in treating common illnesses, make it a significant global public health concern. Combating this growing issue necessitates the discovery of new antibiotics, stricter regulatory policies, enhanced diagnostic capabilities, and greater public awareness to curb the spread of resistance.

Keywords Antibiotic discovery · Mechanisms of action · Antimicrobial resistance · Resistance mechanisms · Public health impact

S. K. Abbas (✉) · S. Qadeer
COMSATS University, Islamabad, Pakistan

R. U. Khan, S. Naz (eds.), *Green Alternatives to Antibiotics in Animal Health*,
https://doi.org/10.1007/978-981-92-0309-3_2

1 Introduction

Animals and humans alike suffer from a variety of illnesses brought on by bacteria. Some are so infectious and lethal that they may destroy entire cities in a matter of days, for example. Bubonic plague (Tenover 2006a, b). Due to a lack of appropriate treatment, germs killed an enormous number of people in ancient times. Even a slight injury or simple illness could have deadly consequences (Harper 2019). People have been attempting to fight these illnesses for a very long time. However, they were unaware of the sorts of bacteria and the causal agent at the time, thus they were unable to treat patients (Mohr 2016). They typically used a variety of substances, particularly plants and their extracts, and treated patients by trial and error. In rare instances, they may be effective, but because each kind of bacteria reacts differently, treatments typically fail (Cowan 1999).

Hook's and Leeuwenhoek's discoveries of the microscope and bacteria, respectively, greatly improved our comprehension of these tiny life forms (Gest 2004). The germ theory considerably improved our comprehension of the association between microorganisms and diseases. Subsequently, many systemic approaches pursued a cure (Tan and Tatsumura 2015). Numerous antibiotics were discovered as a result of intensive research efforts. Early in the twentieth century, Selman Waksman, a renowned actinomycetes researcher, described antibiotics as substances made by microorganisms that prevent or eliminate other germs (Davies and Davies 2010). Most antibiotics originate from the Actinobacteria phylum, with the genus Streptomyces, a type of soil bacterium, responsible for approximately 80% of these compounds (Barka et al. 2016). Chemotherapeutic compounds like salvarsan, sulfa medications, and quinolones were essential in antimicrobial treatment before the discovery of antibiotics (Aminov 2010). When a culture plate was contaminated with the fungus Penicillium in 1928, Alexander Fleming discovered penicillin, the first naturally occurring antibiotic. But it wasn't until the late 1930s that it was created for clinical usage (Hopwood 2007). Penicillin works against Gram-positive bacteria by stopping the bacterial cell walls from developing. But because Gram-negative bacteria are shielded by an outer layer, its effectiveness against them is far less, and against tuberculosis-causing bacteria, which possess unusually thick cell walls (Hopwood 2007).

Observations showed that some bacteria could enzymatically break down penicillin even before it was widely used (Abraham and Chain 1940). Surprisingly, this remains true for Treponema pallidum (Cha et al. 2004), however, a large number of other disease-causing bacteria have become resistant to semi-synthetic penicillins, cephalosporins, and even the more sophisticated carbapenems in addition to natural penicillin (penicillin G) (Rollo and Williamson 1952). Modern antibiotics are known to target nearly every essential bacterial process. They are divided into at least seven main classes according to their chemical makeup and mode of action. Among these are β-lactams, which prevent the formation of cell walls; aminoglycosides, macrolides, and tetracyclines, which prevent the formation of proteins; daptomycin, which interferes with the function of cell membranes; platensimycin, which targets the biosynthesis of fatty acids; and glycopeptides, which also prevent the formation of cell walls (Kumarasamy et al. 2010; Surette and Wright 2017; Peterson and Kaur 2018).

2 Discovery of Antibiotics

Vuillemin coined the term "antibiotic" in the late nineteenth century to describe substances that are harmful to living things, particularly microorganisms, and it comes from the French words antibiose and antibiotique (Halawa et al. 2023). Selman A. Waksman later provided a more thorough explanation in 1947, defining antibiotics as chemicals produced by microorganisms that either eradicate or inhibit the growth of bacteria and other germs (Waksman 1947; Kaur Sodhi and Singh 2022). When Antonie van Leeuwenhoek discovered tiny organisms in 1676, he named them "animalcules," which laid the groundwork for the development of antibiotics (Gould 2016). Joseph Lister found in 1971 that the growth of bacteria was inhibited by *Penicillium glaucum*. He used *P. glaucum* extract to successfully cure a nurse's wound, confirming the idea that infections were caused by bacteria (Durand et al. 2019). German physician Robert Koch and French bacteriologist Louis Pasteur independently studied microbes in the second part of the 1800s. While Pasteur focused on *Bacillus anthracis* and established the connection between particular bacterial species and diseases, Koch studied Mycobacterium tuberculosis. The discovery of antibiotics and microbiology was greatly aided by their pioneering work (Nicolaou and Rigol 2018). The first antibiotic, mycophenolic acid, was discovered in 1893 by Italian scientist Bartolomeo Gosio. It was discovered to suppress *Bacillus anthracis* growth after being isolated from P. glaucum (Mohr 2016). Paul Ehrlich and his colleagues identified Salvarsan (arsphenamine) in 1909 as a synthetic antibiotic based on arsenic that effectively combats the syphilis-causing bacteria *Treponema pallidum* (Halawa et al. 2023). Neosalvarsan was a safer and more effective alternative and was introduced in 1913 (DiGiandomenico and Sellman 2015). However, Prontosil, a broad-spectrum sulfonamide antibiotic created in 1930 by German bacteriologist Gerhard Domagk, finally replaced Salvarsan and Neosalvarsan due to the potential risks associated with arsenic-based drugs. In World War I, it treated war-injured soldiers and marked the beginning of the history of antibiotics (Golkar et al. 2014). Methicillin-resistant *Staphylococcus aureus* (MRSA), the first "superbug," was eventually discovered, highlighting the growing issue of resistance to antibiotics (Halawa et al. 2023).

In 1939, French researcher René Dubos achieved a significant step in antibiotic production when he purified tyrothricin, a mixture of gramicidin D and tyrocidine, from the soil bacterium Bacillus brevis. The substance exhibited strong antibacterial activity against Gram-positive bacteria. However, due to its high toxicity to humans, gramicidin is presently applied only externally (Mohr 2016). Selman Waksman conducted extensive studies on the antibacterial action of soil bacteria, especially Streptomyces species, in the 1940s. He proposed the Waksman system that emphasized the antagonistic interactions between bacterial species (da Cunha et al. 2019). He purified and created a succession of noteworthy antibiotics and antifungal compounds on the basis of this research platform, such as fumigacin (from *Aspergillus fumigatus*), clavacin (from *Aspergillus clavatus*), actinomycin (from *Streptomyces* spp.), neomycin (from *Streptomyces fradiae*), and streptomycin (from *Streptomyces griseus*) (da Cunha et al. 2019). The second half of the twentieth century, known as the "Golden Era" of antimicrobial drug discovery, was characterized by the discovery of several new antimicrobial drugs and was triggered by Waksman's discoveries (Durand et al. 2019).

Paul Ehrlich hypothesized early in the twentieth century that dyes potentially could kill microbes by selectively adhering to them based on their different affinities for diverse tissues. He and Alfred Bertheim and Sachachiro Hata tested 606 chemical compounds in 1909 and isolated arsphenamine, later called Salvarsan, as useful against syphilis (Mitsuhashi 1993). When Alexander Fleming left a bacterial culture out accidentally in 1928, he isolated the first natural antibiotic. He saw mould growing on the culture a few days later, after the surrounding bacteria had been eradicated. However, the active ingredient causing this antimicrobial activity could not be isolated or purified by Fleming, making it useless (Aminov 2010; Fleming 1929). As a result, no chemists were ready to assist him, and his discoveries were initially ignored. In 1939, Ernst Chain, Howard Florey, and Norman Heatley demonstrated the effectiveness of penicillin in vivo, but they also encountered difficulties in its commercialization (Chain et al. 1940). An important breakthrough was made when Andrew Moyer and Robert Coghill of the USDA's Northern Regional Research Laboratory in Illinois were able to synthesize penicillin in large numbers (Moyer and Coghill 1946).

Subsequently, Raper and Fennell discovered high-yielding strains of Penicillium chrysogenum, further enhancing production efficiency (Raper and Fennell 1946). During World War II, penicillin proved invaluable, significantly reducing infection-related fatalities from battlefield wounds. Later, chemist John C. Sheehan produced penicillin synthetically, due to structural research done by Dorothy Crowfoot Hodgkin (Sheehan and Henery-Logan 1959). This finding made it possible to chemically alter the 6-aminopenicillanic acid (6-APA) nucleus, which resulted in the creation of several penicillin derivatives with enhanced medicinal qualities.

Prontosil Red, a red dye associated with a sulfonamide molecule, was created in 1932 by scientists Mietzsch and Klarer while they were employed at Bayer. Prontosil Red was shown to be effective in curing hemolytic streptococci-infected mice by Gerhard Domagk in 1935 (Domagk 1935). In vitro, the chemical did not, however, demonstrate antibacterial activity. When scientists found that Prontosil Red was broken down in the body into two parts—an inert dye and an active antibacterial drug called sulfanilamide—the cause of this disparity was finally explained (Trefouel et al. 1935). Sulfanilamide had already been patented in 1906, but its patent had expired, placing it in the public domain. As a result, numerous pharmaceutical companies began large-scale production of sulfa drugs. However, in 1937, the use of sulfa drugs was temporarily halted when over 100 people died due to the toxic effects of diethylene glycol, which was used as a solvent in one of the formulations. Sulfa medicines had already saved many lives, including that of British Prime Minister Winston Churchill, in spite of this setback. At this point, penicillin started to gain traction as a viable substitute (Trefouel et al. 1935).

After that, a wave of antibiotic discoveries started, which quickly resulted in the identification of several antibiotics, many of which are still in use today. While researching soil bacteria in 1943, Albert Schatz, a pupil of Selman Waksman, made the discovery of streptomycin (Schatz et al. 1944). Streptomycin showed exceptional effectiveness against Gram-negative bacteria, particularly Mycobacterium TB, the causal agent of tuberculosis. By 1946, it had successfully treated both animal models and human patients with tuberculosis. Initially, Merck held the rights to streptomycin; however, due to the urgent need for the drug, the company allowed other manufacturers to produce it on a large scale. In 1952, Waksman won the Nobel Prize, was named the "Father of Antibiotics," and was widely attributed with the discovery, even though Albert Schatz was the original discoverer.

Minge Benjamin In 1945, 71-year-old retired professor Duggar used the bronze-colored actinomycete Streptomyces aureofaciens to find a chemical. Duggar (2011) stated that this material showed significant antibacterial action against both Gram-positive and Gram-negative bacteria. It was later discovered to be chlortetracycline, which proved to be very successful in curing diseases that were previously incurable, including Rocky Mountain spotted fever, *Streptococcus pneumoniae*, *β-hemolytic streptococci*, typhus, and typhoid fever (*Salmonella typhi*). Due to its remarkable efficacy, chlortetracycline quickly earned the nickname "wonder drug."

James McGuire and his associates at Eli Lilly discovered a substance in 1952 while researching a *Streptomyces erythreus* strain (Bunch and McGuire 1953). This compound, initially called illotycin, demonstrated strong activity against Gram-positive bacteria. It was later renamed erythromycin and became the prototype for the macrolide class of antibiotics.

Antibiotics have been essential in improving surgery and medicine in addition to curing serious illnesses (Gould and Bal 2013). They have been crucial in managing chronic conditions treatment and preventing infections in patients receiving chemotherapy, and assisting with intricate surgical procedures like joint replacements, organ transplants, and heart surgeries (Piddock 2012). Antibiotics have had a significant effect on human longevity as well (Rossolini et al. 2014). The average life expectancy in the United States was only 56.4 years in 1920; today, it has increased to nearly 80 years (Shrestha 2005). Similarly, antibiotics have significantly improved health outcomes worldwide. In developing countries, where sanitation remains a challenge, they have helped reduce morbidity and mortality from foodborne and other poverty-related infections (Wright 2014).

3 Mode of Actions of Antibiotics

The effectiveness and spectrum of activity of any antibiotic is determined by its mode of action (Fig. 1). Although numerous antibiotics have been discovered to date, they target only a limited number of bacterial processes. Usually, antibiotics function by preventing basic bacterial processes that are necessary for survival, like the formation of cell walls, cell membrane integrity, synthesis of folic acid, transcription, DNA replication, or protein synthesis (Dowling et al. 2017).

β-Lactam antibiotics hinder the synthesis of cell walls. They specifically target penicillin-binding proteins (PBPs), which are required for cross-linking peptidoglycan precursors, during the bacterial cell wall building process. Bacterial mortality and cell wall disintegration result from β- lactams' disruption of peptidoglycan production by binding to PBPs. Glycopeptides are a significant class of inhibitors of cell wall formation. These antibiotics impede the development of cell walls by

Fig. 1 Mode of actions of antibiotics

directly binding to peptidoglycan subunits and blocking the action of the enzymes transglycosylase and transpeptidase (Tipper 1979).

Drugs such as sulfonamides (sulfa) act by interfering with the production of folic acid in bacterial cells. It is a necessary process for the synthesis of DNA and proteins. The enzyme dihydropteroate synthetase (DHPS) in this pathway transforms pteridine and para-aminobenzoic acid (PABA) into dihydropteroic acid. Then it is utilized to create dihydrofolic acid. This intermediate is subsequently reduced to tetrahydrofolic acid by dihydrofolate reductase (DHFR). Sulfamethoxazole acts by inhibiting DHPS, but trimethoprim acts by inhibiting DHFR and, in the process, collectively stops folic acid synthesis and hence bacterial life (Masters et al. 2003).

The rest of the antibiotics, daptomycin and polymyxins, destroy the bacterial cell membrane through the action of making it to burst. Polymyxins chelate to the lipopolysaccharide lipid moiety and lead to disruption and enhanced permeability, causing cell death (Bialvaei and Samadi 2015). Daptomycin, a calcium lipopeptide antibiotic, causes depolarization of the bacterial membrane, inhibition of macromolecular synthesis, and disruption of the membrane (Alborn et al. 1991).

Bacterial transcription and DNA replication are crucial for cellular survival and growth. Antibiotics interfere with these by interfering with the function of enzymes essential for DNA synthesis. Quinolones, for example, inhibit the function of the helicase enzyme, which plays a key role in unzipping the double helix structure during replication (Chen et al. 1996). Bacterial DNA replication and DNA repair processes are severely impaired due to helicase activity inhibition (Gale et al. 1972).

The other mechanism crucial for bacterial survival is protein synthesis. The two subunits constituting bacterial ribosomes are the 50S and 30S ribosomes. The ribosomes are protein-making factories. Both these ribosomal subunits can be inhibited by antibiotics to stop protein-making. The macrolides (such as erythromycin), lincosamides (such as clindamycin and lincomycin), phenicols (such as chloramphenicol), and oxazolidinones (such as linezolid) bind to the 50S ribosomal subunit, inhibiting the formation of peptide bonds and elongation of the polypeptide chain (Douthwaite 1992; Katz and Ashley 2005). Some antimicrobials inhibit the 30S ribosomal subunit, including spectinomycin, tetracyclines, and aminoglycosides. By blocking aminoacyl-tRNA from binding to the ribosome, these antibiotics inhibit translation and disrupt the synthesis of proteins (Chopra and Roberts 2001; Hong et al. 2014).

4 Mechanism of Resistance

Bacteria began to resist antibiotics shortly after their commercial release (Cantas et al. 2013). They employ a range of strategies to counteract antibiotics, and the resistant mechanisms often differ according to the mode of action of the antibiotic (Fig. 2). Multidrug-resistant (MDR) bacteria emerge due to some highly efficient processes that enable germs to be able to resist multiple drugs simultaneously (Tenover 2006a). Drug inactivation or degradation, target alteration, efflux pumps,

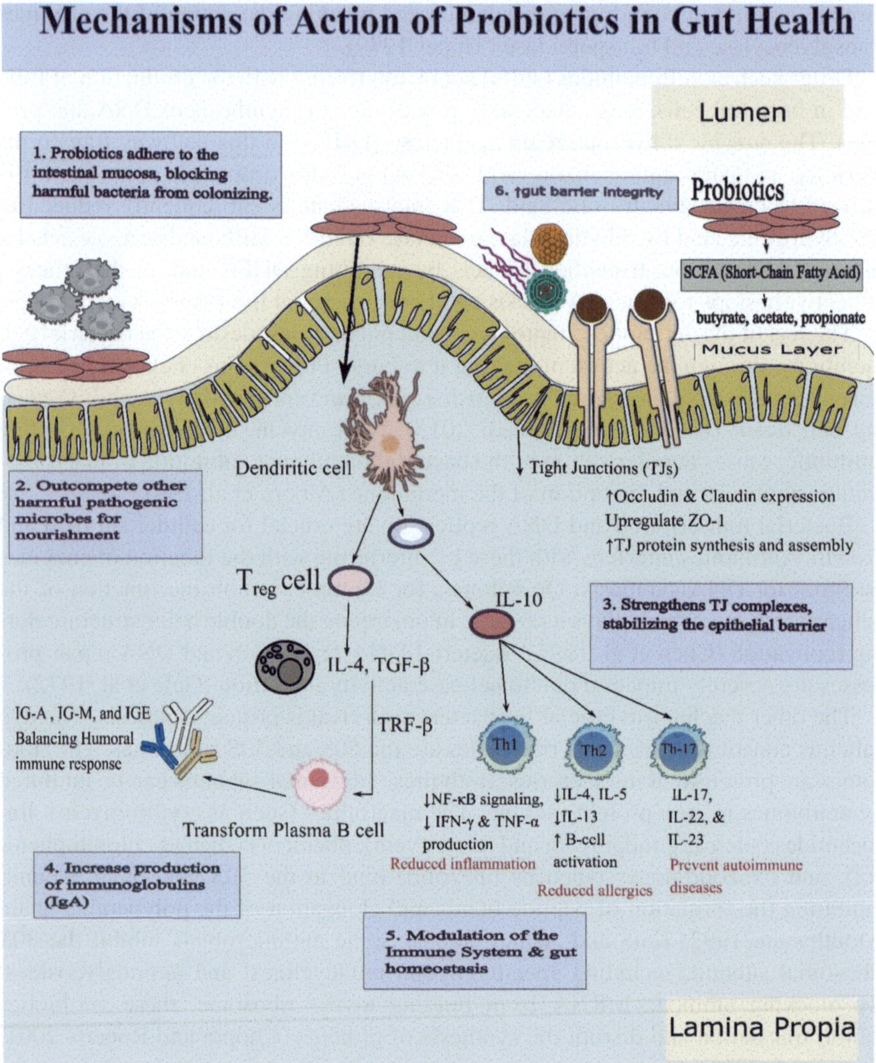

Fig. 2 General mechanism of resistance in bacteria

altered membrane permeability, and metabolic pathway modifications are the main resistance mechanisms.

β-lactam antibiotics target penicillin-binding proteins (PBPs); however, some Gram-positive bacteria, such as methicillin-resistant Staphylococcus aureus (MRSA), change their PBPs to PBP2a, which has a significantly decreased affinity for β-lactams (Rice 2012). However, Gram- negative bacteria often generate β-lactamases, which hydrolyze β-lactam antibiotics and render them useless (Poole 2004).

Drug modification is another resistance mechanism wherein bacteria chemically alter antibiotics, preventing them from binding to their targets. For instance, Staphylococcus aureus and Escherichia coli inactivate chloramphenicol by transferring an acetyl group from acetyl-CoA to the medication (Schwarz et al. 2004).

Modifying the permeability of membranes is another way that bacteria resist. When exposed to antibiotics, some bacteria downregulate porin channels in their outer membrane, reducing drug uptake and preventing the antibiotic from reaching an effective intracellular concentration (Nikaido 2003). As seen in Escherichia coli, this mechanism works especially well against hydrophilic antibiotics like β-lactams and aminoglycosides (Denyer and Maillard 2002).

One of the main causes of bacterial multidrug resistance is efflux pumps. The purpose of these incredibly varied pumps is to remove harmful materials, such as antibiotics, from bacterial cells (Poole 2005). Efflux pumps are often non-specific and may confer resistance to many antibiotics because they are a natural aspect of bacterial physiology. For example, efflux pump function has enabled Salmonella enterica to become multidrug-resistant to tetracyclines, quinolones, and chloramphenicol (Nishino et al. 2006).

Target site modification is another significant mechanism of resistance. Antibiotics are made ineffective when some bacteria modify their binding sites. Gene mutations that affect the RNA polymerase enzyme, the primary target of rifampicin, lead to resistance against the antibiotic in *E. coli* and Mycobacterium tuberculosis (Campbell et al. 2001).

A resistance mechanism called "target bypass" happens when bacteria evade or inactivate a specific enzyme or metabolic pathway targeted by drugs. Many sulfonamide- and trimethoprim-resistant bacteria have been discovered to use this strategy. Bacteria use these resistant bacteria to bypass the enzyme processes involving dihydrofolate reductase (DHFR) and dihydropteroate synthase (DHPS) that are needed for the production of folic acid (Flensburg and Sköld 1987; Eliopoulos and Huovinen 2001).

Both acquired and intrinsic bacterial resistance are possible (Olivares et al. 2013). Some bacteria have resistance genes in their DNA even before they are exposed to drugs. Because bacteria and organisms that produce antibiotics live together in the same environment, protective mechanisms encoded in their genomes—known as the resistome—have evolved, giving rise to this intrinsic resistance (Wright 2007). Proto-resistance genes are also present in the resistome and may develop into functional resistance elements (Morar and Wright 2010). For instance, the multidrug efflux transporter gene mexF provides resistance to tetracycline and chloramphenicol in laboratory settings and improves environmental fitness in the sediment-dwelling bacterium *Shewanella oneidensis* (Groh et al. 2007).

Mutations or horizontal gene transfer (HGT) can result in acquired resistance. Transduction, in which bacteriophages mediate DNA transfer; transformation, in which bacteria absorb free DNA from the environment; and conjugation, in which genetic material, frequently in plasmids, is transferred between bacteria, are the three main ways that HGT enables bacteria to acquire new genetic material (McManus 1997; Tenover 2006b). According to Ogawara et al. (1999), streptomycetes have long been known to produce β-lactamases, which are believed to be the precursors of clinical β-lactam resistance. Furthermore, antibiotic resistance genes can be produced by chance mutations and subsequently disseminated through horizontal gene transfer (Gniadkowski 2008).

The spread of antibiotic resistance has been greatly accelerated by human activity, despite the fact that bacteria naturally carry resistance genes. Antibiotic manufacturing and usage have skyrocketed since the beginning of their commercial use. Because of improper antibiotic use and inaccurate prescriptions, microorganisms have become resistant (Davies and Davies 2010). According to studies, between 30% and 50% of instances involve mistakes in the treatment indication, agent choice, or length of antibiotic therapy (Kollef and Fraser 2001).

Every year, pharmaceutical companies generate a lot of antibiotics, some of which need to be disposed of because they expire before being used. Antibiotic misuse can contaminate the environment and cause microbial communities to become resistant. For example, the development of antibiotic-resistant environmental bacteria has been connected to the pharmaceutical firms in Hyderabad, India, dumping more than 50 kg of ciprofloxacin into rivers per day (Fick et al. 2009).

Furthermore, animal feed contains almost half of the antibiotics manufactured worldwide. Although this approach lowers healthcare expenses and increases the efficiency of cattle production, it also greatly increases antibiotic resistance (Bush et al. 2011). Research labs that frequently test different microorganisms against antibiotics are also dangerous if appropriate biosafety and waste management procedures are not followed. In a similar vein, hospital trash has significant levels of resistant bacteria and antibiotics. Such garbage is frequently disposed of improperly in nations with limited resources, which exacerbates the issue of antibiotic resistance.

The misuse of antibiotics in human populations is another important factor for resistance. Antibiotics are freely accessible in many low- and middle-income nations, which encourages self- medication and improper usage, especially for viral diseases for which they are ineffective (Bush et al. 2011; Michael et al. 2014). Furthermore, antibiotics are constantly present in the gut microbiota, which is made up of commensal (nonpathogenic) microorganisms. As demonstrated by some strains of Escherichia coli, these gut bacteria have the capacity to develop resistance genes over time, transfer them horizontally to pathogenic bacteria, or even turn harmful themselves.

5 Impact of Antibiotic Resistance

It is difficult to fully analyze the effects of antibiotic resistance in bacteria because of its extensive and complex effects. Its danger to the health of people and animals is among its most serious consequences. Once-curable diseases have the potential to become lethal, posing serious concerns to public health. For instance, because of modern treatment, pandemics like the 1918 Spanish flu, which killed almost

50 million people, are no longer seen as serious risks. However, comparable catastrophic epidemics could result from antibiotic-resistant illnesses, making it challenging to treat even common ailments (Michael et al. 2014).

Additionally, surgical procedures will be riskier due to the higher chance of incurable postoperative infections. The financial burden of antibiotic resistance is also substantial. Studies have shown that treating methicillin-resistant Staphylococcus aureus (MRSA) infections is significantly more costly and deadly than treating methicillin-susceptible S. aureus (MSSA) infections. A study conducted in New York City found that, in comparison to MSSA infections, MRSA infections resulted in a 3× increase in mortality (21% versus 8%) and a 22% increase in treatment costs (Rubin et al. 1999).

6 Conclusion

Antibiotic resistance cannot be totally eliminated, although significant actions can be taken to mitigate its consequences. Antibiotics are less effective against bacterial strains that have evolved resistance since many of the medications used in developing countries today were created in the 1950s and 1960s. There is an urgent need for both new, reasonably priced antibiotics and modifications to existing antibiotics to increase their efficacy. Combining older antibiotics with compounds that inhibit resistance may potentially boost their effectiveness.

To control resistance, antibiotic use must be strictly regulated. Only when absolutely required should antibiotics be taken, and quick, precise diagnostic methods should be used to make sure that only the right medicines are given. To stop the spread of resistant diseases, hospital patients with resistant germs should be kept in isolation until the infection is completely treated.

Fighting antibiotic resistance requires raising public awareness. People need to be informed about the causes, effects, and their part in preventing resistance by using antibiotics responsibly. Strong surveillance systems are required both domestically and internationally to track the development of novel resistance mechanisms and bacterial drug susceptibility. Global connectivity between these surveillance systems would allow for a prompt and well-coordinated reaction to stop the spread of resistant germs (Levy and Marshall 2004).

References

Abraham EP, Chain E (1940) An enzyme from bacteria able to destroy penicillin. Nature 146:837

Alborn WE, Allen NE, Preston DA (1991) Daptomycin disrupts membrane potential in growing Staphylococcus aureus. Antimicrob Agents Chemother 35:2282–2287. https://doi.org/10.1128/AAC.35.11.2282

Aminov RI (2010) A brief history of the antibioticera: lessons learned and challenges for the future. Front Microbiol 1. https://doi.org/10.3389/FMICB.2010.00134/FULL

Barka EA, Vatsa P, Sanchez L et al (2016) Taxonomy, physiology, and natural products of actinobacteria. Microbiol Mol Biol Rev 80:1–43. https://doi.org/10.1128/mmbr.00019-15
Bialvaei AZ, Samadi Kafil H (2015) Colistin, mechanisms and prevalence of resistance. Curr Med Res Opin 31(4):707–21
Bunch RL, McGuire JM (1953) Erythromycin, its salts and method of preparation. Eli Lilly, USA, 1–12, Patent No. US2653899A
Bush K, Courvalin P, Dantas G, Davies J, Eisenstein B, Huovinen P et al (2011) Tackling antibiotic resistance. Nat Rev Microbiol 9:894–896
Campbell EA, Korzheva N, Mustaev A et al (2001) Structural mechanism for rifampicin inhibition of bacterial RNA polymerase. Cell 104:901–912. https://doi.org/10.1016/S0092-8674(01)00286-0
Cantas L, Shah SQA, Cavaco LM et al (2013) A brief multi-disciplinary review on antimicrobial resistance in medicine and its linkage to the global environmental microbiota. Front Microbiol 4. https://doi.org/10.3389/FMICB.2013.00096/FULL
Cha JY, Ishiwata A, Mobashery S (2004) A novel β-lactamase activity from a penicillin-binding protein of Treponema pallidum and why syphilis is still treatable with penicillin. J Biol Chem 279. https://doi.org/10.1074/jbc.M400666200
Chain E, Florey HW, Adelaide MB et al (1940) Penicillin as a chemotherapeutic agent. Lancet 236:226–228. https://doi.org/10.1016/S0140-6736(01)08728-1
Chen CR, Malik M, Snyder M, Drlica K (1996) DNA gyrase and topoisomerase IV on the bacterial chromosome: quinolone- induced DNA cleavage. J Mol Biol 258:627–637. https://doi.org/10.1006/jmbi.1996.0274
Chopra I, Roberts M (2001) Tetracycline antibiotics: mode of action, applications, molecular biology, and epidemiology of bacterial resistance. Microbiol Mol Biol Rev 65:232–260. https://doi.org/10.1128/MMBR.65.2.232-260.2001
Cowan MM (1999) Plant products as antimicrobial agents. Clin Microbiol Rev 12:564–582. https://doi.org/10.1128/CMR.12.4.564
da Cunha BR, Fonseca LP, Calado CRC (2019) Antibiotic discovery: where have we come from, where do we go? Antibiotics 8. https://doi.org/10.3390/antibiotics8020045
Davies J, Davies D (2010) Origins and evolution of antibiotic resistance. Microbiol Mol Biol Rev 74:417–433
Denyer SP, Maillard JY (2002) Cellular impermeability and uptake of biocides and antibiotics in gram-negative bacteria. J Appl Microbiol 92(Supply):35–45
DiGiandomenico A, Sellman BR (2015) Antibacterial monoclonal antibodies: the next generation? Curr Opin Microbiol 27:78–85. https://doi.org/10.1016/j.mib.2015.07.014
Domagk G (1935) A contribution to the chemotherapy of bacterial infections. DMW – Dtsch Med Wochenschr 61:250–253. https://doi.org/10.1055/S-0028-1129486
Douthwaite S (1992) Interaction of the antibiotics clindamycin and lincomycin with Escherichia coli 23S ribosomal RNA. Nucleic Acids Res 20:4717–4720
Dowling A, O'Dwyer J, Adley C (2017) Antibiotics: mode of action and mechanisms of resistance. In: Antimicrobial research: Novel bioknowledge and educational programs, vol 1. Formatex Research Center, pp 536–545
Duggar BM (2011) Aureomycin: a product of the continuing search for new antibiotics. Ann N Y Acad Sci 1241:163–169. https://doi.org/10.1111/j.1749-6632.2011.06254.x
Durand GA, Raoult D, Dubourg G (2019) Antibiotic discovery: history, methods and perspectives. Int J Antimicrob Agents 53:371–382
Eliopoulos GM, Huovinen P (2001) Resistance to trimethoprim-sulfamethoxazole. Clin Infect Dis 32:1608–1614
Fick J, Söderström H, Lindberg RH et al (2009) Contamination of surface, ground, and drinking water from pharmaceutical production. Environ Toxicol Chem 28:2522–2527. https://doi.org/10.1897/09-073.1
Fleming A (1929) On the antibacterial action of cultures of a penicillium, with special reference to their use in the isolation of B. Influenzae. Br J Exp Pathol 10:226

Flensburg J, Sköld O (1987) Massive overproduction of dihydrofolate reductase in bacteria as a response to the use of trimethoprim. Eur J Biochem (Wiley Online Libr) 162:473–476. https://doi.org/10.1111/j.1432-1033.1987.tb10664.x
Gale E, Cundliffe E, Reynolds PE, Richmond MH, Waring MJ (1972) The molecular basis of antibiotic action, 2nd edn. Wiley, New York, p 670
Gest H (2004) The discovery of microorganisms by Robert Hooke and Antoni van Leeuwenhoek, fellows of the Royal Society. Notes Rec R Soc 58:187–201
Gniadkowski MJCM (2008) Evolution of extended-spectrum β-lactamases by mutation. Clin Microbiol Infect 14:11–32
Golkar Z, Bagasra O, Pace DG (2014) Bacteriophage therapy: a potential solution for the antibiotic resistance crisis. J Infect Dev Countries 8(02):129–136
Gould K (2016) Antibiotics: from prehistory to the present day. J Antimicrob Chemother 71:572–575. https://doi.org/10.1093/jac/dkv484
Gould IM, Bal AM (2013) New antibiotic agents in the pipeline and how hey can help overcome microbial resistance. Virulence 4:185–191. https://doi.org/10.4161/viru.22507
Groh JL, Luo Q, Ballard JD, Krumholz LR (2007) Genes that enhance the ecological fitness of Shewanella oneidensis MR-1 in sediments reveal the value of antibiotic resistance. Appl Environ Microbiol 73:492–498. https://doi.org/10.1128/AEM.01086-06
Halawa EM, Fadel M, Al-Rabia MW et al (2023) Antibiotic action and resistance: updated review of mechanisms, spread, influencing factors, and alternative approaches for combating resistance. Front Pharmacol 14. https://doi.org/10.3389/FPHAR.2023.1305294/FULL
Harper K (2019) Invisible environmental history: infectious disease in late antiquity. Environ Soc long late Antiq 12:298–313. https://doi.org/10.1163/9789004392083_019
Hong W, Zeng J, Xie J (2014) Antibiotic drugs targeting bacterial RNAs. Acta Pharm Sin B 4:258–265. https://doi.org/10.1016/j.apsb.2014.06.012
Hopwood D (2007) Streptomyces in nature and medicine: the antibiotic makers. Oxford University Press, New York
Katz L, Ashley GW (2005) Translation and protein synthesis: macrolides. Chem Rev 105:499–527. https://doi.org/10.1021/CR030107F
Kaur Sodhi K, Singh CK (2022) Recent development in the sustainable remediation of antibiotics: a review. Total Environ Res Themes 3–4:100008. https://doi.org/10.1016/j.totert.2022.100008
Kollef MH, Fraser VJ (2001) Antibiotic resistance in the intensive care unit. Ann Intern Med 134:298–314. https://doi.org/10.7326/0003-4819-134-4-200102200-00014
Kumarasamy KK, Toleman MA, Walsh TR et al (2010) Emergence of a new antibiotic resistance mechanism in India, Pakistan, and the UK: a molecular, biological, and epidemiological study. Lancet Infect Dis 10. https://doi.org/10.1016/S1473-3099(10)70143-2
Levy SB, Marshall B (2004) Antibacterial resistance worldwide: causes, challenges and responses. Nat Med 10:S122–S129
Masters PA, O'Bryan TA, Zurlo J, Miller DQ, Joshi N, Masters PA, O'Bryan TA et al (2003) Trimethoprim-sulfamethoxazole revisited. Arch Intern Med 163(4):402–410. https://doi.org/10.1001/archinte.163.4.402
McManus MC (1997) Mechanisms of bacterial resistance to antimicrobial agents. Am J Heal Pharm 54:1420–1433
Michael CA, Dominey-Howes D, Labbate M (2014) The antimicrobial resistance crisis: causes, consequences, and management. Front Public Health 2. https://doi.org/10.3389/FPUBH.2014.00145/FULL
Mitsuhashi S (1993) Drug resistance in bacteria: history, genetics and biochemistry. J Int Med Res 21:1–14. https://doi.org/10.1177/030006059302100101
Mohr KI (2016) History of antibiotics research. Curr Top Microbiol Immunol 398:237–272. https://doi.org/10.1007/82_2016_499
Morar M, Wright GD (2010) The genomic enzymology of antibiotic resistance. Annu Rev Genet 44:25–51. https://doi.org/10.1146/ANNUREV-GENET-102209-163517

Moyer AJ, Coghill RD (1946) Penicillin; production of penicillin in surface cultures. J Bacteriol 51:57–78
Nicolaou KC, Rigol S (2018) A brief history of antibiotics and select advances in their synthesis. J Antibiot (Tokyo) 71:153–184
Nikaido H (2003) Molecular basis of bacterial outer membrane permeability revisited. Microbiol Mol Biol Rev 67:593–656. https://doi.org/10.1128/MMBR.67.4.593-656.2003
Nishino K, Latifi T, Groisman EA (2006) Virulence and drug resistance roles of multidrug efflux systems of salmonella enterica serovar Typhimurium. Mol Microbiol 59:126–141. https://doi.org/10.1111/j.1365-2958.2005.04940.x
Ogawara H, Kawamura N, Kudo T et al (1999) Distribution of β-lactamases in actinomycetes. Antimicrob Agents Chemother 43:3014–3017. https://doi.org/10.1128/aac.43.12.3014
Olivares J, Bernardini A, Garcia-Leon G et al (2013) The intrinsic resistome of bacterial pathogens. Front Microbiol 4. https://doi.org/10.3389/FMICB.2013.00103/FULL
Peterson E, Kaur P (2018) Antibiotic resistance mechanisms in bacteria: relationships between resistance determinants of antibiotic producers, environmental bacteria, and clinical pathogens. Front Microbiol 9. https://doi.org/10.3389/fmicb.2018.02928
Piddock LJV (2012) The crisis of no new antibiotics-what is the way forward? Lancet Infect Dis 12:249–253. https://doi.org/10.1016/S1473-3099(11)70316-4
Poole K (2004) Resistance to β-lactam antibiotics. Cell Mol Life Sci 61:2200–2223
Poole K (2005) Efflux-mediated antimicrobial resistance. J Antimicrob Chemother 56(1):20–51
Raper KB, Fennell DI (1946) The production of penicillin X in submerged culture. J Bacteriol 51:761–777. https://doi.org/10.1128/JB.51.6.761-777.1946
Rice LB (2012) Mechanisms of resistance and clinical relevance of resistance to β-lactams, glycopeptides, and fluoroquinolones. Mayo Clin Proc 87:198–208. https://doi.org/10.1016/j.mayocp.2011.12.003
Rollo IM, Williamson J (1952) Acquired resistance to penicillin and to neoarsphenamine in Spirochaeta recurrentis. Br J Pharmacol Chemother 7:33–41. https://doi.org/10.1111/j.1476-5381.1952.tb00686.x
Rossolini GM, Arena F, Pecile P, Pollini S (2014) Update on the antibiotic resistance crisis. Curr Opin Pharmacol 18:56–60. https://doi.org/10.1016/j.coph.2014.09.006
Rubin RJ, Harrington CA, Poon A et al (1999) The economic impact of Staphylococcus aureus infection in New York City hospitals. Emerg Infect Dis 5:9–17. https://doi.org/10.3201/eid0501.990102
Schatz A, Bugle E, Waksman SA (1944) Streptomycin, a substance exhibiting antibiotic activity against gram-positive and gram-negative bacteria. Proc Soc Exp Biol Med 55:66–69. https://doi.org/10.3181/00379727-55-14461
Schwarz S, Kehrenberg C, Doublet B, Cloeckaert A (2004) Molecular basis of bacterial resistance to chloramphenicol and florfenicol. FEMS Microbiol Rev 28:519–542. https://doi.org/10.1016/J.FEMSRE.2004.04.001
Sheehan JC, Henery-Logan KR (1959) The total synthesis of penicillin V. J Am Chem Soc 81:3089–3094. https://doi.org/10.1021/JA01521A044
Shrestha LB (2005) Life expectancy in the United States. Library of Congress. Congressional Research Service
Surette MD, Wright GD (2017) Lessons from the environmental antibiotic resistome. Ann Rev Microbiol 71. https://doi.org/10.1146/annurev-micro-090816-093420
Tan SY, Tatsumura Y (2015) Alexander Fleming (1881–1955): discoverer of penicillin. Singapore Med J 56:366–367. https://doi.org/10.11622/smedj.2015105
Tenover FC (2006a) Mechanisms of antimicrobial resistance in bacteria. Am J Infect Control 34. https://doi.org/10.1016/j.ajic.2006.05.219
Tenover FC (2006b) Mechanisms of antimicrobial resistance in bacteria. Am J Med 119. https://doi.org/10.1016/j.amjmed.2006.03.011
Tipper DJ (1979) Mode of action of β-lactam antibiotics. Rev Infect Dis 1:39–53

Trefouel J, Nitti F, Bovet D (1935) The activity of p-aminophenylsulfamide on experimental streptococcus infections of the mouse and the rabbit. C R Seances Soc Biol Fil 120:756–758
Waksman SA (1947) What is an antibiotic or an antibiotic substance? Mycologia 39:565–569. https://doi.org/10.1080/00275514.1947.12017635
Wright GD (2007) The antibiotic resistome: the nexus of chemical and genetic diversity. Nat Rev Microbiol 5:175–186. https://doi.org/10.1038/nrmicro1614
Wright GD (2014) Something old, something new: revisiting natural products in antibiotic drug discovery. Can J Microbiol 60:147–154. https://doi.org/10.1139/cjm-2014-0063

Antibiotics: Mechanisms, Classification, and Resistance

Iqra Farzeen, Zubair Ahmed, Asma Ashraf, Derya Karatas Yeni, Muhammad Umar Ijaz, and Siddiqua Gulzar

Abstract Antibiotics are transformed into modern medicines that are used against infectious diseases. They have been used against bacterial infections for more than 70 years. In the anti-bacterial fight, millions of lives have been saved by antibiotics, which have undoubtedly benefited human civilization. Multidrug-resistant (MDR) bacterial illnesses are increasing worldwide. According to their chemical structures, mode of action, administration method (oral, topical, or injectable), and range of activity, carbapenems are the most widely used categorization for antibiotics. The efficacy of antibiotics depends upon the mechanism of action. Bacteria show resistance through four different mechanistic pathways, such as membrane inactivation, membrane remodeling, pump efflux, and alteration of the target site. This study highlighted mainly the classification and resistance pathway in microbial resistance.

Keywords Microbial · Antibiotics · Microbial resistance · Beta-lactams and gene transfer

1 Introduction

Antibiotics are revolutionary drugs with broad specificities against a variety of harmful bacteria that efficiently kill or stop development and multiplication (bacteriostatic) without endangering the host, regardless of the infection location (Cook

---

I. Farzeen · Z. Ahmed · A. Ashraf (✉) · S. Gulzar
Department of Zoology, Government College University Faisalabad, Faisalabad, Pakistan
e-mail: asmaashraf@gcuf.edu.pk

D. K. Yeni
Department of Microbiology, Faculty of Veterinary Medicine, Necmettin Erbakan University, Ereğli, Konya, Turkey

M. U. Ijaz
Department of Zoology, Wildlife and Fisheries, University of Agriculture, Faisalabad, Pakistan

R. U. Khan, S. Naz (eds.), *Green Alternatives to Antibiotics in Animal Health*,
https://doi.org/10.1007/978-981-92-0309-3_3

and Wright 2022). The discovery of penicillin, which brought antibiotics into clinical use, was one of the twentieth-century advances that encouraged the pursuit of other antimicrobial medicines (Hutchings et al. 2019). In addition to treating serious infections, antibiotics have contributed to the advancement of healthcare by making uncomplicated cardiac surgery, organ transplantation, and cancer treatment possible (Abushaheen et al. 2020). Unfortunately, overuse and abuse of these life-saving medications over the past few decades have resulted in the evolution of disease resistance due to their low cost, ease of access, and fewer side effects. Furthermore, antibiotic resistance is becoming a bigger problem due to the widespread use of antibiotics as growth enhancers in various industries, including fish farming, poultry, animal husbandry, and agriculture (Hutchings et al. 2019).

When an antibiotic is first administered, it affects strongly; nevertheless, occasionally a tiny bacterial population develops resistance over time and finally survives (Blair et al. 2015). Through a variety of processes, Bacteria can develop antibiotic resistance, such as hydrolysis through enzymes, outflow pump amplification, downregulation, and resilience gene acquisition. As a result of these resistance genes being passed on to their offspring, resistance to that particular antibiotic develops (Frieri et al. 2017). This study was designed to elaborate on antibiotic classification, mechanism of action, and microbial resistance pathways.

1.1 Classification of Antibiotics

Antibiotics may be divided into a number of kinds according to their origin, structure, response, and molecular mechanism of action against infections.

Based on the origin, antibiotics can be divided into two classes: natural and synthetic antibiotics. Natural antibiotics are produced by the bacteria themselves as a byproduct of secondary metabolic pathways, meaning that they are not essential for their survival and are only produced when necessary. Synthetic antibiotics are thought to have a higher toxic effect on pathogens and work more quickly than natural antibiotics. This trend has increased as knowledge about the structure and mode of action of antibiotics has grown.

1.2 Based on Bacterial Response

Bactericidal Antibiotics: These antibiotics inhibit the growth if bacteria attack by various way.

Bacteriostatic Antibiotics: In certain situations, these classes of antibiotics are ineffective in treating diseases because they do not kill the bacteria; rather, they treat the infection by stopping the growth of microorganisms and preventing them from dividing or reproducing. The same antibiotics can be bacteriostatic for certain bacteria and bactericidal for others.

2 Classification of Antibiotics

According to their chemical structures, mode of action, administration method (oral, topical, or injectable), and range of activity, carbapenems are the most widely used categorization for antibiotics (Etebu and Arikekpar 2016).

2.1 Beta-Lactams

The majority of antibiotics fall into the beta-lactam category. Penicillin, cephalosporins, carbapenems, and monobactams are the most well-known members of the beta-lactam class (Etebu and Arikekpar 2016). Every member of a beta-lactam group has a beta-lactam ring. The 3-carbon and 1-nitrogen ring of the highly reactive beta-lactam ring interferes with the bacterial cell wall's protein production, which either kills the cell or stops growth. However, by inhibiting a penicillin-binding protein (PBP), this ring also prevents the formation of peptidoglycan in the bacterial cell wall, which causes cell lysis and death (Smith et al. 2018). Antibiotics can be produced more economically via biochemical microbial fermentation than by synthesis from basic materials (De Rosa et al. 2021).

Penicillin It consists of a thiazolidine ring and a beta-lactam ring attached to side chains (Fig. 1). From *Penicillium notatum*, the first penicillin was prepared. The most prevalent members of the Penicillin class include Penicillin G, Penicillin V, Oxacillin (dicloxacillin), Methicillin, Nafcillin, Ampicillin, Amoxicillin, Carbenicillin, Piperacillin, Meslocillin, and Ticarcillin (Etebu and Arikekpar 2016).

Cephalosporin An antibiotic of the beta-lactam class that shares structural and mechanistic similarities with penicillin (Fig. 1). Guiseppe Brotzu distinguished it from the mould *Cephalosporium acremonium* in 1945 (Petri 2006). It has a side-chain with 3,6-dihydro-2 H-1,3- thiazane rings and a nucleus of 7-aminocephalosporanic acid (Fig. 1). The most widely used classifications for cephalosporins are based on their chemical structure, clinical pharmacology, beta-lactamase resistance, antibacterial spectrum, and generations (Begum et al. 2021).

Monobactams A 2-oxoazetidine-1-sulfonic acid moiety is included in these monocyclic β-lactam drugs. The FDA-approved monobactam antibiotic Aztreonam is sold commercially and exhibits limited spectrum action (Begum et al. 2021). *Chromobacterium violaceum*, a Gram-negative bacterium, was the source of aztreonam. Its acts against Gram-negative bacteria, including Neisseria and Pseudomonas. It has been used to treat urinary tract infections, septicemia, and pneumonia brought on by these bacterial species.

Carbapenems are common antibiotics used in clinical settings in the US. Carbapenems have shown broad-spectrum activity against both types of

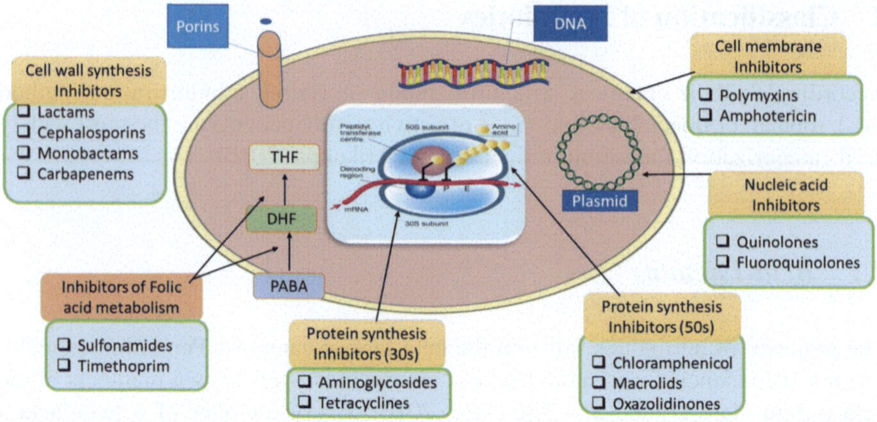

Fig. 1 Classification of antibiotics and mechanism of action

bacteria. The four carbapenems that are most often used are doripenem, ertapenem, meropenem, and ipenem. Carbapenems may easily penetrate bacterial outer membranes and have a high affinity for a variety of penicillin-binding proteins (PBPs). They are frequently called "antibiotics of final resort" and are administered to patients who are believed to have resistant bacteria or disease (Torres et al. 2007).

2.2 *Macrolides*

In 1952, J. M. McGuire discovered and extracted the first macrolides from the soil living fungus (Etebu and Arikekpar 2016). It is an antibiotic with a wide range of properties that mostly treats infections of the respiratory system, soft tissues, and skin. It also has wide-ranging antibacterial properties. The macrolides consist of erythromycin, azithromycin, and clarithromycin. By preventing elongation, macrolides can reversibly attach to the bacterial 23S rRNA and prevent protein production (Waseem et al. 2020).

2.3 *Tetracyclines*

In 1945, Benjamin Duggar isolated tetracycline from a bacterium. The Streptomyces class initial member was the chlorotetracycline. The synthesis approach may be used to categorize the tetracyclines into the first, second, and third generations. The biosynthesis-derived tetracycline, chlortetracycline, oxytetracycline, and demeclocycline are members of the first generation of tetracyclines (Rusu and Buta 2021).

2.4 *Quinolones*

Lesher's accidental discovery of nalidixic acid as a by-product of the synthesis of the antimalarial compound chloroquine in 1962 led chemists to discover the quinolone class of antibiotics. It consists of two rings, though more rings are present in the parent structure of more recent generations of quinolones to increase their spectrum of antimicrobial activity (Begum et al. 2021).

2.5 *Aminoglycosides*

The first antibiotic of the aminoglycoside class, streptomycin, was discovered in 1943 in soil Actinomycetes. Among the antibiotics in this family, streptomycin was the first drug to be identified; it was isolated in 1943 (Mahajan and Balachandran 2012). A family of broad-spectrum antibiotics called aminoglycosides works by preventing the creation of new proteins. They exhibit effectiveness and reduced side effects similar to β-lactam antibiotics, and they attach to the ribosome's A-site with great affinity (Krause et al. 2016).

3 Mechanisms of Antimicrobial Action

The range of action of antibiotics might be either extensive or limited. Antibiotics respond to Gram-positive and Gram-negative bacteria in a comparable way, regardless of their spectrum. The main chemical classes of antibiotics include oxazolidinones, macrolides, β-lactams, tetracyclines, sulfonamides, aminoglycosides, and glycopeptides. Antibiotics can be classified as disruptors of cell membranes, proteins, nucleic acids (DNA/RNA) production, cell wall synthesis, and metabolic pathways based on their method of action. The various modes of action displayed by several antibiotic classes are demonstrated in Fig. 1. The method by which each type of antimicrobial drug kills or inhibits bacteria varies. Antimicrobial agents' modes of action can be divided into the following categories (Abushaheen et al. 2020).

3.1 *Interfering with the Synthesis of Cell Wall*

The bulk of antibacterial agents are primarily attracted to the bacterial cell wall. Composed of poly-[Nacetylglucosamine-N-acetylmuramic acid (Mur-NAc)] cross-linked by brief peptide units, it consists of peptidoglycan (PG), a mesh-like structure. PG synthesis occurs on the cytoplasm's outer surface through the cytoplasmic production of the lipid-linked disaccharide-pentapeptide precursor lipid II (Saikia and Chetia 2024).

Penicillins and their derivatives, cephalosporins, carbapenems, and glycopeptides are among the β-lactam antibiotics that target the formation of peptidoglycans in the bacterial cell wall, leaving the cell susceptible to autolysis and osmotic pressure. Table 1 shows the main targets and mechanisms of certain clinically significant antibiotics. β-lactams (penicillins, cephalosporins, carbapenems, and monobactams) and glycopeptides (vancomycin and teicoplanin) are antibacterial medications that prevent the formation of bacterial cell walls (Saikia and Chetia 2024).

Glycopeptides Vancomycin and teicoplanin are the two main families of glycopeptide antibiotics that are utilized in therapeutic settings. Both are bactericidal against gram-positive bacteria, such as Enterococcus species and methicillin-resistant *Staphylococcus aureus* (MRSA). By blocking the creation of connections between the peptidoglycan layers, glycopeptides affect transpeptidase and cause structural instability. Cell death will ensue as a result of the cell content moving outside the cell (Abushaheen et al. 2020).

3.2 Inhibitors of Protein Synthesis

Messenger RNA (mRNA) contains genetic information that is transformed into a polypeptide chain by ribosomes, which are very complex and conserved cellular machinery (Katz and Ashley 2005). The two uneven ribonucleoprotein subunits, 30S and 50S, that make up the bacterial 70S ribosome are essential for protein synthesis (Diekema and Jones 2001). Targeting the 30S ribosomal subunit, aminoglycosides and tetracyclines prevent the translation process (Krause et al. 2016).

Blocking 30S: By binding to the 30S subunit, macrolides, aminoglycosides, and tetracyclines prevent the synthesis of new proteins. Possessing carbohydrate groups, which attach to the negatively charged plasma membrane of the bacterial cell and then diffuse into it. Once within the bacterial cell, they bind to the 30S subunit of the ribosome's A-site and change the translation from intra-helical to extra-helical, generating an erroneous mRNA-tRNA pairing that makes protein synthesis mistranslated (Garneau-Tsodikova and Labby 2016).

Blocking 50S subunit: Amino acids are linked together to form a polypeptide chain in the peptidyl transferase center (PTC) of the bacterial ribosome's 50S subunit. Furthermore, the polypeptide chain was able to exit the 50S subunit thanks to the nascent peptide exit tunnel (NPET). To prevent newly synthesized polypeptides from passing through the tunnel that links the NPET and the PTC, chloramphenicol may bind to the 50S subunit. Thus, inhibiting the elongation step leads to the inhibition of protein synthesis. A family of artificial antibiotics called oxazolidinones and chloramphenicol functions by blocking the 50S subunit (Abushaheen et al. 2020).

Table 1 Classification of antibiotics with target pathogen and mechanism of action

Class of microbe	Name	Target pathogen	Mechanism
β-lactams combinations	Avibactam	Gram negative bacteria	3rd generation cephalosporin and avibactam
	Aztreonam	Gram negative bacteria	For treatment of MBL producing gram-negative bacteria
	Tazobactam	Gram-negative bacteria + carbapenem-resistant bacteria	5th generation cephalosporin + tazobactm
	Enmetazobactam + cefepime	Entero bacteriaceae PRODUCING ESBL	Combination of β-lactams + cyclosporins
	Recarbrio	MDR Gram-negative bacteria	Imipenem-cilastatin and relebactam combination
β-lactams inhibitor	Durlobactam	Combination with subactam	*Acinetobacter baumannii-calcoaceticus* complex and CRAB
	Exblifeb	cUTIS + pneumonia	Inhibition of peptide synthesis + inhibition of hydrolysis of cefepime
Monobactem	Relebactam	CRE and carbapenem-resistant bacteria	Combination of imipenem and cilastatin
Carbapenem	Carbavance	Gram negative bacteria with β- lactamases	meropenem and vaborbactam
	Vabomere	β-lactamase producers	meropenem + tazobactam
	Ben apenem	cUTI and hospital-acquired pneumonia	Inhibition of bacterial cell wall
Cephalosporins	Cefiderocol	MDR gram-negative bacteria	β-lactamase + cephalosporins
Aminoglycosides	Plazomicin	MDR Entrobacteriaceae	Aminoglycosides antibiotics
Tetracycline	Omadacycline	gram-negative bacteria + Gram positive bacteria	Semi-synthetic tetracyclin
	Eravacycline	MDR gram-negative bacteria	Tetracycline antibiotics
Macrolides	Solithromycin	Azithromycin resistant	Next -generation + active against telithromycin resistant bacteria
	Nafithromycin	Bacterial Pneumonia	Protein synthesis inhibition
Fluoroquinolones	Delafloxacin	Bacterial Pathogen	DNA gyrase inhibitors
	Nemonoxacin	Gram-positive and gram-negative pathogen	Efficient antibiotic
	Zabofloxacin	Respiratory tract pathogen	DNA Gyrase inhibitor
	Belofloxacin	Gram-positive and gram-negative pathogen	DNA Gyrase inhibitor
Macrocyclic peptides	Zosurabalpin	Carbapenem-resistant *A. baumannii*	Inhibition of the LptB2FGC complex

3.3 *Folic Acid Metabolism Inhibitors*

3.3.1 DHFR Inhibitors and Sulfonamides

Dihydrofolate reductase (DHFR) inhibitors are frequently used to treat protozoal illnesses, such as malaria. According to Fernandes and Martens (2017), they also possess antifungal, antibacterial, and anticancer properties (Fernandes and Martens 2017). Synthetic antibiotics that include sulfur are connected to PABA are called sulfonamides, or sulfa medications (Venkatesan et al. 2023). The mechanisms of action of DHFR inhibitors and sulfonamides are similar since they both target the same pathway. As a result, they reduce resistance.

The mechanisms of action of DHFR inhibitors and sulfonamides are similar since they both target the same pathway. As a result, they inhibit the development of resistance. For example, severe skin and urinary tract infections are treated with a combination of trimethoprim and sulfamethoxazole. When used alone, trimethoprim is bacteriostatic; however, when coupled with sulfamethoxazole, it is thought to be bactericidal serious side effects of these medications include severe skin rashes and photosensitivity (Venkatesan et al. 2023).

3.3.2 Inhibitors of DNA Replication

Quinolones Quinolones are broad-spectrum synthetic antibiotics that were first created mainly to treat Gram-negative infections. Because of their remarkable efficacy, they were subsequently altered to cure both Gram-positive and Gram-negative bacteria (Heeb et al. 2011). Fluoroquinolones are a common subtype of quinolones that have an extra fluorine at position 6 and nitrogen at position 8 in the naphthyridine nucleus rather than carbon (Sato et al. 2013). Since the late 1980s, quinolones and fluoroquinolones have been utilized to treat bacterial infections over the past 35 years (Afsharikhah et al. 2023).

Fluoroquinolones and quinolones target gyrase and topoisomerase enzymes, causing incomplete replication and apoptosis. Levofloxacin and ciprofloxacin are the two commonly used wide- ranging fluoroquinolones. Target protection, efflux pump overexpression, and target site mutation are the main causes of multifactorial resistance (Dube et al. 2023).

4 Mechanisms of Antibiotic Resistance

Antibiotic -resistant mechanisms help bacteria survive and spread despite antimicrobial exposure and represent a global health problem. There are different pathways of antibiotic resistance (Prestinaci et al. 2015).

4.1 Horizontal Gene Transfer: Plasmids, Transposons, and Integrons

Horizontal Gene Transfer (HGT) is an important just-in-time mechanism of acquisition and dissemination of antibiotic resistance genes. Unlike vertical gene transfer, where genetic material is passed down from parent to progeny, HGT occurs through horizontal gene transfer, the lateral exchange of genetic material between bacterial cells of different, or the same, species (Collineau et al. 2019).

Plasmids Extrachromosomal, double-stranded DNA molecules capable of autonomous replication within bacterial cells are plasmids. Often these mobile genetic elements carry many resistance genes in the same plasmid, which are resistance plasmids (R-plasmids) that provide bacteria the means to resist different antibiotic classes simultaneously (Bennett 2008).

Conjugative plasmids are quite important since they carry transfer genes that lead to the formation of a pilus between donor and recipient cells. This pilus also facilitates the transfer of circular plasmid DNA, which accelerates the spread of resistance genes both populationally and generically (Lipszyc et al. 2022). For example, the spread of extended-spectrum-beta-lactamases (ESBLs) in Enterobacteriaceae is almost entirely mediated by conjugative plasmids (Bennett 2008).

Transposons DNA sequences that are able to move from one genomic location to another, one that is within a chromosome to another one, or between a chromosome and various plasmids, or even between two plasmids, are commonly called transposons and sometimes called jumping genes (Lipszyc et al. 2022). Most of these elements carry antibiotic resistance genes, and their mobility enables them to become integrated into other genetic elements that enhance their dissemination potential (Cain et al. 2020).

Integrons Integrons are genetic platforms to capture and express gene cassettes, a number of which encode antibiotic resistance genes. Integrons are unlike plasmids and transposons, being mobile by association with these elements. (Bhat et al. 2023). Integrons encode the integrase enzyme that catalyzes the insertion and excision of gene cassettes in bacteria, causing the integration of multiple resistance genes into a single genetic structure. This ability offers a great adaptive and lends itself to responding quickly to changing selective pressures between individuals (Sabbagh et al. 2021).

Integrons tend to be associated with multidrug-resistant pathogens including clinical isolates of *Pseudomonas aeruginosa* and *Klebsiella pneumoniae*. Integrons harboring gene cassettes encoding resistance to beta-lactams, aminoglycosides, and many other important antibiotic classes are common (Bhat et al. 2023).

4.2 *Mutation and Selection*

A number of mutations and selection are also fundamental in the development of antibiotic resistance and in its spread. Such processes include the genetic alterations that make bacteria adapt to antibiotic pressure and new generations of such populations. Mutations and selection taken together are the foundations of adaptive evolution and help bacteria to survive in extreme conditions, including those containing concentrations of antimicrobial substances (Souque et al. 2021).

4.2.1 Genetic Mutations

Bacterial genome alterations are permanent changes in a mutation that may result spontaneously or be induced by environmental factors, for example, ultraviolet (UV) radiation or chemical mutagens. And it can affect the structure and function of bacterial proteins, including those targeted by antibiotics, and make it less effective to fight (Kumar et al. 2023).

Consequently, point mutations in the genes for DNA gyrase (gyrA) and topoisomerase IV (parC) encode enzymes which confer resistance to fluoroquinolones by interfering with drug binding. Clinical isolates of *Escherichia coli* and *Staphylococcus aureus* often contain these mutations (Melander et al. 2023).

4.2.2 Natural Selection

Amplification of the prevalence for advantageous mutations within bacterial populations is driven by natural selection (Kumar et al. 2023). Whenever antibiotics are introduced, all the bacteria possessing the lucky gene, which makes them immune to antibiotics, are put at a clear advantage. They reproduce so fast, while the weak, helpless, and vulnerable are wiped out, leaving the world filled with resistant strain (Buelow et al. 2021).

4.2.3 Adaptive Evolution

It is important to understand that adaptive evolution is the evolution gained in increments of mutations and natural selection, which help bacterial populations to enhance their interaction with antibiotic existences. This process usually begins with stepwise mutation, whereby every single change bestows approximately equal resistance, and is then followed by the accumulation of multiple resistance determinants (Belay et al. 2024).

For example, if bacteria are subjected to near levels that cause them to die, they may acquire mutations that will eventually make them more resistant. These small changes accumulate and over time create strains of *Mycobacterium tuberculosis* such as extensively drug-resistant *Mycobacterium tuberculosis* (XDR-TB) (Subramaniam and Girish 2020). Phenotypic adaptations, including changes in membrane permeability or upregulation of efflux pumps, as well as adaptive evolution, are also counted among adaptive evolution. The causes are generally genetic mutations, and these changes often lead to multidrug resistance and complicate treatment strategies (Belay et al. 2024).

4.3 Enzymatic Inactivation: Beta-Lactamases, Aminoglycoside-Modifying Enzymes, and Chloramphenicol Acetyltransferases

One of the main ways bacteria counteract antibiotics is to become enzymatically inactivated, or in other words, destroy the antibiotics themselves to survive in hostile environments. Bacterial production of enzymes that break (degrade) or chemically modify (inactivate) a molecule causes this mechanism: antibiotics are deactivated in the host (Kumar et al. 2023).

4.3.1 Beta-Lactamases: Hydrolysis of the Beta-Lactam Ring

Beta-lactamases lead to resistance to beta-lactam antibiotics by hydrolyzing the beta-lactam ring of beta lactam antibiotics, that is, the component required for infections to inhibit bacterial cell wall synthesis. Disruption of the interaction of beta-lactam antibiotics with penicillin-binding proteins (PBPs) by the hydrolytic mechanism directly enables bacterial survival (Mellor et al. 2019).

Mechanism of Action

The bacterial defense against antibiotics is created by the production of enzymes that inactivate the antibiotics enzymatically, breaking them or otherwise chemically altering them so they are no longer effective. Beta-lactamases are one of the most important enzymes for resistance to beta-lactam antibiotic (Akhtar et al. 2022). First, the beta-lactamase enzyme binds to the antibiotic's beta-lactam ring at that enzyme's active site. The binding is highly specific, facilitated by structural complementarity between the enzyme and the antibiotic molecule (Mellor et al. 2019). The enzyme is once bound to initiate hydrolysis of the beta lactam ring (Akhtar et al. 2022).

Once hydrolyzed, the enzyme will release the inactivated antibiotic and regenerate itself for another reaction. They achieve this rapid turnover, allowing beta-lactamase enzymes to continuously degrade beta lactam antibiotics (penicillins, cephalospors, and carbapenems) [15].

4.3.2 Aminoglycoside-Modifying Enzymes (AMEs): Chemical Modification of Antibiotics

Aminoglycoside-modifying enzymes destroy aminoglycosides by modifying the groups of the functionality of aminoglycosides. The usual mode of action of aminoglycosides is to bind to the 30S ribosomal subunit to inhibit protein synthesis, but enzymatic modification prevents this binding (Belay et al. 2024).

Mechanisms of Modification

An important mechanism of resistance to aminoglycosides is the modification of the aminoglycoside by bacterial enzymes. And modifying these changes the structure and makes aminoglycosides unable to bind to bacterial ribosomes and stop protein synthesis. Of these modifications, aminoglycoside acetyltransferases (AACs) are one such. These enzymes are acetyltransferring enzymes capable of transferring acetyl groups from acetyl-CoA to the amino groups of aminoglycosides. Then this chemical alteration renders the antibiotic useless because it dramatically reduces its affinity for ribosomal targets (El-Khoury et al. 2022). An important mechanism of resistance to aminoglycosides is the modification of the aminoglycoside by bacterial enzymes (El-Khoury et al. 2022). A third enzymatic approach to aminoglycoside inactivation is adenylation, or nucleotidylation, also used by bacteria. Aminoglycoside nucleotidyltransferases (ANTs) add adenylyl groups to the hydroxyl groups of the molecule of antibiotics. In combination, these enzymatic modifications (acetylation, phosphorylation, and adenylation) represent a robust resistance mechanism allowing bacteria to evade the aminoglycosides' inhibitory effects (Fig. 2) (AL-Tameemi et al. 2023).

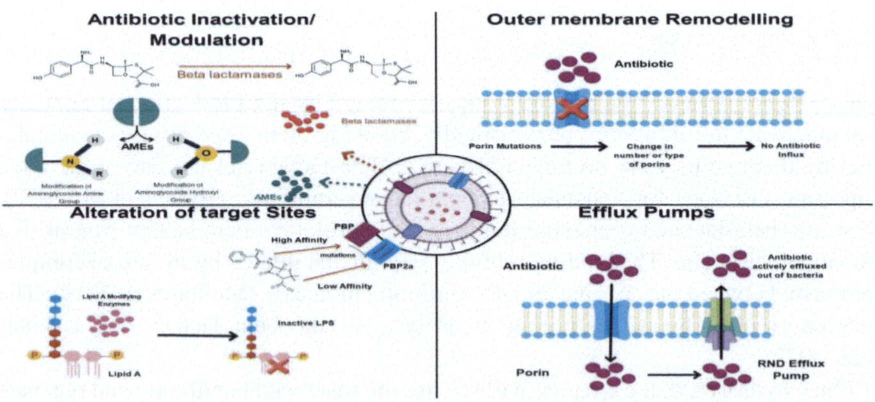

Fig. 2 A brief overview of resistance mechanisms in Gram-negative bacteria is displayed in this figure. It focuses on four main ways used by these bacteria to avoid being killed by antibiotics

4.3.3 Chloramphenicol Acetyltransferases (CATs): Acetylation of Chloramphenicol

Chloramphenicol acetyltransferases modify chloramphenicol by the transfer of acyl groups to its hydroxyl groups. Chloramphenicol's antimicrobial activity depends on interference with peptide bond formation of the 50S ribosomal subunit by means of binding between the ligand and the subunit. This interaction is blocked by acetylation, making the antibiotic ineffective (Liao et al. 2024).

Mechanism of Action

Chloramphenicol acetyltransferases (CATs; enzymes that chemically modify the antibiotic molecule) are important in bacterial resistance to chloramphenicol. At the enzyme active site, both chloramphenicol and acetyl-CoA bind. This precise alignment tightly couples the substrates positioning to the following catalytic reaction, which is a most critical step for the enzyme functionality (Alcala et al. 2020).

After the substrates were aligned, CATs transferred the acetyl group from acetyl-CoA to the hydroxyl group of chloramphenicol. The modification of this chemical makes the antibiotic untouchable to the ribosomal peptidyl transferase center, or it changes the molecular structure of the antibiotic. Without this interaction, chloramphenicol ceases to inhibit bacterial protein synthesis (rendering it a neutralizing force for its antibiotic activity) (AL-Tameemi et al. 2023).

This is illustrated by the fact that after the acetylation reaction, the enzyme releases the chloramphenicol it has acquired, thus making it ineffective. The acetylated form cannot bind to bacterial ribosomes anymore, and so they can produce proteins and replicate. This enzyme thing is an example that demonstrates how wily bacteria can be to get around the consequences of chloramphenicol and continue to develop in the presence of the antibiotic (Liao et al. 2024).

4.4 Efflux Pumps: Mechanisms and Examples

An efflux pump is a type of transmembrane proteins, which aids bacteria to expel the acting molecules, including antibiotics, from the bacterial cell. This process starts with the pump, and it starts to acknowledge and allow the antibiotics to bind to some marked sites on the inner membrane (Mellor et al. 2019). Some pumps, however, show broad specificity, hitting several classes of antibiotics, while others are very specific. Structural changes in the pump occur in response to binding triggers, and these must occur before transport can occur (Belay et al. 2024).

4.4.1 Mechanism of Action

Substrate transport functions require energy (transporting energy) and are obtained either from the proton motive force (PMF) or hydrolysis of ATP. PMF powers pumps such as those in the Resistance Nodulation Divisions (RND) and Major Facilitator Superfamily (MFS), and ABC motors require ATP hydrolysis to drive the process. Conformational changes in the pump are driven to allow the antibiotic to perform the powered movement through the membrane (Fig. 2) (Mohanty et al. 2021).

When the antibiotic reaches the extracellular side, the compound is transported back into the periplasm to maintain a low intracellular concentration of the compound. For instance, in Gram-negative bacteria the efflux takes place through a structure known as the AcrAB–TolC and efficient efflux takes place in both membranes. As soon as the antibiotic is expelled, the transport system reverts to normal to transport other molecules (Belay et al. 2024).

4.5 Biofilm Formation: Role in Antibiotic Resistance

Biofilm formation is a defense mechanism used by bacteria to adapt to a hostile environment; antibiotic exposure included. Biofilms are complex, multicellular assemblies of bacteria, embedded in a matrix of anonymous polymers referred to as exopolysaccharides. This matrix also functions as a protective filter of mechanical strength, antimicrobial resistant and stress protection. It is also involved in chronic infections and is ranked highest in promoting antibiotic resistance in clinical and environment (Table 2) (Rather et al. 2021).

Table 2 Antibiotic resistance mechanisms across bacterial species (Wang et al. 2022)

Resistance mechanism	Details
Beta-Lactamase Production	Includes enzymes such as TEM, CTX-M, OXA variants (e.g., OXA- 23, -48, -181), KPC-like enzymes, NDM (1–5), VIM, SIM, and PER. These enzymes hydrolyze beta-lactam antibiotics.
Modification of Aminoglycosides	Aminoglycoside-modifying enzymes (AMEs) like AAC(3)-II/IV, AAC(6′)-Ib, ANT(2″), APH(3′)-VI, and APH(6′)-Id alter drug structure, reducing efficacy.
Efflux Pump Activation	Efflux systems such as MexAB-OprM, MexCD-OprJ, MexEF-OprN, MacAB, AdeABC, and AcrAB-TolC remove drugs from cells, lowering intracellular drug concentrations.
Target Site Alterations	Mutations in DNA gyrase genes (gyrA, parC) and ribosomal proteins (e.g., 16S rRNA methylation) reduce drug binding. Mutations in genes like pmrB can lead to resistance via overexpression.
Outer Membrane Remodeling	Alterations include porin modifications (OmpK36, OmpF, OprD) and changes to lipopolysaccharide (LPS) or lipid A, affecting drug penetration.

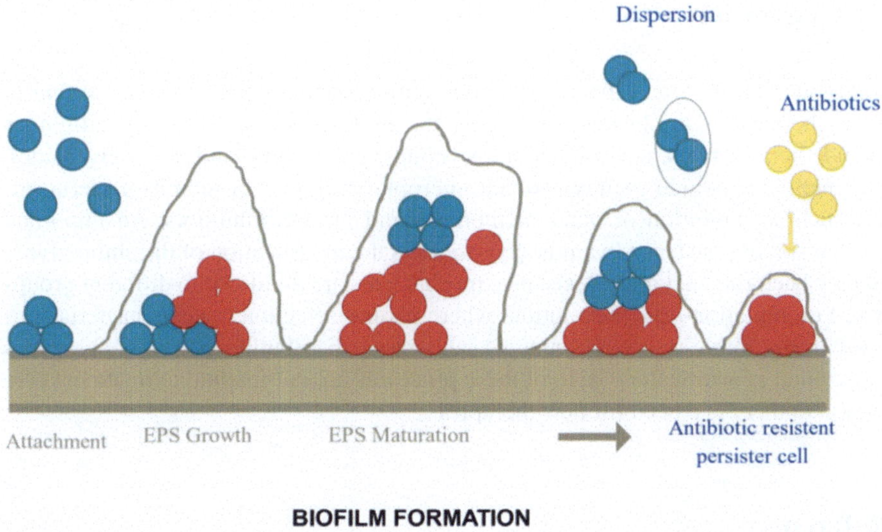

Fig. 3 Antibiotics can be interacted with by the extracellular matrix or enzymes which neutralize them

4.5.1 Mechanisms of Biofilm-Mediated Antibiotic Resistance

Biofilms EPS matrix physically prevents the penetration of antibiotics. The dense matrix, consisting of polysaccharides, is made up of proteins, lipids, and extracellular DNA (eDNA) that simultaneously traps or inactivates the antibiotic preventing it from reaching the bacteria in deeper layers (Fig. 3). In particular, biofilms are not particularly susceptible to antibiotics such as beta-lactams because these antibiotics cannot uniformly diffuse through the protective layer (Abebe 2020). By the generation of gradients of nutrients, oxygen, and pH, the biofilm environment is less metabolically active than the center. Such deeply situated bacteria can go dormant and are not as susceptible to antibiotics, which target actively growing cells. (Liao et al. 2024). Biofilm-forming bacteria also upregulate efflux pumps, which actively remove antibiotics from cells, effectively in concert with the EPS matrix to reduce the intracellular antibiotic concentrations (Belay et al. 2024). To this resistance, persister cells, nonreplicating, dormant cells within the biofilm tolerant of high doses of antibiotics, are added. Once antibiotic pressure eases, these cells can survive treatment and reenter active states, playing a major role in biofilm-associated infections that do not resolve with treatment (Abebe 2020).

5 Conclusion

Antibiotics are a crucial line of defense against bacterial infection. According to their chemical or molecular structures, the most prevalent types of antibiotics include beta-lactams, macrolides, tetracyclines, quinolones, and aminoglycosides. The method by which each type of antimicrobial drug kills or inhibits bacteria varies, such as inhibition of outer membranes and protein inhibitors. An important mechanism of resistance to aminoglycosides is the modification of the aminoglycoside by bacterial enzymes. Resistance mechanisms are divided into different groups based on mutation-driven evolution, where natural selection benefits bacterial survival in antibiotic-rich environments, and rapid distribution of resistance genes (horizontal gene transfer). Both of these processes are detrimental to treatment efficacy, and there is a need for new therapies.

References

Abebe GM (2020) The role of bacterial biofilm in antibiotic resistance and food contamination. Int J Microbiol 2020(1):1705814

Abushaheen MA, Fatani AJ, Alosaimi M, Mansy W, George M, Acharya S, Rathod S, Divakar DD, Jhugroo C, Vellappally S (2020) Antimicrobial resistance, mechanisms and its clinical significance. Dis Mon 66(6):100971

Afsharikhah S, Ghanbarpour R, Mohseni P, Adib N, Bagheri M, Jajarmi M (2023) High prevalence of β-lactam and fluoroquinolone resistance in various phylotypes of Escherichia coli isolates from urinary tract infections in Jiroft city, Iran. BMC Microbiol 23(1):114

Akhtar A, Fatima N, Khan HM (2022) Beta-lactamases and their classification: an overview. In: Beta- Lactam resistance in gram-negative bacteria: threats and challenges, pp 25–33

AL-Tameemi ZA, Almanaseer ND, Nasser NA (2023) Systematic study of aminoglycoside group antibiotics drug resistance. World 2(6)

Alcala A, Ramirez G, Solis A, Kim Y, Tan K, Luna O, Nguyen K, Vazquez D, Ward M, Zhou M (2020) Structural and functional characterization of three type B and C chloramphenicol acetyltransferases from Vibrio species. Protein Sci 29(3):695–710

Begum S, Begum T, Rahman N, Khan RA (2021) A review on antibiotic resistance and way of combating antimicrobial resistance. GSC Biol Pharm Sci 14(2):087–097

Belay WY, Getachew M, Tegegne BA, Teffera ZH, Dagne A, Zeleke TK, Abebe RB (2024) Mechanism of antibacterial resistance, strategies and next-generation antimicrobials to contain antimicrobial resistance. Front Pharmacol 15(16 August). https://doi.org/10.3389/fphar.2024.1444781

Bennett PM (2008) Plasmid encoded antibiotic resistance: acquisition and transfer of antibiotic resistance genes in bacteria. Br J Pharmacol 153(S1):S347–S357

Bhat BA, Mir RA, Qadri H, Dhiman R, Almilaibary A, Alkhanani M, Mir MA (2023) Integrons in the development of antimicrobial resistance: critical review and perspectives. Front Microbiol 14:1231938

Blair JM, Webber MA, Baylay AJ, Ogbolu DO, Piddock LJ (2015) Molecular mechanisms of antibiotic resistance. Nat Rev Microbiol 13(1):42–51

Buelow E, Ploy M-C, Dagot C (2021) Role of pollution on the selection of antibiotic resistance and bacterial pathogens in the environment. Curr Opin Microbiol 64:117–124

Cain AK, Barquist L, Goodman AL, Paulsen IT, Parkhill J, van Opijnen T (2020) A decade of advances in transposon-insertion sequencing. Nat Rev Genet 21(9):526–540

Collineau L, Boerlin P, Carson CA, Chapman B, Fazil A, Hetman B, McEwen SA, Parmley EJ, Reid-Smith RJ, Taboada EN (2019) Integrating whole-genome sequencing data into quantitative risk assessment of foodborne antimicrobial resistance: a review of opportunities and challenges. Front Microbiol 10:1107

Cook MA, Wright GD (2022) The past, present, and future of antibiotics. Sci Transl Med 14(657):eabo7793

De Rosa M, Verdino A, Soriente A, Marabotti A (2021) The odd couple (s): an overview of Beta-Lactam antibiotics bearing more than one pharmacophoric group. Int J Mol Sci 22(2):617

Diekema DJ, Jones RN (2001) Oxazolidinone antibiotics. The Lancet 358(9297):1975–1982

Dube PS, Legoabe LJ, Beteck RM (2023) Quinolone: a versatile therapeutic compound class. Mol Divers 27(3):1501–1526

El-Khoury C, Mansour E, Yuliandra Y, Lai F, Hawkins BA, Du JJ, Sundberg EJ, Sluis-Cremer N, Hibbs DE, Groundwater PW (2022) The role of adjuvants in overcoming antibacterial resistance due to enzymatic drug modification. RSC Med Chem 13(11):1276–1299

Etebu E, Arikekpar I (2016) Antibiotics: classification and mechanisms of action with emphasis on molecular perspectives. Int J Appl Microbiol Biotechnol Res 4(2016):90–101

Fernandes P, Martens E (2017) Antibiotics in late clinical development. Biochem Pharmacol 133:152–163

Frieri M, Kumar K, Boutin A (2017) Antibiotic resistance. J Infect Public Health 10(4):369–378

Garneau-Tsodikova S, Labby KJ (2016) Mechanisms of resistance to aminoglycoside antibiotics: overview and perspectives. Medchemcomm 7(1):11–27

Heeb S, Fletcher MP, Chhabra SR, Diggle SP, Williams P, Cámara M (2011) Quinolones: from antibiotics to autoinducers. FEMS Microbiol Rev 35(2):247–274

Hutchings MI, Truman AW, Wilkinson B (2019) Antibiotics: past, present and future. Curr Opin Microbiol 51:72–80

Katz L, Ashley GW (2005) Translation and protein synthesis: macrolides. Chem Rev 105(2):499–528

Krause KM, Serio AW, Kane TR, Connolly LE (2016) Aminoglycosides: an overview. Cold Spring Harb Perspect Med 6(6):a027029

Kumar V, Yasmeen N, Pandey A, Ahmad Chaudhary A, Alawam AS, Ahmad Rudayni H, Islam A, Lakhawat SS, Sharma PK, Shahid M (2023) Antibiotic adjuvants: synergistic tool to combat multi-drug resistant pathogens. Front Cell Infect Microbiol 13:1293633

Liao J, Qi Q, Kuang L, Zhou Y, Xiao Q, Liu T, Wang X, Guo L, Jiang Y (2024) Chloramphenicol binding sites of Acinetobacter baumannii Chloramphenicol acetyltransferase CatB8. ACS Infect Dis 10(3):870–878

Lipszyc A, Szuplewska M, Bartosik D (2022) How do transposable elements activate expression of transcriptionally silent antibiotic resistance genes? Int J Mol Sci 23(15):8063

Mahajan GB, Balachandran L (2012) Antibacterial agents from actinomycetes-a review. Front Biosci-Elite 4(1):240–253

Melander RJ, Mattingly AE, Nemeth AM, Melander C (2023) Overcoming intrinsic resistance in gram-negative bacteria using small molecule adjuvants. Bioorg Med Chem Lett 80:129113

Mellor KC, Petrovska L, Thomson NR, Harris K, Reid SW, Mather AE (2019) Antimicrobial resistance diversity suggestive of distinct Salmonella Typhimurium sources or selective pressures in food-production animals. Front Microbiol 10:708

Mohanty H, Pachpute S, Yadav RP (2021) Mechanism of drug resistance in bacteria: efflux pump modulation for designing of new antibiotic enhancers. Folia Microbiol 66(5):727–739

Petri W (2006) Penicillins, cephalosporins, and other β-lactam antibiotics. In: Goodman and gilman's the pharmalogical basis of therapeutics, 11st edn. Mc Graw Hill, pp 1127–1152

Prestinaci F, Pezzotti P, Pantosti A (2015) Antimicrobial resistance: a global multifaceted phenomenon. Pathogens Glob Health 109(7):309–318

Rather MA, Gupta K, Mandal M (2021) Microbial biofilm: formation, architecture, antibiotic resistance, and control strategies. Braz J Microbiol:1–18

Rusu A, Buta E (2021) The development of third-generation tetracycline antibiotics and new perspectives. Pharmaceutics 13:2085. In: s Note: MDPI stays neutral with regard to jurisdictional claims in published ….

Sabbagh P, Rajabnia M, Maali A, Ferdosi-Shahandashti E (2021) Integron and its role in antimicrobial resistance: a literature review on some bacterial pathogens. Iran J Basic Med Sci 24(2):136

Saikia S, Chetia P (2024) Antibiotics: from mechanism of action to resistance and beyond. Indian J Microbiol:1–25

Sato T, Yokota S-I, Uchida I, Okubo T, Usui M, Kusumoto M, Akiba M, Fujii N, Tamura Y (2013) Fluoroquinolone resistance mechanisms in an Escherichia coli isolate, HUE1, without quinolone resistance-determining region mutations. Front Microbiol 4:125

Smith PW, Zuccotto F, Bates RH, Martinez-Martinez MS, Read KD, Peet C, Epemolu O (2018) Pharmacokinetics of β-lactam antibiotics: clues from the past to help discover long-acting oral drugs in the future. ACS Infect Dis 4(10):1439–1447

Souque C, Escudero JA, MacLean RC (2021) Integron activity accelerates the evolution of antibiotic resistance. elife 10:e62474

Subramaniam G, Girish M (2020) Antibiotic resistance—A cause for reemergence of infections. Indian J Pediatr 87(11):937–944

Torres JA, Villegas MV, Quinn JP (2007) Current concepts in antibiotic-resistant gram-negative bacteria. Expert Rev Anti-Infect Ther 5(5):833–843

Venkatesan M, Fruci M, Verellen LA, Skarina T, Mesa N, Flick R, Pham C, Mahadevan R, Stogios PJ, Savchenko A (2023) Molecular mechanism of plasmid-borne resistance to sulfonamide antibiotics. Nat Commun 14(1):4031

Wang Z, Koirala B, Hernandez Y, Zimmerman M, Park S, Perlin DS, Brady SF (2022) A naturally inspired antibiotic to target multidrug-resistant pathogens. Nature 601(7894):606–611

Waseem T, Zargaham MK, Shahid F, Rajput TA, Ibrahim B, Babar MM (2020) New approaches to antimicrobial discovery: current development and future prospects. In: New and future developments in microbial biotechnology and bioengineering. Elsevier, pp 67–77

Mechanism of Antibiotics and Potential Replacements

Awais Hameed and Tayyaba Ali

Abstract The introduction of antibiotics into the field of medicine drastically lowered mortality rates caused by bacterial infections. Since the discovery of penicillin in the year 1928, various antibiotics have been developed that target protein synthesis, nucleic acid replication, metabolism, and the bacterial cell wall. The use of antibiotics, however, has come under threat due to the escalating phenomenon of antimicrobial resistance (AMR), primarily due to the bacteria's self-adaptive processes like enzymatic degradation, modification of efflux pumps and target sites, and biofilm formation. Throughout the world, the misuse and overuse of antibiotics in healthcare and agriculture have enabled the propagation of multidrug-resistant pathogens, presenting a serious issue for public health. To tackle the issue of AMR, new therapeutic alternatives are being constructed, including, but not limited to, bacteriophages, nanotechnology-based antimicrobials, enzymes, and antimicrobial peptides. Immunomodulatory agents coupled with vaccines should decrease the dependence on antibiotics. Furthermore, anti-resistance factors can be supplemented through more advanced synthetic biology tools and CRISPR–cas9 gene editing technologies. To truly tackle the problem, collaboration of countries along with strict regulations regarding antibiotics, continuous funding in research, and policies is the key to ensuring that effective solutions do not lose their effectiveness.

Keywords Antibiotics · Antimicrobial Resistance (AMR) · Antibiotic mechanism of action · Multidrug resistance · Global health policy · Alternative to antibiotics

A. Hameed · T. Ali (✉)
Department of Bioinformatics and Biotechnology, Government College University, Faisalabad, Pakistan

R. U. Khan, S. Naz (eds.), *Green Alternatives to Antibiotics in Animal Health*,
https://doi.org/10.1007/978-981-92-0309-3_4

1 Introduction

Antibiotics play a major role in modern medicine by improving illness treatment and lowering mortality rates. Alexander Fleming's 1928 discovery of penicillin was celebrated in the medical community since antibiotics were a revolutionary development. All infections, from minor bacterial invasions to life-threatening cases of sepsis and tuberculosis, require life-saving antibiotic treatment. Scientists developed various antibiotic families consisting of aminoglycosides, macrolides, and quinolones to effectively treat individual bacterial infections. Medicinal antibiotics extend beyond therapeutic applications for human beings (Zinner 2007). Antibiotics provide advantages to the veterinary fields and agricultural industries, as well as biotechnological applications. These medical products have multiple fields of use, which demonstrates their importance. Antibiotics remain crucial medicines, while antimicrobial resistance (AMR) threatens their effectiveness severely. We need to investigate the present ability of antibiotics, along with researching novel treatment strategies without delay (Alós 2015). Knowledge about antibiotic function structures through research serves as the foundation to achieve better tromping antibiotic resistance throughout the lasting problem of antibiotic resistance. Antibiotics destroy bacterial processes through attacks on cell wall formation, protein synthesis, nucleic acid reproduction, and cellular metabolic activities. The mechanism of penicillin beta-lactams prevents bacterial cell wall generation before causing bacterial cell death through wall destruction (Denakpo et al. 2023). Tetracyclines, along with macrolides, prevent protein synthesis by obstructing ribosomes. These antibiotic actions exploit the base structural differences between bacteria and eukaryotic cells to achieve successful removal of the infectors and eukaryotic cells. Bacteria possess rapid evolutionary capacity, which allows them to create more sophisticated self-defence systems against these processes. Beta-lactamase enzymes produced by bacteria lead to antibiotic breakdown, while target site modifications and antibiotic expulsion mechanisms are some additional forms of bacterial resistance. The speed of multi-drug–resistant pathogen dissemination becomes worse because bacteria transfer these resistance genes through horizontal gene transfer. Examining antibiotics alongside bacterial treatment-resistance mechanisms enables the development of strategies which preserve vital medications (Garner et al. 2020).

Recent outcomes of antibiotic resistance call for revolutionary methods to combat bacterial infections. The recognition of resistance as a global health emergency imposes serious threats to the success of antibiotic treatment for standard infections, along with vital medical procedures such as surgical operations and chemotherapeutic treatments along with organ transplant protocols (Tarín-Pelló et al. 2022). The World Health Organization (WHO) asserts that failure to take immediate action will lead the world toward a future without available treatments for routine infections as well as stressed outcomes from typical injuries (Cosentino et al. 2023). This crisis emerged from frequent and incorrect antibiotic distribution within medical facilities and agricultural domains, insufficient supervisory protocols, and limited innovation in antibiotic development due to market restrictions. The urgent situation calls for alternative treatment methods to replace traditional antibiotics because they assure our ability to fight bacterial infections (Stocco et al. 2020).

2 Antimicrobial Mechanisms of Action

The mechanism of antimicrobial agents includes cell wall synthetic inhibition by beta-lactams and protein synthesis disruption by tetracyclines and aminoglycosides and their impact on nucleic acid replication through quinolones. The bacteria experience leakage of their vital contents when some antimicrobials disrupt the cell membranes (e.g., polymyxin) and sulphonamides stop folic acid from being synthesised. Antimicrobial agents work by attacking structures and processes which exist only in pathogens without affecting the structures of host cell tissue.

2.1 Molecular Mechanisms of Action

Specific molecular targets among bacterial structures, along with enzymes, form the fundamental behaviour of antibiotics. The active site of penicillin-binding proteins (PBPs) remains bound to β-lactams as they imitate these enzymes' natural substrates, therefore preventing their enzymatic functions (Epand et al. 2016). The penicillin-like antibiotics disable peptidoglycan cross-linking functions and activate destructive autolytic enzymes, which degrade the bacterial cell wall. Protein synthesis inhibitors make use of bacterial ribosomes' distinct structural characteristics to prevent translation in bacteria without affecting the translational machinery of eukaryotic ribosomes. The A-site of the 30S subunit provides binding space for tetracyclines, which inhibit amino acid placement in the growing polypeptide chain (Hancock Robert and Chapple Daniel 1999).

These molecular processes produce immense effects on bacteria's physiological functions. The essential cellular functions of antibiotics get blocked, which leads cells to experience stress reactions and then deplete energy supplies until they die. When bacteria receive treatment with DNA gyrase inhibitor fluoroquinolones, their DNA replication stops while double-strand breaks occur, which forces bacterial repair pathways into overdrive until the cell dies. Membrane disruptors within the polymyxin category rapidly harm bacterial membranes to produce fatal outcomes for bacteria (Evans et al. 1999).

2.2 Classification of Antibiotics by Target

The classification system for antibiotics depends on which bacterial cell components they target. Beta-lactams that include penicillins, along with glycopeptides like vancomycin, belong to the category of cell wall synthesis inhibitors, which stop bacteria from generating their rigid cell walls. The group of antibiotics, which includes tetracyclines, macrolides and aminoglycosides, prevents bacterial ribosome function through protein synthesis inhibition, while the group of antibiotics, including fluoroquinolones and rifamycins, interrupts both DNA replication and transcription processes (Table 1). Polymyxins serve as cell membrane disruptors, while sulphonamides, together with trimethoprim, inhibit metabolic pathway functions in bacteria.

Table 1 Antibiotics mechanism of action, with examples and target bacterial species

Antibiotic mechanism of action	Examples	Bacterial species targeted
Inhibition of cell wall synthesis	Penicillins (e.g., Penicillin G, Amoxicillin), Cephalosporins (e.g., Ceftriaxone, Cefazolin), Vancomycin	Gram-positive bacteria (e.g., *Staphylococcus aureus*, *Streptococcus pneumoniae*, *Enterococcus faecalis*), some Gram-negative bacteria (e.g., *Neisseria gonorrhoeae*, *Haemophilus influenzae*)
Disruption of protein synthesis	Macrolides (e.g., Erythromycin, Azithromycin), Aminoglycosides (e.g., Gentamicin, Tobramycin), Tetracyclines (e.g., Doxycycline, Minocycline)	Gram-positive bacteria (e.g., *Streptococcus pneumoniae*, *Staphylococcus aureus*, *Listeria monocytogenes*), Gram-negative bacteria (e.g., *Escherichia coli*, *Haemophilus influenzae*, *Neisseria gonorrhoeae*), Atypical bacteria (e.g., *Chlamydia trachomatis*, *Mycoplasma pneumoniae*)
Inhibition of nucleic acid synthesis	Fluoroquinolones (e.g., Ciprofloxacin, Levofloxacin), Rifamycins (e.g., Rifampin, Rifabutin)	Gram-negative bacteria (e.g., *Pseudomonas aeruginosa*, *Escherichia coli*, *Neisseria gonorrhoeae*), Gram-positive bacteria (e.g., *Streptococcus pneumoniae*, *Staphylococcus aureus*), Mycobacteria (e.g., *Mycobacterium tuberculosis*)
Disruption of cell membrane integrity	Polymyxins (e.g., Polymyxin B, Colistin), Daptomycin	Gram-negative bacteria (e.g., *Pseudomonas aeruginosa*, *Klebsiella pneumoniae*, *Acinetobacter baumannii*), Gram-positive bacteria (e.g., *Staphylococcus aureus*, *Enterococcus faecium*)
Inhibition of metabolic pathways	Sulphonamides (e.g., Sulphamethoxazole), Trimethoprim (often in combination as Cotrimoxazole)	Broad-spectrum (e.g., *Escherichia coli*, *Streptococcus pneumoniae*, *Enterococcus faecalis*, *Shigella dysenteriae*, *Haemophilus influenzae*)

2.2.1 Cell Wall Synthesis Inhibitors

Many antibiotic classes aiming to block bacterial cell wall synthesis operate at distinct stages of this essential biological process. β-Lactams that include penicillins (amoxicillin, penicillin G) as well as cephalosporins (ceftriaxone, cefepime) and carbapenems (imipenem, meropenem) and monobactams (aztreonam) all interact with penicillin-binding proteins (PBPs). Cross-linking of peptidoglycan strands is the responsibility of these enzymes (Walsh 2000). The permanent attachment of β-lactams to PBPs obstructs proper cell wall construction, thereby making bacterial cell walls sensitive to osmotic pressure until they fragment and lyse. Glycopeptides work against *methicillin-resistant Staphylococcus aureus* (MRSA) with vancomycin and Enterococcus infections with teicoplanin because they directly link with the D-Ala-D-Ala termination of peptidoglycan precursors. The antibiotics prevent transglycosylase and transpeptidase enzymes from performing their essential functions in peptidoglycan elongation (Blair et al. 2015).

Bacitracin, used in topical ointments, blocks bacterial growth of Gram-positive infections caused by *Staphylococcus aureus* by preventing the reuse of bactoprenol lipid molecules needed for peptidoglycan precursor transmembrane transport. Bactoprenol activation is required for peptidoglycan precursors to enter the cell wall, so bacteria cannot grow anymore (Kohanski et al. 2010). Fosfomycin represents an effective medication for treating urinary tract infections originating from multidrug-resistant *Escherichia coli* and *Klebsiella pneumoniae* through its unique method of action. Fosfomycin denotes its activity against MurA to prevent early peptidoglycan precursor synthesis, thus causing the breakdown of bacterial cell wall formation (Wright 2005).

During antibacterial therapy, lipoglycopeptides dalbavancin and oritavancin seek and bind to bacterial membranes through their lipophilic side chains in addition to blocking the D-Ala-D-Ala terminal, thus achieving enhanced effect duration and increased potency against vancomycin-resistant *Enterococcus* (VRE) (Delcour 2009). The antibacterial agent daptomycin treats MRSA and VRE bacteraemia infections through membrane destruction, even though it does not stop cell wall formation directly. The process of rapid depolarisation due to this action creates instability in the cell wall by preventing the synthesis of DNA, RNA and proteins (Wright 2005).

2.2.2 Protein Synthesis Inhibitors

Antibiotics that belong to the Protein synthesis inhibitors category function by attacking the bacterial ribosome, which performs essential functions such as mRNA protein translation. The antibiotics use ribosome structure differences between 70S prokaryotic and 80S eukaryotic ribosomes to target bacterial cells safely passing human cells. The 30S small subunit and 50S large subunit constitute bacterial ribosomes, while these ribosome subunits provide specific antibiotic targets that block different stages of protein synthesis from initiation through elongation and termination (Kohanski et al. 2010).

Macrolides (e.g., erythromycin, azithromycin, and clarithromycin) bind reversibly to the 23S rRNA of the 50S ribosomal subunit, specifically at the peptide exit tunnel. The antibiotic binds at a specific location in the ribosome to prevent the passage of the expanding polypeptide chain and causes ribosome stalling, which halts protein chain elongation (Walsh 2000). Macrolide antibiotics prove highly beneficial against pneumococcal *Streptococcus pneumoniae* and *Haemophilus influenzae* among Gram-positive bacteria. The presence of resistance mechanisms, which include ribosomal methylation through erm genes along with efflux pumps, diminishes the effectiveness of these antibiotics.

Aminoglycosides (e.g., gentamicin, streptomycin, tobramycin, and amikacin) bind irreversibly to the 16S rRNA of the 30S ribosomal subunit. The binding activity disrupts the decoding operation of the ribosome, which results in incorrect translation of mRNA codons, thus producing improper, toxic, functional, or nonfunctional proteins (Mingeot-Leclercq and Tulkens Paul 1999). Aminoglycosides prevent the ribosome from moving forward along the decoded mRNA sequence during translation. Aminoglycosides function as bactericidal agents, which demonstrate superior effectiveness against the aerobic Gram-negative bacteria *Escherichia coli* and *Pseudomonas aeruginosa* and belong to the group of protein synthesis inhibitors, but differ from most others in their killing ability. Drug resistance emerges primarily through the 16S rRNA mutating and the enzymatic acetylation or phosphorylation of drugs (Blair et al. 2014).

Tetracyclines (e.g., doxycycline, tetracycline, and minocycline) bind reversibly to the 16S rRNA of the 30S ribosomal subunit at the A-site. Since aminoacyl-tRNA cannot attach to the ribosome, the polypeptide chain development ends, and bacterial growth stops. Tetracyclines are broad-spectrum antibiotics active against both Gram-positive and Gram-negative bacteria, as well as atypical pathogens such as Chlamydia and Mycoplasma. However, resistance mechanisms, including efflux pumps and ribosomal protection proteins, have reduced their clinical effectiveness (Xu et al. 2014). Resistance mechanisms, which consist of efflux pumps and ribosomal protection proteins, have reduced their effectiveness (Xu et al. 2014).

The 50S ribosomal subunit features the 23S rRNA, to which Chloramphenicol binds to block the essential peptide bond formation mechanism known as peptidyl transferase. This inhibition halts the elongation process of protein synthesis. The antibiotic chloramphenicol proves successful against *Haemophilus influenzae* and *Salmonella typhi,* but doctors typically limit its use because it causes severe adverse effects, including aplastic anaemia when treating fatal infections of typhoid fever and meningitis (Poole 2000). Resistances toward chloramphenicol occur due to enzymatic processes that lead to acetylation of chloramphenicol.

2.2.3 Nucleic Acid Synthesis Inhibitors

The process of DNA replication and transcription needed for bacterial survival and proliferation gets disrupted by antibiotics that target nucleic acid synthesis. These antibiotics take advantage of specific bacterial enzymes to treat various infectious bacteria with reduced side effects on human cells. The growth of resistance strategies against antibiotics reduces their effectiveness in the future (Blair et al. 2015).

The DNA gyrase, along with topoisomerase IV enzymes, form the primary targets for Fluoroquinolones, which include ciprofloxacin, levofloxacin, and moxifloxacin. DNA gyrase creates negative supercoils in DNA to reduce replication strains, and topoisomerase IV releases tension in duplicated DNA. Fluoroquinolones cause the inhibition of DNA enzyme activities by binding to these enzymes, which results in improper DNA supercoiling and decatenation processes. The impediment of bacterial replication causes DNA break accumulation alongside DNA replication blockage, which ends in bacterial death. Because fluoroquinolones show effectiveness against the Gram-negative *E. Coli* and Gram-positive *S. Pneumoniae* and *P. Aeruginosa* pathogens, the drugs effectively treat urinary tract infections, respiratory infections and gastroenteritis (Ehmann and Lahiri 2014). The main factor contributing to growing fluoroquinolone drug resistance stems from mutations within the quinolone resistance-determining regions (QRDRs) of DNA gyrase and topoisomerase IV. Bacterial cells successfully remove antibiotics through the resistance mechanism of efflux pump overexpression.

The Rifamycins group, which includes rifampin as well as rifabutin and rifaximin, block bacterial RNA polymerase functions so that DNA transcription into RNA cannot occur. The β-subunit of bacterial RNA polymerase has rifamycins bound to ıt, which block transcription initiation, thus preventing mRNA synthesis and necessary bacterial protein production (Wehrli 1983). Such drug activity demonstrates special efficacy against *Mycobacterium tuberculosis* and related mycobacterial microorganisms, which cause tuberculosis. The antibiotic properties of rifampin extend to treating *Staphylococcus aureus* and *Neisseria meningitidis,* along with Gram-positive and negative bacterial species. Rifamycins occupy a primary position as tuberculosis treatment drugs because health professionals use them as the first option, combined with other antibiotics, to stall resistance formation (Pang et al. 2013). The development of rifamycin resistance originates from mutations occurring in the rpoB gene that produces the β-subunit of RNA polymerase, which modifies the binding site for the drug to cause reduced therapeutic effect. Rifamycin resistance demonstrates why it is essential to use rifamycins alongside other antibiotics during treatment, because it blocks the speedy development of resistant bacterial strains (Appelbaum and Hunter 2000).

2.2.4 Metabolic Pathway Inhibitors

The therapeutic agents from metabolic pathway inhibitors disrupt bacterial-specific biosynthetic pathways that maintain essential biological processes for bacterial growth and survival. The bacterial pathways in human cells show essential differences from these pathways because of which these antibiotics maintain their bacterial-specificity (Biswas et al. 2012). The folate biosynthesis pathway becomes the target of two common metabolic pathway inhibitors, sulphonamides, and trimethoprim, leading to DNA and RNA synthesis slowdown in bacterial cells.

Bacterial production of folic acid requires para-aminobenzoic acid (PABA) as a substrate, but sulphonamides occupy this position to prevent folic acid synthesis. The production of nucleotides requires folic acid since nucleotides act as building blocks for DNA and RNA. Sulphonamides function as competitors with PABA to bind dihydropteroate synthase, which constitutes an enzyme responsible for integrating PABA into the folic acid synthesis pathway. The blockade of dihydropteroate synthase through sulphonamide action interferes with folic acid precursor production and results in bacterial DNA and RNA synthesis blockage due to folic acid deficiency (Epand et al. 2016). Sulphonamide drugs effectively eliminate a wide range of Gram-positive and Gram-negative bacteria as a result of their blocking mechanism at dihydropteroate synthase. This inhibition prevents PABA from binding to the enzyme and stops folic acid synthesis. Sulphonamides prove effective against *Escherichia coli* along with *Streptococcus pneumoniae* and *Haemophilus influenzae* (Evans et al. 1999).

The antibacterial substance trimethoprim stops dihydrofolate reductase from functioning properly in the folate biosynthesis pathway, where it converts dihydrofolic acid into tetrahydrofolic acid. The synthesis of pyrimidines and purines requires tetrahydrofolic acid to build nucleotides, which produces bacterial DNA and RNA (Hancock Robert and Chapple Daniel 1999). The blockade of dihydrofolate reductase by trimethoprim interrupts folate metabolic functions alongside nucleotide production, which results in inhibited bacterial cell multiplication. Trimethoprim can effectively combat bacterial infections from two major species, including *Staphylococcus aureus* and *Enterococcus species* (Kosmidis and Levine 2010).

Sulphonamides and trimethoprim combined produce enhanced antibacterial results since they block separate yet cooperative processes within folate synthesis. Medical practitioners often refer to the combined drugs TMP–SMX as cotrimoxazole for treating infections from *Escherichia coli, Staphylococcus aureus*, and *Pneumocystis jirovecii*. The joint action between these medications boosts their antibacterial work while minimising bacterial adaptability to drugs. The dual drug treatment shows effectiveness when treating urinary tract infections and respiratory infections, as well as particular forms of pneumonia (Mohapatra et al. 2021).

2.2.5 Membrane Disruptors

Membrane disruptors destroy the bacterial cell membrane structure, which protects the cell while regulating ion flow through its barrier. These antibiotics break down the membrane through which cells eventually perish by causing depolarisation alongside essential content leakage. Their effectiveness against Gram-negative bacterial species is high since these bacteria possess an outer membrane which blocks many antibiotic types (Walsh 2000).

The antibiotic group known as polymyxins (including polymyxin B along with colistin) specifically attacks the lipid A portion of lipopolysaccharide molecules located in Gram-negative bacteria's outer membranes. The bacterial membrane depends on lipopolysaccharides for both structural maintenance and membrane stability purposes. Polymyxins attach themselves to the lipid A part of LPS, which destabilises the compound structure while enlarging bacterial membrane permeability. The membrane disruption creates a passage through which bacterial contents such as ions and proteins escape, leading to fatal damage to the bacterial cell (Evans et al. 1999). Polymyxins show strong efficacy against three significant multidrug-resistant Gram-negative bacteria, including *Pseudomonas aeruginosa,* along with *Acinetobacter baumannii* and *Klebsiella pneumoniae*. Polymyxins serve purposes in treating serious multidrug-resistant organism infections, as their toxicity, especially renal damage, makes them suitable only for cases with no other available treatment options (Mohapatra et al. 2021).

Daptomycin, as a lipopeptide antibiotic, creates pores in bacterial membranes by a mechanism that depends on calcium. Through its insertion into bacterial cell membranes, Daptomycin forms pores, which results in excessive potassium passage alongside other ions through the membrane channels. The cellular process of membrane potential loss (depolarisation) harms ATP production and enzyme activity because of membrane disruption (Kosmidis and Levine 2010). Bacterial cells lose their ability to preserve their internal conditions and die because of this condition. Daptomycin demonstrates exceptional efficacy against *Staphylococcus aureus* and *Enterococcus faecalis* among Gram-positive bacteria, so healthcare providers use it for treating complicated skin and soft tissue infections and both bacteraemia and endocarditis. The therapeutic range of this drug applies only to Gram-positive organisms, while the treatment creates muscle toxicities in patients.

3 Broader Impacts of Antimicrobials

Medical treatment with antibiotics affects both the microbial ecosystem and the host organism through many important changes. The microbes that reside inside the human body experience alterations with antibiotic treatment since this shifts the helpful bacterial population structure toward opportunistic pathogens (Coates et al. 2020). The disturbance in bacterial composition, which doctors refer to as dysbiosis, connects with different illnesses like gastrointestinal problems, metabolic diseases and immune system problems. The use of broad-spectrum antibiotics enables the growth of *Clostridioides difficile* until it causes severe colitis (Denakpo et al. 2023).

The administration of antibiotics leads to additional modifications in both host immunity and disease results. The decrease in bacterial numbers reduces inflammation, so the immune system rebuilds itself. Long-term antibiotic exposure, together with wrong medication timing, leads to damaged immune function since the microbiome directly influences how the body controls its immune response (Majalekar and Shirote 2020). The extensive use of antibiotics in agriculture, combined with animal husbandry practices, causes environmental contamination through which resistance genes propagate further in soil and water microbial communities.

Knowledge about antibiotic working mechanisms, along with their wider effects, enables proper utilisation in medicine and helps scientists create resistance-mitigation approaches. The global challenge of antimicrobial resistance can only be addressed through maintaining current antibiotic effectiveness and developing new therapeutic options because resistance mechanisms continue to form Cosentino et al. (2023).

4 Resistance Pathways and Mechanisms

Worldwide concern for antibiotic-resistant microbes grows because these microorganisms reduce the success of antibiotic treatments while increasing footprints on infection management systems. Microorganisms develop antibiotic resistance through diverse, complex systems, which include genetic modifications and obtaining resistance genes.

4.1 Genetic Basis of Antibiotic Resistance

Bacteria obtain and distribute genetic changes through which they develop survival abilities against antimicrobial agents. Resistance development results from genetic changes that occur either through chromosomal mutations or horizontal gene transfer (HGT). The ability of bacteria to resist antibiotic pressure leads to strain development that becomes resistant to standard therapeutic drugs.

4.1.1 Chromosomal Mutations

Bacterial DNA carries changes called chromosomal mutations when alterations appear in its sequence. Genetic alterations can occur naturally during DNA replication or develop because of environmental conditions that include radiation chemicals, and various types of stress. Occurring only rarely, chromosomal mutations produce major effects on bacterial antibiotic susceptibility when they take place. The mutations affect drug targets by making antibiotics less effective through diminished drug uptake and increased drug displacement properties (Chaplin et al. 2021).

Point mutations function as one of the frequent forms of chromosomal mutations by replacing individual base pairs within DNA. Point mutations can transform essential protein and enzyme amino acids, which end up modifying antibiotic binding point structures. *Mycobacterium tuberculosis* acquires genetic changes in the gyrA gene, which modify DNA gyrase as a fluoroquinolone receptor site (Hong et al. 2015). The mutational modification weakens the antibiotic-binding capacity of fluoroquinolones, thus making the drugs ineffective. A single gene mutation in *Streptococcus pneumoniae* targeting penicillin-binding protein (PBP) weakens the antibiotic binding capacity, causing penicillin resistance (Hussain et al. 2021).

Gene duplication functions as a major mutation mechanism that generates multiple gene copies, which increases the amount of protein expression. The increased amount of target protein due to gene duplication weakens antibiotic potency because the drug has more compounds to bind to and reduces its effectiveness. The abnormal increase of β-lactamase enzymes in *Pseudomonas aeruginosa* leads to antibiotic resistance because it destroys the β-lactam antibiotic before it can perform its effect (Jian et al. 2021).

Target modification contributes substantially to resistance development. Bacteria achieve drug resistance by adapting their molecular targets in which antibiotics would normally bind, including enzymes as well as ribosomal subunits, through mutating their shape and functional properties (Medakina et al. 2023). Bacterial survival occurs since the modified target site stops allowing effective antibiotic binding. *Staphylococcus aureus* develops penicillin-binding protein 2a (PBP2a), which exhibits weak methicillin and β-lactam antibiotic affinity while living in drug environments, hence allowing methicillin-resistant *Staphylococcus aureus* (MRSA) strains to survive (Poyart et al. 2003).

Increasing the amount of antibiotic target proteins within bacterial cells contributes to drug resistance. When *Escherichia coli* cells produce increased levels of ribosomal RNA (rRNA), it results in tetracycline antibiotics having reduced affinity to bacterial ribosomes, therefore allowing the bacteria to live through elevated drug concentrations.

4.1.2 Horizontal Gene Transfer (HGT)

Through Horizontal Gene Transfer (HGT), bacteria efficiently obtain new genetic material with resistance genes from additional bacteria. HGT distinguishes itself from chromosomal mutations because it enables the bacterial population to get resistance traits and exchange genes among different species (Munita and Arias 2016). Resistance genes spread through various mechanisms that include plasmids, transposons and integrons, as each element serves independently in the spread of antibiotic resistance in Table 2.

Table 2 Genetic basis of antibiotic resistance of different species

Mechanism of antibiotic resistance	Example	Bacterial species
Chromosomal mutations	Point mutations in gyrA gene leading to reduced affinity for fluoroquinolones in *Mycobacterium tuberculosis* Gene duplication in *Pseudomonas aeruginosa*, leading to overproduction of β-lactamase enzymes	*Mycobacterium tuberculosis* (fluoroquinolone resistance) *Pseudomonas aeruginosa* (β-lactam resistance)
Point mutations	gyrA mutation in *Escherichia coli* resulting in fluoroquinolone resistance. PBP mutation in *Streptococcus pneumoniae* leading to reduced affinity for penicillin	*Escherichia coli* (fluoroquinolone resistance) *Streptococcus pneumoniae* (penicillin resistance)
Gene duplication	Gene duplication in *Streptococcus pneumoniae*, leading to reduced effectiveness of penicillin by producing more PBPs that are less sensitive to β-lactams	*Streptococcus pneumoniae* (penicillin resistance)
Target modifications	Modified PBP2a in MRSA (Methicillin-resistant *Staphylococcus aureus*) causing resistance to methicillin and other β-lactams. Altered rRNA in Streptococcus pneumoniae conferring resistance to macrolides like erythromycin	*Staphylococcus aureus* (MRSA) *Streptococcus pneumoniae* (macrolide resistance)
Overproduction of target	Overproduction of ribosomal RNA in *Escherichia coli* making the bacterium less susceptible to tetracycline	*Escherichia coli* (tetracycline resistance)
Horizontal gene transfer (HGT)	Plasmid-mediated transfer of bla gene, causing β-lactam resistance in *Klebsiella pneumoniae*. Tn4001 transposon in *Staphylococcus aureus* carrying aminoglycoside resistance genes. Class 1 integrons in *Enterobacteriaceae* carrying resistance to aminoglycosides and fluoroquinolones	*Klebsiella pneumoniae* (β-lactam resistance) *Staphylococcus aureus* (aminoglycoside resistance) *Enterobacteriaceae* (multidrug resistance)
Plasmids	bla gene on plasmids in *Escherichia coli* resulting in β-lactam resistance. Transfer of carbapenem resistance genes on plasmids from *Klebsiella pneumoniae* to *Escherichia coli*	*Escherichia coli* (β-lactam resistance) *Klebsiella pneumoniae* (carbapenem resistance)
Transposons	Tn4001 transposon in *Staphylococcus aureus* carrying resistance to aminoglycosides like gentamicin and tobramycin Tn21 transposon in *Escherichia coli* conferring resistance to sulphonamides and trimethoprim	*Staphylococcus aureus* (aminoglycoside resistance) *Escherichia coli* (sulphonamide and trimethoprim resistance)
Integrons	Class 1 integrons in *Enterobacteriaceae* carrying resistance to aminoglycosides, fluoroquinolones, and sulphonamides Class 2 integrons in *Pseudomonas aeruginosa* enabling resistance to β-lactams and aminoglycosides	*Enterobacteriaceae* (multidrug resistance) *Pseudomonas aeruginosa* (β-lactam and aminoglycoside resistance)

The DNA structure known as a plasmid maintains its circular form as independent genetic information outside bacterial chromosome sequences. Antibiotic resistance genes exist on multiple plasmids of bacteria that spread between bacterial cells through conjugation, which uses direct cell-to-cell contact. Self-transferring plasmids enable antibiotic resistance gene transmission from one bacterial species to another, thus rapidly spreading antibiotic-resistant bacteria throughout entire bacterial communities. In *Klebsiella pneumoniae* bacteria, the bla gene, which produces β-lactamase, usually exists on plasmids. These plasmids enable bacterial transfer between different sorts of microorganisms while granting β-lactam antibiotic resistance to such organisms as *Escherichia coli* (Poirel et al. 2011).

Transposons function as "jumping genes" which possess the capability to move between genomic areas of bacteria or between plasmids and chromosomes. The internal movement capabilities of transposons make it easy for them to transport antibiotic resistance genes between bacterial organisms. The transposon Tn4001, located in *Staphylococcus aureus* contains resistance to gentamicin and tobramycin, which can transfer resistance genes to different bacterial strains (Pontes and Groisman 2020). The transposons generate integration spaces that allow resistance genes to enter both bacterial genomes and plasmids, thus enlarging the antibiotic resistance genetic material.

HGT contains integrons as genetic elements that show the capability to capture and integrate resistance gene cassettes. Integrons enhance the quick acquisition of resistance through their antibiotic-resistance gene-containing gene cassettes. Class 1 integrons occur frequently within *Enterobacteriaceae* bacteria while simultaneously containing resistance genes against aminoglycosides and fluoroquinolones alongside various other antibiotic substances. Integrons help bacteria exchange genetic material, during which process they speed up the development of drug-resistant microorganisms in hospital environments (Salyers 2002).

Plasmid-mediated conjugation represents an HGT mechanism which swiftly spreads resistance genes throughout distinct bacterial populations and makes their treatment challenging.

Antibiotic resistance poses particular risks in healthcare facilities and hospitals because resistance spreads rapidly among pathogenic organisms (Wang et al. 2023).

4.2 *Biochemical Mechanisms of Antibiotic Resistance*

4.2.1 Enzymatic Degradation

The enzymatic breakdown mechanism stands as one of the main strategies through which bacteria become impervious to antibiotic treatment. Specific enzymes produced by bacteria cause the breakdown or chemical alteration of antibiotics, thus making them useless. Bacteria manifest this resistance against β-lactam antibiotics and aminoglycosides together with some additional antibiotic classes (Darby et al. 2023).

Among all enzymes involved in enzymatic degradation, β-lactamases have earned the status of being the most prominent ones. β-lactamase enzymes degrade different classes of antibiotics, including penicillins, by breaking down their β-lactam band to stop antibiotic binding with penicillin-binding proteins engaged in bacterial cell wall synthesis. β-lactamases fall under different structural groups in Table 3. Plasmids carry TEM and SHV enzymes of Class A, but OXA-type enzymes under Class D are responsible for carbapenem resistance. β-lactamases with carbapenemase activity belong to carbapenemases as a β-lactamase subgroup, and they can break down carbapenem antibiotics that health professionals save to treat drug-resistant infections. The most recognised carbapenemase variants consist of KPC (*Klebsiella pneumoniae* carbapenemase), NDM (New Delhi metallo-β-lactamase), and VIM (Verona integron-encoded metallo-β-lactamase) (Abajo 2024).

Bacterial resistance to aminoglycosides, including gentamicin, tobramycin, and amikacin, is significantly affected by the activity of aminoglycoside-modifying enzymes (AMEs) in addition to β-lactamases. The bacterial enzymes add functional groups such as acetyl, phosphate and adenyl groups to aminoglycosides, following which these antibiotics become unable to bind to ribosomes, thus inhibiting protein

Table 3 Antibiotic resistance mechanisms, corresponding drugs, bacterial species impacted, and WHO-recommended dosing strategies for treatment

Resistance mechanism	Antibiotics	Bacterial species	WHO recommendations and dosing
Enzymatic degradation	Penicillins, Cephalosporins, Gentamicin, Erythromycin, Ticarcillin, Piperacillin	*E. coli, Klebsiella, P. aeruginosa, Enterobacter*	Use β-lactamase inhibitors, e.g., penicillin 500 mg q6h
Alteration of target sites	Penicillins, Macrolides, Fluoroquinolones, Vancomycin, Tetracyclines, Clindamycin	*MRSA, S. pneumoniae, Enterococcus faecalis*	Treat MRSA with vancomycin (15 mg/kg q12h)
Efflux pumps	Fluoroquinolones, Aminoglycosides, Tetracyclines, Chloramphenicol, Ciprofloxacin	*P. aeruginosa, A. baumannii, E. coli, Streptococcus* spp.	Use combination therapy, e.g., ciprofloxacin 500 mg BID
Reduced permeability	Carbapenems, Penicillins, Tetracyclines, Ciprofloxacin, Colistin	*Klebsiella, A. baumannii, P. aeruginosa, Enterococcus*	Early use of meropenem 500 mg q8h
Biofilm formation	Aminoglycosides, Ciprofloxacin, Vancomycin, Clindamycin, Imipenem	*P. aeruginosa, S. aureus, Enterococcus* spp.	Gentamicin (3–5 mg/kg/day), combined with other agents
Persister cells	Fluoroquinolones, Beta-lactams, Rifampin, Vancomycin	*S. aureus, E. coli, P. aeruginosa*	Use ciprofloxacin 500 mg q12h for 7–14 days
Stress response mechanisms	Fluoroquinolones, Sulphonamides, Trimethoprim, Beta-lactams	*S. pneumoniae, E. coli, Klebsiella pneumoniae*	Complete course, e.g., Sulphamethoxazole/trimethoprim 160/800 mg q12h

synthesis (Alekshun and Levy 2007). The different types of AMEs include acetyltransferases, together with phosphotransferases and adenyltransferases. The drug modification diminishes treatment effectiveness so that bacteria succeed in producing proteins.

Healthcare professionals can use chloramphenicol acetyltransferase (CAT) to deactivate chloramphenicol through its ability to add an acetyl group to the antibiotic molecule. Bacterial ribosomes cannot attach to chloramphenicol when it has this modification, which makes the antibiotic ineffective as a protein synthesis inhibitor.

Bacterial species such as *Escherichia coli* and *Staphylococcus aureus* exhibit resistance through enzymatic degradation. E. coli expresses the β-lactamase enzymes TEM and SHV that break down penicillins and cephalosporin antibiotics. The strains of *E. coli* resistant to carbapenems can generate either NDM-1 or KPC carbapenemases (Hernández 2016). The bacterial species *Staphylococcus aureus*, including methicillin-resistant strains (MRSA), generates penicillinase enzymes as β-lactamase types, which disable penicillin drugs alongside other β-lactam antibiotic medications. KPC carbapenemases produced by *Klebsiella pneumoniae* pathogen enable this bacterium to develop resistance toward carbapenems as well as other β-lactams.

The resistance mechanism provides major effects on bacterial survival by giving bacteria the power to break down potent antibiotics. Bacteria survive antibiotic treatment because they make degrading enzymes continuously, which enable them to live through medicine-based resistance mechanisms. The spread of resistance traits in bacterial populations becomes increasingly difficult due to horizontal gene transfer mechanisms that occur through plasmids or transposons (Si et al. 2023).

4.2.2 Alteration of Target Sites

Mutations in specific bacterial proteins or enzymes cause target site alteration, which makes the antibiotic unable to connect to its destination or generate toxic effects. Antibiotics experience failure because several genetic alterations modify their target, making them ineffective.

Bacteria that are resistant to penicillin and cephalosporin antibiotics like β-lactams become impervious to treatment through genomic mutations, which alter penicillin-binding protein (PBP) targets, thus preventing antibiotic attachment. The β-lactams inhibit PBPs during cell wall synthesis, while these enzymes are vital proteins for bacterial cell wall synthesis (Egan 2018). Cell lysis occurs as a consequence. *Staphylococcus aureus* with methicillin resistance, such as MRSA, contains the mecA gene that produces a modified PBP called PBP2a, which shows a minimal affinity for both methicillin and additional β-lactams, so these bacteria can maintain their cell wall production in the antibiotic presence (Oliveira and Reygaert 2025).

Antibiotics in the groups known as macrolides and aminoglycosides attack bacterial ribosomes as their protein synthesis inhibition targets. The binding affinity of these antibiotics becomes weaker due to bacteria altering either ribosomal RNA (rRNA) or changes to ribosomal proteins. *Streptococcus pneumoniae* bacteria that show erythromycin resistance carry a methylation of their 23S rRNA compound at the macrolide attachment site that decreases erythromycin accessibility and renders the drug useless (Kumar and Schweizer 2005).

The antibiotic resistance mechanism of fluoroquinolones stems from their interaction with DNA gyrase and topoisomerase IV, which perform vital DNA supercoiling functions. Fluoroquinolone antibiotic effectiveness decreases when the genes gyrA and parC undergo mutations that decrease their affinity to bind the antibiotic. The resistance mechanism operates frequently in multidrug-resistant *Mycobacterium tuberculosis* populations (Laborda et al. 2022).

4.2.3 Efflux Pumps and Reduced Permeability

The bacterial cell employs membrane proteins called efflux pumps, which actively pump antibiotics outside the cell until reaching sub-therapeutic drug levels. The cell membrane maintains restricted permeability, which creates another essential mechanism that blocks antibiotic entry into bacterial cells (Fletcher et al. 2023). The membrane proteins known as efflux pumps operate inside different bacterial types, including *Pseudomonas aeruginosa, Escherichia coli, and Acinetobacter baumannii*. Multidrug resistance through these membrane pumps extends to the removal of fluoroquinolones together with β-lactams, aminoglycosides, and tetracyclines. The *Pseudomonas aeruginosa* MexAB–OprM efflux system functions to transport away different antibiotic categories, such as scarbapenems, cephalosporins, and quinolones and enhances multidrug resistance in this bacterium.

The process of acquiring resistance includes bacteria adjusting their outer membrane permeability (Zgurskaya et al. 2021). The protective membrane surrounding gram-negative bacteria prevents numerous antibiotics from entering the bacterial cell. Mutations affecting or downregulation of outer membrane porin protein channels slow down the penetration of antibiotics through bacterial membranes. *Klebsiella pneumoniae* strains resistant to carbapenems show mutations or disappear OmpK35 and OmpK36 porins, which block the entry of meropenem and imipenem antibiotics (Hamzaoui et al. 2018). Outer membrane porin expression in *Acinetobacter baumannii* diminishes when this bacterium attempts to restrict penetration of β-lactam antibiotics (Fig. 1).

4.2.4 Adaptive Resistance Mechanisms

The protective cellular responses activated by bacteria through adaptive resistance mechanisms enable them to survive hostile conditions, which include antibiotic treatments. The survival tactics developed by bacteria for antibiotic tolerance do not depend on evolutionary changes but operate through organismal changes in phenotype.

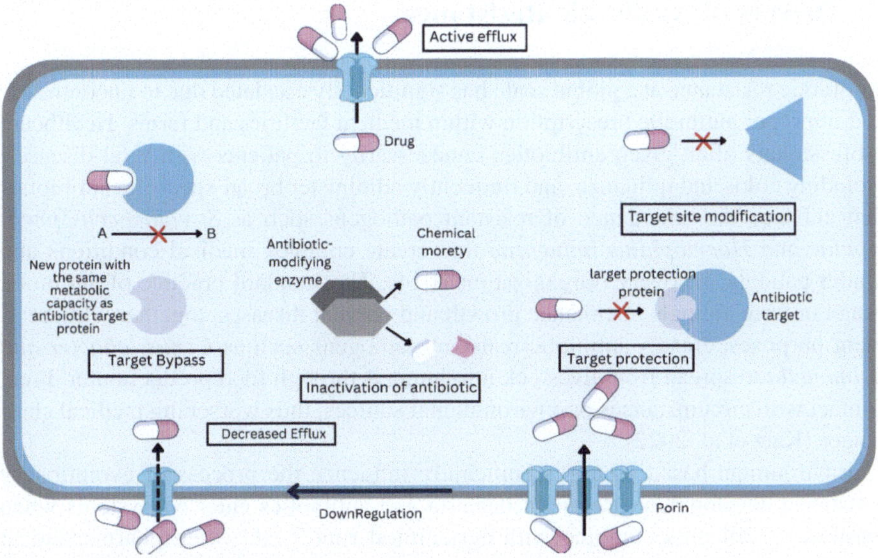

Fig. 1 Antibiotic resistance mechanisms, including active efflux, target site modification, target protection, target bypass, antibiotic inactivation, decreased efflux, and porin regulation

The formation of biofilms occurs when bacteria create clusters that hide within an extracellular matrix formed by polysaccharides, together with proteins and DNA. The drug penetration barrier of the biofilm matrix and the slow cell growth speed protect biofilm bacteria from antibiotic exposure. The bacteria inside biofilm structures demonstrate stronger drug elimination capabilities, and at the same time, they succeed in higher gene exchange during horizontal gene transfer events. Biofilms created by *Pseudomonas aeruginosa* protect these bacteria from antibiotic agents during infections like cystic fibrosis because they develop innate resistance to tobramycin and ciprofloxacin (Fletcher et al. 2023).

Microbial persister cells become dormant at both biofilm locations and antibiotic exposure situations. The reduced cell activity of persistent cells causes them to develop antibiotic resistance because antibiotics primarily affect active cellular growth and metabolism. The cells maintain survival even when exposed to elevated antibiotic amounts, yet they can quickly recover their growth capacity following antibiotic removal to cause reinfections. Persistent infection development results from persister cells formed by *Staphylococcus aureus* and *Escherichia coli*, which survive antibiotic treatments

Bacteria enact stress response pathway activation after antibiotic exposure, so they develop resistance. The SOS bacterial stress response works as a DNA damage activation system triggered by medications in the fluoroquinolone class. When SOS response activation occurs, it enables DNA repair gene expression while simultaneously producing high mutation frequencies, which promote the development of resistant bacterial strains.

5 Drivers of Antibiotic Resistance

Antibiotic resistance at a global scale has significantly escalated due to uncontrolled and improper antibiotic prescription within medical facilities and farms. Healthcare professionals often given antibiotics unnecessarily to patients with viral diseases, including colds and influenza, and frequently administer broad-spectrum antibiotics that enhance the emergence of resistant pathogens such as *Streptococcus pneumoniae* and *Haemophilus influenzae* that create complex medical conditions and hinder patients' recovery (Organization 2014). The standard practice of antibiotic usage in farm animals to enhance growth and prevent diseases, together with treatment purposes, creates antibiotic-resistant pathogens such as *Campylobacter and Salmonella* to spread from livestock into humans through food production or direct contact with circumstances or environmental sources, thus worsening medical challenges (Katz et al. 2022).

Environment-based factors significantly influence the process of evolutionary resistance development. Resistant bacteria and antibiotics enter ecosystems when wastewater effluents combine with agricultural runoff and when pharmaceutical substances are discarded incorrectly. Water sources contaminated by environmental chemicals serve as natural locations where resistance genes accumulate in rivers and lakes as well as drinking water reservoirs. The survival and spread of resistant bacteria thrive within wastewater treatment facilities and poor disposal procedures because these environments allow resistant bacteria to thrive and multiply, thus creating a more dangerous resistance problem that increases infection risks for human beings as well as animals and the wider environmental ecosystem (Van Boeckel et al. 2015).

5.1 Role of Global Health Policies and Regulation

Antibiotic resistance spread can be effectively controlled by global health policies together with regulatory mechanisms. Many countries lack effective rules for antibiotic sales, which promotes excessive utilisation and improper distribution of these drugs. The practice of selling antibiotics without medical prescriptions in several low- and middle-income countries enables improper medication by self-treating consumers. Attempts to control resistance face erosion because of this behaviour, which allows resistant pathogens to continue existing (Organization 2014).

The World Health Organization (WHO), together with the Centers for Disease Control and Prevention (CDC), has executed substantial global initiatives to inform people about antibiotic resistance and related outcomes. The WHO Global Action Plan on Antimalarial Resistance (2015) focuses on antibiotic surveillance systems as well as antibiotic management and infection control standards to support antibiotic discovery and development. Some regions face difficulties in deploying these strategies because they lack the necessary infrastructure, together with minimal political support.

Healthcare regulations have intensified in countries with high incomes, resulting in enhanced antibiotic management practices. The European Union has established regulations that prohibit antibiotics in agriculture for livestock growth enhancement. The new government standards work to decrease antibiotic-related environmental harm, together with limiting antibiotic resistance spread. Universal adoption of these initiatives remains a challenge because more countries need to participate in tight regulatory enforcement with international partners (Martinez 2009).

Policies for infection prevention and control must also form part of the strategy to minimise antibiotic use. Mass vaccination programs combined with advanced diagnostic tests and enhanced hygiene practices help cut down antibiotic use by either blocking infection development or implementing accurate antibiotic prescriptions. The control of antibiotic resistance requires strict campaigns to enhance antibiotic usage responsibility among human and veterinary medicine practitioners.

6 Alternatives to Antibiotics

6.1 Non-traditional Antimicrobials

Research scientists are developing non-traditional antimicrobials to combat resistant infections because the antibiotic resistance threat keeps escalating. The alternative approaches use different functions than traditional antibiotics, thus enabling the medical treatment of antibiotic-resistant bacterial infections Fig. 2.

6.1.1 Bacteriophages and Phage Therapy

The great potential of bacteriophages as antibiotic substitutes exists because these viruses exclusively infect bacterial cells. The infection mechanism of these viruses targets particular bacterial species in a precise manner, which allows beneficial bacteria to remain unaffected. The replication process of phages within bacterial cells produces bacterium lysis, which releases numerous phages responsible for additional bacterial infections. The therapeutic application of phage therapy shows successful results in treating *Pseudomonas aeruginosa,* together with *Staphylococcus aureus* and *Escherichia coli* bacterial infections that include drug-resistant bacterial strains. Phage therapy operates with obstacles involving phage resistance formation and regulatory restrictions, together with the necessity for

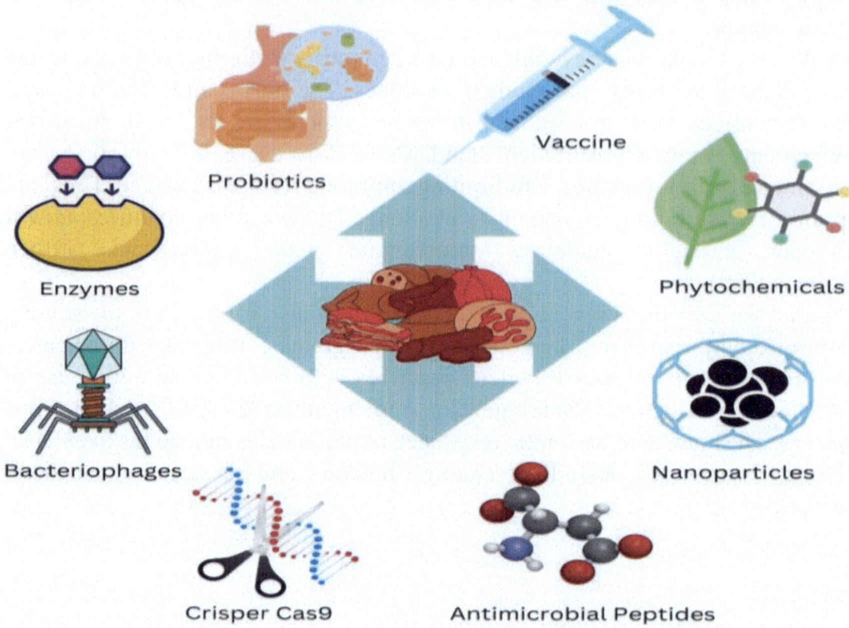

Fig. 2 Various alternatives to antibiotics, such as probiotics, vaccines, phytochemicals, nanoparticles, antimicrobial peptides, CRISPR–Cas9, bacteriophages, and enzymes, which offer innovative ways to fight bacterial infections

tailored therapeutic approaches based on bacterial response patterns (Zhang et al. 2015).

6.1.2 Antimicrobial Peptides (AMPs)

The cellular defence system of diverse organisms produces antimicrobial peptides (AMPs) through small protein compounds that humans maintain. The peptide compounds demonstrate universal anti-microbial properties by fighting against bacteria, fungi, and viruses. The bacterial cell membrane becomes compromised when AMPs function, which creates cellular content leakage that ultimately ends in bacterial mortality (Chung and Khanum 2017). AMPs that occur naturally, such as defensins, cathelicidins, and cecropins, together with synthetic peptides like Daptomycin, serve present clinical uses, especially to treat MRSA and Gram-positive infections. The barriers to applying AMPs in clinical use include their expensive production methods, as well as toxicity risks at elevated dose levels, and bacterial resistance formation because of altered membrane structures.

6.1.3 Enzyme-Based Strategies

Lysins stand as promising antimicrobial tools, which make up another alternative approach to conventional antibiotics because they emerge from bacteriophage enzymes. Bacterial cell walls are the exclusive targets of these enzymes, which initiate a mechanism that causes bacterial death by cell lysis. Lysins function independently from bacterial multiplication because of their ability to produce instant results. The lysin enzymes LysK and ClyA represent two therapeutic components under investigation because LysK targets *Staphylococcus aureus* while ClyA targets *Escherichia coli* and other Gram-negative bacteria (Marshall and Arenas 2003). Experiments with animal subjects indicate that lysins successfully fight both MRSA and *Klebsiella pneumoniae* infections. The application of lysins in clinical settings requires solving their limited bacterial spectral scope, together with their possible breakdown in human environments.

6.1.4 Plant-Derived Antimicrobials

Scientists have renewed their examination of plant-derived compounds for treatment alternatives to antibiotics due to their extensive traditional medicinal antimicrobial effects. Plant-derived compounds fight bacterial infections through the destruction of cellular membranes, inhibition of enzyme functions, and interruption of DNA/RNA synthetic processes. Laboratory experiments demonstrate that allicin from garlic neutralises *E. coli*, Salmonella spp. and Staphylococcus aureus, as well as tea tree oil decimates MRSA (O'Neill 2016). The use of plant-derived antimicrobials as natural alternatives comes with problems because their strength varies unpredictably, dosage guidelines are not standardised and allergic reactions to them may occur.

6.1.5 Nanotechnology-Based Antimicrobials

Infection treatment through nanotechnology utilises nanoparticles to deal with bacteria using their minute particle structure that engages with cells according to multiple mechanisms. Among all antimicrobial nanoparticles, silver stands out because it damages bacterial structures while causing cellular oxidative damage. Microscopic copper and zinc oxide particles, together with others, demonstrate effective antimicrobial abilities (Ruh et al. 2017). The effective delivery of antimicrobial agents to bacterial cells becomes more potent when nanoparticles serve as delivery vehicles. Research is needed to address issues about nanoparticle toxicity in human cells, along with their environmental effects and possible bacterial resistance to nanoparticle therapies (Gupta et al. 2017).

6.2 *Immune-Based Interventions: Harnessing the Immune System Against Resistant Pathogens*

Immune-based interventions represent hopeful alternative strategies when dealing with bacterial infections that resist conventional antibiotic treatments because antibiotic resistance affects global health on a significant scale. The interventions seek to strengthen natural defence capabilities and directly destroy immune-resistant pathogens. Two major immune-based strategies include immunomodulatory agents alongside vaccine development, which create modern methods for decreasing antibiotic medication dependence (Laborda et al. 2022).

6.2.1 Immunomodulatory Agents

Immunomodulatory agents function as engineered substances that strengthen or regulate the immune system's infection reactions to help the body fight resistant pathogens better. The agents activate immune cell populations, including T cells and macrophages, alongside neutrophils, to improve their capability of detecting bacterial enemies and destroying them. The substances can boost immune responses through cytokine production while generating pathogen-neutralising antibodies. The use of immunomodulatory agents helps the immune system identify bacteria that have developed antibiotic resistance so the immune system can destroy them. The treatment with Granulocyte colony-stimulating factor (G-CSF strengthens the production of white blood cells to enhance infection resistance. The strategy of Immunoglobulin therapy gives passive immunity through direct antibody administration to fight infections in susceptible patients with suppressed immune systems (Marshall and Arenas 2003).

6.2.2 Vaccine Development Targeting Resistant Pathogens

Vaccine development stands as a fundamental immune-based prevention method for obstructing bacterial infections from resistant pathogens. Through vaccination, the body learns to detect particular pathogens so it can generate quick and powerful protection when exposed to these germs again. The creation of solid vaccines that target antibiotic-resistant pathogens will help decrease antibiotic prescriptions while eliminating the risk of emerging resistance in microorganisms. PCV13 stands as an effective Pneumococcal vaccine that has shown success in cutting down infections from drug-resistant *Streptococcus pneumoniae* strains (Laborda et al. 2022). Scientists conduct vaccine development research to protect against *Staphylococcus aureus*, including MRSA, together with *Escherichia coli*, particularly multidrug-resistant strains. These vaccines activate immune responses to identify hostile pathogens before they establish infections, helping decrease the international scale of antibiotic resistance effect.

6.3 *Synthetic and Bioengineered Solutions: Innovating the Fight Against Antibiotic Resistance*

The scientific community uses synthetic biology and bioengineering technologies to create novel solutions against resistant bacteria because antibiotic resistance presents itself as a critical worldwide health crisis. Scientists identify CRISPR–Cas bacterial procedures and synthetic bio-based antimicrobial manufacturing techniques as the best approaches in this field (Alkhnbashi et al. 2020). Advanced technology presents the possibility to focus on bacterial resistance mechanism elimination while developing better treatments that demonstrate promise against traditional antibiotics' deficiencies.

6.3.1 CRISPR–Cas Systems for Bacterial Gene Editing

The CRISPR–Cas system operates as a ground-breaking gene-editing technology that enables new opportunities to fight antibiotic resistance. Bacterial genome modifications become possible through this technology for scientists who wish to focus on resistance-causing genes. Through CRISPR–Cas technology, researchers can disable or delete resistance genes, which enables them to restore bacterial sensitivity to antibiotics, thus creating a new method to treat untreatable infections (Hossain 2021).

CRISPR-Cas systems achieve their DNA-modification function through targeted bacterial DNA breaking, which generates repair activity that deletes targeted genes. Scientists employ CRISPR-Cas9 technology to target both NDM-1 and mcr-1 genes, which lead bacteria to resist beta-lactam antibiotics and colistin antibiotics, respectively (Hryhorowicz et al. 2017). Researches demonstrate that CRISPR-Cas9 shows great potential as a tool to address antimicrobial resistance in *Klebsiella pneumoniae* and *Escherichia coli*, together with multiple resistant bacterial pathogens. Widespread therapeutic application of CRISPR needs resolution of multiple issues, which include distribution approaches along with unintended consequences and ethical considerations (Liang et al. 2024).

6.3.2 Synthetic Biology Approaches for Tailored Antimicrobials

An innovative framework for developing new antimicrobials uses methods of synthetic biology. New biological systems are engineered and designed using biological engineering principles along with chemical and biological approaches to manufacture antimicrobial products. These medically created substances can fight bacterial resistance through strategic designs or attack unfamiliar microorganism targets. Synthetic biology obtains its potential against antibiotic resistance by making possible the development of antimicrobials that target specific resistant bacteria without harming the host (Jiang et al. 2015).

Science-based genetic engineering allows researchers to develop synthetic antimicrobial peptides that model natural organism defence capabilities but incorporate optimised properties to fight antibiotic-resistant bacteria. Research shows that synthetic AMP Plectasin successfully defeats methicillin-resistant *Staphylococcus aureus* (MRSA) bacteria, which are known to be extremely resistant to treatment. Through synthetic biology, researchers can manufacture bacteriophage strains that target specific antibiotic-resistant bacteria while providing an antibiotic-like treatment that exceeds broad-spectrum coverage. The creation of bacterial detection sensors constitutes one of many biosensing applications because these devices help identify bacterial strains and their resistance markers to enable better treatment planning (Shmakov et al. 2017).

6.4 Combination Therapies: Enhancing Efficacy Against Resistant Pathogens

The worldwide progression of antibiotic resistance prompts healthcare professionals to develop combination antibiotic therapies as an effective treatment strategy for existing drugs. Antibiotic treatment success rates rise when multiple drugs are combined or when medications use adjuvants, thus providing simultaneous treatment benefits and resistance prevention. The combination therapy employs two successful methods, including synergistic drug pairings and existing drug repurposing assisted by adjuvants to defeat drug resistance while enhancing therapeutic results (Jiang et al. 2015).

6.4.1 Synergistic Drug Combinations

Using two or more antibiotics together with antimicrobial agents produces favourable drug interaction effects, which lead to better treatment outcomes than each drug would achieve on its own. Drugs given together can attack different bacterial processes, which decreases the probability of bacteria becoming resistant while delivering better treatment results. Various drug cooperative effects emerge from mechanisms that target different bacterial growth phases, increase drug absorption, and block bacterial expulsion pumps responsible for their elimination from cells.

A significant drug synergy exists between piperacillin β-lactams paired with tazobactam β-lactamase inhibitors that help combat bacterial β-lactamase enzymes causing resistance. The treatment of infections due to methicillin-resistant *Staphylococcus aureus* (MRSA) utilises the drug combination of rifampin and vancomycin. The combined antibiotic treatments simultaneously improve effectiveness while decreasing bacterial resistance in the treatment of multiple-drug resistance pathogens (Wehrli 1983).

6.4.2 Repurposing Existing Drugs with Adjuvants

Doctors apply existing medical treatments to bacterial resistance by creating new drug pairings from previously uninvolved medications. Drug adjuvants applied together with antimicrobial agents allow healthcare professionals to break through resistance mechanisms, thus achieving better treatment results. The mechanism of action behind adjuvants includes blocking bacterial efflux pumps and modifying bacterial cell membranes for drug uptake, but it also includes stopping bacterial enzyme activity that breaks down antibiotics.

The combination of outdated antibiotic colistin with either clindamycin or tigecycline effectively treats *Acinetobacter baumannii* and *Klebsiella pneumoniae* resistance infections. Avibactam, as a β-lactamase inhibitor, when paired with ceftazidime, provides a third-generation cephalosporin treatment against Carbapenem-resistant *Enterobacteriaceae* (CRE).

Medical investigations have demonstrated that antibiotics and non-antibiotic drugs like chloroquine and azithromycin provide combined therapy against resistant strains of Mycobacterium tuberculosis bacteria and additional resistant bacterial pathogens (Kohanski et al. 2010).

7 Challenges and Future Directions in Combating Antibiotic Resistance

Antibiotic resistance represents a worldwide medical emergency because scientists lack full comprehension of resistance mechanisms, face economic limitations in finding new medications, and require global multinational partnerships. The emergence of three resistance mechanisms and the lack of development in alternative drugs like bacteriophages and antimicrobial peptides create challenges for medical treatment. The Worldwide Organization for Health sets research initiatives on three specific resistant pathogens, including *Klebsiella pneumoniae,* together with *Pseudomonas aeruginosa* and *Mycobacterium tuberculosis* (Jose and Meyer Thomas 2007).

The process of developing new antibiotics faces economic barriers alongside logistical challenges because pharmaceutical firms concentrate on profitable chronic disease drugs, yet developing countries encounter problems with antibiotic distribution. As a global health organisation, the WHO supports innovations in markets to promote new antibiotic developments while promoting responsible use of antibiotics.

The WHO Global Action Plan on Antimicrobial Resistance demands international cooperation because it ensures monitoring programs and enhances both infection prevention measures and develops new tests, treatments, and diagnostic methods. The fight against antibiotic resistance in all countries requires worldwide collaboration to build sustainable antibiotic practices (Ehmann and Lahiri 2014).

References

Abajo T (2024) Beta-lactam resistance in 'Pseudomonas Geruginosa'and its detection in the microbiology laboratory. Aust J Med Sci 45(3):126–145

Alekshun MN, Levy SB (2007) Molecular mechanisms of antibacterial multidrug resistance. Cell 128(6):1037–1050

Alkhnbashi OS, Meier T, Mitrofanov A, Backofen R, Voß B (2020) CRISPR-Cas bioinformatics. Methods 172:3–11. https://doi.org/10.1016/j.ymeth.2019.07.013

Alós JI (2015) Antibiotic resistance: A global crisis. Enferm Infecc Microbiol Clin 33(10):692–699. https://doi.org/10.1016/j.eimc.2014.10.004. (Resistencia bacteriana a los antibióticos: una crisis global)

Appelbaum PC, Hunter PA (2000) The fluoroquinolone antibacterials: past, present and future perspectives. Int J Antimicrob Agents 16(1):5–15. https://doi.org/10.1016/S0924-8579(00)00192-8

Biswas S, Brunel J-M, Dubus J-C, Reynaud-Gaubert M, Rolain J-M (2012) Colistin: an update on the antibiotic of the 21st century. Expert Rev Anti-Infect Ther 10(8):917–934. https://doi.org/10.1586/eri.12.78

Blair JMA, Richmond GE, Piddock LJV (2014) Multidrug efflux pumps in gram-negative bacteria and their role in antibiotic resistance. Future Microbiol 9(10):1165–1177. https://doi.org/10.2217/fmb.14.66

Blair JMA, Webber MA, Baylay AJ, Ogbolu DO, Piddock LJV (2015) Molecular mechanisms of antibiotic resistance. Nat Rev Microbiol 13(1):42–51. https://doi.org/10.1038/nrmicro3380

Chaplin AV, Korzhanova M, Korostin DO (2021) Identification of bacterial antibiotic resistance genes in next-generation sequencing data (review of literature). Klin Lab Diagn 66(11):684–688. https://doi.org/10.51620/0869-2084-2021-66-11-684-688. (Выявление генов антибиотикорезистентности бактерий в данных полногеномного секвенирования (обзор литературы))

Chung PY, Khanum R (2017) Antimicrobial peptides as potential anti-biofilm agents against multidrug-resistant bacteria. J Microbiol Immunol Infect 50(4):405–410

Coates ARM, Hu Y, Holt J, Yeh P (2020) Antibiotic combination therapy against resistant bacterial infections: synergy, rejuvenation and resistance reduction. Expert Rev Anti-Infect Ther 18(1):5–15. https://doi.org/10.1080/14787210.2020.1705155

Cosentino F, Viale P, Giannella M (2023) MDR/XDR/PDR or DTR? Which definition best fits the resistance profile of Pseudomonas aeruginosa? Curr Opin Infect Dis 36(6):564–571. https://doi.org/10.1097/qco.0000000000000966

Darby EM, Trampari E, Siasat P, Gaya MS, Alav I, Webber MA, Blair JMA (2023) Molecular mechanisms of antibiotic resistance revisited. Nat Rev Microbiol 21(5):280–295. https://doi.org/10.1038/s41579-022-00820-y

Delcour AH (2009) Outer membrane permeability and antibiotic resistance. Biochim Biophys Acta 1794(5):808–816. https://doi.org/10.1016/j.bbapap.2008.11.005

Denakpo E, Naas T, Iorga BI (2023) An updated patent review of metallo-β-lactamase inhibitors (2020–2023). Expert Opin Ther Pat 33(7–8):523–538. https://doi.org/10.1080/13543776.2023.2262763

Egan AJF (2018) Bacterial outer membrane constriction. Mol Microbiol 107(6):676–687. https://doi.org/10.1111/mmi.13908

Ehmann DE, Lahiri SD (2014) Novel compounds targeting bacterial DNA topoisomerase/DNA gyrase. Curr Opin Pharmacol 18:76–83. https://doi.org/10.1016/j.coph.2014.09.007

Epand RM, Walker C, Epand RF, Magarvey NA (2016) Molecular mechanisms of membrane targeting antibiotics. Bioch Biophys Acta 1858(5):980–987. https://doi.org/10.1016/j.bbamem.2015.10.018

Evans ME, Feola DJ, Rapp RP (1999) Polymyxin B sulfate and colistin: old antibiotics for emerging multiresistant gram-negative bacteria. Ann Pharmacother 33(9):960–967. https://doi.org/10.1345/aph.18426

Fletcher M, McCormack A, Parcell BJ, Coote PJ (2023) Combination therapy with ciprofloxacin and pentamidine against multidrug-resistant Pseudomonas aeruginosa: assessment of in vitro and in vivo efficacy and the role of resistance-nodulation-division (RND) efflux pumps. Antibiotics (Basel) 12(8). https://doi.org/10.3390/antibiotics12081236

Garner MR, Sethuraman SA, Schade MA, Boateng H (2020) Antibiotic prophylaxis in open fractures: evidence, evolving issues, and recommendations. J Am Acad Orthop Surg 28(8):309–315. https://doi.org/10.5435/jaaos-d-18-00193

Gupta D, Singh A, Khan AU (2017) Nanoparticles as efflux pump and biofilm inhibitor to rejuvenate bactericidal effect of conventional antibiotics. Nanoscale Res Lett 12(1):454. https://doi.org/10.1186/s11671-017-2222-6

Hancock Robert EW, Chapple Daniel S (1999) Peptide Antibiotics. Antimicrob Agents Chemother 43(6):1317–1323. https://doi.org/10.1128/aac.43.6.1317

Hamzaoui Z, Ocampo-Sosa A, Fernandez Martinez M, Landolsi S, Ferjani S, Maamar E, Saidani M, Slim A, Martinez-Martinez L, Boutiba-Ben Boubaker I (2018) Role of association of OmpK35 and OmpK36 alteration and blaESBL and/or blaAmpC genes in conferring carbapenem resistance among non-carbapenemase-producing Klebsiella pneumoniae. Int J Antimicrob Agents 52(6):898–905. https://doi.org/10.1016/j.ijantimicag.2018.03.020

Hernández IA (2016) Aislamiento e identificación de bacterias de origen ambiental por MALDI-TOF MS.

Hong J, Kim Y, Kim J, Heu S, Kim SR, Kim KP, Roh E (2015) Genetic diversity and antibiotic resistance patterns of staphylococcus aureus isolated from leaf vegetables in Korea. J Food Sci 80(7):M1526–M1531. https://doi.org/10.1111/1750-3841.12909

Hossain MA (2021) CRISPR-Cas9: a fascinating journey from bacterial immune system to human gene editing. Prog Mol Biol Transl Sci 178:63–83. https://doi.org/10.1016/bs.pmbts.2021.01.001

Hryhorowicz M, Lipiński D, Zeyland J, Słomski R (2017) CRISPR/Cas9 immune system as a tool for genome engineering. Arch Immunol Ther Exp 65(3):233–240. https://doi.org/10.1007/s00005-016-0427-5

Hussain HI, Aqib AI, Seleem MN, Shabbir MA, Hao H, Iqbal Z, Kulyar MF, Zaheer T, Li K (2021) Genetic basis of molecular mechanisms in β-lactam resistant gram-negative bacteria. Microb Pathog 158:105040. https://doi.org/10.1016/j.micpath.2021.105040

Jian Z, Zeng L, Xu T, Sun S, Yan S, Yang L, Huang Y, Jia J, Dou T (2021) Antibiotic resistance genes in bacteria: occurrence, spread, and control. J Basic Microbiol 61(12):1049–1070. https://doi.org/10.1002/jobm.202100201

Jiang Y, Chen B, Duan C, Sun B, Yang J, Yang S (2015) Multigene editing in the Escherichia coli genome via the CRISPR-Cas9 system. Appl Environ Microbiol 81(7):2506–2514. https://doi.org/10.1128/aem.04023-14

Jose J, Meyer Thomas F (2007) The autodisplay story, from discovery to biotechnical and biomedical applications. Microbiol Mol Biol Rev 71(4):600–619. https://doi.org/10.1128/mmbr.00011-07

Katz MJ, Tamma PD, Cosgrove SE, Miller MA, Dullabh P, Rowe TA, Ahn R, Speck K, Gao Y, Shah S (2022) Implementation of an antibiotic stewardship program in long-term care facilities across the US. JAMA Netw Open 5(2):e220181–e220181

Kohanski MA, Dwyer DJ, Collins JJ (2010) How antibiotics kill bacteria: from targets to networks. Nat Rev Microbiol 8(6):423–435. https://doi.org/10.1038/nrmicro2333

Kosmidis C, Levine DP (2010) Daptomycin: pharmacology and clinical use. Expert Opin Pharmacother 11(4):615–625. https://doi.org/10.1517/14656561003598893

Kumar A, Schweizer HP (2005) Bacterial resistance to antibiotics: active efflux and reduced uptake. Adv Drug Deliv Rev 57(10):1486–1513. https://doi.org/10.1016/j.addr.2005.04.004

Laborda P, Hernando-Amado S, Martínez JL, Sanz-García F (2022) Antibiotic resistance in pseudomonas. Adv Exp Med Biol 1386:117–143. https://doi.org/10.1007/978-3-031-08491-1_5

Liang R, He Z, Zhao KT, Zhu H, Hu J, Liu G, Gao Q, Liu M, Zhang R, Qiu JL, Gao C (2024) Prime editing using CRISPR-Cas12a and circular RNAs in human cells. Nat Biotechnol 42(12):1867–1875. https://doi.org/10.1038/s41587-023-02095-x

Majalekar PP, Shirote PJ (2020) Fluoroquinolones: blessings or curses. Curr Drug Targets 21(13):1354–1370. https://doi.org/10.2174/1389450121666200621193355

Marshall SH, Arenas G (2003) Antimicrobial peptides: a natural alternative to chemical antibiotics and a potential for applied biotechnology. Electron J Biotechnol 6(3):271–284

Martinez JL (2009) Environmental pollution by antibiotics and by antibiotic resistance determinants. Environ Pollut 157(11):2893–2902

Medakina I, Tsapkova L, Polyakova V, Nikolaev S, Yanova T, Dekhnich N, Khatkov I, Bordin D, Bodunova N (2023) Helicobacter pylori antibiotic resistance: molecular basis and diagnostic methods. Int J Mol Sci 24(11). https://doi.org/10.3390/ijms24119433

Mingeot-Leclercq M-P, Tulkens Paul M (1999) Aminoglycosides: nephrotoxicity. Antimicrob Agents Chemother 43(5):1003–1012. https://doi.org/10.1128/aac.43.5.1003

Mohapatra SS, Dwibedy SK, Padhy I (2021) Polymyxins, the last-resort antibiotics: mode of action, resistance emergence, and potential solutions. J Biosci 46(3):85. https://doi.org/10.1007/s12038-021-00209-8

Munita JM, Arias CA (2016) Mechanisms of antibiotic resistance. Microbiol Spectrum 4(2). https://doi.org/10.1128/microbiolspec.VMBF-0016-2015

O'Neill J (2016). Tackling drug-resistant infections globally: final report and recommendations

Oliveira J, Reygaert WC (2025) Gram-negative bacteria. In StatPearls. StatPearls Publishing. Copyright © 2025, StatPearls Publishing LLC

Organization WH (2014) Antimicrobial resistance: global report on surveillance. World Health Organization

Pang Y, Lu J, Wang Y, Song Y, Wang S, Zhao Y (2013) Study of the rifampin monoresistance mechanism in Mycobacterium tuberculosis. Antimicrob Agents Chemother 57(2):893–900. https://doi.org/10.1128/aac.01024-12

Poirel L, Bonnin RA, Nordmann P (2011) Genetic basis of antibiotic resistance in pathogenic Acinetobacter species. IUBMB Life 63(12):1061–1067. https://doi.org/10.1002/iub.532

Pontes MH, Groisman EA (2020) A physiological basis for nonheritable antibiotic resistance. mBio 11(3). https://doi.org/10.1128/mBio.00817-20

Poole K (2000) Efflux-mediated resistance to fluoroquinolones in gram-positive bacteria and the mycobacteria. Antimicrob Agents Chemother 44(10):2595–2599. https://doi.org/10.1128/aac.44.10.2595-2599.2000

Poyart C, Jardy L, Quesne G, Berche P, Trieu-Cuot P (2003) Genetic basis of antibiotic resistance in Streptococcus agalactiae strains isolated in a French hospital. Antimicrob Agents Chemother 47(2):794–797. https://doi.org/10.1128/aac.47.2.794-797.2003

Ruh C, Banjade R, Mandadi S, Marr C, Sumon Z, Crane JK (2017) Immunomodulatory effects of antimicrobial drugs. Immunol Investig 46(8):847–863

Salyers AA (2002) An overview of the genetic basis of antibiotic resistance in bacteria and its implications for agriculture. Anim Biotechnol 13(1):1–5. https://doi.org/10.1081/abio-120005766

Shmakov S, Smargon A, Scott D, Cox D, Pyzocha N, Yan W, Abudayyeh OO, Gootenberg JS, Makarova KS, Wolf YI, Severinov K, Zhang F, Koonin EV (2017) Diversity and evolution of class 2 CRISPR-Cas systems. Nat Rev Microbiol 15(3):169–182. https://doi.org/10.1038/nrmicro.2016.184

Si Z, Pethe K, Chan-Park MB (2023) Chemical basis of combination therapy to combat antibiotic resistance. JACS Au 3(2):276–292

Stocco G, Lucafò M, Decorti G (2020) Pharmacogenomics of antibiotics. Int J Mol Sci 21(17). https://doi.org/10.3390/ijms21175975

Tarín-Pelló A, Suay-García B, Pérez-Gracia MT (2022) Antibiotic resistant bacteria: current situation and treatment options to accelerate the development of a new antimicrobial arsenal. Expert Rev Anti-Infect Ther 20(8):1095–1108. https://doi.org/10.1080/14787210.2022.2078308

Van Boeckel TP, Brower C, Gilbert M, Grenfell BT, Levin SA, Robinson TP, Teillant A, Laxminarayan R (2015) Global trends in antimicrobial use in food animals. Proc Natl Acad Sci 112(18):5649–5654

Walsh C (2000) Molecular mechanisms that confer antibacterial drug resistance. Nature 406(6797):775–781. https://doi.org/10.1038/35021219

Wang WJ, Yu LM, Shao MY, Jia YT, Liu LQ, Ma XH, Zheng Y, Liu YF, Zhang YZ, Luo XX, Li FM, Zheng H (2023) Research review on the pollution of antibiotic resistance genes in livestock and poultry farming environments. Ying Yong Sheng Tai Xue Bao 34(5):1415–1429. https://doi.org/10.13287/j.1001-9332.202305.032. (畜禽养殖环境中抗生素抗性基因污染的研究进展)

Wehrli W (1983) Rifampin: mechanisms of action and resistance. Rev Infect Dis 5(Supplement_3):S407–S411. https://doi.org/10.1093/clinids/5.Supplement_3.S407

Wright GD (2005) Bacterial resistance to antibiotics: enzymatic degradation and modification. Adv Drug Deliv Rev 57(10):1451–1470. https://doi.org/10.1016/j.addr.2005.04.002

Xu Z-Q, Flavin MT, Flavin J (2014) Combating multidrug-resistant gram-negative bacterial infections. Expert Opin Investig Drugs 23(2):163–182. https://doi.org/10.1517/13543784.2014.848853

Zgurskaya HI, Walker JK, Parks JM, Rybenkov VV (2021) Multidrug efflux pumps and the two-faced Janus of substrates and inhibitors. Acc Chem Res 54(4):930–939. https://doi.org/10.1021/acs.accounts.0c00843

Zhang J, Li Z, Cao Z, Wang L, Li X, Li S, Xu Y (2015) Bacteriophages as antimicrobial agents against major pathogens in swine: a review. J Anim Sci Biotechnol 6:1–7

Zinner SH (2007) Antibiotic use: present and future. New Microbiol 30(3):321–325

Phytogenic Feed Additives

Role of Herb, Spices, and Plant Extracts in Animal Health and Productivity

Sayydah Suryya Sani, M. Shahid Mahmood, Shabana Naz, and Azhar Rafique

Abstract Herbs, spices, and several other plant derivatives are examples of phytogenic feed additives, which are utilized as animal feed supplements. One of the biggest challenges facing animal nutritionists is figuring out how to enhance animal output while preventing and curing disease in an environmentally responsible manner. Herbs, spices, and plant extracts have been studied more and employed in animal nutrition and production due to their possible health benefits, which include improving growth, boosting immunity, and increasing productivity. They provide natural alternatives to synthetic additives, promoting sustainability and animal welfare. Herbs, spices, and plant extracts include a variety of antioxidants, including flavonoids, polyphenols, and vitamins C and E. They may aid in digestion. Numerous plant and herb extracts are well known for their anti-inflammatory and immunomodulatory properties. Furthermore, it has been asserted that several spices, herbs, or their constituents promote the production of intestinal mucus. This effect could account for better production performance after adding these phytogenic feed additives. According to the research now accessible, the list of non-antibiotic growth stimulants for cattle, which includes probiotics and organic acids, may be enhanced by phytogenic feed additions like herbs and spices.

Keywords Herbs · Spices · Plant extract · Anti-inflammatory · Immunomodulatory · Anti-oxidant

S. S. Sani · S. Naz · A. Rafique (✉)
Department of Zoology, Government College University, Faisalabad, Pakistan

M. S. Mahmood
Institute of Microbiology, University of Agriculture Faisalabad, Faisalabad, Pakistan

R. U. Khan, S. Naz (eds.), *Green Alternatives to Antibiotics in Animal Health*,
https://doi.org/10.1007/978-981-92-0309-3_5

1 The Acquaint

In animal nutrition, products referred to as feed additives are used to improve the quality of food and feed made from animals, as well as the well-being and productivity of the animals. The production of nutrient-dense animal products with the necessary organoleptic qualities can only be ensured by high-quality feed in conjunction with appropriate management, potable water, and hygiene methods (Gupta 2013). To produce healthy animal products, farm animals must be kept in good health. Over the past decade, there has been a push to incorporate natural ingredients in animal and human nutrition. Several investigations focused on elucidating the physiological functions and biochemical composition of several feed additives, including plant extracts, probiotics, organic acids, and prebiotics. Thousands of years ago, spices, herbs, and their extracts were already being used in China, Mesopotamia, Egypt, ancient Greece, and India. There, they were prized for their distinct aroma and range of therapeutic applications (Greathead 2003). We can barely rely solely on outdated notions about the health benefits of particular herbs and spices or the active ingredients they contain when talking about their use as feed additives. To support their use, we require scientific evidence of their positive impacts on the general health and performance of the animals. We can now easily identify the components and roles of as-yet-unidentified active molecules of plant origin because of technical advancements.

2 Spice and Herbs

Spices and herbs can be utilized in food in the form of dried plants, plant segments, or extracts to further increase their beneficial effects (Table 1). The extraction process and the characteristics of the extraction solvent determine the makeup of extracts from the same plant. In the EU, it is presently unlawful to use antibiotics as growth promoters for purposes other than therapeutic ones. Antibiotic-resistant forms of *Salmonella*, *Escherichia coli*, *Enterococcus*, and *Campylobacter* bacteria have been linked to the application of antibiotics in food animal production. According to US and European studies, human infections caused by resistant bacteria are not treatable with conventional antibiotics. In light of these practices and associated issues, a number of organizations, including the American Public Health Association (APHA), American Medical Association (AMA), and American Society for Microbiology (ASM), have called for limitations on the use of antibiotics in food animal production as well as the abolition of all non-therapeutic uses. The growth performance of animals will be impacted in ways that can be avoided, particularly in the poultry industry (Greathead 2003).

Table 1 Often used plants, its active components, and functions (Loo and Richard 1992)

Plants	Used parts	Mayor active components	Function
Rosemay	Leaves	Cineol	Anti-inflammatory, digestion stimulant, antioxidant
Thyme	Whole plant	Thymol	Digestion Stimulant, antibacterial, antiseptic
Laurel	Leaves	Cineol	Digestion, appetite,
Nutmeg	Seed	Sabinene	Digestion stimulant
Cumin	Seed	Cuminaldehyde	Caminative, digestive
Pepper	Fruit	Piperine	Digestion Stimulant
Garlic	Bulb	Allicin	Antiseptic, methane inhibitor, immunity enhancer
Coriander	Seed, Leaves	Linalol	Digestion Stimulant, carminative
Celery	Fruit, leaves	Phtalides	Antiseptic, appetite stimulant
Mustard	Seed	Allyl izotiocianat	Digestion stimulant
Ginger	Rizhom	Zingerone	Methane reducing captivity, gastric stimulant
Cloves	Cloves	Eugenol	Antiseptic, insecticidal
Parsley	Leaves	Apiol	Appetite stimulant
Capsicum	Fruit	Capsaicin	Digestion stimulant

3 Significance of Using Herbal Products in Livestock Production

Long-term use of synthetic pharmaceuticals raises concerns about their unfavorable effects, such as the existence of toxic metabolites and antibiotic residues in meat and byproducts, which can contribute to the issue of antibiotic resistance in humans. Furthermore, the majority of Indian livestock owners cannot afford this medical care. Sometimes the remoteness of villages makes it difficult to receive health care. These problems have highlighted the usage of herbal remedies, which rural residents believe to be somewhat safe and reasonably priced (Gupta et al. 2013). Globally, the utilization of herbal products has significantly increased. About 70% of the animals injected with herbal treatments are cattle, horses, sheep, goats, and pigs. Poultry (9.1%), dogs (5.3%), and rabbits (4.3%) follow (Viegi et al. 2003; Amna et al. 2024). This is because there is a lot of research supporting the effectiveness of herbal therapies in addition to the broader trend towards using natural items to treat ailments. The importance of using natural goods is growing. In actuality, individuals are choosing natural products over synthetic ones more and more because they believe—rightly or incorrectly—that natural products have less negative effects and side effects (Viegi et al. 2003). When used properly, herbal solutions can offer a believable solution to the negative effects of prescription medications. These can reduce the needless usage of chemical medications, such as antibiotics, to address residual issues and the increasing resistance of microorganisms. Herbal goods are essential because herbs have medicinal properties. The items are made from herbs that are rich in magical qualities that enhance wellbeing without having any negative effects. Herbal goods have many benefits over non-herbal ones, making the supplements or medications far more beneficial. Herbal products are readily available, well-tolerated, and more effective for chronic health conditions. They are affordable, readily available, and simple to administer and prepare. Because they don't leave residues in animal products, herbal growth promotants are suitable for human consumption. In the long run, herbs are also less expensive and could result in lower feed expenses (Alimon 2009). Herbs, spices, and different plant extracts have the ability to stimulate digestion and hunger. Plants produce a variety of secondary metabolites that act as defenses against pathogenic microbes, predators, and environmental and physiological stressors (Klevenhusen et al. 2011). Together with other beneficial plants, the herbs and spices can be used as an anti-inflammatory, anti-oxidative, immunostimulant, and natural alternative to antibiotics (Klevenhusen et al. 2011; Aljohani 2024). Many pharmacologically active substances can be found in herbs, and each one has a special combination and set of characteristics. Herbs that are utilized whole or in their leaves, along with other herbs, have active components that operate in harmony and synergy to produce the desired results naturally. This isn't the case with contemporary medicine, though (Alimon 2009). Large volumes of free radicals are produced by the rapid metabolic rate of developing tissues (Koivula and Eeva 2010), and oxidative stress may result if antioxidants are unable to properly eliminate them. Antioxidant supplements may therefore aid in mitigating the adverse consequences of oxidative stress linked to growth. Growth is certainly influenced by an individual's total oxidative condition because it extends to multiple organs. It is logical to assume that polyphenols, the antioxidants with the largest environmental abundance and antioxidant potency, would be crucial for growth (Koivula and Eeva 2010). According to a report, broiler growth could be considerably enhanced by feed supplied with 2% *Thymus vulgaris* (Ocak et al. 2008). In broilers, the application of herbal extracts, particularly *Allium sativum*, increased feed conversion rate when compared to the antibiotic virginiamycin. The improvement in the release of digestive enzymes may be the cause of this effect (Rahimi et al. 2011). Coneflower and garlic were the most successful in enhancing immune function and lowering the serum lipid profile, but they are not a good substitute for antibiotics as feed additives to increase broiler growth (Rahimi et al. 2011). The impact of *T. vulgaris* (thyme) and *Cinnamonuim zeylanicum* (cinnamon) on broiler performance was investigated by Yalçin et al. (2020), and improvements in feed intake, feed conversion ratio (FCR), live weight gain, and poultry health were noted. When the feed was enhanced with Chinese herbal medicines, which exhibited a growth-promoting impact comparable to that of antibiotics in the growth of broiler chickens, a higher growth rate was also noted in the chicks. In both normal and immune-inhibition-treated rats and chicks, polysaccharides from mushrooms and herbs promoted the growth of immunological organs such the spleen, bursa, and thymus (Demir et al. 2003), compared the use of probiotics, antibiotics, and two herbal preparations and found that the natural preparations were successful in promoting the growth of broiler chicks in place of virginiamycin, and that using essential oils instead of antibiotics was effective in promoting the growth of broiler chicks (Demir et al. 2003).

4 Effect of Herbs and Herbal Preparation on Animal Products

Meat, milk, and egg quality are directly impacted by feed and feed supplements, either positively or negatively. Numerous secondary plant metabolites are produced by plants; some of these metabolites may be toxic to animals, while others may not. Numerous phytochemicals from a variety of plants, including flavonoids, essential oils, tannins, and saponins, have been discovered to have potential uses in rumen manipulation and increasing animal production as substitutes for chemical feed additives (McIntosh et al. 2003; Benchaar et al. 2007).

5 Milk Production

The interaction of many hormones regulates milk production. The process of producing milk is actually a complicated one that involves both mental and physical factors as well as the collaboration of several hormones, the most significant of which is thought to be prolactin. Progesterone levels drop and the entire milk supply begins with parturition and placenta evacuation (Neville et al. 2001). Dopamine agonists and antagonists interact with the anterior pituitary gland and hypothalamus to increase and decrease prolactin secretion and, consequently, milk production (endocrine regulation). After that, a local feedback mechanism (autocrine regulation) maintains or increases the milk production as the prolactin level steadily declines. Thus, to enhance but not sustain the milk supply, a rise in prolactin levels is required (Lawrence and Lawrence 2004). Farm animals have been using a variety of herbs and their mixtures to improve and maintain their milk production for generations. Among the herbs used as galactagogues include *Nigella sativa*, *Leptadenia reticulate*, and *Asparagus recemosus*. It is thought that galactagogues help to start, sustain, or increase the production of milk (Abascal and Yarnell 2008). Goats, sheep, cows, and buffaloes have all been demonstrated to exhibit a significant galactopoitic response to *Leptadenia reticulata* (Morsy et al. 2018). Galog is a herbal substance that has been proposed to increase the quantity of milk (Saleh et al. 2023). Saleh et al. (2023). found that dairy cows given Lectovet (herbal combination) had higher milk yields and higher milk fat percentages. A non-hormonal herbal supplement called galactin dramatically increased dairy cow milk production, which in turn improved milking economics. Cows with subclinical mastitis produced more milk while using indigenous herbal formulations, which also successfully preserved the changed milk contents (Bhatt et al. 2009). A combination of herbs called

Ruchamax improves digestive processes, enhances salivary secretions, increases the numbers of beneficial bacteria and protozoa, and eventually aids in feed digestion and metabolism (Bhatt et al. 2009). By boosting milk production and halting the decline phase, herbal preparations extend the breastfeeding period and produce a gradual but gradual post-peak drop. The overall percentage of financial gains from animals fed either Ruchamax or Paya pro (a combination of more than 10 herbs) was 50.37% and 23.49%, respectively, in comparison to the control group (Bhatt et al. 2009). Supplementing dairy cows with Ilex paraguarensis increased their milk output (Celi and Raadsma et al. 2009). Herbs can be added to dairy cows' feed to alter the flavor of their milk. When the dried herbs were given to nursing dairy cows, the distinctive scent of the milk was reduced because the plants' unique components were transferred into the milk (Ando et al. 2001). When some herbal galactagogue was added, Preciado et al. (2011) found that the flavor of milk increased.

6 Meat Production and Quality

Meat microbiological deterioration is caused by lipid oxidation, which takes place during the production and preservation of meat and meat products. Meat's color, flavor, texture, and nutritional content are all negatively impacted by lipid peroxidation products (Kanatt et al. 2007). The production of nutrient-dense animal products with the appropriate organoleptic qualities can be guaranteed by using high-quality feed in conjunction with good hygiene, drinkable water, and management (Wu et al. 2014). In addition to causing harm to live cells, free radicals oxidize animal products, lowering their quality. One of the most destructive reactive oxygen species is the superoxide anion radical, which worries animal scientists and feed producers about product quality. Animal fat can be improved by feeding n−3 PUFA-enriched diets (Wood et al. 2004), although this technique has been linked to an increase in plasma lipoperoxidation (Scislowski et al. 2005). As a result, new bioefficient antioxidants—especially natural antioxidants—are a fresh way to address consumer concerns about toxicity and safety. Because they are readily obtainable from natural sources and effectively stop lipid oxidation in food products, plant extracts high in polyphenols are excellent candidates. In rats given a PUFA-rich diet with purified polyphenols (catechin+quercetin), a study showed the bioefficiency of polyphenols. Another study found that rats fed a diet high in polyunsaturated fats (PUFAs) added with 2 grams of isoflavone-rich powder per kilogram of food did not see a decrease in plasma malondialdehyde (MDA) under comparable experimental settings (Kawakami et al. 2004). Recent in vitro research has shown that polyphenols can preserve a high amount of tocopherol in lipid structure by either recycling these oxidized tocopherol molecules or by preserving it in LDL-based models (Deckert et al. 2002) or SDS micelles (Zhou et al. 2005). Nevertheless, vitamin E kinetic studies showed that tocopherol is quickly eliminated and that the plasma pool is continuously modified by tocopherol reserves in the tissue or from the food (Traber et al. 1994). Due to its abundance of polyphenolic chemicals, the genus Mentha may have potent antioxidant qualities (Ali et al. 2022). Mentha is currently grown as industrial crops in a number of nations because of its volatile oils, which are of significant economic value and are utilized by the food, beverage, pharmaceutical, cosmetic, and liquor sectors.

The antioxidant properties of mentha have been reported (Dorman et al. 2003), but their use in the preparation of meat has not been investigated. Supplementing turkeys with 200 mg/kg of essential oil of oregano not only preserved the quality of poultry meat during freezing (Botsoglou et al. 2003b) but also successfully decreased the lipid peroxidation of both cooked and fresh meat during refrigeration (Botsoglou et al. 2003a). Adding vitamins C and E to extracts from herbs and spices helps reduce lipid peroxidation in tissues (Young et al. 2003). Studies comparing the carcass features and meat quality of lambs raised on mountain pastures and low-land pastures revealed notable differences in the meat's color, flavor, and fat level and fatty acid composition. Furthermore, they stated that lambs raised on mountain pastures without any additional nutrition or care are frequently thought to provide meat of the highest caliber. This was due to the fact that the meat of lambs raised on mountain pastures is frequently delicious, and they have access to an abundance of wild bushes and magical herbs.

7 Stability of Animal Products

The use of herbs and spices as antioxidants is essential for the animals' excellent health and the oxidative stability of their products. The effects of oregano essential oil on the oxidative stability of chicken and turkey meat have been well studied. Supplementing turkeys with 200 mg/kg of oregano essential oil significantly decreased the amount of lipid peroxidation that occurred in both cooked and raw meat while it was refrigerated (Botsoglou et al. 2003b). Additionally, during frozen storage, the quality of chicken flesh was effectively maintained by the volatile oil of oregano (Botsoglou et al. 2003a). Lipid peroxidation in chicken and turkey tissues is considerably more successfully prevented by extracts from herbs and spices when combined with vitamins C and E (Young et al. 2003). In one study, goat feed was supplemented with *Andro graphis paniculata*, turmeric, and vitamin E to assess the meat's antioxidant content. Higher antioxidant activities and better sensory qualities were found in the flesh of supplemented goats. Jang et al. (2008) assessed the quality and antioxidative potential of the breast meat of broiler chickens fed a dietary medicinal herb extract mix (MHEM), which included goldthread, Japanese honeysuckle, and *Morus nigraleaf* in a ratio of 48.5:48.5:3.0. They found that while MHEM had no effect on the proximate composition of the breast meat, the supplemented group's phenol content was twice that of the control group.

The treated diets' levels of tribarbituric acid (2-TBA) were lower than those of the control and did not rise over time. They came to the conclusion that this herbal blend improved the overall flavor and oxidative capability while it was being stored cold. Plant extracts from oregano inhibited lipoperoxidation in muscle tissues and may be complementary to vitamin E, according to feeding tests done on chickens (Young et al. 2003).

8 Immuno-Stimulation

After an illness, a number of therapeutic herbs are used to strengthen the immune system or restore it to normal. Herbs and spices high in flavonoids, vitamin C, and carotenoids often strengthen the immune system. *Allium sativum, Glycyr rhiza glabra, Uacanria tomentosa,* and *Echinacea purpurea* are plants that contain compounds with immunostimulatory qualities. These plants can enhance NK cells, macrophages, and lymphocyte function; they can promote interferon generation or phagocytosis (Craig 1999). According to Roussel et al. (2009), who looked at how *C. zeylanicum* essential oil affected the antioxidant status of hens, it has a strong antioxidant effect on fattening them and can be added to dietary supplements as an antioxidant source. The essential oil of *C. zeylanicum* exhibited stimulatory effects on macrophages, which are the body's main defense against infections and are crucial for the start and control of the immune response. Without causing negative side effects or leaving toxic residues in the bodies of animals, preparations from herbs and spices aid in the inhibition and treatment of a variety of health issues. Cows with endometritis, or inflammation of the endometrium, can benefit from the use of herbs and spices. The impact of extracts of garlic (*Allium sativum*, L.), eucalyptus (*Eucalyptus globulus*, Labill.), and *Gnaphalium onoideum* on acute endometritis in Holstein cows was assessed by Esparza-Borges and Ortiz-Márquez (1995). *Allium sativum* extract was the most efficient of all the extract however, eucalypt also had positive effects. Tannins have antihelmitic qualities and stop rumen bloat (Barry and McNabb 1999). In hypercholemic rats, asparagus root enhances antioxidant status and controls cholesterol metabolism. *Ayurvedic scriptures* recommend *Asparagus racemosus* (Shatavari) as a galactagogue, a therapy for dyspepsia, and a preventative measure against stomach ulcers.

9 Active Compounds and Their Function Phytochemicals

Chemical substances are synthesized by plants as part of their metabolic activities. Secondary metabolites, also known as phytochemicals, are non-essential substances that are found in a small number of plants, some of which are unique to specific genera or species. All plants have primary metabolites like lipids and carbohydrates. Pigments exhibit colors, absorb light, and shield the organism from radiation to attract pollinators. Plants naturally contain organic molecules called phytochemicals. Generally, the term refers to substances that may have health effects but are not considered essential nutrients. A medicinal plant's biological or therapeutic activity is directly related to its chemical components. Because phytochemicals, also known as plant secondary metabolites (PSM), are so diverse, it is impossible to provide a thorough summary of their biochemistry, chemistry, and bioactivity in a short document (Hashemi and Davodi 2010).

10 Phytogenics Classification

In addition to standard botanical classification, classification is based on the part used (whole plant, root, bark, stem, seed, leaf, fruit, and flower), habit (grasses, sedges, herbs, shrubs, climbers, and trees), habitat (temperate, tropical and subtropical), therapeutic value (antifungal, anti-inflammatory, antioxidant, antibacterial, anticancer, anti-ulcer, immunostimulant, antiviral etc.), and routes of administration (inhalation, tincture, maceration, syrup, decoction, and tisanes).

11 Phytogenics Properties

Historically, both humans and animals have benefited from the healing properties of herbal plants. Information about the usage of ethnoveterinary medications is scarce. The primary ways that phytogenics improve animal growth performance and health are through their antimicrobial properties (Guo et al. 2004a), antioxidant effects (Roussel et al. 2009), anti-stress capabilities (Chattopadhyay et al. 2005), gut microbiota management, and nutrigenomics effects (Franco-Jimenez et al. 2007), and immune enhancement (Guo et al. 2004b).

12 Herbal Plants as Antibiotics

Numerous studies have documented the antimicrobial activity of various herbal extracts (in vitro) across diverse global regions (Sudhakar et al. 2006; Yano et al. 2006). Antimicrobial phytochemicals can be categorized into numerous groups: phenolics and polyphenols, alkaloids, lectins and polypeptides, terpenoids and essential oils (Windisch et al. 2008).

13 Use of Herbal Plants to Support Health and Growth

Animal species, age, and production purpose are some of the factors that affect the practical use of herbal supplements in diet or extraction. Each herb contains unique biological substances that produce a variety of effects through different mechanisms. The main mechanism of phytogenic feed additives is their capacity to positively impact the gut microflora ecosystem by regulating potential pathogens. One possible indirect effect of phytogenic chemicals stabilizing microbial eubiosis in the gastrointestinal system is the enhancement of digestion in the small intestine. By improving the intestinal availability of vital nutrients for absorption and encouraging growth within their genetic potential, phytogenics help host animals cope with immune defense stress in stressful settings (Hashemi and Davoodi 2010). In the last few years, herbal plant feed additives have concerned a lot of interest as a possible substitute for growth stimulants made from antibiotics in feeding plans. The impact of phytobiotic feed additives on poultry production performance was established by Hashemi and Davoodi (2010). It was found that extracts from some herbal plants, including sage (*Salvia officinalis*), thyme (*T. vulgaris*), and rosemary (*Rosmarinus officinalis*), together with a combination of cinnamonaldehyde, carvacrol, and capsaicin, increased the digestibility of broiler feed (Hernandez et al. 2004).

14 Spices and Herbs as Digestion and Appetite Enhancers

Understanding the taste favorites of different animal species is crucial for adding herbs, spices, or their extracts to animal feed in an efficient manner that stimulates hunger Fig. 1. According to Janz et al. (2007), pigs preferred a diet enriched with garlic or thyme over feed supplied with ginger or oregano. It was found that adding oregano or thyme to the diet of weaned pigs resulted in a noteworthy decrease in feed consumption. Pigs in this experiment showed a preference for unsupplemented feed over feed that had the previously listed spices in it. Laurel, mint, cardamom,

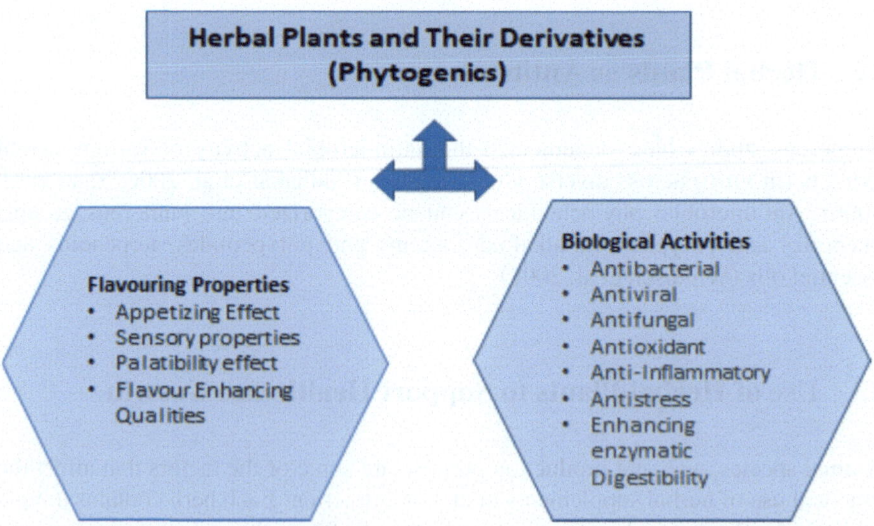

Fig. 1 Herbal plants and their activities

cloves, and cinnamon are known to stimulate hunger (Suresh and Srinivasan 2007). Different herbs and spices have different active ingredients that have different effects on the digestive system. Many of them encourage the production of saliva, ginger, mint, anise, curcuma, cumin, Cayenne pepper, onions, and fenugreek all help the liver produce bile acids and excrete them in bile, which improves lipid absorption and digestion. While some of the prelisted spices also improve the performance of digestive enzymes in the gastric mucosa, many of them increase the activity of pancreatic enzymes, such as amylases, lipases, and proteases (Srinivasan 2005). Extraction from herbs and spices improves digestion and shortens the time that food or feed passes through the digestive tract, in addition to affecting bile production and enzyme activity (Suresh and Srinivasan 2007).

15 The Antimicrobial Properties of Spices and Herbs

Growth-promoting feed additives improve feed stability and have a positive impact on the gastrointestinal environment, mostly by inhibiting the growth of harmful microorganisms. Animals are not as exposed to microbiologically derived poisons because of their increased digestive system health. According to Windisch et al. (2008), herbs and spices therefore aid in enhancing the animals' growth by increasing their tolerance to various stressors and their absorption of vital nutrients. Many of the secondary metabolites that plants produce act as defenses against pathogenic microbes, predators, and environmental and physiological stressors. Numerous in vitro investigations have shown that specific plant extracts have powerful antibacterial activity against both Gram-positive and Gram-negative bacteria; *E. coli* membranes cultured in the presence of limonene or cinnamonaldehyde showed a modification in the long-chain fatty acid profile, according to Di Pasqa et al. (2006). Comparable results were obtained with *Bronchotrix thermosphacta* cultivated with either limonene, cinnamonaldehyde, carvacrol, or eugenol, and with *Salomonella enterice* grown with either eugenol or carvacrol. No phytochemical that was tested altered the fatty acid profile of Pseudomonas fluorescens in *Staphylococcus aureus*. The ability of microbes to survive can be impacted by changes in the content of fatty acids.

16 Anti-Inflammatory Activity

Rat studies have demonstrated the anti-inflammatory effects of extracts of black pepper, red pepper, curcuma, cloves, cumin, cinnamon, nutmeg, mint, and ginger (Srinivasan 2005). The primary active components responsible for this action are flavonoids and terpenoids, which inhibit the metabolism of inflammatory prostaglandins. Notable plants and spices with anti-inflammatory properties in this area include licorice, anise, marigold, and chamomile (Craig 1999).

17 Anti-oxidant Activity

Phenolic compounds, including hydrolyzable tannins, phenolic acids, flavonoids, phenolic terpenes, and proanthocyanidins, as well as particular vitamins A, C, and E, are the primary components that give herbs and spices their antioxidative qualities (Meng et al. 2024). Green tea, chamomile, dandelion, marigold, oregano, sage, thyme, rosemary, and ginkgo are common herbs that are high in phenolics. These active substances can stop lipid peroxidation by either scavenging free radicals or triggering antioxidant enzymes such as glutathione reductase, catalase, superoxide dismutase, and glutathione peroxidase (Fasseas et al. 2008). During storage, herbs and spices help protect feed against oxidative deterioration. The pet and human food businesses both make extensive use of this tactic. The herb rosemary (*Rosmarinus officinalis*) is frequently used to preserve food and feed. It can be taken alone or in combination with synthetic antioxidants or tocopherols (Jacobsen et al. 2008).

18 Improved Digestive Tract Functions

Herbs improve the flavor of animal feed, which affects feeding habits, the release of digestive juices, and total feed consumption. The primary location of activity is the digestive tract. Different herbs and spices have different active ingredients that have different effects on the digestive system. The animal body can use herbs for a variety of purposes (Hernandez et al. 2004; Cho et al. 2024). Many of these compounds have sialagogue properties, which encourage salivation and make swallowing easier. Broiler feed digestibility was improved by extracts from *Rosmarinus officinalis, T. vulgaris*, and *S. officinalis* as well as a mixture of carvacrol, cinnamaldehyde, and capsaicin (Hernandez et al. 2004). Additionally, noted were the beneficial effects of plant extracts on antibacterial qualities, hunger stimulation, digestion improvement, and nutritional digestibility. Animals receiving phytobiotic-supplemented feed exhibit increased feed intake and digestive secretions (Windisch et al. 2008).

19 Immunostimulant Function

Spices and herbs rich in flavonoids, carotenoids, and vitamin C strengthen the immune system. Certain plants, such as cat's claw, garlic, liquorice, and echinacea, contain compounds that have immunostimulatory qualities. These herbs either boost phagocytosis, induce interferon generation, or increase the activity of lymphocytes, NK cells, and macrophages (Craig 1999).

20 Limitations of the Study

It is difficult to achieve consistent results of herbs and plant extracts due to variations in plant species, growing conditions, harvesting methods, and the preparation process. Some herbs, spices and plant extracts can be toxic to animals when consumed in large quantities or over extended periods. Extracts can sometimes be more expensive than conventional feed additives or pharmaceuticals, particularly for large-scale operations. Lack of clear regulatory guidelines can result in inconsistent quality or contamination. Many farmers and veterinarians may not be sufficiently trained in the proper use of herbs and plant extracts. This knowledge gap can lead to improper usage, limiting the potential benefits.

21 Recommendations

- Use plant-based products as a complement to, not a replacement for, good husbandry and conventional management practices.
- Consult with professionals to identify appropriate herbs, spices, and extracts for your specific animal species and production system.
- Educate farmers, farm managers, and workers about the safe and effective use of herbs, spices, and plant extracts.
- Keep up-to-date with new research findings and regulatory developments related to the use of herbs and plant extracts.

22 Conclusion

Achieving high productivity, preserving animal health, and creating high-quality, stable, and processing-ready animal products are the major goals of animal production. In addition to stimulating appetite and digestion, herbs and spices have a good impact on a number of physiological processes, improving an animal's general health and welfare and performance. Additionally, there are a number of benefits to using herbs instead of antibiotics. Understanding the chemical structures, values, and properties of useful herbs or their derivatives, as well as the physiological requirements and general health of animals, as well as—most importantly—consumer tastes and expectations, will be crucial to the future of using herbs in animal feed.

References

Abascal K, Yarnell E (2008) Botanical galactagogues. Altern Complement Ther 14:288–294. https://doi.org/10.1089/act.2008.14602

Ali MS, Saleem M, Ahmad W, Parvez M, Yamdagni R (2022) A chlorinated monoterpene ketone, acylated β-sitosterol glycosides and a flavanone glycoside from Mentha longifolia (Lamiaceae). Phytochem 59:889–895. https://doi.org/10.1016/S0031-9422(01)00490-3

Alimon AR (2009) Alternative raw materials for animal feed. Indonesian Bull Ani Vet Sci 19:117–124. http://medpub.litbang.pertanian.go.id/index.php/wartazoa/article/view/917

Aljohani AS (2024) Phenolics of botanical origin for the control of Coccidiosis in poultry. Pak Vet J 44(2). https://doi.org/10.29261/pakvetj/2024.179

Amna D, Islam MR, Farooq A, Munawar I (2024) Unveiling the functional implications and complex interplay between bound phenolic compounds and phenolics in food: a comprehensive review. Agrobio Rec 13:70–81. https://doi.org/10.47278/journal.abr/2023.027

Ando S, Nishida T, Ishida M, Kochi Y, Kami A, Se S (2001) Transmission of herb essential oil to milk and change of milk flavor by feeding dried herbs to lactating cows (in Japanese). Nippon Shok Kagaku Kaishi 48:142–145. https://doi.org/10.3136/nskkk.48.142

Barry TN, McNabb WC (1999) The implications of condensed tannins on the nutritive value of temperate forages fed to ruminants. Br J Nutr 81:263–272. https://doi.org/10.1017/S0007114599000501

Benchaar C, Petit HV, Berthiaume R, Ouellet DR, Chiquette J, Chouinard PY (2007) Effects of essential oils on digestion, ruminal fermentation, rumen microbial populations, milk production, and milk composition in dairy cows fed alfalfa silage or corn silage. J Dairy Sci 90:886–897. https://doi.org/10.3168/jds.S0022-0302(07)71572-2

Bhatt N, Singh M, Ali A (2009) Effect of feeding herbal preparations on milk yield and rumen parameters in lactating crossbred cows. Int J Agric Biol 11:721–726. http://www.fspublishers.org/

Botsoglou NA, Fletouris DJ, Florou-Paneri P, Christaki E, Spais AB (2003a) Inhibition of lipid oxidation in long-term frozen stored chicken meat by dietary oregano essential oil and α- tocopheryl acetate supplementation. Food Res Int 36:207–213. https://doi.org/10.1016/S0963-9969(02)00095-9

Botsoglou NA, Grigoropoulou SH, Botsoglou E, Govaris A, Papageorgiou G (2003b) The effects of dietary oregano essential oil and α-tocopheryl acetate on lipid oxidation in raw and cooked Turkey during refrigerated storage. Meat Sci 65:1193–1200. https://doi.org/10.1016/S0309-1740(03)00029-9

Celi P, Raadsma HW (2009) The effects of Yerba Mate (Ilex paraguarensis) supplementation on the productive performance of lambs. In: Ruminant physiology. Wageningen Academic, pp 804–805. https://doi.org/10.3920/9789086866830_375

Cho HY, Lee GY, Ali MS, Park SC (2024) Effects of dietary intake of heat-inactivated Limosilactobacillus reuteri PSC102 on the growth performance, immune response, and gut microbiota in weaned piglets. Pak Vet J 44(3). https://doi.org/10.29261/pakvetj/2024.224

Craig WJ (1999) Health-promoting properties of common herbs. Am J Clin Nutr 70:491S–499S. https://doi.org/10.1093/ajcn/70.3.491s

Deckert V, Desrumaux C, Athias A, Duverneuil L, Palleau V, Gambert P, Masson D, Lagrost L (2002) Prevention of LDL α-tocopherol consumption, cholesterol oxidation, and vascular endothelium dysfunction by polyphenolic compounds from red wine. Atherosclerosis 165:41–50. https://doi.org/10.1016/S0021-9150(02)00189-2

Demir E, Sarica Ş, Özcan MA, Sui Mez M (2003) The use of natural feed additives as alternatives for an antibiotic growth promoter in broiler diets. Br Poult Sci 44:44–45. https://doi.org/10.1080/713655288

Di Pasqua R, Hoskins N, Betts G, Mauriello G (2006) Changes in membrane fatty acids composition of microbial cells induced by addiction of thymol, carvacrol, limonene, cinnamaldehyde, and eugenol in the growing media. J Agric Food Chem 54:2745–2749. https://doi.org/10.1021/jf0527221

Dorman HD, Koşar M, Kahlos K, Holm Y, Hiltunen R (2003) Antioxidant properties and composition of aqueous extracts from Mentha species, hybrids, varieties, and cultivars. J Agric Food Chem 51:4563–4569. https://doi.org/10.1021/jf034108k

Esparza-Borges H, Ortiz-Márquez A (1995) Therapeutic efficacy of plant extracts in the treatment of bovine endometritis. In: International symposium on medicinal and aromatic plants, vol 426, pp 39–46. https://doi.org/10.17660/ActaHortic.1996.426.3

Fasseas MK, Mountzouris KC, Tarantilis PA, Polissiou M, Zervas G (2008) Antioxidant activity in meat treated with oregano and sage essential oils. Food Chem 106:1188–1194. https://doi.org/10.1016/j.foodchem.2007.07.060

Franco-Jimenez DJ, Scheideler SE, Kittok RJ, Brown-Brandl TM, Robeson LR, Taira H, Beck MM (2007) Differential effects of heat stress in three strains of laying hens. J Appl Poult Res 16:628–634. https://doi.org/10.3382/japr.2005-00088

Greathead H (2003) Plants and plant extracts for improving animal productivity. Proc Nutr Soc 62:279–290. https://doi.org/10.1079/PNS2002197

Guo FC, Kwakkel RP, Williams BA, Parmentier HK, Li WK, Yang ZQ, Verstegen MW (2004a) Effects of mushroom and herb polysaccharides on cellular and humoral immune responses of Eimeria tenella-infected chickens. Poult Sci 83:1124–1132. https://doi.org/10.1093/ps/83.7.1124

Guo FC, Williams BA, Kwakkel RP, Li HS, Li XP, Luo JY, Li WK, Verstegen MW (2004b) Effects of mushroom and herb polysaccharides, as alternatives for an antibiotic, on the cecal microbial ecosystem in broiler chickens. Poult Sci 83:175–182. https://doi.org/10.1093/ps/83.2.175

Gupta GT (2013) Prevalence of musculoskeletal disorders in farmers of Kanpur-rural. India J Commun Med Health Educ 3. https://doi.org/10.4172/2161-0711.1000249

Hashemi SR, Davoodi H (2010) Herbal plants and their derivatives as growth and health promoters in animal nutrition. Vet Res Commun 35:169–180. https://doi.org/10.1007/s11259-010-9458-2

Hernandez F, Madrid J, Garcia V, Orengo J, Megias MD (2004) Influence of two plant extracts on broilers performance, digestibility, and digestive organ size. Poult Sci 83:169–174. https://doi.org/10.1093/ps/83.2.169

Jacobsen C, Let MB, Nielsen NS, Meyer AS (2008) Antioxidant strategies for preventing oxidative flavour deterioration of foods enriched with n-3 polyunsaturated lipids: a comparative evaluation. Tr Food Sci Technol 19:76–93. https://doi.org/10.1016/j.tifs.2007.08.001

Jang A, Liu XD, Shin MH, Lee BD, Lee SK, Lee JH, Jo C (2008) Antioxidative potential of raw breast meat from broiler chicks fed a dietary medicinal herb extract mix. Poult Sci 87:2382–2389. https://doi.org/10.3382/ps.2007-00506

Janz JA, Morel PC, Wilkinson BH, Purchas RW (2007) Preliminary investigation of the effects of low-level dietary inclusion of fragrant essential oils and oleoresins on pig performance and pork quality. Meat Sci 75:350–355. https://doi.org/10.1016/j.meatsci.2006.06.027

Kanatt SR, Chander R, Sharma A (2007) Antioxidant potential of mint (Mentha spicata L.) in radiation-processed lamb meat. Food Chem 100:451–458. https://doi.org/10.1016/j.foodchem.2005.09.066

Kawakami Y, Tsurugasaki W, Yoshida Y, Igarashi Y, Nakamura S, Osada K (2004) Regulative actions of dietary soy isoflavone on biological antioxidative system and lipid metabolism in rats. J Agric Food Chem 52:1764–1768. https://doi.org/10.1021/jf0345898

Klevenhusen F, Duval S, Zeitz JO, Kreuzer M, Soliva CR (2011) Diallyl disulphide and lovastatin: effects on energy and protein utilisation in, as well as methane emission from, sheep. Arch Anim Nutr 65:255–266

Koivula MJ, Eeva T (2010) Metal-related oxidative stress in birds. Environ Pollut 158:2359–2370. https://doi.org/10.1016/j.envpol.2010.03.013

Lawrence RM, Lawrence RA (2004) Breast milk and infection. Clin Perinatol 31:501–528. https://doi.org/10.1016/j.clp.2004.03.019

Loo A, Richard H (1992) Nature, origine et propriétés des épices et des aromates bruts. In: Richard H (ed) Épices et Aromates. Lavoisier, Paris, pp 18–22
McIntosh FM, Williams P, Losa R, Wallace RJ, Beever DA, Newbold CJ (2003) Effects of essential oils on ruminal microorganisms and their protein metabolism. Appl Environ Micro 69:5011–5014. https://doi.org/10.1128/AEM.69.8.5011-5014.2003
Meng A, Zhang X, Pubu P, Ali M, Wang J, Xu C, Almutairi MH, Li K (2024) Protective effect of Lentinan against LPS-induced injury in mice via influencing antioxidant enzyme activity, inflammatory pathways and gut microbiota. Pak Vet J 44(3). https://doi.org/10.29261/pakvetj/2024.225
Morsy TA, Kholif AE, Matloup OH, Elella AA, Anele UY, Caton JS (2018) Mustard and cumin seeds improve feed utilisation, milk production and milk fatty acids of Damascus goats. J Dairy Res 85:142–151. https://doi.org/10.1017/S0022029918000043
Neville MC, Morton J, Umemura S (2001) Lactogenesis: the transition from pregnancy to lactation. Pediatr Clin N Am 48:35–52. https://doi.org/10.1016/S0031-3955(05)70284-4
Ocak N, Erener G, Ak F, Sungu M, Altop A, Ozmen A (2008) Performance of broilers fed diets supplemented with dry peppermint (Mentha piperita L.) or thyme (Thymus vulgaris L.) leaves as growth promoter source. Czeh J Anim Sci 53:169–175. https://doi.org/10.17221/373-cjas
Preciado T, Orozco-Hernández J-R, Carranza A, Mora V, Chávez G (2011) Use of an herbal galactogogue on milk quality and yield. Asian J Anim Vet Adv 6:297–300. https://doi.org/10.3923/ajava.2011.297.300
Rahimi S, Teymori Zadeh Z, Torshizi K, Omidbaigi R, Rokni H (2011) Effect of the three herbal extracts on growth performance, immune system, blood factors and intestinal selected bacterial population in broiler chickens. J Agric Sci Technol 13:527–539. https://dorl.net/dor/20.1001.1.16807073.2011.13.4.6.5
Roussel AM, Hininger I, Benaraba R, Ziegenfuss TN, Anderson RA (2009) antioxidant effects of a cinnamon extract in people with impaired fasting glucose that are overweight or obese. J Am Coll Nutr 28(1):16–21. https://doi.org/10.1080/07315724.2009.10719756
Saleh AA, Soliman MM, Yousef MF, Eweedah NM, El-Sawy HB, Shukry M, Wadaan MA, Kim IH, Cho S, Eltahan HM (2023) Effects of herbal supplements on milk production quality and specific blood parameters in heat-stressed early lactating cows. Front Vet Sci 10:1180539. https://doi.org/10.3389/fvets.2023.1180539
Scislowski V, Bauchart D, Gruffat D, Laplaud PM, Durand D (2005) Effects of dietary n-6 or n-3 polyunsaturated fatty acids protected or not against ruminal hydrogenation on plasma lipids and their susceptibility to peroxidation in fattening steers. J Anim Sci 83:2162–2174. https://doi.org/10.2527/2005.8392162x
Srinivasan K (2005) Spices as influencers of body metabolism: an overview of three decades of research. Food Res Int 38:77–86. https://doi.org/10.1016/j.foodres.2004.09.001
Sudhakar M, Rao CV, Rao PM, Raju DB, Venkateswarlu Y (2006) Antimicrobial activity of Caesalpinia pulcherrima, Euphorbia hirta and Asystasia gangeticum. Fitoterapia 77:378–380. https://doi.org/10.1016/j.fitote.2006.02.011
Suresh D, Srinivasan K (2007) Studies on the in vitro absorption of spice principles–curcumin, capsaicin and piperine in rat intestines. Food Chem Toxicol 45:1437–1442. https://doi.org/10.1016/j.fct.2007.02.002
Traber MG, Ramakrishnan R, Kayden HJ (1994) Human plasma vitamin E kinetics demonstrate rapid recycling of plasma RRR-alpha-tocopherol. Proc Natl Acad Sci 91:10005–10008. https://doi.org/10.1073/pnas.91.21.10005
Viegi L, Pieroni A, Guarrera PM, Vangelisti R (2003) A review of plants used in folk veterinary medicine in Italy as basis for a databank. J Ethnopharmacol 89:221–244. https://doi.org/10.1016/j.jep.2003.08.003
Windisch W, Schedle K, Plitzner C, Kroismayr A (2008) Use of phytogenic products as feed additives for swine and poultry. J Anim Sci 86:140–148. https://doi.org/10.2527/jas.2007-0459

Wood JD, Richardson RI, Nute GR, Fisher AV, Campo MM, Kasapidou E, Sheard PR, Enser M (2004) Effects of fatty acids on meat quality: a review. Meat Sci 66:21–32. https://doi.org/10.1016/S0309-1740(03)00022-6

Wu G, Fanzo J, Miller DD, Pingali P, Post M, Steiner JL, Thalacker-Mercer AE (2014) Production and supply of high-quality food protein for human consumption: sustainability, challenges and innovations. Ann N Y Acad Sci 1321:1–9. https://doi.org/10.1111/nyas.12500

Yalçin S, Eser H, Onbaşilar İ, Yalçin S (2020) Effects of dried thyme (Thymus vulgaris L.) leaves on performance, some egg quality traits and immunity in laying hens. Ankara Üniv Vet Fak Derg 67:303–311. https://doi.org/10.33988/auvfd.677150

Yano Y, Satomi M, Oikawa H (2006) Antimicrobial effect of spices and herbs on Vibrio parahaemolyticus. Int J Food Microbiol 111:6–11. https://doi.org/10.1016/j.ijfoodmicro.2006.04.031

Young JF, Stagsted J, Jensen SK, Karlsson AH, Henckel P (2003) Ascorbic acid, alpha-tocopherol, and oregano supplements reduce stress-induced deterioration of chicken meat quality. Poult Sci 82:1343–1351. https://doi.org/10.1093/ps/82.8.1343

Zhou BO, Wu LM, Yang LI, Liu ZL (2005) Evidence for α-tocopherol regeneration reaction of green tea polyphenols in SDS micelles. Free Radic Biol Med 38:78–84. https://doi.org/10.1016/j.freeradbiomed.2004.09.023

Essential Oils as Natural Feed Additives: Functional Properties and Emerging Role in Sustainable Animal Nutrition

Ali Aygun, Gülşah Kanbur, Suleyman Dogu, and Zeliha Ustun-Argon

Abstract Essential oils are complex mixtures of volatile, aromatic compounds derived from various plant parts through processes such as distillation or solvent extraction. In recent years, they have attracted growing scientific interest due to their diverse applications in sectors including medicine, cosmetics, food technology, agriculture, and textiles. These bioactive substances are recognized for their multifaceted biological properties, including anti-inflammatory, antimicrobial (encompassing antibacterial, antifungal, and antiviral), antioxidant, and immunomodulatory activities. The extraction technique employed plays a critical role in determining the chemical profile and yield of essential oils. While conventional methods such as hydrodistillation remain widely used, advanced techniques like microwave-assisted extraction and supercritical fluid extraction have been developed to enhance efficiency and preserve bioactivity. Additionally, the phytochemical composition and functional quality of essential oils are influenced by variables such as botanical species, plant part used, harvesting stage, and post-harvest processing conditions. In the context of animal nutrition, essential oils have emerged as promising alternatives to synthetic growth promoters and antibiotic feed additives. Their incorporation into animal diets has been associated with improved feed conversion efficiency, enhanced nutrient digestibility, modulation of gut microbiota, and stimulation of digestive enzyme secretion. Moreover, their antimicrobial properties contribute to the maintenance of gut health by inhibiting pathogenic microorganisms, thereby reducing the risk of enteric infections and improving overall animal performance. Given these functional attributes, essential oils are increasingly regarded as natural feed additives with the potential to support sustainable

A. Aygun (✉) · G. Kanbur
Agriculture Faculty, Department of Animal Science, Selçuk University, Konya, Turkey
e-mail: aaygun@selcuk.edu.tr; gkanbur@selcuk.edu.tr

S. Dogu · Z. Ustun-Argon
Meram Vocational School, Department of Plant and Animal Production, Necmettin Erbakan University, Konya, Türkiye

Medical and Cosmetic Plants Application and Research Center, Konya, Turkey
e-mail: sdogu@erbakan.edu.tr; zargon@erbakan.edu.tr

R. U. Khan, S. Naz (eds.), *Green Alternatives to Antibiotics in Animal Health*,
https://doi.org/10.1007/978-981-92-0309-3_6

livestock production, enhance animal health, and contribute to food safety by minimizing the use of synthetic compounds.

Keywords Essential oil · Alternative to antibiotics · Animal health

1 Introduction

Essential Oils are "aromatic products with a mixture of compounds derived from plant raw material, either separated by steam, dry distillation, or by a suitable mechanical technique without heating" (EPC 2004). Essential oil is separated from its liquid phase without changing its chemical composition by any physical method.

The exploration and study of essential oils have gained substantial traction and considerable interest in recent years, particularly within the dynamic realms of health, cosmetics, food, pharmaceuticals, wearable devices, leather, agriculture, and textiles (Belahcene et al. 2023; Hudz et al. 2023; Ke et al. 2023; Liu et al. 2022; Rout et al. 2022; Zhang et al. 2023; Zhou et al. 2023) industries. Essential oils, precisely defined as concentrated and hydrophobic liquids that contain a rich array of volatile aroma compounds derived from multiple parts of various plants, are intricate and complex mixtures. These remarkable substances can exhibit beneficial properties, which include, but are not limited to, notable anti-inflammatory, antibacterial, amoebicidal, antifungal, larvicidal, anticancer, nematocidal, antiviral, antioxidant, antidiabetic, antimutation, and anti-spasmodic effects activities (Anacarso et al. 2019; Berenji Ardestani et al. 2016; Drioiche et al. 2023; Gogoi et al. 2023; Huo et al. 2024; Jadhav et al. 2023; Kalaiselvi et al. 2019; Kamaruddin et al. 2023; Kant and Kumar 2022; Li et al. 2025; Shin et al. 2021; Taglienti et al. 2023; Xu et al. 2022).

As a result of the developments in extraction methods, essential oils are increasingly being considered for a wide range of practical applications, primarily in the food industry (Wang et al. 2023), therapeutic uses in an effort to enhance well-being and support the overall health of human beings, and to replace the chemical drugs of animals with natural alternatives (Miguel 2010).

The extraction methods for essential oils are critical, as they affect both the yield and the chemical composition of the final product, along with plant species, growth conditions, environmental factors, and harvesting time (Aguele et al. 2023). Traditional techniques such as hydrodistillation and steam distillation have been widely utilized, but innovations in extraction technology, including microwave-assisted extraction and supercritical fluid extraction, have emerged as promising alternatives that enhance efficiency and reduce processing times (Lainez-Cerón et al. 2021).

Acknowledging the inherent limitations of conventional essential oil extraction methodologies, contemporary research has focused on the development and implementation of alternative extraction techniques. These methodologies are predicated on several key principles. These are the utilization of non-toxic and readily separable solvents, such as supercritical fluid extraction (SFE), to mitigate solvent residue contamination; the application of microwave irradiation to activate cellular disruption and uniform heating, facilitating EO efflux without supplementary solvent application through gravitational forces, thereby circumventing hydrolytic degradation of EO constituents, as exemplified by solvent-free microwave extraction (SFME) and microwave water diffusion gravity methods; the employment of techniques that rapidly compromise cellular integrity, such as flash vacuum expansion, to expedite EO extraction and substantially reduce processing duration; the integration of novel energy sources, including solar energy collection systems, to promote energy efficiency, and environmental sustainability within the extraction process; and the implementation of post-extraction purification protocols to enhance EO quality and efficacy through advanced processing. Collectively, these advancements signify substantial progress in refining EO extraction, addressing the deficiencies associated with traditional methods, and establishing a foundation for the integration of renewable energy sources in this domain (Huo et al. 2024).

For the efficiency of the essential oil yield from plants, the processes before the extraction should be carefully planned. Harvesting is the initial process of collecting the intended plants from the field. The harvesting time depends on maturity, planned end use, cultivar differences, and the target part of the plant, such as flowers, buds, leaves, stems, roots, rhizomes, bark, wood, and resins. Transportation of the harvested plants should be done carefully in dry and clean conditions. Threshing is the separation process of the plant from unwanted or damaged parts. The next step is drying, one of the most important processes. It is necessary to produce different essential oils since they decrease the water content of the plants and enable better extraction and easier storage. Different techniques of drying can be applied, such as direct sun exposure, drying rooms, ovens, solar drying systems, and freeze-drying methods. Cleaning and classification of the plant materials would help to increase the essential oil quality. If the extraction process is not completed after the drying process, cleaned and classified material should be packed with proper packaging material to ensure controlling the moisture, heat, light exposure, and microbiological growth to protect the essential oil content. Plants such as nutmegs and cardamom need a grinding process to increase the yield and to create uniform products (Bayrak 2023; Lego et al. 2021; UNIDO 2005).

Extraction methods of essential oils can be classified as conventional and advanced extraction methods.

1.1 Conventional Methods

1.1.1 Hydrodistillation

Hydrodistillation is the most economical method used to obtain essential oils from spices and aromatic plants. The main advantage of distillation is that it can be carried out with simple equipment close to where the plant is produced. Even in remote settlements, large amounts of material can be processed in a short time, and fewer human forces are used in distillation.

Water distillation is the preferred method for distilling small pieces of plants and flowers. Steam is not successful in effectively distilling this type of material because steam can create channels in powdered material, but this is not so easy among flowers. For all other materials, both high and low steam pressure distillation methods are preferred.

The distillation process steps are as follows:

1. The plant is placed in a distiller and filled with water to the top.
2. Heat is applied until steam is formed.
3. At the beginning of the distillation, clear water is obtained.
4. When the water loses its clarity and takes on a milky appearance, distillation of the aromatic substance begins.
5. Distillation ends when the color of the water becomes completely clear again as it was at the beginning.

It does not require qualified workers as in the solution reaction, but using the simplest and most economical method can cause low economic value due to reasons such as low efficiency, low oil quality, and low market value. Distillation with water is the simplest of the three types of distillation. In this method, the plant material is mixed directly with water in the distiller. In this method, the plant should not settle to the bottom and should not come into direct contact with heat. Water distillation is the simplest and cheapest method to obtain essential oils, but there is a loss of quality in the oil due to the contact of the material with heat and water (Fig. 1) (Kant and Kumar 2022).

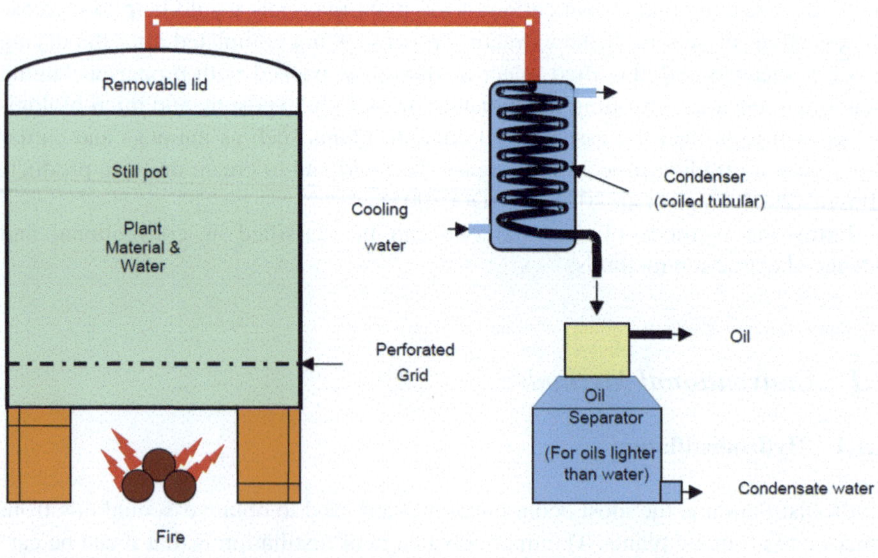

Fig. 1 Water distillation unit. (UNIDO 2005)

Water distillation is similar to steam distillation in many ways; the difference is that in water distillation, the plant is in contact with water. In this type of distillation, if the boiler is directly heated, the odor of the distiller is a constant problem. The burnt nature of the volatile oils is called the odor of the distiller, and this odor decreases over time during the storage process of the oils. Water and steam distillation methods have some common aspects; plant parts are separated from the boiling water with a perforated metal grid at the bottom of the boiler, and do not come into contact with water. If the boiler is heated carefully, the danger of odor in the distiller is significantly reduced thanks to this grid. Hydro diffusion is a separate event in the normal steam distillation process. In this system, steam enters from the top of the boiler, and the oil and water mixture accumulates at the bottom of the boiler. This method reduces the distillation time and is especially suitable for the distillation of seeds and fruits (Lainez-Cerón et al. 2021).

The disadvantages of this method are as follows:

1. Time loss in excessive loading and unloading of the distillation still.
2. Direct contact of the plant with the distillation tank and formation of unpleasant odors due to high temperature.
3. The layer formed due to calcium on the inner wall of the boiler prevents heat transfer to some extent.

The oils obtained by water distillation are generally darker in color and have a stronger odor than other methods, so they have the lowest value. The disadvantages of water distillation are greater than its advantages (Bayrak 2023).

To increase the effectiveness of this method, turbo-hydrodistillation, where the extractor in the Clevenger system is replaced with a stainless-steel stirrer to fasten the process and also with sharp blades, extending the exposure area of the plant, which helps accelerate the extraction process of essential oils. This technique has been used to extract essential oils of the Alliaceae family and lavender. Adding salt to the water also helps to increase the essential oil yield in plants such as lavender (Filly et al. 2016). Pretreating the plants with enzymes such as pectinase, cellulase, xylanase also develop the effectiveness of the oil yield while decreasing the duration of extraction (Kashyap et al. 2001).

1.1.2 Steam Distillation

Steam distillation is the distillation of plant material with steam produced in a separate steam generator outside the distiller. Essential oils are produced by various methods, the most commonly used of which is steam distillation. In this method, steam is produced in a steam boiler separate from the system and is brought into direct or indirect contact with the plant material in the distiller (Fig. 2). Essential oil producers usually use large distillation systems heated directly or with steam. In the steam distillation system, the plants are placed on a grate located in the distiller and not in contact with water, i.e., just above the water level, not in water. As the steam passes through the plant parts, it takes the volatile substances and multiplies one

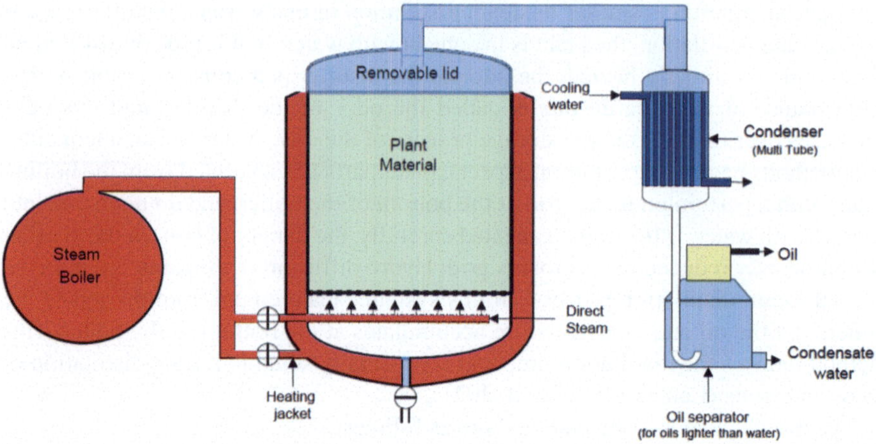

Fig. 2 Steam distillation (UNIDO 2005)

after another, becomes saturated, and condenses in an effective cooler and turns into liquid. The oil is separated from the water by decantation. The steam distillation method is preferred to the water distillation method performed with direct heating due to the rapid and short-term oil production. As a result, since the plant is boiled for a short time, the quality of the oil obtained is not very low. Steam distillation is also done by giving steam directly into the distiller or by using double-walled distillation distillers. Good results are obtained by circulating unused steam in double-walled steam boilers. In this type of distillation, the plant is protected from contact with the extremely hot metallic part of the distiller. However, this method is a less-used method due to the shortness of time and high cost compared to the distillation method, which is carried out by sending steam directly into the steam boiler. In steam distillation, hydrolysis, that is, the disintegration caused by the effect of water, is limited and even decreases if high temperatures are applied (Bayrak 2023; Kant and Kumar 2022; Lego et al. 2021).

The advantages and disadvantages of the steam distillation method are as follows:

1. The amount and quality of steam can be controlled.
2. The risk of thermal deterioration is less because the temperature generally does not exceed 100 ° C.
3. It is another of the most used methods in the production of essential oils, and a higher quality oil is obtained compared to the other two methods.
4. It is an extraction method used especially for the industry producing aroma and flavor substances.
5. It requires a lot of capital and can take more than 10 years to amortize in low-cost oils.
6. It requires high technical skills, fabrication, maintenance, and installation.
7. There are many types of this application: batch system, hydrodiffusion, maceration distillation, mobile distillers, and continuous distillation process.
8. This method is a more expensive method than the other two distillation methods (Bayrak 2023).

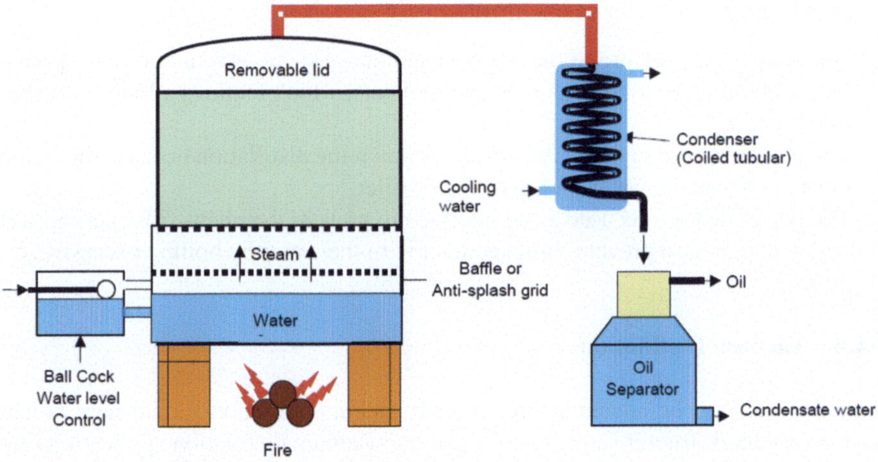

Fig. 3 Water–steam distillation (UNIDO 2005)

1.1.3 Water–Steam Distillation

In this distillation method, the plant material is placed on a grid or a perforated tray placed slightly above the bottom of the distiller (Fig. 3). The empty space under the grid is filled with water, but the water should not touch the grid. The water is heated, and low-pressure and saturated wet steam rises through the plant material. The typical feature of this method is:

1. The steam is always saturated and wet, but not extremely hot.
2. The plant material is in contact only with the steam, not with boiling water.

In this method, the distiller capacity decreases in volume, but at the end of the process, high-density material is obtained. Because the plant material is not suspended in water.

The advantages of water steam distillation are quite high compared to water distillation. These are:

1. High oil yield.
2. Very little change in oil components due to the heat source, thermal conductivity, and wetness in the distiller.
3. Ease of rapid processing and energy saving.
4. The components of the essential oil are less sensitive to hydrolysis and polymerization.
5. If reflux is controlled, the loss of polar components is minimized.
6. The quality of the oil obtained is better.
7. Water–steam distillation is faster than water distillation, therefore energy is used more effectively.
8. It is the method applied in many essential oil extractions today. Water steam distillation is widely used in rural areas (Bayrak 2023).

Disadvantages of water steam distillation:

1. Due to the low pressure of the rising steam, more steam is required to evaporate oils containing high-boiling-point components, which requires a longer distillation time.
2. The plant material gets wet, which slows down the distillation because the steam must condense the water on top of the distiller.
3. The perforated grid is used as an interface to prevent the plants from getting wet with water, and to prevent direct contact with the water by boiling excessively.

1.1.4 Vacuum Distillation

It may not always be advantageous to apply steam distillation for substances that have a very hard structure and change easily, so vacuum distillation applications are made under low pressure to obtain some flavor components, especially in essential oils. This application has advantages such as saving fuel, preventing drying and wetting due to applications in airless conditions, reducing the possibility of reactions between components at higher temperatures, and reducing losses. Distillation and fermentation can be applied together for mustard and bitter almond essential oils and some essences.

Fractional distillation is based on the principle of separating medium and high molecular weight components into fractions, purifying, concentrating, and refractionating them. This method has been successful in solving important problems in different sectors (Bayrak 2023).

Molecular distillation is an important method based on the evaporation of components for the distillation and separation of products with high boiling points.

Dry distillation is used to obtain a very small number of essential oils, such as Juniper oil, which are generally included in secretory oils.

1.1.5 Cold Pressing

This method, called squeezing or cold pressing, is only used in the production of citrus oil. The term squeezing describes a physical process that involves crushing or breaking the volatile oil pockets inside the peel to release the oil. Citrus essential oils are usually obtained by cold pressing the peels of these fruits. With this method, the peels are crushed, and the oil is separated as a watery emission. This separated emission is then centrifuged to obtain the oil. Citrus essential oils obtained by pressing have superior odor properties compared to oils obtained by distillation.

1.1.6 Enfleurage

Freshly picked flower batches are spread on a glass surface with specially prepared oil, kept for a while and taken by scooping, then replaced with fresh flowers. After a while, the fat absorbs the essential oil of the flowers and becomes saturated. Then this volatile or solid oil is separated by treating it with alcohol. This is a method used for jasmine and tuberose flowers. For this extraction, lard, coconut oil, or shea butter can be used (Bayrak 2023).

1.1.7 Hot Fat Extraction

Hot fat extraction is the process of treating the plants with hot fat and thus obtaining extra moisture. In other words, it is the process of treating the flowers in batches with hot fat and processing the same batch of fresh flowers several times until the oil is saturated with the scent of the flower. The flowers that are used are removed, and the fragrant oil is sold under names such as orange pomade, rose pomade, etc. (Bayrak 2023).

1.1.8 Solvent Extraction

In this method, solvents such as benzene, acetone, or hexane are used to dissolve the essential oil in the plants. Evaporation or a slight heating process can separate the solvents from the extracted oil–solvent mixture (Chakravarty et al. 2023). Percolation and Soxhlet extractions are the methods mainly preferred with the solvents. In the percolation system, dried plant material should be powdered and is placed in the percolator. The solvent should be added at the top of the plant samples, and it should be left for a determined period of time, depending on the plant material. The valve at the bottom of the percolator should be in the closed position during the extraction. At the end of the process, the solvent and essential oil mixture should be separated through the valve. Fresh solvent could be added from the top several times until the extraction is completed (Jones and Kinghorn 2005).

Soxhlet extraction is the other method, including a condenser, extractor, and distillation flasks. The basic principle of the system is based on the concentrated liquid solvent passing through the plant material to extract the essential oil. The essential oil is collected in the distillation flask, and the solvents are reused in the cycle of evaporation and condensing several times until the extraction is reached to the end (Tandon and Rane 2008).

Extraction systems usually have 3–4 extractors, 4–5 metal tanks containing solvents and solutions, and an evaporator to concentrate the essential oils. The extractors are classified as fixed and rotary types. Extraction is done systematically with each batch of plants and solvent three times. This process is done systematically with consecutive washing processes. It takes approximately 4.5–5 h to complete the extraction of each batch of plants, but each plant type requires different working methods, and each producer applies their own method (Bayrak 2023).

1.2 Advanced Extraction Methods

1.2.1 Microwave-Assisted Extraction (MAE)

Microwave energy is used to heat the solvent and plant material mixture. Since the weak hydrogen bonds and the cell structure are disrupted in this method, the essential oil extraction is accelerated and helps to reduce the time by more than 50% compared with the traditional methods. Energy consumption is also reduced in this method while the quality of the essential oil is not changed (Bendahou et al. 2008; Flamini et al. 2007). To produce more effective and high-quality products with this method, microwave applications are used, such as solvent-free (Lucchesi et al. 2004), vacuum hydro-distillation (Zill-e-Huma et al. 2011), microwave-assisted solvent extraction (Tomaniova et al. 1998), microwave hydrodiffusion and gravity (Vian et al. 2008), and microwave-assisted simultaneous distillation solvent extraction (Ferhat et al. 2007; Gaikwad et al. 2024; Lego et al. 2021).

1.2.2 Supercritical Fluid Extraction (SFE)

This method is applied effectively with temperature-sensitive compounds such as *Echinophora platyloba* (Sodeifian and Sajadian 2017) and *Crocus sativus* petals (Ahmadian-Kouchaksaraie and Niazmand 2017). Carbon dioxide is heated and compressed to obtain a supercritical fluid and is passed through grounded plant material, and the extract mix with CO2 and plant material are sent to a separator to discharge the CO2 by decompression (Kant and Kumar 2022). SFE method covers high-pressure working conditions between 10 and 50 MPa, and temperatures between 35 and 80 °C with a determined CO2 flow (Gaikwad et al. 2024). Since there is no solvent residue in the extracted product and discharged CO2 is circulated to the storage tanks and released to the atmosphere at the end of the process, it is called green technology (Akdag et al. 2025; Argon et al. 2023). The effectiveness of the SFE process can be increased with the integration of mechanical pre-treatment processes such as ultrasound, microwave, microfluidics, cold pressing, and milling. The SFE system can also be assisted with non-mechanical methods such as enzymes, nano or micro encapsulation, adsorbents, subcritical water, molecular and steam-distillation, pressurized liquid extraction, and supercritical antisolvent additions. Offline and online fractionation processes and integrating the SFE systems with chromatography methods also increase the quality of the essential oil obtained (Albuquerque et al. 2022; Dashtian et al. 2024).

1.2.3 Subcritical Extraction Liquid (SEL)

This method uses subcritical water to extract the essential oil under high-pressure and high-temperature (130–240 °C) conditions. Subcritical water is a liquid material that is below its critical point, and during the extraction process, it acts like solvents such as ethanol or methanol (Mondor and Hernández-Álvarez 2023). The extraction process takes approximately 15 min, compared to 3–6 h of extraction time with traditional methods (Aziz et al. 2018; Özel et al. 2006). Subcritical water extraction was found to be the best extraction option for essential oils such as rosemary oil in terms of speediness, oil quality, effectiveness, and clean technology (Díaz-Reinoso et al. 2023). Subcritical extraction can be applied to a wide range of sample matrices such as seasoning and medicinal herbs, plants, food byproducts, organic pollutants, marine algae, and fungi (Gaikwad et al. 2024; Kant and Kumar 2022).

1.2.4 Ultrasound-Assisted Extraction (UAE)

In this method, ultrasonic waves (20–100 kHz) are used to extract the essential oil from plant materials, which are submerged in solvent or water. The high-frequency sound waves create micro-sized bubbles and increase the solvent penetration into the cells of the plant. The mechanical vibration of cell walls with the effect of temperature and the high pressure during the process, plant cells release its components in an effective way. This method can also be applied efficiently at room temperature on the temperature-sensitive plant materials with simple equipment and in an economical way. The effectiveness of the extraction depends on pressure, temperature, water content of the plant, type of the solvent, particle size of the plant, and time length of the process (Gaikwad et al. 2024; Lego et al. 2021).

1.2.5 Pulsed Electric Field (PEF) Technology

Pulsed electric field technology is generally preferred to accelerate the extraction process and pretreatment before extraction to increase the oil yield with its short operation time and low-level power consumption (Naliyadhara et al. 2022). Traditional extraction methods can yield a low quantity of the high-value oils, applying the PEF technology can help to increase the oil yield, the quality, and the amount of antioxidant components (Yang et al. 2024). The studies showed that the addition of electricity in 200 pulses, 2 kV/cm, before 60 min of distillation of rosemary leaves, eucalyptus, and mint leaves increased the oil yields 11%, 17%, and 68,3%, respectively (Ferraz and Silva 2024).

Traditional systems have their own limitations, and therefore, researchers have developed novel methods for the extraction of essential oils. These inventions focused on using non-toxic, environmentally friendly, easily separable solvents, sustainable energy sources, more rapid and effective processes with increased and high-quality oil yield, systems providing more purified products through extensive processing steps. These systems can be classified as concentrated solar thermal systems-based extraction, flash vacuum expansion, vacuum distillation technology, solvent-free microwave extraction, microwave hydro-diffusion and gravity, molecular distillation, sequential extraction, and ohmic-assisted hydro distillation methods (Huo et al. 2024; Katekar et al. 2023).

1.3 Factors Affecting Essential Oil Extraction

The particle size of the plant material will affect the extraction effectiveness. The size of the plant material should be reduced before the extraction to increase the surface area which contacted with the extraction solvent. Various methods, such as grinding and milling, can be applied to the extracted material based on the type of essential oil (Tischer et al. 2017).

Microwave and ultrasound pre-treatments are other techniques for increasing essential oil yield. The heating induced by microwave and reducing the intracellular moisture disrupts the cells. The important point of this pre-treatment is to optimize the exposure time of the irradiation, the ratio of solvent and the solid, the microwave power level, and to create specially designed recovery units for evaporated oils. Ultrasound pretreatment is also a very effective way to enhance the oil quality and yield when it is combined and optimized with other extraction methods (Chakravarty et al. 2023; Radzi and Kasim 2020; Yu et al. 2021).

Temperature is another critical point determining the oil quality, viscosity, yield, and solvent diffusion. It should be optimized to maximize the oil yield and to protect the temperature-sensitive contents while maintaining the oil quality (Florczak 2015; Wang et al. 2021).

Pressure enhances the oil yield and efficiency by increasing the solvent exposure to the plant material. Increasing pressure will reduce the process duration by fastening the extraction and improving the solubility of volatiles. With some extraction methods, such as supercritical fluid extraction, the application of different levels of pressure will allow extract targeted essential oil components selectively without any degradation of medicinal and aromatic components (Gaikwad et al. 2024).

2 Compounds of Essential Oils

The chemical composition of essential oils is directly related to the extracted part of the plant. The volatiles and the essential oil concentrations vary depending on tissues from leaves, flowers, roots, resins, and wood. Extraction methods, applied techniques, environmental factors, harvest conditions, and integrated technologies also affect the essential oil quality, yield, efficiency, and the chemical composition (Gaikwad et al. 2024; Ngcangatha-Maqhude 2017; Orchard and van Vuuren 2017).

Essential oils containing aldehydes, phenols, esters, alcohols, acids, and terpenes are particularly susceptible to oxidation, decomposition, and isomerization when exposed to elevated temperatures and humidity. These processes can lead to a marked decline in the biological activity of essential oils and may even result in the formation of harmful byproducts. Consequently, the conditions during the extraction of EOs are crucial in preserving their quality and effectiveness (Huo et al. 2024).

Essential oils extracted from various plant parts are complex mixtures of natural components, including terpenes, aromatic compounds, and terpenoids (Reddy et al. 2025).

2.1 Terpenes

Terpenes include monoterpenes (C10) and sesquiterpenes (C15).

2.1.1 Monoterpenes

Monoterpene (C10) hydrocarbons such as α and β-pinenes, limonene, and myrcene are found in complex mixtures in many essential oils, especially in the leaf oils of plants, while seed and flower oils contain more specific monoterpenes. The effects of oils rich in these components in skin applications can be summarized as bactericidal and antiviral activity, preventing the initiation, development, and progression of cancer cells, and activities that increase skin penetration (Eslahi et al. 2017).

2.1.2 Sesquiterpenes

They are 15-carbon hydrocarbons (C15) components that are less volatile and more viscous than monoterpenes. Many woody oils contain high levels of sesquiterpenes and sesquiterpenols. Sesquiterpenols are oxidized in storage and eventually polymerize to form long-chain resins. Unlike monoterpenes, their oxidation products are not skin irritants, but shelf life and storage conditions are important factors in preserving their therapeutic activity. They are widely used in skin care and have anti-inflammatory effects (Bayrak, 2023).

2.2 Alcohols

Terpenoid alcohols have a hydroxyl group attached to one of the carbon atoms in the monoterpene and sesquiterpene, or diterpene chain. They are classified as primary alcohols, secondary alcohols, and tertiary alcohols (Bayrak 2023).

2.3 Monoterpenols

Primary monoterpene alcohols, such as geraniol, are slowly oxidized to aldehydes or acids in the presence of air and light to form resins. Monoterpenols have proven antibacterial and antibiotic activity and are generally well tolerated in the skin and mucous membranes (Zuzarte and Salgueiro 2015).

2.4 Sesquiterpenols

Those compounds are complex molecules with two or three closed rings. Its skin penetration is slow due to the large size of the molecule, and irritation is not expected. Sesquiterpenol-rich oils are viscous, the viscosity of which affects the size of the drop from a standard dropper. These compounds have skin healing, anti-inflammatory, wound-scab-facilitating, and virucidal effects (Eslahi et al. 2017).

2.5 Aromatic Alcohols

The most common aromatic alcohol is 2 phenyl ethyl alcohol and is found in distilled rose oil. It is used in the perfume industry because of its slightly floral rose scent (Bayrak 2023).

2.6 Phenols

Phenols in essential oils have a hydroxyl group attached to the aromatic benzene ring. They act as weak acids and react easily with other polar or positively charged molecules in the skin, including proteins, to form ions. Therefore, they have greater antimicrobial activity compared to alcohols. However, some phenols, such as eugenol, are irritating to the dermal and mucous membranes. Some also have the potential to trigger liver damage when taken regularly and in high concentrations (Eslahi et al. 2017).

2.7 Aldehydes

Aldehydes characteristically have the HC=O structure on the terminal carbon atom. They are highly reactive and unstable. Associated with dermal sensitivity, some aldehydes have antibacterial, antifungal, and antiviral effects as a treatment (Zuzarte and Salgueiro 2015).

2.8 Ketones

The carbon atom has a carbonyl (C=O) group, so they are less reactive than aldehydes. Many ketones, especially the common monoterpenoid ketones, have a strong odor. Ketones are known to have wound healing properties and inhibit wound growth. High ketone concentrations in some oils have a potential risk of toxicity. Single dermal doses are not uncommon, but the high stability of ketones and their resistance to liver metabolism can lead to cumulative chronic toxicity over long periods of time due to repeated daily doses (Shinde et al. 2023).

2.9 Carboxylic Acids

The carboxylic acid group (-COOH) represents most of the acids in volatile oils. These acids are very soluble in water and are found in trace amounts in oils, although they are not found anywhere else. Although this acid is frequently found in commercial cosmetics and preparations, it is not found in oils obtained by steam distillation (Bayrak 2023).

2.10 Esters

Plant acids and alcohol can easily combine to form esters through the separation of water. They are well tolerated on the skin but can trigger sensitivity when used for long periods. Some esters have proven therapeutic effects in aroma dermatology, including sedative and anti-inflammatory effects, which are particularly useful in itchy and painful skin conditions caused by sleep disturbances and restlessness (Khan et al. 2023).

2.11 Oxides

Oxides are formed by removing hydrogen atoms from the hydroxyl group (-OH) and bonding the oxygen atom to another carbon atom to form a closed ring. In topical use of essential oils, 1,8-cineole has been observed to disrupt the lipid bilayers of the skin. Neurological side effects have been reported in children after oral administration of eucalyptus oil. For this reason, eucalyptus oil containing 1,8-cineole should not be applied to the face and nose of children and babies (Bayrak 2023).

2.12 *Biological Activity of Essential Oils*

Essential oils show antimicrobial, antibacterial, antifungal, antivirus, anticancer, antioxidant, anti-inflammatory, antihyperpigmentation, antidiabetic, cardioprotective, neuroprotective, hepatoprotective, and tooth protective effects (Bertella et al. 2024; Khan et al. 2023; Miguel 2010; Radu et al. 2023; Shinde et al. 2023).

3 Essential Oil Applications in Livestock Systems

The One Health approach has gained prominence in livestock production, as in all agricultural systems. The demand for more sustainable livestock practices and consumer preference for natural products have driven increased research into non-synthetic, natural-based feed additives for nutrition and treatment. Recently, in addition to synthetic feed additives, many plant-based natural components have been used to enhance production performance, improve product quality, and safeguard animal health. Some products, such as prebiotics and herbal extracts, have even been commercialized as feed additives to enhance production and influence the sensory qualities or overall quality of animal products. Essential oils (EOs) are mixtures of low-weight molecules, including terpenes (monoterpenes and sesquiterpenes), alcohols, aldehydes, and ketones. Besides their bioactive properties, these compounds also contribute to the characteristic aromatic fragrance of essential oils (Jean 2008; Nehme et al. 2021). EOs are frequently used for their therapeutic, odoriferous, and flavor properties, in an extensive selection of products like cosmetics, foods, and medicines (Hanif et al. 2019). On the other hand, it has been reported that EOs stimulate digestive secretions, improve blood circulation, exhibit antioxidant properties, reduce pathogenic bacteria, and may enhance immune function (Brenes and Roura 2010). Given their beneficial properties, research has increasingly focused on using essential oils in animal nutrition to evaluate their effects on performance and product quality. Previous studies on the use of essential oils as feed additives have shown that they can positively affect growth parameters in broilers, including cecal microflora balance, nutrient digestibility, antioxidant activity, and the fatty acid profile of broiler meat (Chowdhury et al. 2018; Mohebodini et al. 2021; Ruan et al. 2021; Su et al. 2021). Although fewer studies have examined the use of essential oils in laying hen diets, it has been reported that EO supplementation enhances feed efficiency, improves egg quality, and significantly boosts egg performance traits (Laptev et al. 2021; Ramirez et al. 2021). Similarly, experimental studies on ruminants have shown that adding essential oils to diets can increase feed intake and weight gain (Dorantes-Iturbide et al. 2022). Comprehensive reviews suggest that EOs can enhance production and health performance in ruminants, though their effectiveness may vary depending on factors such as plant sources, extraction methods, dosage, and diet composition across different studies and conditions (Caroprese et al. 2023; Nehme et al. 2021; Ramdani et al. 2023). Previous research on the utilization of essential oils as natural feed supplements in livestock nutrition has demonstrated encouraging outcomes in terms of performance metrics. Moreover, essential oils have been extensively investigated for their antimicrobial properties, highlighting their potential as effective alternatives to conventional additives. In livestock production systems, safeguarding animal health and preventing diseases of microbiological origin are critical to ensuring production efficiency and maintaining global food security. Notably, about 50% of zoonotic diseases are transmitted to humans from vertebrate animals (Klous et al. 2016; Taylor et al. 2001), with a significant proportion of these infections originating from livestock and spreading through animal-derived food products. Therefore, natural feed additive alternatives with antimicrobial properties, such as essential oils (EOs), have garnered significant attention in recent years. These alternatives are increasingly recognized for their potential to enhance animal health, improve productivity, and reduce the reliance on conventional antimicrobial agents, thereby addressing concerns related to antibiotic resistance and promoting sustainable livestock practices. Numerous EOs have been extensively evaluated in vitro for their antimicrobial properties, and research has identified several bioactive compounds within these oils that exhibit significant efficacy (Chouhan et al. 2017; Lv et al. 2011; Romeo et al. 2008; Škrinjar and Nemet 2009). These findings highlight the potential of EOs as viable natural alternatives for combating microbial pathogens, paving the way for their application in various fields, including livestock nutrition and health management. For instance, oregano EO has demonstrated effectiveness against various microorganisms, including fungi like Candida albicans and both Gram-positive and Gram-negative bacteria (Adame-Gallegos et al. 2016; Andrade-Ochoa et al. 2021; Leyva-López et al. 2017). The exact mechanism of action of EOs remains unclear due to their complex composition. Their antimicrobial activity is likely not attributed to a single mode of action but rather involves multiple cellular targets. Antimicrobial effectiveness of the essential oils is reported to depend on factors such as pH, chemical structure, concentration, and the specific bioactive compound, as well as the population and type of microorganisms affected. The antimicrobial mechanisms involve various actions, including membrane disruption by terpenoids and phenolics, metal chelation by phenols and flavonoids, and effects on genetic material by coumarins and alkaloids, which are believed to inhibit microbial growth (Krishan and Narang 2014). Terpenes, terpenoids, and phenylpropanoids, whether individually or as components of essential oils, exhibit varying biological activity across several biological systems. The variety of antimicrobial evaluation methods presents an overwhelming array of facts that are challenging to correlate. The various experimental parameters (methods for assessing activity, growing medium, and incubation conditions) can affect the antibacterial efficacy of the essential oils and their components (Andrade-Ochoa et al. 2021). Although in vitro studies have demonstrated that essential oils can inhibit pathogens, research on their antimicrobial properties—particularly thymol, eugenol, and carvacrol—has primarily been conducted in vitro. These compounds have shown strong activity against pathogenic bacteria such as *Escherichia coli* and *Salmonella typhimurium*, both of which pose a risk for enteric infections (Zhai et al. 2018). Despite the promising results from laboratory studies, the in vivo application of EOs as antimicrobial agents in livestock requires further comprehensive investigation. For this purpose, in this chapter, previous studies regarding the application of different EOs in livestock production and their effects on microorganisms have been examined, and results have been summarized.

3.1 Antimicrobial Effects of Essential Oils in Poultry Production

Poultry meat and eggs continue to be a common source of animal protein for humans globally, and it is estimated that demand will increase as the world population rises. According to OECD/FAO (2024), world egg production reached 97 million tons in 2023, and poultry meat will account for 43% of the protein consumed from all meat sources in 2033. The intense production pressure in the poultry sector has driven new efforts to enhance efficiency, especially following the ban on antimicrobial growth promoters in poultry production in 2006 in many countries, including Europe, because of antimicrobial resistance. Antimicrobial resistance has emerged in zoonotic enteropathogens (e.g., *Salmonella* spp., *Campylobacter* spp.), commensal bacteria (e.g., *Escherichia coli*, enterococci), and pathogens affecting animals (for instance, *Pasteurella* spp. *and Actinobacillus* spp.), with varying levels of prevalence (McEwen and Fedorka-Cray 2002). Contaminated chicken meat remains a major global source of *Salmonella* and *Campylobacter* infections in humans and has occasionally contributed to worldwide salmonellosis outbreaks due to egg contamination with Salmonella Enteritidis. Moreover, the gastrointestinal tract is vital for poultry health, serving as the first line of defense against pathogens while facilitating nutrient absorption. Additionally, maintaining homeostasis and nutrient processing, the gastrointestinal tract hosts a diverse microbiota, including *Lactobacillus, Clostridium, Ruminococcus, Salmonella enterica serovars, Enterococcus* spp., and *Escherichia coli* (Mak et al. 2022; Pan and Yu 2014). Combating pathogens and promoting beneficial bacteria in poultry production is a key concern. Research has focused on natural feed additives that enhance gut health, inhibit pathogen colonization, and improve production performance. In recent years, with consumers' attempts to be non-synthetic and antibiotic-free, poultry products, aromatic plants, and their extracts have gained attention as growth and health promoters in poultry production. Although EOs have come forward as a feed additive in poultry production in vivo studies, the related antimicrobial activity of EOs in poultry species is scarce. This section reviews the results of previous studies on the use of essential oils (EOs) as feed additives in poultry and their effects on bacteria. Table 1 presents findings from studies that examine EOs as feed additives in broilers and their impact on relevant bacterial populations. The outcomes of in vivo research indicate that the effectiveness of EOs on bacteria varies depending on the usage level and herbal source. As shown in Table 1, *Nigella sativa* oil as a feed additive increases *Lactobacillus* while reducing *Clostridium perfringens* and *Escherichia coli*, with no

Table 1 Study results on the use of essential oils as a feed additive in broiler and their effects on related bacteria

Essential oil	Application dose (mg/kg feed)	Target microorganisms	Effects	Reference
Nigella sativa	0, 30, 40, 50	*Lactobacillus*, Total Coliform, *Enterobacteriaceae*, *C. perfringens*, *E. coli*	↑[a] *Lactobacillus*, no effect on Total Coliform & *Enterobacteriaceae*, ↓[a] *C. perfringens* & *E. coli*	Elbaz et al. (2025)
Oregano	0, 500	*Escherichia coli*, *Salmonella* spp., *Lactobacillus* spp.	↓ *E. coli* & *Salmonella*, ↑ *Lactobacillus* spp.	Khukhodziinai et al. (2024)
Carvacrol & Cinnamaldehyde	0, 100	*E. coli*, Total Microbial Count	↓ *E. coli* & Total Viable Count	Chaurasia (2024)
Clove	0, 300	*Lactobacillus*, *E. coli*, *Enterococcus*, Total Coliforms	↓ *E. coli*, ↑ *Lactobacillus*, No effect on Total Coliforms & *Enterococcus*	Elbaz et al. (2023)
Cinnamon	0, 500, 1000, 1500	Total Bacterial Count, Total Yeast & Mold Count, *E. coli*, *Salmonella*, Lactic Acid Bacteria	↓ Total Bacterial Count, Yeast & Mold Count, *E. coli*, *Salmonella*; ↑ Lactic Acid Bacteria	Saied et al. (2022)
Oregano, Sage, & Lavender Blend	0, 100, 200, 400	Total Aerobic Bacteria, Total Anaerobic Bacteria, *Enterobacteriaceae*, *Enterococcus* spp., *Lactobacillus* spp.	↓ Total Aerobic & Anaerobic Bacteria, ↑ *Lactobacillus* spp., No effect on *Enterobacteriaceae* & *Enterococcus* spp.	Bonos et al. (2022)
Coriander, Ajwain, & Dill Seed	0, 200, 400, 600	*Salmonella* spp., *E. coli*, *Lactobacillus* spp.	↓ *Salmonella* spp. & *E. coli*, No effect on *Lactobacillus* spp.	Ali et al. (2024)
Garlic, Lemon, & Garlic Mix	0, 200	Total Lactic Acid Bacteria, *Lactobacillus*, *E. coli*	↑ Cecal *Lactobacillus*, ↓ *E. coli*, No effect on Total Lactic Acid Bacteria Count	Elbaz et al. (2022)

[a] ↑ indicates increase, ↓indicates decrease

effect on total coliforms or *Enterobacteriaceae*. Oregano oil and a combination of carvacrol & cinnamaldehyde significantly reduce *E. coli* and *Salmonella*, while increasing *Lactobacillus* populations. Clove and cinnamon oils reduce *E. coli* and total bacterial counts, with cinnamon oil also decreasing yeast and mold counts. A blend of oregano, sage, and lavender reduces both aerobic and anaerobic bacteria but has no effect on *Enterobacteriaceae* or *Enterococcus* spp. Coriander, ajwain, and dill seed oils lower *Salmonella* and *E. coli* levels but do not affect *Lactobacillus*. Garlic, lemon, and their combination increase *Lactobacillus* while reducing *E. coli*, with no impact on total lactic acid bacteria. The results suggest that EOs as feed additives in broiler diets tend to enhance *Lactobacillus* strains, which are generally considered beneficial for gut health, while suppressing pathogenic bacteria. Overall, essential oils demonstrate potential as natural antimicrobial agents in broiler nutrition by selectively promoting beneficial bacteria and inhibiting pathogens. However, to establish EOs as effective antibacterial agents in broiler production, further in vivo studies are required to determine the most effective herbal EO and optimal dosage. On the other hand, studies on the use of EOs as antimicrobial agents in laying hens are more limited, mainly because zoonotic bacteria are less prevalent in laying hens. Table 2. Shows the study results on the use of essential oils as a feed additive in laying hens and their effects on related bacteria. The researchers reported that Carvacrol, thyme, and cinnamaldehyde increased beneficial bacteria like *Firmicutes, Intestinimonas*, and *Megamonas* while reducing *Spirochaeta.* Oregano and cinnamaldehyde increased *Actinobacteriota* and decreased *Desulfobacterota.* Oregano essential oil increased *Cyanobacteria, Burkholderiales*, and *Comamonadaceae*. Black cumin and cinnamon essential oils reduced *Escherichia coli* on eggshells but had no effect on *Enterococcus*. The results support that EOs may reduce bacterial populations when used as feed additives in laying hens, but their effects depend on the application level, herbal source, and bacterial strain. However, the number of in vivo studies in laying hens is insufficient to draw a definitive conclusion.

The egg is one of the carriers of pathogenic bacteria from the sheds to the table, and the bacterial load of the egg is important for food safety. Freshly laid eggs are

Table 2 Study results on the use of essential oils as a feed additive in laying hens and effects on related bacteria

Essential oil	Application dose	Evaluated parameter	Effects	Reference
Carvacrol, Thyme, and Cinnamaldehyde	0, 300 mg/kg	Microbial composition of cecal contents	↑ *Firmicutes*, I*ntestinimonas*, and *Megamonas*; ↓ *Spirochaetota*	Xiao et al. (2022)
Oregano & Cinnamaldehyde	0, 100 oreganos +20 cinnamaldehyde 200 oreganos +20 cinnamaldehyde 300 oreganos +20 cinnamaldehyde mg/kg	*Actinobacteriota, Desulfobacterota*	↑ *Actinobacteriota*; ↓ *Desulfobacterota*	Gao et al. (2022)
Oregano	0, 200 mg/kg	*Cyanobacteria, Burkholderiales, Comamonadaceae*	↑ *Cyanobacteria, Burkholderiales*, and *Comamonadaceae*	Feng et al. (2021)
Black Cumin & Cinnamon	0, 0.5 (ml/kg)	Eggshell *Escherichia coli* & *Enterococcus* contamination	↓ *Escherichia coli* (with Black Cumin EO); No effect on *Enterococcus*	Yalçın and Denli (2021)

[a] ↑ indicates increase, ↓indicates decrease

typically free of microorganisms. However, exposure to environmental factors such as soil, dust, and dirty nesting materials can lead to contamination with various microorganisms. Additionally, these microorganisms can enter the egg contents either by penetrating the shell through its pores or via the trans-ovarian route (Ghosh et al. 2021). *Salmonella* Enteritidis is considered the only pathogen currently posing a major risk of egg-borne diseases in the European Union (Hazards 2014). However, it is not only *Salmonella* that poses a concern. The eggshell, which can act as a potential breeding ground for bacteria, is especially susceptible to colonization by Gram-positive bacteria such as *Bacillus cereus* and *Staphylococcus aureus* (Zhu et al. 2024). On the other hand, cleaning hatching eggs is important. It has been noted that failing to sanitize hatching eggs before incubation may result in significant bacterial contamination and subsequent growth, which can decrease hatchability, poor chick quality, hinder growth and performance, and raise mortality rates (Cadirci 2009). To provide bacteria-free eggs for table and incubation, disinfection procedures may be used. A diverse array of disinfectant formulations is available on the market, and although there are significant variations in their effectiveness, aldehydes tend to be more efficient than other disinfectants used in poultry houses (Carrique-Mas et al. 2009; McDonnell and Russell 1999). Even though chemicals such as formaldehyde are very effective for egg sanitization, it has been mentioned that it may be harmful to human health as it is a toxic gas, a strong irritant, a potent sensitizer, and a suspected carcinogen (Cadirci 2009). With concerns about the chemicals that are used for disinfectants, natural source antimicrobials such as EOs have gained importance, especially for food safety. Previous studies regarding the use of EOs as egg sanitizers with different application methods have been reviewed, and the results are presented in Table 3. In those studies, clove, cinnamon, rosemary, peppermint, oregano, thymol, carvacrol, and cinnamaldehyde EOs were tested on hen, quail, and broiler eggs using spraying or immersion methods. Most EOs significantly reduced bacterial counts, including total aerobic bacteria, mesophilic bacteria, and *Enterobacteriaceae*. Clove, peppermint, and rosemary oils were effective in reducing aerobic bacteria, while cinnamon had no effect. Oregano and a combination of thymol, carvacrol, and cinnamaldehyde also showed antibacterial properties. Overall, EOs demonstrate potential as natural egg disinfectants, improving microbial safety and reducing the microbial load on eggs. However, further research is needed to determine the optimal application method, dosage, and type of EO for effective egg sanitization.

3.2 Antimicrobial Effects of Essential Oils in Ruminants

Ruminants, including cattle, sheep, and goats, are key livestock species in modern systems, providing essential sources of meat and milk for the human diet. The rumen microbiome enables ruminants to convert low-quality feed into high-quality protein and has been studied for over six decades. This unique habitat hosts diverse microorganisms, including at least 50 bacterial genera, 25 ciliate protozoan genera,

Table 3 Effects of essential oils on egg bacteria

Essential oil	Application dose	Application method	Egg type	Effects	References
Clove, Cinnamon, Rosemary, Peppermint	0.39 mg/mL for clove, 1.56 mg/mL for cinnamon, 1.56 mg/mL for rosemary, and 12.5 mg/mL for peppermint EO.	Spraying	Hen table eggs	↓ Aerobic bacteria with clove, peppermint, and rosemary EOs. No effect on aerobic bacteria with cinnamon EO.	El-Soufi et al. (2025)
Clove	0.5, 1, or 2%	Spraying	Broiler hatching eggs	↓Total bacterial count	Hassan et al. (2023)
Tahiti Lemon + Green Banana Flour	1% v/v	Immersing	Quail table eggs	↓ Total Aerobic Mesophilic Bacteria	Oliveira et al. (2023)
Clove	0.39%	Spraying	Hen hatching eggs	↓ Total aerobic mesophilic bacteria and Enterobacteriaceae	Oliveira et al. (2020)
Thymol, Carvacrol, Cinnamaldehyde	1% (v/v)	Spraying	Quail hatching eggs	↓Total bacteria	Ulucay and Yildirim (2010)
Oregano	0 and 0.2 ml/L	Spraying	Quail hatching eggs	↓Total bacteria and coliforms.	Yildirim et al. (2003)

[a] ↑indicates increase, ↓indicates decrease

six fungal genera, methanogenic archaea, and bacteriophages (Cobellis et al. 2016). Ruminant digestion relies on anaerobic microbial fermentation in the rumen, and dietary carbohydrates are converted into volatile fatty acids (acetate, propionate, butyrate), CO_2, and CH_4, providing a key energy source. The major fibrolytic rumen bacteria include *Ruminococcus flavefaciens, Fibrobacter succinogenes, Ruminococcus albus, Prevotella ruminicola, Butyrivibrio fibrisolvens, Eubacterium cellulosolvens,* and *Eubacterium ruminantium*. However, *F. succinogenes, R. albus, and R. flavefaciens* have been considered representative cellulolytic bacteria species in the rumen (Koike and Kobayashi 2009). Maintaining the natural rumen microflora is essential for optimal production performance and rumen health. However, the diversity of the rumen microbiome can be significantly influenced not only by diet composition but also potentially by host genetics and environmental factors (Chaucheyras-Durand and Ossa 2014). As alternatives to non-nutritive chemical and antibiotic feed additives, a wide range of phytochemicals (e.g., plant secondary metabolites, or PSM) derived from various herbs and spices have been utilized to modulate ruminal fermentation, suppress methanogenesis, and improve livestock performance (Kholif and Olafadehan 2021). EOs' antimicrobial properties have sparked interest in their potential to enhance ruminal fermentation and feed efficiency. While some tested EOs show promise, many remain unexplored. Most studies have been in vitro, highlighting the need for in vivo research to confirm their effects, mechanisms, and optimal concentrations for modifying ruminal fermentation (Benchaar et al. 2008). On the other hand, in vitro studies that consider the antimicrobial effects of EO's in rumen microorganisms are very limited. In this part, results on the effects of EO's on rumen bacteria of studies that used EO's feed supplement in ruminants are reviewed and summarized in Table 4. Results showed that Thyme and cinnamon essential oils, administered at 5 g/day/calf, decreased the populations of *Ruminococcus albus* and *R. flavefaciens* in cattle, while having no effect on *Fibrobacter succinogenes*. Oregano essential oil at 52 mg/day increased the abundance of *Bifidobacterium, Ruminococcus*, and *Enterococcus* in sheep, and higher doses in cattle enhanced populations of *Parabacteroides distasonis*, *Bacteroides thetaiotaomicron*, and *Tannerella forsythia*. The findings indicate that in ruminants, the dosage of administration of EO's as feed additives and the determined bacteria strains are effective factors in the antimicrobial action of EOs. However, the overall findings of these studies indicate that essential oils (EOs) in ruminant diets may enhance fibrolytic bacteria, thereby benefiting rumen health. However, more in vivo studies are needed to draw definitive conclusions.

Table 4 Study results on the use of essential oils as a feed additive in ruminants on rumen bacteria

Essential oil	Application dose	Characteristics	Species	Effects	References
Thyme & Cinnamon	0 & 5 g/d/calf	*Ruminococcus albus, R. flavefaciens, & Fibrobacter succinogenes*	Cattle	↓[a] *R. albus & R. flavefaciens*; No effect on *F. succinogenes*	Khorrami et al. (2024)
Oregano	52 mg/head/day	*Bifidobacterium, Ruminococcus, & Enterococcus*	Sheep	↑*Bifidobacterium, Ruminococcus, & Enterococcus*	Jia et al. (2022)
Oregano	0, 130, 260 mg/kg feed	*Parabacteroides distasonis, Bacteroides thetaiotaomicron, & Tannerella forsythia*	Cattle	↑ *P. distasonis, B. thetaiotaomicron, & T. forsythia*	Zhang et al. (2021)
Thymol, Eugenol, Vanillin & Limonene	0, 1000 mg/day/cattle	*C. sticklandii, P. anaerobius, & Clostridium aminophilum*	Cattle	✕ *C. sticklandii & P. anaerobius*; No effect on *C. aminophilum*	McIntosh et al. (2003)

[a]: ↑ indicates increase, ↓ indicates decrease, ✕ indicates inhibition

4 Conclusions

Essential oils represent a valuable resource in various industries due to their diverse biological activities and health benefits. Their antimicrobial properties make them particularly promising as natural alternatives to synthetic additives in both human health and animal nutrition. Advances in extraction techniques have enhanced the efficiency and quality of essential oils, allowing for broader applications. As research continues to uncover the potential of these natural compounds, essential oils may play a crucial role in promoting sustainable practices and improving product quality across multiple sectors. Emphasizing their use can lead to healthier alternatives in food production and contribute to overall well-being, highlighting the importance of integrating essential oils into modern health and agricultural practices.

References

Bayrak, A (2023) Uçucu Yağlar Bilimi, Kimyası, Teknolojisi ve Tedavide Kullanımı. Sidas Yayınevi, İzmir, 304 pp

Adame-Gallegos JR, Andrade-Ochoa S, Nevarez-Moorillon GV (2016) Potential use of Mexican oregano essential oil against parasite, fungal and bacterial pathogens. J Essent Oil Bear Plants 19:553–567

Aguele FO, Oke EO, Abam FI, Nnabodo D, Agbana AS (2023) Biological and antioxidant activities, extraction methodology and prospects of essential oil from Hyptis suaveolens (L.): a review. Cleaner Eng Technol 17:100685

Ahmadian-Kouchaksaraie Z, Niazmand R (2017) Supercritical carbon dioxide extraction of antioxidants from Crocus sativus petals of saffron industry residues: optimization using response surface methodology. J Supercrit Fluids 121:19–31

Akdag T, Keskınkaya HB, Argon-Ustun Z, Dogu S, Kocak M, Güneş E (2025) Characterization and bioactivity of carrot (Daucus carota L. subsp. sativus (Hoffm.) Arcang by supercritical liquid CO 2 extraction. Modern Phytomorphology, 19(1)

Albuquerque PM, Azevedo SG, de Andrade CP, D'Ambros NCS, Pérez MTM, Manzato L (2022) Biotechnological applications of nanoencapsulated essential oils: a review. Polymers 14:5495

Ali U, Qaisrani SN, Mahmud A, Hayat Z, Toyomizu M (2024) Effects of supplemented coriander, Ajwain, and dill seed essential oils on growth performance, carcass characteristics, gut health, meat quality, and immune status in broilers. J Poult Sci 61:2024006

Anacarso I, Sabia C, de Niederhäusern S, Iseppi R, Condò C, Bondi M, Messi P (2019) In vitro evaluation of the amoebicidal activity of rosemary (Rosmarinus officinalis L.) and cloves (Syzygium aromaticum L. Merr. & Perry) essential oils against Acanthamoeba polyphaga trophozoites. Nat Prod Res 33:606–611

Andrade-Ochoa S, Chacón-Vargas KF, Sánchez-Torres LE, Rivera-Chavira BE, Nogueda- Torres B, Nevárez-Moorillón GV (2021) Differential antimicrobial effect of essential oils and their main components: insights based on the cell membrane and external structure. Membranes 11:405

Keskinkaya HB, Ustun-Argon Z, Doğu S, Akdag T (2023) Phenolic content and in-vitro antioxidant activity of Olea europaea L. subs. oleaster leaves by supercritical CO_2 extraction. Ereğli Tarım Bilim Derg 3:75–85

Aziz ZA, Ahmad A, Setapar SHM, Karakucuk A, Azim MM, Lokhat D, Rafatullah M, Ganash M, Kamal MA, Ashraf GM (2018) Essential oils: extraction techniques, pharmaceutical and therapeutic potential-a review. Curr Drug Metab 19:1100–1110

Belahcene S, Kebsa W, Omoboyowa DA, Alshihri AA, Alelyani M, Bakkour Y, Leghouchi E (2023) Unveiling the chemical profiling antioxidant and anti-inflammatory activities of algerian Myrtus communis L. essential oils, and exploring molecular docking to predict the inhibitory compounds against cyclooxygenase-2. Pharmaceuticals 16:1343

Benchaar C, Calsamiglia S, Chaves AV, Fraser G, Colombatto D, McAllister TA, Beauchemin KA (2008) A review of plant-derived essential oils in ruminant nutrition and production. Anim Feed Sci Technol 145:209–228

Bendahou M, Muselli A, Grignon-Dubois M, Benyoucef M, Desjobert J-M, Bernardini A-F, Costa J (2008) Antimicrobial activity and chemical composition of Origanum glandulosum Desf. essential oil and extract obtained by microwave extraction: comparison with hydrodistillation. Food Chem 106:132–139

Berenji Ardestani S, Sahari MA, Barzegar M (2016) Effect of extraction and processing conditions on anthocyanins of barberry. J Food Process Preserv 40:1407–1420

Bertella A, Gavril G-L, Wrona M, Pezo D, Sidaoui A, Benlahcen K, Kihal M, Olewnik-Kruszkowska E, Salafranca J, Nerín C (2024) Analysis of bioactive aroma compounds in essential oils from Algerian plants: implications for potential antioxidant applications. Foods 13:749

Bonos E, Skoufos I, Giannenas I, Fotou K, Sidiropoulou E, Tsiftsoglou O, Lazari D, Galamatis D, Stylianaki I, Tzora A (2022) Augmenting broiler diets with essential oils affects growth, intestinal microbiota and morphology, and meat quality. South Afr J Anim Sci 52:506–520

Brenes A, Roura E (2010) Essential oils in poultry nutrition: main effects and modes of action. Anim Feed Sci Technol 158:1–14

Cadirci S (2009) Disinfection of hatching eggs by formaldehyde fumigation–a review. Arch Geflügelk 73:116–123

Caroprese M, Ciliberti MG, Marino R, Santillo A, Sevi A, Albenzio M (2023) Essential oil supplementation in small ruminants: a review on their possible role in rumen fermentation, microbiota, and animal production. Dairy 4:497–508

Carrique-Mas JJ, Marin C, Breslin M, McLaren I, Davies R (2009) A comparison of the efficacy of cleaning and disinfection methods in eliminating Salmonella spp. from commercial egg laying houses. Avian Pathol 38:419–424

Chakravarty I, Parmar VM, Mandavgane SA (2023) Current trends in essential oil (EO) production. Biomass Convers Biorefinery 13:15311–15334

Chaucheyras-Durand F, Ossa F (2014) Review: the rumen microbiome: composition, abundance, diversity, and new investigative tools. Prof Anim Sci 30:1–12

Chaurasia J (2024) Effect of dietary supplementation of Phytosome conjugated Carvacrol and Cinnamaldehyde essential oil on performance and gut health of broiler chicken. Maharashtra Animal and Fishery Sciences University, Doktoral

Chouhan S, Sharma K, Guleria S (2017) Antimicrobial activity of some essential oils—present status and future perspectives. Medicine 4:58

Chowdhury S, Mandal GP, Patra AK (2018) Different essential oils in diets of chickens: 1. Growth performance, nutrient utilisation, nitrogen excretion, carcass traits and chemical composition of meat. Anim Feed Sci Technol 236:86–97

Cobellis G, Trabalza-Marinucci M, Yu Z (2016) Critical evaluation of essential oils as rumen modifiers in ruminant nutrition: a review. Sci Total Environ 545:556–568

Dashtian K, Kamalabadi M, Ghoorchian A, Ganjali MR, Rahimi-Nasrabadi M (2024) Integrated supercritical fluid extraction of essential oils. J Chromatogr A 1733:465240

Díaz-Reinoso B, Rivas S, Rivas J, Domínguez H (2023) Subcritical water extraction of essential oils and plant oils. Sustain Chem Pharm 36:101332

Dorantes-Iturbide G, Orzuna-Orzuna JF, Lara-Bueno A, Mendoza-Martínez GD, Miranda-Romero LA, Lee-Rangel HA (2022) Essential oils as a dietary additive for small ruminants: a meta-analysis on performance, rumen parameters, serum metabolites, and product quality. Vet Sci 9:475

Drioiche A, Ailli A, Handaq N, Remok F, Elouardi M, Elouadni H, Al Kamaly O, Saleh A, Bouhrim M, Elazzouzi H (2023) Identification of compounds of Crocus sativus by GC-MS and HPLC/

UV-ESI-MS and evaluation of their antioxidant, antimicrobial, anticoagulant, and antidiabetic properties. Pharmaceuticals 16:545

Elbaz AM, Ashmawy ES, Salama AA, Abdel-Moneim A-ME, Badri FB, Thabet HA (2022) Effects of garlic and lemon essential oils on performance, digestibility, plasma metabolite, and intestinal health in broilers under environmental heat stress. BMC Vet Res 18:430

Elbaz AM, Ashmawy ES, Ali SA, Mourad DM, El-Samahy HS, Badri FB, Thabet HA (2023) Effectiveness of probiotics and clove essential oils in improving growth performance, immuno-antioxidant status, ileum morphometric, and microbial community structure for heat-stressed broilers. Sci Rep 13:18846

Elbaz AM, Ashmawy ES, Farahat MAA, Abdel-Maksoud A, Amin SA, Mohamed ZS (2025) Dietary Nigella sativa nanoparticles enhance broiler growth performance, antioxidant capacity, immunity, gene expression modulation, and cecal microbiota during high ambient temperatures. Sci Rep 15:861. https://doi.org/10.1038/s41598-024-82725-9

El-Soufi A, Al Khatib A, Khazaal S, El Darra N, Raafat K (2025) Evaluation of essential oils as natural antibacterial agents for eggshell sanitization and quality preservation. Processes 13:224

EPC (2004) European pharmacopoeia 5. Council of Europe

Eslahi H, Fahimi N, Sardarian AR (2017) Chemical composition of essential oils. In: Hashemi SMB, Khaneghah AM, Sant'Ana AS (eds) Essential oils in food processing: chemistry, safety and applications. Wiley-Blacwell, pp 119–171

Feng J, Lu M, Wang J, Zhang H, Qiu K, Qi G, Wu S (2021) Dietary oregano essential oil supplementation improves intestinal functions and alters gut microbiota in late-phase laying hens. J Anim Sci Biotechnol 12:1–15

Ferhat MA, Meklati BY, Chemat F (2007) Comparison of different isolation methods of essential oil from citrus fruits: cold pressing, hydrodistillation and microwave 'dry' distillation. Flavour Fragr J 22:494–504

Ferraz LP, Silva EK (2024) Pulsed electric field-assisted extraction techniques for obtaining vegetable oils and essential oils: Recent progress and opportunities for the food industry. Sep Purif Technol:128833

Filly A, Fabiano-Tixier AS, Louis C, Fernandez X, Chemat F (2016) Water as a green solvent combined with different techniques for extraction of essential oil from lavender flowers. C R Chim 19:707–717

Flamini G, Tebano M, Cioni PL, Ceccarini L, Ricci AS, Longo I (2007) Comparison between the conventional method of extraction of essential oil of Laurus nobilis L. and a novel method which uses microwaves applied in situ, without resorting to an oven. J Chromatogr A 1143:36–40

Florczak MC (2015) Extraction and characterisation of the essential oil from Centella Asiatica (Pennywort) growing in South Africa

Gaikwad RK, Mondal IH, Dash KK, Béla K (2024) Effectiveness of sustainable oil extraction techniques: a comprehensive review. J Agric Food Res:101546

Gao F, Zhang L, Li H, Xia F, Bai H, Piao X, Sun Z, Cui H, Shi L (2022) Dietary oregano essential oil supplementation influences production performance and gut microbiota in late-phase laying hens fed wheat-based diets. Animals 12:3007

Ghosh N, Bhattacharyya S, Pati RR, Raj A, Banik A (2021) Study of colonization of pathogenic bacteria on raw eggshells. The Indian J Nutr Diet:92–98

Gogoi R, Sarma N, Begum T, Chanda SK, Lekhak H, Sastry GN, Lal M (2023) Agarwood (Aquilaria malaccensis L.) a quality fragrant and medicinally significant plant based essential oil with pharmacological potentials and genotoxicity. Ind Crop Prod 197:116535

Hanif MA, Nisar S, Khan GS, Mushtaq Z, Zubair M (2019) Essential oils. In: Essential oil research: trends in biosynthesis, analytics, industrial applications and biotechnological production. Springer, Cham, pp 3–17

Hassan AS, Morsy EA, El Moustafa K-EM, Ibrahim FA, Elmenawey MA (2023) Effects of clove essential oil on eggshell bacterial load, antibacterial sensitivity, and hatchability. Egypt Pharm J 22:650–658

Hazards EPoB (2014) Scientific opinion on the public health risks of table eggs due to deterioration and development of pathogens. EFSA J 12:3782

Hudz N, Kobylinska L, Pokajewicz K, Horčinová Sedláčková V, Fedin R, Voloshyn M, Myskiv I, Brindza J, Wieczorek PP, Lipok J (2023) Mentha piperita: essential oil and extracts, their biological activities, and perspectives on the development of new medicinal and cosmetic products. Molecules 28:7444

Huo Y, Deng W, Sun X, Zhou L, Zhang Q, Hu J (2024) Extract toolkit for essential oils: state of the art, trends, and challenges. Food Chem 461:140854

Jadhav H, Raina I, Gogate P, Annapure U, Casanova F (2023) Sonication as a promising technology for the extraction of triacylglycerols from fruit seeds—a review. Food Bioprocess Technol 16(2023):1625–1651

Jean B (2008) Pharmacognosy, phytochemistry, medicinal plants (2e ed.-retirage broch"). Lavoisier

Jia L, Wu J, Lei Y, Kong F, Zhang R, Sun J, Wang L, Li Z, Shi J, Wang Y (2022) Oregano essential oils mediated intestinal microbiota and metabolites and improved growth performance and intestinal barrier function in sheep. Front Immunol 13:908015

Jones WP, Kinghorn AD (2005) Extraction of plant secondary metabolites. In: Natural products isolation. Humana Press, New York, pp 323–351

Kalaiselvi D, Mohankumar A, Shanmugam G, Thiruppathi G, Nivitha S, Sundararaj P (2019) Altitude-related changes in the phytochemical profile of essential oils extracted from Artemisia nilagirica and their nematicidal activity against Meloidogyne incognita. Ind Crop Prod 139:111472

Kamaruddin MSH, Chong GH, Daud NM, Putra NR, Salleh LM, Suleiman N (2023) Bioactivities and green advanced extraction technologies of ginger oleoresin extracts: a review. Food Res Int 164:112283

Kant R, Kumar A (2022) Review on essential oil extraction from aromatic and medicinal plants: techniques, performance and economic analysis. Sustain Chem Pharm 30:100829

Kashyap D, Vohra P, Chopra S, Tewari R (2001) Applications of pectinases in the commercial sector: a review. Bioresour Technol 77:215–227

Katekar VP, Rao AB, Sardeshpande VR (2023) A hydrodistillation-based essential oils extraction: a quest for the most effective and cleaner technology. Sustain Chem Pharm 36:101270

Ke Q, Ma K, Zhang Y, Meng Q, Huang X, Kou X (2023) Antibacterial aroma compounds as property modifiers for electrospun biopolymer nanofibers of proteins and polysaccharides: a review. Int J Biol Macromol 253:126563

Khan S, Sahar A, Tariq T, Sameen A, Tariq F (2023) Essential oils in plants: plant physiology, the chemical composition of the oil, and natural variation of the oils (chemotaxonomy and environmental effects, etc.). In: Essential oils. Elsevier, pp 1–36

Kholif AE, Olafadehan OA (2021) Essential oils and phytogenic feed additives in ruminant diet: chemistry, ruminal microbiota and fermentation, feed utilization and productive performance. Phytochem Rev 20:1087–1108

Khorrami B, Vakili SA, Danesh Mesgaran M (2024) The effect of thyme and cinnamon essential oils on performance, rumen fermentation profile and rumen microbial population under in vitro and in vivo conditions. J Anim Product, Print ISSN: 2008-6776., Online ISSN: 2382-994X 26: 263–277. https://doi.org/10.22059/jap.2024.375614.623792

Khukhodziinai JS, Das PK, Mukherjee J, Banerjee D, Ghosh PR, Das AK, Samanta I, Jas R, Mondal S, Patra AK (2024) Effect of dietary benzoic acid and oregano essential oil as a substitute for an anti-coccidial agent on growth performance and physiological and immunological responses in broiler chickens challenged with Eimeria species. Animals 14:3008

Klous G, Huss A, Heederik DJ, Coutinho RA (2016) Human–livestock contacts and their relationship to transmission of zoonotic pathogens, a systematic review of literature. One Health 2:65–76

Koike S, Kobayashi Y (2009) Fibrolytic rumen bacteria: their ecology and functions. Asian-Australas J Anim Sci 22:131–138

Krishan G, Narang A (2014) Use of essential oils in poultry nutrition: a new approach. J Adv Vet Anim Res 1:156–162

Lainez-Cerón E, Jiménez-Munguía MT, López-Malo A, Ramírez-Corona N (2021) Effect of process variables on heating profiles and extraction mechanisms during hydrodistillation of eucalyptus essential oil. Heliyon 7:e08234

Laptev GY, Yildirim EA, Ilina LA, Filippova VA, Kochish II, Gorfunkel EP, Dubrovin AV, Brazhnik EA, Narushin VG, Novikova NI, Novikova OB, Dunyashev TP, Smolensky VI, Surai PF, Griffin DK, Romanov MN (2021) Effects of essential oils-based supplement and salmonella infection on gene expression, blood parameters, Cecal microbiome, and egg production in laying hens. Animals 11:360

Lego N, Shadap A, Momin KC (2021) Essential oils and the methods of extraction. In: Sharangi AB (ed) Aromatic plants the technology, human welfare and beyond. Nova Science Publishers, New York, p 376

Leyva-López N, Gutiérrez-Grijalva EP, Vazquez-Olivo G, Heredia JB (2017) Essential oils of oregano: biological activity beyond their antimicrobial properties. Molecules 22:989

Li H, Qiao S, Zhang S (2025) Essential oils in grain storage: a comprehensive review of insecticidal and antimicrobial constituents, mechanisms, and applications for grain security. J Stored Prod Res 111:102537

Liu H, Ni Y, Hu J, Jin Y, Gu P, Qiu H, Chen K (2022) Self-healing and antibacterial essential oil-loaded mesoporous silica/polyacrylate hybrid hydrogel for high-performance wearable body-strain sensing. ACS Appl Mater Interfaces 14:21509–21520

Lucchesi ME, Chemat F, Smadja J (2004) Solvent-free microwave extraction of essential oil from aromatic herbs: comparison with conventional hydro-distillation. J Chromatogr A 1043:323–327

Lv F, Liang H, Yuan Q, Li C (2011) In vitro antimicrobial effects and mechanism of action of selected plant essential oil combinations against four food-related microorganisms. Food Res Int 44:3057–3064

Mak PH, Rehman MA, Kiarie EG, Topp E, Diarra MS (2022) Production systems and important antimicrobial resistant-pathogenic bacteria in poultry: a review. J Anim Sci Biotechnol 13:148

McDonnell G, Russell AD (1999) Antiseptics and disinfectants: activity, action, and resistance. Clin Microbiol Rev 12:147–179

McEwen SA, Fedorka-Cray PJ (2002) Antimicrobial use and resistance in animals. Clin Infect Dis 34:S93–S106

McIntosh F, Williams P, Losa R, Wallace R, Beever D, Newbold C (2003) Effects of essential oils on ruminal microorganisms and their protein metabolism. Appl Environ Microbiol 69:5011–5014

Miguel MG (2010) Antioxidant and anti-inflammatory activities of essential oils: a short review. Molecules 15:9252–9287

Mohebodini H, Jazi V, Ashayerizadeh A, Toghyani M, Tellez-Isaias G (2021) Productive parameters, cecal microflora, nutrient digestibility, antioxidant status, and thigh muscle fatty acid profile in broiler chickens fed with Eucalyptus globulus essential oil. Poult Sci 100:100922

Mondor M, Hernández-Álvarez AJ (2023) Emerging solvent extraction technologies for plant protein extraction: aqueous two-phase extraction; deep eutectic solvent; subcritical water extraction. In: Green protein processing technologies from plants: novel extraction and purification methods for product development. Springer, pp 111–130

Naliyadhara N, Kumar A, Girisa S, Daimary UD, Hegde M, Kunnumakkara AB (2022) Pulsed electric field (PEF): Avant-garde extraction escalation technology in food industry. Trends Food Sci Technol 122:238–255

Nehme R, Andrés S, Pereira RB, Ben Jemaa M, Bouhallab S, Ceciliani F, López S, Rahali FZ, Ksouri R, Pereira DM, Abdennebi-Najar L (2021) Essential oils in livestock: from health to food quality. Antioxidants 10:330

Ngcangatha-Maqhude B (2017) The influence of extraction techniques on the chemical composition of essential oils from Pelargonium cv. rose in the Eastern Cape

Oliveira GS, Nascimento ST, dos Santos VM, Silva MG (2020) Clove essential oil in the sanitation of fertile eggs. Poult Sci 99:5509–5516. https://doi.org/10.1016/j.psj.2020.07.014

Oliveira GS, McManus C, Salgado CB, Pires PGS, de Figueiredo Sousa HA, da Silva ER, dos Santos VM (2023) Antimicrobial coating based on Tahiti lemon essential oil and green Banana flour to preserve the internal quality of quail eggs. Animals 13:2123

Orchard A, van Vuuren S (2017) Commercial essential oils as potential antimicrobials to treat skin diseases. Evid Based Complement Alternat Med 2017:4517971

Özel M, Göğüş F, Lewis A (2006) Comparison of direct thermal desorption with water distillation and superheated water extraction for the analysis of volatile components of Rosa damascena Mill. using GCxGC-TOF/MS. Anal Chim Acta 566:172–177

Pan D, Yu Z (2014) Intestinal microbiome of poultry and its interaction with host and diet. Gut Microbes 5:108–119

Radu C-M, Radu CC, Bochiş S-A, Arbănaşi EM, Lucan AI, Murvai VR, Zaha DC (2023) Revisiting the therapeutic effects of essential oils on the oral microbiome. Pharmacy 11:33

Radzi NC, Kasim FA (2020) Effect of microwave pretreatment on gaharu essential oil using hydrodistillation method. Indones J Chem 20:960–966

Ramdani D, Yuniarti E, Jayanegara A, Chaudhry AS (2023) Roles of essential oils, polyphenols, and Saponins of medicinal plants as natural additives and Anthelmintics in ruminant diets: a systematic review. Animals 13:767

Ramirez SY, Peñuela-Sierra LM, Ospina MA (2021) Effects of oregano (Lippia origanoides) essential oil supplementation on the performance, egg quality, and intestinal morphometry of Isa Brown laying hens. Vet World 14:595

Reddy BHV, Hussain SMS, Hussain MS, Kumar RN, Gupta J (2025) Essential oils in cosmetics: antioxidant properties and advancements through nanoformulations. Pharm Res Natl Prod:100192

Romeo FV, De Luca S, Piscopo A, Poiana M (2008) Antimicrobial effect of some essential oils. J Essent Oil Res 20:373–379

Rout S, Tambe S, Deshmukh RK, Mali S, Cruz J, Srivastav PP, Amin PD, Gaikwad KK, de Aguiar Andrade EH, de Oliveira MS (2022) Recent trends in the application of essential oils: the next generation of food preservation and food packaging. Trends Food Sci Technol 129:421–439

Ruan D, Fan Q, Fouad AM, Sun Y, Huang S, Wu A, Lin C, Kuang Z, Zhang C, Jiang S (2021) Effects of dietary oregano essential oil supplementation on growth performance, intestinal antioxidative capacity, immunity, and intestinal microbiota in yellow- feathered chickens. J Anim Sci 99. https://doi.org/10.1093/jas/skab033

Saied A, Attia A, El-Kholy M, Reda F, Nagar AE (2022) Effect of cinnamon oil supplementation into broiler chicken diets on growth, carcass traits, haemato- biochemical parameters, immune function, antioxidant status and caecal microbial count. J Anim Feed Sci 31:21–33

Shin J, Seo S-M, Park I-K, Hyun J (2021) Larvicidal composite alginate hydrogel combined with a Pickering emulsion of essential oil. Carbohyd Polym 254:117381

Shinde SS, Sarkate AP, Nirmal NP, Sakhale BK (2023) Bioactivity, medicinal applications, and chemical compositions of essential oils: detailed perspectives, recent frontiers of phytochemicals. Elsevier, pp. 353–367

Škrinjar MM, Nemet NT (2009) Antimicrobial effects of spices and herbs essential oils. Acta periodica technologica:195–209

Sodeifian G, Sajadian SA (2017) Investigation of essential oil extraction and antioxidant activity of Echinophora platyloba DC. using supercritical carbon dioxide. J Supercrit Fluids 121:52–62

Su G, Wang L, Zhou X, Wu X, Chen D, Yu B, Huang Z, Luo Y, Mao X, Zheng P (2021) Effects of essential oil on growth performance, digestibility, immunity, and intestinal health in broilers. Poult Sci 100:101242

Taglienti A, Donati L, Dragone I, Ferretti L, Gentili A, Araniti F, Sapienza F, Astolfi R, Fiorentino S, Vecchiarelli V (2023) In vivo antiphytoviral and aphid repellency activity of essential oils and hydrosols from Mentha suaveolens and Foeniculum vulgare to control zucchini yellow mosaic virus and its vector Aphis gossypii. Plants 12:1078

Tandon S, Rane S (2008) Decoction and hot continuous extraction techniques. Extraction technologies for medicinal and aromatic plants 93:12–18

Taylor LH, Latham SM, Woolhouse ME (2001) Risk factors for human disease emergence. Philos Trans R Soc London Ser B Biol Sci 356:983–989

Tischer B, Vendruscolo RG, Wagner R, Menezes CR, Barin CS, Giacomelli SR, Budel JM, Barin JS (2017) Effect of grinding method on the analysis of essential oil from Baccharis articulata (Lam.) Pers. Chem Pap 71:753–761

Tomaniova M, Hajšlová J, Pavelka J Jr, Kocourek V, Holadova K, Klımova I (1998) Microwave-assisted solvent extraction—a new method for isolation of polynuclear aromatic hydrocarbons from plants. J Chromatogr A 827:21–29

Ulucay I, Yildirim I (2010) Hatching traits of quail (Coturnix coturnix japonica) eggs disinfected with carvacrol, cinnamaldehyde or thymol. J Appl Anim Res 38:139–142

UNIDO F (2005) Herbs, spices and essential oils: post-harvest operations in developing countries. Agro-Industries and Sectoral Support Branch, UNIDO, Vienna

Vian MA, Fernandez X, Visinoni F, Chemat F (2008) Microwave hydrodiffusion and gravity, a new technique for extraction of essential oils. J Chromatogr A 1190:14–17

Wang J, Li Y, Lu Q, Hu Q, Liu P, Yang Y, Li G, Xie H, Tang H (2021) Drying temperature affects essential oil yield and composition of black cardamom (Amomum tsao-ko). Ind Crop Prod 168:113580

Wang Y, Li L, Hu J (2023) Development of biobased multifunctional films incorporated with essential oils@ polydopamine nanocapsules for food preservation applications. Int J Biol Macromol 253:127161

Xiao G, Zheng L, Yan X, Gong L, Yang Y, Qi Q, Zhang X, Zhang H (2022) Effects of dietary essential oils supplementation on egg quality, biochemical parameters, and gut microbiota of late-laying hens. Animals 12:2561

Xu R, Deng W, Dai Y, Hu J (2022) pH-responsive citral microcapsules with tannic acid-FeIII coordination complexes. Food Chem 397:133715

Yalçın S, Denli M (2021) Effect of dietary supplementation of black cumin (Nigella sativa L.) and cinnamon (Cinnamomum zeylanicum L.) essential oils on performance and egg quality of laying hens. Middle East J Sci 7:56–63

Yang S, Li S, Li G, Li C, Li W, Bi Y, Wei J (2024) Pulsed electric field treatment improves the oil yield, quality, and antioxidant activity of virgin olive oil. Food Chem X 22:101372

Yildirim I, Ozsan M, Yetisir R (2003) The use of oregano (Origanum vulgare L) essential oil as alternative hatching egg disinfectant versus formaldehyde fumigation in quails (Coturnix coturnix japonica). Rev Med Vet-Toulouse 154:367–370

Yu F, Wan N, Zheng Q, Li Y, Yang M, Wu Z (2021) Effects of ultrasound and microwave pretreatments on hydrodistillation extraction of essential oils from Kumquat peel. Food Sci Nutr 9:2372–2380

Zhai H, Liu H, Wang S, Wu J, Kluenter A-M (2018) Potential of essential oils for poultry and pigs. Anim Nutr 4:179–186

Zhang R, Wu J, Lei Y, Bai Y, Jia L, Li Z, Liu T, Xu Y, Sun J, Wang Y (2021) Oregano essential oils promote rumen digestive ability by modulating epithelial development and microbiota composition in beef cattle. Front Nutr 8:722557

Zhang T, Sun H, Hu S, Ding S, Zhang P, Wang L, Fan W, Liu F, Mu W, Pang X (2023) Self- assembly of eugenol-loaded particles to regulate the adhesion of carriers on leaves for efficient foliar applications and ecotoxicological safety. Ecotoxicol Environ Saf 267:115602

Zhou L, Huo Y, Cheng Y, Guang W, Sun X, Deng W, Hu J (2023) One-pot controllable synthesis of silica nanocapsules with encapsulated osmanthus fragrance for durable odor improvement of leather. ACS Sustain Chem Eng 11:11799–11810

Zhu J, Liu Q, Wang Y, Zhu K, Guo J, Jin Y, Liu Y (2024) Mangosteen extract reduces the bacterial load of eggshell and improves egg quality. Heliyon 10:e35857

Zill-e-Huma Z-e-H, Abert-Vian M, Elmaataoui M, Chemat F (2011) A novel idea in food extraction field: study of vacuum microwave hydrodiffusion technique for by-products extraction. J Food Eng 105(2):351–360

Zuzarte M, Salgueiro L (2015) Essential oils chemistry. In: Bioactive essential oils and cancer. Springer, Cham, pp 19–61

Tannins and Saponins: Natural Defense Compounds in Animal Feed

Hira Akbar, Huma Riffat, M. Shahid Mahmood, Bilal Hussain, Shumiala Kiran, and Azhar Rafique

Abstract Rapid advancements in microbial resistance have made it the world's leading cause of mortality. Finding substitutes to combat microbes has become essential. Valuable secondary metabolites, including terpenoids, polyphenols, alkaloids, quinones, and tannins, are abundant in plants. Secondary metabolites found in plants are a source for the discovery of antibiotics. Essential oils, tannins, and saponins are naturally occurring plant compounds that are evaluated as natural substitutes for in-feed antibiotics. Plants contain a number of polyphenolic compounds called tannins that have biological properties such as antimicrobial, anti-parasitic, antiviral, antioxidant, anti-inflammatory, and immunomodulatory effects. Numerous plant and marine animal species produce saponins, which are secondary metabolites. Their microbial, anti-tumor, anti-insect, hepatoprotective, hemolytic, and anti--inflammatory properties have medical applications, used as an adjuvant in vaccines and to lower blood cholesterol levels. The sum of methane that can be decreased by enteric fermentation relies on the levels and sources of saponin and tannin. In order to control ruminal fermentation, saponnins have function in the rumen as membrane permeabilizing, immunostimulating, hypocholesterolaemic, and defaunting agents. Saponins obstruct the digestion of proteins in the gut, interact with cholesterol in the cell membrane, cleave cells, and remove ruminal protozoa exclusively. This upgrades the efficiency of N-use and likely improves ruminant animal performance.

Keywords Bioactive compounds · Saponins · Tannins · Rumen · Properties

H. Akbar · H. Riffat · B. Hussain · A. Rafique (✉)
Department of Zoology, Government College University, Faisalabad, Pakistan

M. S. Mahmood
Institute of Microbiology, University of Agriculture Faisalabad, Faisalabad, Pakistan

S. Kiran
Department of Applied Chemistry, Government College University, Faisalabad, Pakistan

R. U. Khan, S. Naz (eds.), *Green Alternatives to Antibiotics in Animal Health*,
https://doi.org/10.1007/978-981-92-0309-3_7

1 Introduction

Plants have been recognized for their therapeutic benefits for thousands of years. Over millions of years, they have evolved to defend against insects, bacteria, fungi, and environmental factors, producing a wide variety of secondary metabolites with diverse structures. These compounds ethno pharmacological properties have long served as a key inspiration for developing new medications (McRae et al. 2007). Throughout history, plant-derived biological compounds have been the primary source of physiologically active secondary metabolites. Plants have played a vital role in human culture, serving as therapeutic agents, flavors, and scents. Until the commercialization of aspirin in 1899, minerals and plants were essential for maintaining human health and driving economic and social progress, providing the only therapeutic properties and ingredients for preventing and treating illnesses. Saponins and tannins are key components of secondary metabolites. These complex organic substances play vital biological roles. They contribute to various plant functions and interactions (Hapid et al. 2021).

2 Tannins

Seguin coined the name "tannin" in 1796 to describe compounds found in plant extracts that may interact with animal hide protein to stop putrefaction and turn the hides into leather. Plant polyphenols have an astringent, bitter taste that binds and causes proteins to precipitate. Although the word "tan" describes the source of the tannins used to turn animal hides into leather, it can also refer to any large polyphenolic compound that has enough hydroxyls and other appropriate groups (like carboxyls) to form potent complexes with proteins and other macromolecules (Bharadwaj et al. 2021). Conversely, tannins are phenolic chemicals with strong antiviral and antioxidant capabilities that have a large molecular weight (between 500 and 3000 $gmol^{-1}$) present in many plant components, such as the bark, stem, leaf, or root (Romero et al. 2022). Depending on their nature, concentration, animal species, health, and feed composition, tannins can have both negative and positive effects. Tannin-based forages have anti-bloat and anthelmintic properties that improve the quality of meat and milk products, fatty acid composition, and ruminant antioxidant status. Any phenolic compound of sufficiently high molecular weight (Fig. 1) containing sufficient hydroxyls and other suitable groups (i.e., carboxyls) to form effectively strong complexes with protein and other macromolecules under the particular environmental conditions being studied (Blackman et al. 2016).

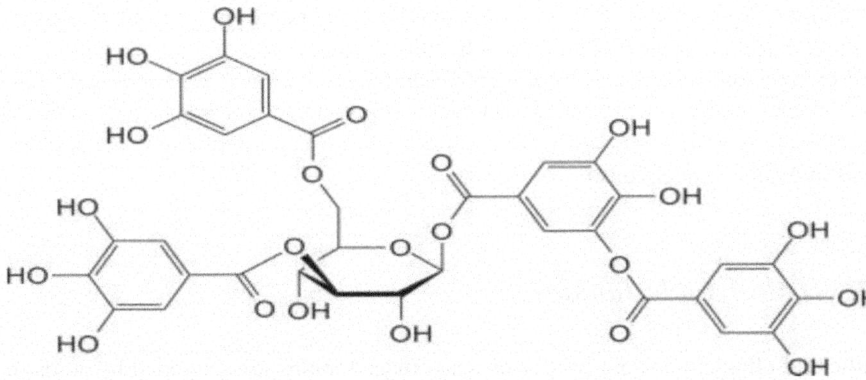

Fig. 1 Tannin Formula

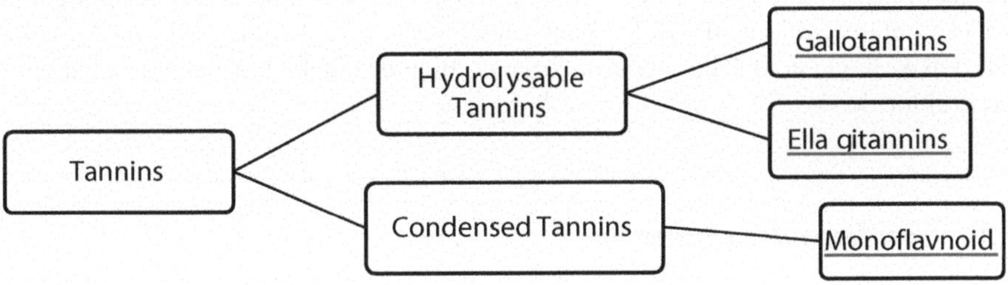

Fig. 2 Types of tannins

3 Types of Tannins

A subgroup of polyphenols, tannins are found in nature in large quantities and are made up of many components with varying molecular weights. Tannins have the ability to precipitate proteins. Tannins are separated based on their molecular structure into hydrolyzable tannin (HTs) (Fig. 2) and condensed tannin (CT). Hydrolyzable tannins include ellagitannins, which can hydrolyze in vivo to release ellagic acid and create urolithin through the metabolism of gut microbes. Generally speaking, the antibacterial activity of tannins has been identified in vitro, but it is not well demonstrated how these tannins affect the colonic ecosystem's abundance in vivo. Tannin-based forages contain anthelmintic and anti-bloat properties that enhance the antioxidant status of ruminants, the composition of fatty acids, and the quality of meat and milk products (Blackman et al. 2016). Furthermore, these plants that contain tannins also help to prevent the release of enteric methane (Widyarini et al. 2021). Depending on their nature, concentration, animal species, health, and feed composition, tannins can have both negative and positive effects. Certain plants that contain tannins make ruminants and monogastric animals poisonous, which lowers their feed intake, protein availability, nutrient digestibility, and production capabilities (Tong et al. 2022). Investigating the many tactics to limit the detrimental impact of dietary tannins on animal health is necessary.

Proteins, amino acids, and polysaccharides can all form compounds with tannins. When CT attaches to protein, it prevents ruminal bacteria from breaking down tannin-protein complexes, which 60% improve the absorption of amino acids in the small intestine (Lawal 2022)

3.1 Hydrolyzable Tannins

Gallic and ellagic acids are produced when these tannins are degraded by acids or enzymes. These are chemically known as gallic acid and ellagic acid, which are esters of phenolic acid. Gallitannins are tannins formed from gallic acid, whereas gallitannins are tannins derived from ellagic acid (Hassen et al. 2024). Rhubarb and cloves contain gallic acid, while eucalyptus leaves, myrobalans, and pomegranate bark have ellagic acid. These tannins became blue or black after being treated with ferric chloride.

3.2 Condensed Tannins

These tannins are resistant to hydrolysis, and they are derived from the flavonols, catechins, and flavan-3, 4-diols. On treatment with acids or enzymes, they are decomposed into phlobaphenes. In dry distillation, condensed tannin produces catechol. These tannins are called catechol tannins. These tannins are found in (Table 1) cinchona bark, male fern, areca seeds, tea leaves, wild cherry bark, bahera fruits, Amla, etc. they produce green colour with ferric chlorides (Hassen et al. 2024).

4 Saponins

Most plants naturally contain saponins, which are non-volatile glycosidic compounds (Vincken et al. 2007). Triterpenoid saponins are secondary metabolites that are found in various parts of plants, including seeds, leaves, flowers, and roots (Table 2). To date, approximately 60 different types of triterpenoid saponins have been identified and extracted (Singh and Chaudhuri 2018). The sapogenin and sugar components (aglycone and glycone) of its large molecule are joined by a glycosidic bond (Fig. 3). Several oligosaccharide chains, such as glucose, galactose, pentose, etc., may make up the carbohydrate portion. Sapogenins, on the other hand, can be triterpenoid or steroid.

They are big molecules with a hydrophobic portion made up of a steroid backbone (27 carbon atoms with a 6-ring spirostane or 5-ring furostane skeleton) or

Table 1 Different tannin sources and concentrations

No	Types	Name of plant	Tannin (%)	References
1.	Leafy vegetables	*Moringa oleifera Dehulled* Whole *Digera arvensis* *Carduus leaves Amaranthus tricolor*	0.08 3.91 2.71 0.079 0.36 0.305	Kumari and Jain (2012)
2.	Fruits	*Acacia Arabica* Ripened banana *Kaki* *Tipo Prsopis cineraria* *Vitis vinifera*	7.27 0.025–0.48 17–32 0.14 0.11	Bueno et al. (2020)
3.	Cereals and millets	*Oryza sativa* Pearl millet Brown finger millet *Triticum astivum* Yellow sorghum grain *Eleusine coracana*	0 0.152 0.12–3.47 0.041 0.2–2.0 0.36	Bueno et al. (2020)
4.	Seeds/ Nutseeds	*Cuminum cyminum* Dry fenugreek Walnut Pistachio Hazelnut Faba beans Pecan	0.23 0.01 0.18–0.34 0.02–0.22 0.04–0.23 0.7–3.8 0.84–0.88	Sriwichai et al. (2016)
5.	Legumes	Pigeon pea Chick pea Bengal gram *Vigna catjang* *Dolichos lablab*	4.3–11.4 1.9–6.1 0.33 0.175 1.024	Mamatha and Prakash (2016)

triterpenoid (30 carbon atoms) and a hydrophobic portion made up of multiple saccharide residues connected to the hydrophobic scaffold by glycose bonds (Dong et al. 2023). Glucosides having foaming properties are called saponins. Polycyclic aglycones joined to one or more sugar side chains make up saponins. The aglycone portion, commonly known as sapogenin, is either a triterpene (C30) or a steroid (C27). The combination of a hydrophilic (water-soluble) sugar component and a hydrophobic (fat-soluble) sapogenin gives saponins their foaming capacity. They are of various types Fig. 4.

The flavor of saponins is bitter. Sapotoxin is the term for certain saponins that are harmful. They display a range of biological functions as well. Triterpenoid and steroidal saponins obtained from plants have been employed in fire extinguishers, culinary additives, and the pharmaceutical industry to produce steroid hormones, among other industrial uses. Additional intriguing biological uses include their employment in immune-stimulating, hypocholesteromelic, and anti-inflammatory medications (Liu et al. 2021).

Table 2 Some common saponin-rich plant sources

No.	Common name	Scientific name	Parts used	References
1.	Shikakai	*Acacia concinna*	Pods and bark	Zhuoyue et al. (2021)
2.	Soap root	*Chlorogalum pomeridia*	Root	Chen et al. (2021)
3.	Soap wort	*Saponaria officinalis*	Roots and Leaves	Chen et al. (2021)
4.	Kanor, barkhor	*Aesulus indica*	Fruits	Dong et al. (2023)
5.	Maguey	*Agave offoyana*	Flowers and Leaves	Singh et al. (2022)
6.	Sisal	*Agave sislana*	Leaves	Bustos et al. (2024)
7.	Tea	*Camellia oleifera*	Seeds	Mohammed (2024)
8.	Chickpeas	*Cicer arietnum*	Seeds	Nguyen et al. (2020)
9.	Yams	*Discorea composite*	Rhizomes and Roots	Nagtode et al. (2023)
10.	Bitter melon	*Momordica charantia*	Fruits and Stem	Moghimipour and Handali (2015)
11.	Asian rice	*Oryza sativa*	Peels	Wisetkomolmat et al. (2019)
12.	Kidney beans	*Phaseolus vulgaris*	Seeds	Nguyen et al. (2020)

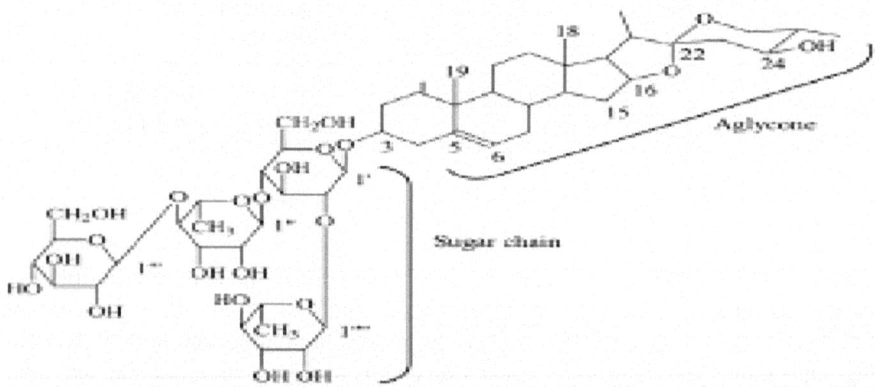

Fig. 3 Sapogenin formula

4.1 *Mechanism of Action of Saponins*

The primary purpose of the plants' secondary metabolites was to protect them from numerous herbivores and infections. To put it another way, plants use secondary metabolites as a defense mechanism. Being a type of secondary metabolite, saponins serve the primary purpose of shielding plants against a variety of diseases and herbivores. The antibacterial, antifungal, antiviral, antihelminthic, insecticidal, larvicidal, and molluscicidal properties of saponins have all been extensively studied (Fig. 5). Saponins produce membrane disruption by creating pores on the

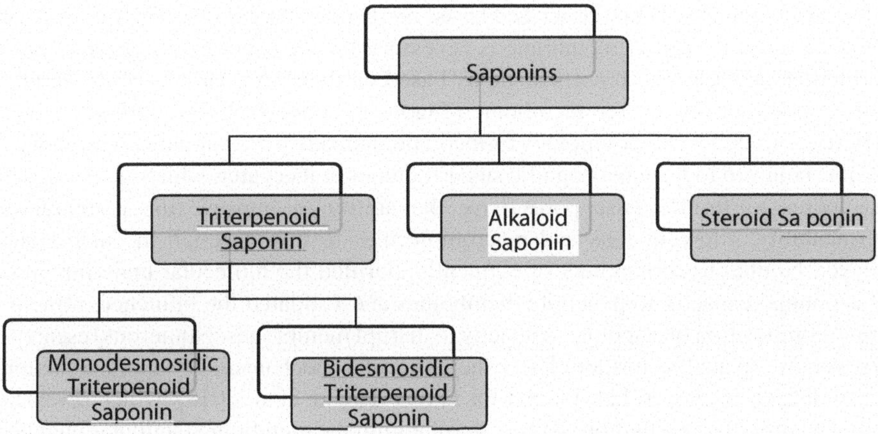

Fig. 4 Types of saponins

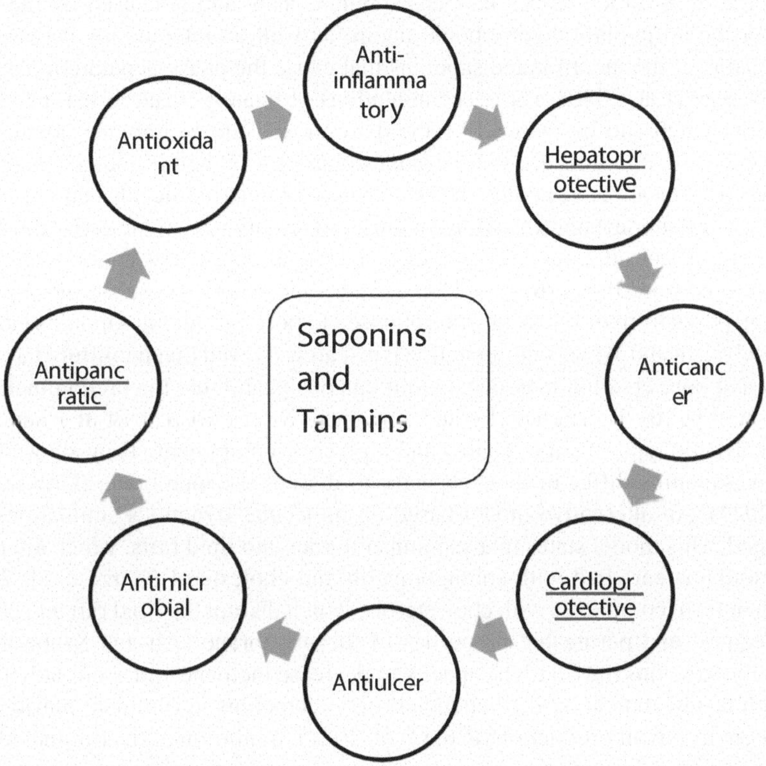

Fig. 5 Saponins and tannins properties

membrane, which was first described based on their investigations. The presence of cholesterol on the target membrane is necessary for the saponins to promote pore formation, according to reports made by based on their observations of the formation of pores or pits on the membrane (Tagousop et al. 2021). According to their reports, a micelle-like compound formed spontaneously by saponins and cholesterol is believed to have hydrophilic sugar moieties at its center, which causes aqueous pores to form. These pores have the ability to improve the membrane's permeability, allowing ions and macromolecules to flow through the membrane bilayer. Numerous researchers subsequently clarified the molecular underpinnings of saponins' capacity to penetrate membranes and validated the influence of membrane composition on saponins' capacity to disrupt membranes. Numerous researchers were motivated to further clarify the molecular mechanism of saponins' action by studies on cholesterol as a target for toxins (Huang et al. 2018), which revealed that the particular orientation of cholesterol within the membrane facilitates channel formation by polyene antibiotics, bacterial protein toxins, and various sterols.

According to the molecular basis of saponins' activity reported that saponins' incorporation into membranes happens spontaneously and is caused by the aglycone portion's lipophilic/hydrophobic nature as well as interactions between the sugar chains of the incorporated saponins that cause the phase-separation phenomenon (Nawab et al. 2020). These accumulations ultimately result in the membrane curvature, which can be induced by the disruption of the membrane domain, the development of pores inside these plaques/matrices, or hemi-tubular changes as protuberances promote vesiculation. The passage of macromolecules up to proteins through the membrane and variations in ion conductivity are explained by the pores. When certain steroidal saponins targeted the membranes, it was also noted that there were no pore-like structures (Nepal and Stine 2023). These steroid saponins even stop pores from developing when exposed to pore-producing saponins later on. Ultimately, Lin and Wang's computational research showed that hemi-tubular vesiculation and pore creation may occur simultaneously, and that the predominant perturbation type is determined by the chemical characteristics of the saponins, specifically the type of sugar chains and a glycone component. They showed how different saponins differ in their capacity to disrupt the membrane (Grzywaczyk et al. 2024). An alternative model based on molecular dynamics simulations was developed. This model states that saponins migrate into lipid rafts, which are membrane domains enriched with sphingomyelin and cholesterol (Garza et al. 2022). They then form complexes with cholesterols, which disrupt the lipid raft and change the structural and permeable properties of the membrane (Ali and Szabo 2023). Generally, saponins have a high capacity to decrease methane emissions at both low and high levels, indicating the beneficial role of saponins in ruminant nutrition for cleaner environment production (Ridla et al. 2021). Inhibition of the ruminal methanogen population by saponins was suggested as a mode of action to lower methane emissions (Singh and Kaur 2020)

The saponins are used as a productive added ingredient to create and maintain oil-in-water and nano-emulsions (Liao et al. 2021). In addition, they have a number of biological characteristics, including anti-inflammatory, anti-tumor, hypocholesterolemic, and hypoglycemic effects. Because saponins have antimicrobial, antifungal, antiparasitic, and molluscicidal properties, they are also linked to defense mechanisms in plants (Shen et al. 2017; Kitagawa et al. 2016). Saponins are useful for treating and preventing cardiovascular and cerebrovascular diseases because of their dioscoresides, while dioscin, gracillin, and pseudoprotodioscin are used to treat rheumatism and are also effective against other chronic conditions (Moghimipour and Handali 2015; Hannan et al. 2024). Saponins were also being used as a starting harbinger for the semi-synthesis of steroidal drugs in the pharmaceutical industry.

5 Antibacterial Activity

One of the major causes of human illness in both developed and developing nations is thought to be bacterial infection. As time passes, pathogens become more resistant to previously identified antibiotics. Concerns regarding the effectiveness of the current antibiotics in treating different infectious diseases have grown among scientists. A lot of research has focused on developing new alternative medications to treat resistant issues. Potential sources for drug discovery include nature. For many years, scientists have been drawn to the phytochemicals found in plants because of their abundance, lack of negative effects, and increased human acceptability. The majority of these phytochemicals are antimicrobial. Disease-causing bacteria can be controlled by using antibacterial agents, especially the drug-resistant ones. Natural plant products are less toxic to both humans and animals, many plants are used as a source of dynamic antibacterial agents (Yao et al. 2014). Accordingly, the application of natural antimicrobial substances is crucial for the prevention of infectious diseases in both humans and plants as well as for food preservation.

5.1 Tannin and Their Antibacterial Activity

Secondary metabolites called tannins are a component of plants' chemical defenses against insect and pathogen invasion. The toxicity of tannins to bacteria, fungi, and yeasts has been reviewed, and their antimicrobial properties have long been known (Singh and Kaur 2020). The inhibition of extracellular microbial enzymes, the deprivation of substrates required for microbial growth, direct action on microbial metabolism through the inhibition of oxidative phosphorylation, and metal ions are some of the mechanisms that have been proposed to explain tannin's antimicrobial activity. Bacterial cell wall morphological changes and increased membrane permeability are caused by deprivation or complex formation with the bacterial cell membrane. Research has demonstrated that tannins primarily inhibit microbial cell membrane function by causing cell aggregation and disrupting cell membranes and functions. In general, it has been reported that tannins have more antimicrobial activity against Gram-positive bacteria than Gram-negative bacteria. The reason behind this is that Gram-negative bacteria have an outer membrane made up of a lipid bilayer structure, which is made up of an inner layer of phospholipids and an outer layer of proteins and lipopolysaccharides. However, it has been demonstrated that tannins, particularly CT, which have been isolated from a number of plants, have potent anti-gram-negative bacterial properties.

Tannins are naturally occurring polyphenolic compounds with antibacterial properties. Additionally, it has been noted that tannins alter the number of microorganisms in chickens' colon and small intestine (Jamroz et al. 2009). Based on the aforementioned findings, tannins may be a viable substitute for antibiotics in animal production. In the treatment groups, tannins decreased the growth of *Clostridium perfringens* and gut damage (Tosi et al. 2013). But the inhibition of intracellular bacteria is also important because tannins work by blocking oxidative phosphorylation and extracellular microbial enzymes, which directly impacts microbial metabolism by increasing membrane permeability and depriving microorganisms of growth-promoting substrates (Liu et al. 2013). According to reports, hydrogen peroxide release and the quantity of hydroxyl groups in tannins are significant markers for assessing the antibacterial qualities, which are positively connected with antioxidant qualities.

5.2 *Effect of Tannins on the Growth of* C. perfringens

Tannins differ in their nutritional value and toxic effects, and they can be divided into two categories: condensed and hydrolyzable (Haslam 1996; Reed 1995). Hydrolyzable tannins are hydrolyzed by acids and microbes to produce simpler phenolics, whereas condensed tannins are not easily broken down in the gut. The use of tannins to improve the health of animals in general and ruminants like sheep, cattle, and deer in particular has a lot of promise (Schneider and Blaut 2000). During feeding, low to moderate tannin concentrations may improve the digestive utilization because they reduce the breakdown of rumen protein and increase the flow of amino acids to the small intestine (Tabacco et al. 2006). Improved animal performance is a direct result of these nutrition-related effects. Ruminant diets supplemented with tannin compounds may help prevent intestinal disorders in addition to serving as a protein substitute. Both quebracho (*Schinopsis lorentzii,* condensed tannins) and chestnut (*Castanea sativa*, hydrolyzable tannins) are among the most widely available tannins in the market.

5.3 *Tannins as an Alternative to Antibiotics*

Numerous plant-based foods and a variety of other plants contain tannins. Plants that contain tannins in higher amounts include eucalyptus, maple trees, acacia, oak, chestnut, quebracho, and mimosa. Tea, coffee, and wine are among the beverages that contain tannins. Food tannins are primarily found in fruits such as grapes, pomegranates, blackberries, cranberries, and persimmons. The comparatively high tannin content of unripe fruits shields them from frugivorous animals before they ripen. Additionally, tannins play a big part in plant defense mechanisms and shield plants from herbivores and insects. The control of plant growth is another function of tannins. Numerous reports discuss the use of tannin-rich extracts in traditional medicinal preparations to treat bacterial infections and other illnesses.

Both Gram-positive and Gram-negative bacteria can be inhibited by tannins, and the majority of tannins have demonstrated bacteriostatic rather than bactericidal activity. Because the bacterial cell envelopes of Gram-positive and Gram-negative bacteria differ structurally, the MIC values of tannins for the former are significantly lower. Tanning agents' antibacterial activity is influenced by their structural characteristics. Antibacterial tannins can be added to livestock to improve the prevention of bacterial infections and promote the growth of bacteria. As a result, tannins may be used in place of or in addition to antibiotics. Because they come from natural sources, are widely available, inexpensive, have very little toxicity, and are easy to extract from plants, tannins are appealing biomolecules. According to recent research, tannins effectively inhibited the growth of various bacteria.

6 Saponins and Their Antibacterial Activity

In order to find new treatments for infectious diseases, numerous researchers have examined the mutual effect that arises from the combination of antibiotics and saponins (Tagousop et al. 2018; Tarhane et al. 2024). According to reports, citrus peel saponins have a variety of pharmacological and therapeutic uses. Flavonoids, glycosides, sitosterol, and volatile oils found in citrus fruit peels can also be effectively utilized as medications or dietary supplements because of their synergistic effects (Haroen et al. 2018).

6.1 *Effects of Saponins Against Clinical* Escherichia coli *Strains and Eukaryotic Cell Line*

The number of sugar units and/or the aglycone's structure determine the biological characteristics of saponins. Because of their surface-active characteristics, saponins may enter the lipid bilayer, attach to cholesterol, create domains that are rich in cholesterol-saponin complexes, and then lyse cells. However, the lack of cholesterol in the membrane structure does not prevent saponins from forming pores (Segal et al. 1974).

6.2 *Saponins and Their Antibacterial Activity Against* **Staphylococcus aureus** *and* **E. coli**

According to reports, 70–80% of Staphylococcus aureus strains globally are resistant to methicillin, and 90–95% are resistant to penicillin (Ayaz et al. 2019). Plant--derived biologically active antimicrobials have attracted more scientific attention as a result of the increase in antibiotic resistance in bacteria, fungi, parasites, and viruses. *Melanthera elliptica* saponin extract has a bactericidal effect on *Shigella flexneri* and *E. coli* (Tagousop et al. 2018).

6.3 *Mechanism of Action*

Saponins have the ability to combat bacteria through a number of different methods. Above all, it has been demonstrated that they interact with the cholesterol in cell membranes to form a pore that ultimately causes the membrane to rupture. Both as antibiotics and as agents to increase bacteria's susceptibility to currently available conventional antibiotics, saponins have been studied. Nevertheless, it might be difficult to convert the synergy between saponins and other traditional antibiotic medications into a finding that is clinically beneficial. Apart from their possible use as antibiotics to treat human infections, saponins have also drawn interest for their potential use as substitute food preservatives. Because the current practice of using artificial chemical additives in food processing is linked to a number of health issues, this area has gained importance.

7 Impact on Gut Health and Performance

High mortality rates are caused by bacterial and fungal infections, which also pose a hazard to cattle, poultry, and animal husbandry. Several antibiotics have been used for decades to address this issue and have shown great efficacy; as a result, the world's livestock and poultry output has improved. However, it is commonly recognized that overuse of antibiotics encourages the development of antibiotic-resistant bacteria in cattle (Das et al. 2024). As a result, plant-based antibacterial agents like tannin and in-feed antibiotics have been developed and promoted, and they have enormous potential for the future. The antimicrobial properties of tannins and their impact as an animal food source have received a lot of interest lately (Serrano et al. 2009; Anju et al. 2024). The usage of tannins with saponins and essential oils as in-feed antibiotics against bacteria, fungi, and yeasts has been determined. Notably, tannins have species-specific antibacterial activity despite being poisonous to bacteria and perhaps inhibiting the growth of Helicobacter pylori, *Salmonella, Shigella, Staphylococcus,* and *Pseudomonas* (Baky et al. 2022). Additionally, feeding cattle fodder that contains tannins helps reduce intestinal parasites, disease-causing bacteria in animal waste, and pasture bloating. Through a variety of processes, including depriving the bacterial cell of nutrients, deactivating essential extracellular enzymes. Additionally, feeding cattle fodder that contains tannins helps reduce intestinal parasites, disease-causing bacteria in animal waste, and pasture bloating. Through a variety of processes, including depriving the bacterial cell of nutrients, deactivating essential extracellular enzymes, inhibiting oxidative phosphorylation, chelating metal ions, and forming complexes with membranes and proteins, tannins can impede the growth of microorganisms (Liu et al. 2021).

Condensed tannins have been shown to be mostly found in trees, shrubs, browse shrubs, and forage legumes; however, the concentration of these tannins varies from species to species and is also impacted by environmental factors (Da silva et al. 2024). Animals have been given tannins from mimosa (HT), chestnut (HT), and quebracho (CT) as in-feed antibiotics (Fraga et al. 2021). However, the main obstacle to using tannins as antibiotics is the dearth of thorough and systematic research on the different aspects, including dosages and side effects on digestion. Prolonged use of tannins can lead to resistance to in-feed antibiotics, just like with regular antibiotics (Liu et al. 2021). Additionally, monogastric mammals, including fowl, are negatively impacted by tannins. Because tannins attach to essential macromolecules in biological systems, they can also function as anti-nutrients in the animal rumen. Numerous negative consequences include nutrient availability, metal ion chelation, protein binding, and hindrance. The gut microbiota of cattle, including fungi, is only weakly impacted by tannins. At the same time, it demonstrates that a high amount of tannin in the diet worked as an antibacterial agent against dangerous protozoa and methanogens without influencing ruminal fermentation (Kumar et al. 2014). How fermentation and rumen microbes are impacted by saponins. By altering the composition and activity of the rumen microbiota (bacteria, fungi, protozoa, and archaea), saponins directly alter rumen fermentation (Gunun et al. 2019). The structure and dosage of saponins, animal adaptability, feed mix, and microbial community structure are the primary factors that determine how they affect rumen microbiota populations.

The enhanced permeability of protozoan cell membranes, which results in the release of cell contents, may be the cause of saponins' antiprotozoal action (Kholif 2023). While cholestanol is the primary sterol in Holotrichs, stigmastanol, campestanol, and cholestanol are the main sterols in Entodiniomorphs. The variance in ruminal protozoa susceptibility to saponins may be explained by the different sterol compositions of ruminal protozoa. On the membrane surface of protozoa, saponins combine with sterols to produce complexes that degrade and ultimately disintegrate the organisms (Kholif 2023). Although they make up a minor fraction of the total ruminal microbial mass, anaerobic ruminal fungus plays a significant role in the digestion of fiber.

Numerous enzymes secreted by rumen fungus are more potent than those produced by bacteria. The impact of saponins on ruminal fungi is not well understood. Defaunation and/or direct impacts may be the means by which ruminal fungi react to saponins. As mentioned before, some studies revealed that adding saponins increased the number of ruminal fungi, which was explained by a decrease in protozoa's engulfment of them.

The metabolism of saponins in sheep's digestive systems was described. As per their assertion, saponins underwent a swift hydrolysis process to yield free sapogenins, some of which underwent oxidation and reduction to yield epismilagenin, smilagenone, smilagenin, and tigogenin. These were subsequently absorbed in the duodenum and subsequently liberated as conjugated and free sapogenins. Furthermore, the colon and caecum continued to epimerize sapogenins (Yanuartono et al. 2024). The main lignocellulosic feedstuff degraders in the rumen include fungi, bacteria, and protozoa. Protozoa only contribute 20% of the fermentative action, whereas bacteria and fungi account for around 80%. The effects of saponins or plants that contain them on feed digestion are not entirely consistent. Perhaps as a result of their impact on particular bacterial populations, saponins have a substrate-dependent effect on nutrient digestibility (Yanza et al. 2024). However, in numerous tests, saponins or plants that contain them had no influence on the digestion of nutrients (Singh et al. 2024).

8 Limitations of the Study

The antimicrobial efficacy of saponins is closely linked to their chemical structure. Saponins with trisaccharide chains have demonstrated significant antifungal activity, whereas those with mono- or disaccharide chains are less effective.

Saponins possess amphiphilic properties, allowing them to interact with bacterial membranes, alter their permeability, and disrupt microbial cell integrity.

Tannins can bind to microbial cell walls, proteins, and enzymes, leading to the disruption of cell membranes, inhibition of oxidative phosphorylation, interference with DNA replication, and overall inhibition of microbial metabolism.

Both saponins and tannins often have limited solubility in water, which can affect their bioavailability and, consequently, their antimicrobial effectiveness.

While saponins and tannins exhibit promising antimicrobial properties, their limitations—including structural dependence, potential toxicity, solubility challenges, and sensory impacts—should be carefully evaluated when considering their use in therapeutic and food preservation applications.

It is difficult to evaluate tannins' and saponins' long-term impacts on gut health and general performance because most research on them concentrates on their immediate effects. It can be difficult to generalize results since different concentrations can yield various outcomes. It can be challenging to identify the effects of tannins and saponins apart because of their interactions with other food substances and gut microorganisms. The accuracy of findings may be impacted by these interactions, which can be tough to account for in research. Even though tannins and saponins have health benefits, too much of them can have harmful effects, like inhibiting the absorption of nutrients.

The precise biological mechanisms by which tannins and saponins affect gut health and function remain unknown, despite signs to the contrary. To completely understand their roles, more mechanistic research is required.

9 Recommendations

1. Specific plants with high saponin content and efficacy should be chosen for the research for better results like *Trifolium* species, which have demonstrated significant antibacterial activity against common pathogens. And same for the tannins, utilize plant sources with substantial tannin concentrations, such as the galls of *Quercus infectoria*, which have been traditionally used for their antibacterial properties.
2. Future research should carry out long-term clinical trials in people in order to better assess the long-term effects of tannins and saponins on gut health and function. Over time, this will shed additional light on these chemicals' safety and sustainability.
3. To make the results more consistent across trials, future research should work to standardize tannin and saponin extraction techniques, doses, and administration forms.
4. Understanding how tannins and saponins affect the gut microbiota is essential to understanding how they affect inflammatory response, digestion, and general health.
5. Techniques for microbiome profiling may be utilized to detect specific shifts in microbial communities.
6. Cooperation between nutritionists, biologists, pharmacologists, and microbiologists could speed learning of how tannins and saponins contribute to gut health and human performance.
7. Future research can offer more solid, dependable, and useful insights into the use of tannins and saponins as natural substances to enhance gut health and function by addressing these shortcomings and putting these proposals into practice.

10 Conclusion

Due to antibacterial resistance towards antibiotics, there is a need for an alternative that can be used against bacterial infection. Saponins and tannins obtained from the plants contain phytochemicals that act as antibacterial agents. These are found in various parts of plants, including seeds, leaves, flowers, and roots. Citrus peel saponins have a variety of pharmacological and therapeutic uses. These saponins are a good source of antibacterial phytochemicals, particularly against *S. aures* and *E. coli.*

Plant tannins, a naturally occurring polyphenolic compound, have been shown to serve a number of biological purposes. Plant tannins have a huge potential for use in animal production as feed additives. Tannins also control plant growth. Both Gram-positive and Gram-negative bacteria can be inhibited by tannins, and the majority of tannins have demonstrated bacteriostatic rather than bactericidal activity. Furthermore, they are eco-friendly and renewable materials that remove dangerous emissions. When used as feed or as feed supplements for ruminants, saponins and plants that contain them—whether entire plants or extracts—show some positive effects. Elevated levels of saponins could function as organic rumen regulators to alter the fermentation and composition of ruminal microbial communities, which could have a beneficial or negative impact on rumen metabolism. In contrast to the microbiological effects, saponins impact rumen metabolism by influencing ammonia adsorption and modifying the ruminal flow of digesta, resulting in minimal physiological reactions. Because saponins have a positive impact on N metabolism, their use can address issues related to ruminants' ineffective or inadequate N retention and utilization

References

Ali O, Szabó A (2023) Review of eukaryote cellular membrane lipid composition, with special attention to the fatty acids. In J Mol Sci 24(21):15693. https://doi.org/10.3390/ijms242115693

Anju T, Saritha GNG, Ramchiary N, Kumar A (2024) Assessing the impact of different cooking methods on nutrients, phytochemicals and antioxidant activity of traditional food plants. Food Chem Adv 4:100677. https://doi.org/10.1016/j.focha.2024.100677

Ayaz M, Ullah F, Sadiq A, Ullah F, Ovais M, Ahmed J (2019) Synergistic interactions of phytochemicals with antimicrobial agents: potential strategy to counteract drug resistance. Chem Int 308:294–303. https://doi.org/10.1016/j.cbi.2019.05.050

Baky MH, Elsaid MB, Farag MA (2022) Phytochemical and biological diversity of triterpenoid saponins from family Sapotaceae: a comprehensive review. Phytochem 202:113345. https://doi.org/10.1016/j.phytochem.2022.113345

Bharadwaj M, Mondal BC, Lata M (2021) Scope of utilization of tannin & saponin to improve animal performance. J Ento Zool Stud 9(1):2168–2179

Blackman A, Foster GD, Zammit G, Rosenberg R, Aronne L, Wadden T, Claudius B, Jensen CB, Mignot E (2016) Effect of liraglutide 3.0 mg in individuals with obesity and moderate or severe obstructive sleep apnea: the SCALE Sleep Apnea randomized clinical trial. Int J Obes 40(8):1310–9. https://doi.org/10.1038/ijo.2016.52

Bueno ICS, Brandi RA, Fagundes GM, Benetel G, Muir JP (2020) The role of condensed tannins in the in vitro rumen fermentation kinetics in ruminant species: feeding type involved? Animals 10(4):635. https://doi.org/10.3390/ani10040635

Bustos KAG, Muñoz SS, da Silva SS, Alarcon MAF, Dos Santos JC, Andrade GJC, Hilares RT (2024) Saponin molecules from quinoa residues: exploring their surfactant, emulsifying, and detergent properties. Molecules 29(20):4928. https://doi.org/10.3390/molecules29204928

Chen C, Li R, Li D, Shen F, Xiao G, Zhou J (2021) Extraction and purification of saponins from Sapindus mukorossi. New J Chem 45(2):952–960. https://doi.org/10.1039/DONJ04047A

Das R, Burnase N, Suryawanshi N, Barapatre A, Kumar B (2024) 7 infliction of antibiotic. Antimicrob Environ 76. https://doi.org/10.1201/9781003268925-6

Dong M, Li J, Yang D, Li M, Wei J (2023) Biosynthesis and pharmacological activities of flavonoids, triterpene saponins and polysaccharides derived from Astragalus membranaceus. Molecules 28(13):5018. https://doi.org/10.3390/molecules28135018

Fraga-Corral M, Otero P, Echave J, Garcia-Oliveira P, Carpena M, Jarboui A, Prieto MA (2021) By-products of agri-food industry as tannin-rich sources: a review of tannins' biological activities and their potential for valorization. Foods 10(1):137. https://doi.org/10.3390/foods10010137

Garza DL, Hanashima S, Umegawa Y, Murata M, Kinoshita M, Matsumori N, Greimel P (2022) Behavior of triterpenoid saponin ginsenoside Rh2 in ordered and disordered phases in model membranes consisting of sphingomyelin, phosphatidylcholine, and cholesterol. Langmuir 38(34):10478–11049. https://doi.org/10.1021/acs.langmuir.2c01261

Grzywaczyk A, Smułek W, Kaczorek E (2024) Saponaria officinalis saponins as a factor increasing permeability of Candida yeasts' biomembrane. World J Microbiol Biotechnol 40(5):152. https://doi.org/10.1007/s11274-024-03961-9

Gunun P, Gunun N, Khejornsart P, Ouppamong T, Cherdthong A, Wanapat M, Kang S (2019) Effects of Antidesma thwaitesianum Muell. Arg. pomace as a source of plant secondary compounds on digestibility, rumen environment, hematology, and milk production in dairy cows. Anim Sci J 90(3):372–381. https://doi.org/10.1111/asj.13147

Hannan A, Du X, Maqbool B, Khan A (2024) Nanoparticles as potent allies in combating antibiotic resistance: a promising frontier in antimicrobial therapy. Pak Vet J 44(4). https://doi.org/10.29261/pakvetj/2024.227

Hapid A, Napitupulu M, Zubair MS (2021) Ethnopharmacology and antioxidant activity studies of woody liana original Wallacea. Int J Design Nat Ecodyn 16(5):495–503. https://doi.org/10.18280/ijdne.160503

Haroen U, Budiansyah A, Nelwida (2018) Phytochemical screening and in vitro antimicrobial effect of orange (Citrus sinensis) ethyl acetate extract silage. Pak J Nutr 17:214–218. https://doi.org/10.3923/pjn.2018.214.218

Haslam E (1996) Natural polyphenol (vegetable tannins) as drugs: possible modes of action. J Nat Prod 59(2):205–215. https://doi.org/10.1021/np960040

Hassen A, Tesfamariam EN, Pepeta BN, Tasfamariam EN (2024) The potential of strategies used to mitigate enteric methane emissions in nutrition studies of ruminants: a review. https://doi.org/10.20944/preprints202401.0284.v1

Huang Q, Liu X, Zhao G, Hu T, Wang Y (2018) Potential and challenges of tannins as an alternative to in-feed antibiotics for farm animal production. Anim Nutr 4(2):137–150. https://doi.org/10.1016/j.aninu.2017.09.004

Jamroz D, Wiliczkiewicz A, Skorupinska J, Orda J, Kuryszko J, Tschirch H (2009) Effect of sweet chestnut tannin (SCT) on the performance, microbial status of intestine and histological characteristics of intestine wall in chickens. Br Poult Sci 50(6):687–699. https://doi.org/10.1080/00071660903191059

Kholif AE (2023) A review of effect of saponins on ruminal fermentation, health and performance of ruminants. Vet Sci 10(7):450. https://doi.org/10.3390/vetsci10070450

Kitagawa N, Morikawa T, Motai C, Ninomiya K, Okugawa S, Nishida A, Muraoka O (2016) The antiproliferative effect of chaka saponins I and II, floratheasaponin A, and epigallocatechin 3-O-gallate isolated from Camellia sinensis on human digestive tract carcinoma cell lines. Int J Mol Sci 17(12):1979. https://doi.org/10.3390/ijms17121979

Kumar K, Chaudhary LC, Kumar S (2014) Exploitation of tannins to modulate rumen ecosystem and ruminant's performance: a review. Indian J Anim Sci 84(6):609–618. https://doi.org/10.56093/ijans.v84i6.41560

Kumari M, Jain S (2012) Tannins: an antinutrient with positive effect to manage diabetes. Res J Recent Sci 1(12):70–73

Lawal IB (2022) Changes in microbial population in the rumen of sheep fed *Acacia Mearnsii* tannin extract for methane reduction. Master's thesis, University of Pretoria (South Africa). https://doi.org/10.1016/j.anifeedsci.2022.115502

Liao Y, Li Z, Zhou Q, Sheng M, Qu Q, Shi Y, Shi X (2021) Saponin surfactants used in drug delivery systems: a new application for natural medicine components. Int J Pharm 15(603):120709. https://doi.org/10.1016/j.ijpharm.2021.120709

Liu XL, Hao YQ, Jin L, Xu ZJ, McAllister TA, Wang Y (2013) Anti-Escherichia coli O157:H7 properties of purple prairie clover and sainfoin condensed tannins. Molecules 18(2):2183–2199. https://doi.org/10.3390/molecules18022183

Liu WH, La ALT, Evans ACO, Gao ST, Yu ZT, Ma L, BDP (2021) Supplementation with Yucca schidigera improves antioxidant capability and immune function and decreases fecal score of dairy calves before weaning. J Dairy Sci 104(4):4317–4325. https://doi.org/10.3168/jds.2020-18980

Mamatha C, Prakash J (2016) Formulation of iron fortified masala powders and assessment of nutritional and sensory qualities. Indian J Nutr Dietetics 53:330–342. https://doi.org/10.21048/ijnd.2016.53.3.5316

McRae J, Yang Q, Crawford R, Palombo E (2007) Review of the methods used for isolating pharmaceutical lead compounds from traditional medicinal plants. Environmentalist 27(1):165–174. https://doi.org/10.1007/s10669-007-9024-9

Moghimipou E, Handali S (2015) Saponin: properties, methods of evaluation and applications. ARRB 5:207–220. https://doi.org/10.1016/j.indcrop.2016.10.008

Mohammed AE, Aldahasi RM, Rahman I, Shami A, Alotaibi M, BinShabaib MS, ALHarthi SS, Aabed K (2024) The antimicrobial activity of tea tree oil (Melaleuca alternifolia) and its metal nanoparticles in oral bacteria. PeerJ 6(12):e17241. https://doi.org/10.7717/peerj.17241

Nagtode VS, Cardoza C, Yasin HK, Mali SN, Tambe SM, Roy P, Singh K, Goel A, Amin PD, Thorat BR, Cruz JN (2023) Green surfactants (biosurfactants): a petroleum-free substitute for sustainability— comparison, applications, market, and future prospects. ACS Omega 8(13):11674–11699. https://doi.org/10.1021/acsomega.3c00591

Nawab A, Tang S, Gao W, Li G, Xiao M, An L, Liu W (2020) Tannin supplementation in animal feeding; mitigation strategies to overcome the toxic effects of tannins on animal health: a review. J Agric Sci 12(4):217. https://doi.org/10.5539/jas.v12n4p217

Nepal B, Stine KJ (2023) Atomic force microscopy study of the complexation of sterols and the glycoalkaloid α-tomatine in Langmuir-Blodgett monolayers. Chem Phys Lipids 252:105293. https://doi.org/10.1016/j.chemphyslip.2023.105293

Nguyen LT, Fărcaş AC, Socaci SA, Tofană M, Diaconeasa ZM, Pop OL, Salanță LC (2020) An overview of saponins-a bioactive group. Food Sci Technol 77. https://doi.org/10.15835/buasvmcn-fst:2019.0036

Reed JD (1995) Nutritional toxicology of tannins and related polyphenols in forage legumes. J Anim Sci 73(5):1516–1528. https://doi.org/10.2527/1995.7351516x

Ridla M, Laconi EB, Jayanegara A (2021) Effects of saponin on enteric methane emission and nutrient digestibility of ruminants: an in vivo meta-analysis. Proc IOP Conf Series: Earth Environ Sci 788:012028. https://doi.org/10.1088/1755-1315/788/1/012028

Romero-Puertas MC, Peláez-Vico MÁ, Pazmiño DM, Rodríguez-Serrano M, Terrón-Camero L, Bautista R, Sandalio LM (2022) Insights into ROS-dependent signalling underlying transcriptomic plant responses to the herbicide 2, 4-D. Plant, Cell Environ 45(2):572–590. https://doi.org/10.1111/pce.14229

Schneider H, Blaut M (2000) Anaerobic degradation of flavonoids by Eubacterium ramulus. Arch Microbiol 173(1):71–75. https://doi.org/10.1007/s002030050010

Segal R, Shatkovsky P, Milo-Goldzweig I (1974) On the mechanism of saponin hemolysis--I. Hydrolysis of the glycosidic bond. Biochem Pharmacol 123(5):973–981. https://doi.org/10.1016/0006-2952(74)90027-6

Serrano J, Puupponen-Pimiä R, Dauer A, Aura AM, Saura-Calixto F (2009) Tannins: current knowledge of food sources, intake, bioavailability and biological effects. Mol Nutr Food Res 53(S2):S310–S329. https://doi.org/10.1002/mnfr.200900039

Shen X, Shi L, Pan H, Li B, Wu Y, Tu Y (2017) Identification of triterpenoid saponins in flowers of four camellia Sinensis cultivars from Zhejiang province: differences between cultivars, developmental stages, and tissues. Ind Crops Prod 1(95):140–147. https://doi.org/10.1016/j.indcrop.2016.10.008

da Silva IAG, Dubeux JCB Jr, Souza CG, Moren MR, Dos Santos MVF, de Oliveira Apolinário VX, Lira MA Jr (2024) Nutritive value and condensed tannins of tree legumes in silvopasture systems. Sci Reports 14(1):18080. https://doi.org/10.1038/s41598-024-68563-9

Singh D, Chaudhuri PK (2018) Structural characteristics, bioavailability and cardioprotective potential of saponins. Integr Med Res 7(1):33–43. https://doi.org/10.1016/j.imr.2018.01.003

Singh AS, Kaur P (2020) Effect of Saponins mitigation of methane—a review. Int J Curr Microbiol App Sci 9(06):3310–3324. https://doi.org/10.20546/ijcmas.2020.906.394

Singh R, Sharma R, Mal G, Varshney R (2022) A comparative analysis of saponin-enriched fraction from Silene vulgaris (Moench) Garcke, Sapindus mukorossi (Gaertn) and Chlorophytum borivilianum (Santapau and Fernandes): an in vitro hemolytic and cytotoxicity evaluation. Anim Biotech 33(1):193–199. https://doi.org/10.1080/10495398.2020.1775627

Singh AK, Ojha L, Kumari P, Choubey M, Chaudhary SK (2024) Phytochemicals as natural feed additives for ruminants. In: Feed additives and supplements for ruminants. Springer Nature Singapore, Singapore, pp 167–196. https://doi.org/10.1007/978-981-97-0794-2-8

Sriwichai W, Berger J, Picq C, Avallone S (2016) Determining factors of lipophilic micronutrient bioaccessibility in several leafy vegetables. J Agric Food Chem 64(8):1695–1701. https://doi.org/10.1021/acs.jafc.5b05364

Tabacco E, Borreani G, Crovetto GM, Galassi G, Colombo D, Cavallarin L (2006) Effect of chestnut tannin on fermentation quality, proteolysis, and protein rumen degradability of alfalfa silage. J Dairy Sci 89:4736–4746. https://doi.org/10.3168/jds.S0022-0302(06)72523

Tagousop CN, Tamokou J-D, Kengne IC, Ngnokam D, Voutquenne-Nazabadioko L (2018) Antimicrobial activities of saponins from Melanthera elliptica and their synergistic effects with antibiotics against pathogenic phenotypes. Chem Cent J 12:1–9. https://doi.org/10.1186/s13065-018-0466-6

Tagousop CN, Feugap LDT, Harakat D, Voutquenne-Nazabadioko L, Ngnokam D (2021) New hemisynthetic oleanane saponin with antimicrobial activities. Adv Biol Chem 11(01):1. https://doi.org/10.4236/abc.2021.111001

Tarhane AK, Tarhane S, Dursun İ, Büyük F, Coşkun MR & Ersoy Y (2024) GC-MS analysis of chemical composition and determination of antimicrobial activity of Laurel leaf extracts prepared by different methods and solvents. Pak Vet J 44(3). https://doi.org/10.29261/pakvetj/2024.207

Tong Z, He W, Fan X, Guo A (2022) Biological function of plant tannin and its application in animal health. Front Vet Sci 8:803657. https://doi.org/10.3389/fvets.2021.803657

Tosi G, Massi P, Antongiovanni M, Buccioni A, Minieri S, Marenchino L (2013) Efficacy test of a hydrolysable tannin extract against necrotic enteritis in challenged broiler chickens. Italian J Anim Sci 12:62. https://doi.org/10.34233/jpr.1317771

Vincken JP, Heng L, Gruppen H (2007) Saponins, classification and occurrence in the plant kingdom. J Phytochem 68(3):275–297. https://doi.org/10.1016/j.phytochem.2006.10.008

Widyarini S, Nagari FS, Hanim C, Bachruddin Z, Muhlisin M, Mira Yusiati L (2021) Effect of Nigella sativa L. as saponin sources on in vitro rumen fermentation, enzyme activity and nutrients digestibility. Adv Anim Vet Sci 9(12):2247–2257. https://doi.org/10.17582/journal.aavs/2021/9.12.2247.2257

Wisetkomolmat J, Suppakittpaisarn P, Sommano SR (2019) Detergent plants of Northern Thailand: potential sources of natural saponins. Resources 8(1):10. https://doi.org/10.3390/resources8010010

Yanuartono Y, Paryuni AD, Soedarmanto I (2024) Rumput Brachiaria decumbens: Antara manfaat dan risiko untuk pakan ruminansia. PETERPAN (Jurnal Peternakan Terapan) 6(2):92–109. https://doi.org/10.25181/peterpan.v6i2.3595

Yanza YR, Irawan A, Jayanegara A, Ramadhani F, Respati AN, Fitri A, Hidayat C, Niderkorn V, Cieslak A, Szumacher-Strabel M (2024) Saponin extracts utilization as dietary additive in ruminant nutrition: a meta-analysis of in vivo studies. Animals 14(8):1231. https://doi.org/10.3390/ani14081231

Yao Y, Yang X, Shi Z (2014) Anti-inflammatory activity of saponins from quinoa (Chenopodium quinoa Willd) seeds in lipopolysaccharide-stimulated RAW 264.7 macrophages cells. J Food Sci 79(5):H1018–H1023. https://doi.org/10.1111/1750-3841.12425

Zhuoyue Z, Ruangaram W, Kato E (2021) Saponins are responsible for the anti-obesogenic activity of *Acacia concinna*. J Nat Med 75(4):1005–1013. https://doi.org/10.1007/s11418-021-01530-0

Probiotics and Prebiotics

Prebiotics: Feeding the Good Microbes

Shahid Nazir, Muhammad Adil, Muhammad Usman, Muhammad Farooq, Mubarik Mahmood, and Aman Ullah Khan

Abstract The restriction on the use of antimicrobial growth promoters in animal production has accelerated the demand for alternative strategies to control and prevent the colonization of pathogenic bacteria while enhancing the growth rate and other production parameters. Modulating the gut microbiota through zootechnical feed additives, including prebiotics and synbiotics, offers a promising approach to improve animal health, welfare, and productivity in modern husbandry systems. Prebiotics are indigestible fibers that promote the growth of beneficial gut microbes, whereas synbiotics combine probiotics and prebiotics to exert a synergistic effect on gut health. Promoting a healthy balance of gut microbiota contributes to reducing the risk of infections and the need for antimicrobial agents. This chapter provides a comprehensive overview of types and functional properties of prebiotics, their roles in microbiota modulation, gut health improvement, and use as sustainable alternatives to antibiotic growth promoters in animal nutrition. Additionally, it highlights the application of synbiotics across different animal species and their role in boosting immunity and improving growth performance.

Keywords Prebiotics · Synbiotics · Gut microbiota · Gut health · Animal health · Antimicrobial resistance (AMR)

S. Nazir · A. U. Khan (✉)
Institute of Microbiology, University of Veterinary and Animal Sciences, Lahore, Jhang, Pakistan
e-mail: 2025-mphil-1002@uvas.edu.pk; amanullah.khan@uvas.edu.pk

M. Adil · M. Usman
University of Veterinary and Animal Sciences, Lahore, Jhang, Pakistan

M. Farooq
Department of Clinical Sciences, University of Veterinary and Animal Sciences, Lahore, Jhang, Pakistan
e-mail: muhammad.farooq@uvas.edu.pk

M. Mahmood
Department of Animal Sciences, University of Veterinary and Animal Sciences, Lahore, Jhang, Pakistan

R. U. Khan, S. Naz (eds.), *Green Alternatives to Antibiotics in Animal Health*,
https://doi.org/10.1007/978-981-92-0309-3_8

1 Introduction

The continuously growing human population is directly associated with an increased demand for both plant and animal-origin food. To address this challenge, it is imperative to implement strategies that allow intensification of food production while minimizing costs and ensuring high standards of quality and safety for both human health and the environment. Animal feed additives play a critical role in improving animal health, increasing the palatability of diet, and boosting the production of high-quality food items like meat, eggs, milk, and fish (Pandey et al. 2019).

The gut microbiota interacts with the animal body, aids in nutrient digestion and absorption, produces bioactive compounds such as amino acids, short-chain fatty acids, and vitamins, and metabolizes the toxic wastes produced in the gut (Lian et al. 2022). Since animal nutrition is closely linked with the health of consumers, and the presence of intestinal pathogens in animals such as *Campylobacter*, *Salmonella,* etc., poses a significant risk leading to food contamination and can eventually result in outbreaks of zoonotic diseases (Markowiak and Śliżewska 2018a). To maintain a healthy gut microbiota, it is essential to use external interventions to regulate the balance between probiotics and pathogenic bacteria. Previously, antibiotics and other medicinal products were widely used to modify the gut microbiota, enhance productivity, and promote animal health. The prolonged use of these substances has contributed to the development of drug-resistant microorganisms, negatively impacting the health of both animals and humans, coupled with a degrading environment. Keeping this in view, the European Union banned the use of antibiotic-based growth promoters on 1 January 2006 (Da Silva et al. 2023). As a result, alternative natural sources providing similar benefits without associated risks have been sought. Among all the available alternatives, prebiotics have emerged as a promising option in animals, as they can pass through the digestive tract, support the symbiotic relationship between the host and the GIT microbiota, and contribute to overall health benefits (Solis-Cruz et al. 2019; Bednarczyk et al. 2016).

Prebiotics are defined as "a group of compounds which act as substrates for commensal gut microbiota, induce specific modulations in their composition, and promote their growth and activity, thus conferring benefits upon host health" (Salminen et al. 2021; Bindels et al. 2015).

Prebiotic compounds include non-absorbable carbohydrates (oligo and polysaccharides), peptides, proteins, and lipids. Natural sources of prebiotics include legumes and cereals. Additionally, prebiotics can also be synthesized through industrial chemical and enzymatic methods (Śliżewska et al. 2013).

Prebiotics lead to increased concentrations of short-chain fatty acids (SCFAs), particularly butyrate, which serves as the preferred energy source for colonocytes and plays a crucial role in maintaining gut integrity (Dankowiakowska et al. 2013). Elevated SCFAs concentrations also lower intestinal pH, creating an environment that suppresses pathogenic bacteria and enhances the solubility of certain nutrients (Van Loo and Vancraeynest 2008). Additionally, the absorption of SCFAs may also modulate certain systemic physiological processes, including glucose metabolism in the liver (Hesta et al. 2006).

Supplementing animals with prebiotic compounds can significantly enhance growth rates, improve feed conversion ratios (FCR), increase gut microbiota diversity, reduce intestinal permeability, and strengthen immune function (Pattanaik et al. 2022; Usman et al. 2025). Despite their benefits, prebiotics in animal nutrition may have certain limitations, such as variable responses among species, potential gastrointestinal disturbances such as diarrhea or bloating at high doses, and the cost of production. Moreover, inconsistent results in field conditions challenge their widespread adoption (Slavin 2013; Gaggìa et al. 2010; Rahman et al. 2022).

A synbiotic is a combination of prebiotics and probiotics that beneficially affects the host's health by enhancing the survival and colonization of live microbial dietary supplements (beneficial microbes) in the GIT. It works by selectively stimulating the growth and activity of a limited number of health-promoting bacteria, resulting in improved gut function and overall host welfare (Gibson and Roberfroid 1995). The principal purpose of this combination is to promote the survival of probiotic microorganisms in the GIT, where probiotics help maintain intestinal balance and constitute a protective barrier, while prebiotics supply the energy and essential nutrients needed for their growth and activity (Le Blay et al. 1999). Synbiotics can inhibit the growth of pathogenic bacteria either by direct antagonism or by competitive exclusion. Additionally, they can combat multidrug-resistant microorganisms (Nguyen et al. 2022; Newman and Arshad 2020). Advancements in DNA-based molecular tools and metagenomics have enabled scientists to characterize and better understand the complex interactions between microbes and their environments.

This chapter aims to explore the different types of prebiotics used as feed additives in livestock, poultry, aquaculture, and companion animals. It highlights their functional properties, including their beneficial effects on the microbiota composition, their ability to control pathogenic infections, and their role in promoting positive changes in intestinal morphology. Additionally, their synbiotic applications in animal health are discussed, emphasizing their ability to enhance productive parameters and immunomodulatory effects as possible mechanisms of action. Moreover, it evaluates current limitations and opportunities in prebiotic application and outlines future directions in precision nutrition and microbiome-targeted interventions. By elaborating on the intricate relationships between prebiotics, probiotics, and gut microbiota, this chapter emphasizes their potential as environmentally friendly, safe, and sustainable alternatives to the traditional antibiotic growth promoters in animal nutrition.

2 Types of Prebiotics

Prebiotics exist in different forms, and the majority of these belong to indigestible carbohydrates (oligo and polysaccharides), whereas others are lipids and proteins in nature (Davani-Davari et al. 2019). Fructo-oligosaccharides (FOS), oligofructose, trans-galacto-oligosaccharides (TOS), gluco-oligosaccharides, galacto-oligosaccharides, lactulose, lactitol, malto-oligosaccharides, xylo-oligosaccharides, stachyose, and raffinose are commonly used prebiotics in animal nutrition (Markowiak and Śliżewska 2018a; Patterson and Burkholder 2003). The classification, source, and general functions of different prebiotics used in animal nutrition are presented in Table 1 and Fig. 1.

2.1 Oligosaccharides (Short-Chain Carbohydrates)

2.1.1 Fructo-Oligosaccharides (FOS)

FOS are naturally found in a variety of plants, including wheat, garlic, chicory, onion, asparagus, and artichoke (Fishbein et al. 1988). FOS consist of linear chains of 2 to 60 units of fructose and often end with a glucose unit. These chains are linked by a β-2,1 bond. Its slight sweetness, non-carcinogenic nature, and soluble dietary fiber-like properties make FOS a good prebiotic that helps improve mineral absorption and reduce serum cholesterol levels (Sabater-Molina et al. 2009). FOS especially stimulates *Bifidobacterium* growth in the colon, stimulating the fermentation of dietary fiber into lactate and SCFAs (Bornet et al. 2002).

2.1.2 Mannan-Oligosaccharides (MOS)

MOS are indigestible oligosaccharides composed of mannose units. Mannose-based polymers contain glycosidic linkages primarily of the α-(1,6) and α-(1,2) types. By-products of *Saccharomyces cerevisiae* are specially used as a source of MOS in animal diets. The action pathway of MOS is different from that of other prebiotics. MOS supplementation helps by promoting beneficial bacteria and inhibiting harmful pathogens. It acts through competitive exclusion, preventing pathogens like *E. coli* from adhering to the gut lining. MOS also enhances gut morphology, immune response, and overall growth performance, making it a promising dietary supplement during the stress of weaning and improving their health and productivity (Halas and Nochta 2012).

2.1.3 Galacto-Oligosaccharides (GOS)

GOS are also non-digestible carbohydrates and support the survival of beneficial microbiota (Souza et al. 2022). They are synthesized from lactose via trans-galactosylation reactions catalyzed by the enzyme β-galactosidase. They resemble milk oligosaccharides, giving them prebiotic effects (Fischer and Kleinschmidt 2018). They are composed of 2–8 sugar units, including glucose and galactose (Wang et al. 2024). The synthesized GOS contained approximately 13%

Table 1 Classification and functional overview of prebiotics in animal nutrition

Prebiotic type	Prebiotic substance	Source	Composition	Functions	Target species	References
Oligosaccharide	*Fructo-oligosaccharides (FOS)*	Wheat, onion, chicory, garlic, asparagus, banana, artichoke	Linear chains (2–60 units) of fructose, ends with glucose; β-2,1 linkage	Stimulates *Bifidobacterium* growth, produces SCFAs (acetate, propionate, butyrate), improves mineral absorption, reduces serum cholesterol, improves gut integrity	Piglets, Poultry, Calves, Young animals	Sabater-Molina et al. (2009), Bornet et al. (2002)
	Mannan-oligosaccharides (MOS)	Yeast (*Saccharomyces cerevisiae*) cell wall	Mannose units with α- (1,6) and α-(1,2) glycosidic linkages	Promotes beneficial bacteria, prevents pathogen adhesion (e.g., *E. coli*), enhances gut morphology, boosts immune response, improves growth, supports gut health	Poultry, Swine, Ruminants, Calves, Pigs	Halas and Nochta (2012)
	Galacto-oligosaccharides (GOS)	Synthesized from lactose using β- galactosidase enzyme	2–8 sugar units (glucose and galactose); includes di-, tri-, tetra-saccharides	Supports beneficial microbiota, resembles milk oligosaccharides, improves taste, texture, shelf life, nutritional value, stimulates gut health	Infants, Piglets, Poultry, Calves, Young animals	Wang et al. (2024), Fischer and Kleinschmidt (2018)
	Xylo-oligosaccharides (XOS)	Agro-residues: wheat straw, corncobs, rice husks, sugarcane bagasse	2–7 xylose units, β- (1,4)-glycosidic bonds	Stimulates *Bifidobacteria* and lactic acid bacteria, increases calcium absorption, enhances gastrointestinal activity, stable at low pH and high temperature	Emerging use in poultry, pigs, and dairy animals	Capetti et al. (2021), Bhatia et al. (2019)
Polysaccharide	*Inulin*	Over 36,000 plant species globally	Soluble fiber; linear fructose chains with terminal glucose; forms fructans	Increases *Bifidobacterium* activity, produces SCFAs, modulates gene expression, boosts immunity, strengthens gut symbiosis, enhances effects when combined with probiotics	Dogs, Pigs, Poultry, Companion animals	Qin et al. (2023), Alexander et al. (2018)

(continued)

Table 1 (continued)

Prebiotic type	Prebiotic substance	Source	Composition	Functions	Target species	References
	Beta-glucans	Plants, fungi, yeast (*Saccharomyces cerevisiae*), bacteria (*Agrobacterium* sp.)	β-(1,3) and β-(1,6) glycosidic bonds in yeast/bacterial forms; cereal forms vary	Improves gut health, strengthens immunity, supports growth in nursery pigs, reduces intestinal viscosity, offers consistent antibiotic alternative when using yeast-based β-glucans	Pigs (nursery pigs), Poultry	Choi and Kim (2023), Ciecierska et al. (2019)
	Pectins	Plant cell walls	High molecular weight acidic polysaccharide; galacturonic acid backbone with diverse side chains	Enhances gut bacteria capable of pectin degradation, supports energy production via specialized enzymes, promotes bacterial diversity, supports gut health	Ruminants, Pigs, Poultry	Dang et al. (2025), Li et al. (2024)
	Arabinoxylans	Grains (barley, corn)	β-(1,4)-linked xylose chains with α-L-arabinose branches	Improves digestion, fermentation increases with bird age, reduces viscosity, supports beneficial microbiota, especially in poultry	Poultry (broilers), Swine	Huang et al. (2024), Bautil et al. (2019)
Resistant Starch	*Resistant Starch (fermentable carbohydrates)*	Foods with varying starch structure	Starch and derivatives resistant to small intestine digestion; amylose/amylopectin ratio impacts content	Produces SCFAs, promotes *Bifidobacterium* and *Lactobacillus*, resists digestion, ferments in colon, selective for beneficial bacteria	Humans, Monogastric (Pigs)	Tekin and Dincer (2023)
Synthetic	*Lactulose*	Synthetic disaccharide	Artificial sugar (galactose + fructose)	Improves egg production and quality, boosts body weight, enhances gut morphology (villi, goblet cells, crypts), improves reproductive gene expression, supports antibiotic-free farming	Poultry (Laying hens)	Elkomy et al. (2023), Sitanggang et al. (2016)

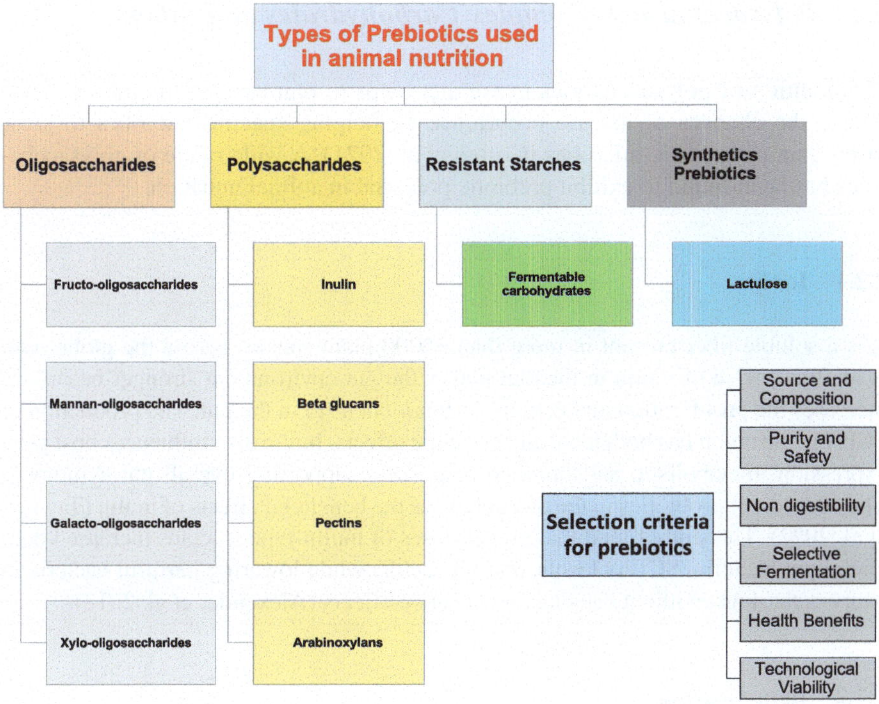

Fig. 1 Types of prebiotics generally used in animal nutrition and selection criteria for prebiotics

non-lactose disaccharides, 17% tri-saccharides, and 4% tetra-saccharides (Gobinath and Prapulla 2014). They are safe, stable, and widely used as food additives for improving taste, texture, shelf life, and nutritional value (Wang et al. 2024).

2.1.4 Xylo-Oligosaccharides (XOS)

XOS are derived from lignocellulosic biomass (LB), which mainly originates from plant cell walls that consist of cellulose, hemicellulose, and lignin. The wheat straw, corncobs, rice husks, and sugarcane bagasse are rich in xylan, the major constituent of hemicellulose (Capetti et al. 2021). XOS are stable at low pH and high temperature and promote the growth of bifidobacteria and lactic acid bacteria, increase the absorption of calcium in the gut, and enhance gastrointestinal activity (Bhatia et al. 2019).

2.2 *Polysaccharides (Complex Carbohydrates and Fibers)*

The addition of polysaccharides in the diet helps to regulate gut microflora. They reduce the chances of disease occurrence by helping specific microbes to grow faster than pathogenic microbes (Corino et al. 2021). A wide range of polysaccharides has been found to exhibit prebiotic potential in animal nutrition.

2.2.1 Inulin

It is a soluble fiber present in more than 36000 plant species across the globe (Qin et al. 2023). Use of inulin in the diet makes the gut environment stronger because it increases the proliferation and activity of *Bifidobacteria* in the gut. The production of SCFA in the colon has both local and systemic effects. Inulin also influences host gene expression, metabolism, and immune responses, supporting overall gut symbiosis. The addition of probiotics in the diet enhances the beneficial effects of inulin (Tawfick et al. 2022). The study found that higher doses of inulin-type fructans increased beneficial gut bacteria, SCFAs levels, and bile acids, while lowering harmful bacteria in overweight dogs, without causing any negative effects (Alexander et al. 2018).

2.2.2 Beta-Glucans

Beta-glucan is present in the cell wall of plants and fungi (Ciecierska et al. 2019). Microbial β-glucans derived from yeast (*Saccharomyces cerevisiae*) and bacteria (*Agrobacterium* sp.) have been shown to reliably improve intestinal health, strengthen immune responses, and support growth in nursery pigs. Conversely, cereal-based β-glucans, although present in greater quantities, often lead to increased intestinal viscosity, hindering nutrient absorption and showing variable outcomes. The effectiveness of microbial β-glucans is linked to their distinct structure and their insoluble characteristics. Supplementing diets with yeast β-glucans yields optimal benefits. Thus, microbial β-glucans represent effective and consistent alternatives to antibiotics for young pigs (Choi and Kim 2023).

2.2.3 Pectins

Pectin is a high molecular weight acidic polysaccharide present primarily in the cell walls of plants (Dang et al. 2025). Its structural complexity requires multiple enzymes for degradation. Bacteria present in the gut break down the pectin through specialized polysaccharide utilization loci (PULs). These loci encode enzymes like polygalacturonases, lyases, and esterases that degrade the pectin backbone and side chains. The coordinated action of these enzymes enables bacteria to utilize pectin as a source of energy (Li et al. 2024).

2.2.4 Arabinoxylans (AXs)

These are present in grains and known for their numerous health advantages, especially their role as prebiotics (Huang et al. 2024). The study found that arabinoxylan digestion in broilers improves with age. In young birds, the microbiota solubilized water-unextractable arabinoxylans, leading to increased intestinal viscosity. By three weeks of age, microbial activity enhanced arabinoxylan solubilization and fermentation (Bautil et al. 2019).

2.3 Resistant Starches (Fermentable Carbohydrates)

Resistant starch includes starch and its derivatives that resist digestion in the small intestine, are fermented in the colon, and produce SCFAs. They stimulate the proliferation of beneficial microbes, including *Bifidobacteria* and *Lactobacilli* (Tekin and Dincer 2023). For a resistant starch to be considered as a prebiotic, it should have these properties: resistant to digestion in the upper GIT, capable of fermentation by intestinal microbiota, and selectively promote the activity or growth of specific beneficial bacteria in the colon (Zaman and Sarbini 2016).

2.4 Synthetic Prebiotics

Lactulose, an artificial disaccharide, has potential prebiotic properties (Sitanggang et al. 2016). It enhanced egg production, improved weight gain, enhanced biochemical markers, and reproductive gene expression. Histological analysis showed improved intestinal morphology, including longer villi, more goblet cells, and deeper crypts, promoting better nutrient absorption and growth in antibiotic-free poultry farming (Elkomy et al. 2023).

2.5 Selection Criteria for Prebiotics

To define and determine a substance as a potential prebiotic, certain important parameters should be considered, including its source, origin, purity, chemical makeup, and molecular structure. Prebiotics also need to comply with international standards of safety. This involves possessing Generally Recognized as Safe (GRAS) status, proper dosage, a side effects profile, and being free of contaminants or impurities. A prebiotic mustn't induce deleterious changes to the gut microbiota. The term prebiotics must only be used when it confers a health benefit via microbiota modulation at a specific site (Sarkar 2007). Five fundamental criteria must be met for a food component to be classified as a prebiotic (Wang 2009). It must resist digestion in the upper GI tract, be specifically fermented by beneficial gut microbiota, and confer beneficial health outcomes to the host via metabolic or immune system enhancement. Additionally, it should selectively stimulate the growth and/or activity of probiotic bacteria and possess technological viability, i.e., the prebiotic material should be manufacturable and available to gut microorganisms metabolically. All these requirements assure that prebiotics are safe and functionally effective for animal and human use in nutrition (Roberfroid 2008).

3 Functional Properties of Prebiotics

Prebiotics are particularly important for achieving effects similar to those provided by antibiotic-based growth promoters (Markowiak and Śliżewska 2018b). Antibiotics harm gut microbiota and are a major cause of dysbiosis. They also contribute to the neural degenerative disorders in animals, and studies show that these effects can be reversed by using prebiotics in the diet (Karakan et al. 2021).

3.1 Selective Stimulation of Beneficial Microbes

There is a mutualistic relationship between animals and microbes that helps animals to extract nutrients from dietary fibres (Jennings and Clavel 2024). Prebiotics reduce the number of pathogenic microbes by promoting the proliferation of beneficial microbes. There is a competition for nutrients in the gut between beneficial and pathogenic microorganisms. Prebiotics help beneficial microbes to proliferate at a higher rate than pathogenic microbes, and they decrease the chances of disease occurrence (Clavijo and Flórez 2018). Prebiotics are beneficial for gut-associated lymphoid tissues. They increase the colony count of *Lactobacillus* and *Bifidobacterium* without any adverse effect and help to decrease *Salmonella* and *Escherichia coli* (Patel and Goyal 2012). All species of *Bifidobacteria* cannot metabolize FOS. The increase in abundance of this specific species is attributed to its ability to ferment FOS more efficiently than others (Gu et al. 2019). The different mechanisms by which prebiotics influence gut microbiota are illustrated in Fig. 2.

3.2 Modulation of the Immune System

A balance between immune protection and tolerance is essential, as the gut is continuously exposed to both dietary antigens and potential pathogens (Schley and Field 2002). Prebiotics modulate innate and acquired immunity by interacting with gut-associated lymphoid tissues (GALT). They strengthen the intestinal barrier by increasing the production of mucin and help in the reduction of pathogen translocation. They increase phagocytic activity to increase pathogen clearance, like

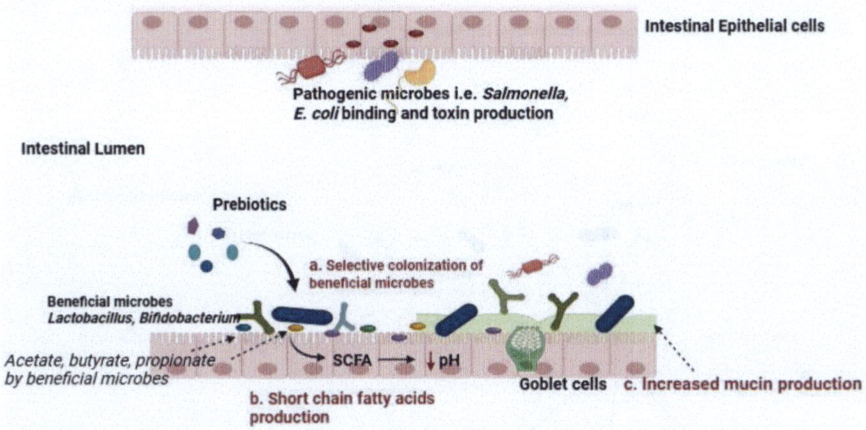

Fig. 2 Schematic diagram depicting mechanisms by which prebiotics influence the gut microbiota

Salmonella, *Clostridium difficile*, and rotavirus infections (Lomax and Calder 2008). Peyer's patches respond to prebiotics by altering gut microbiota, increasing immunoregulatory SCFAs, and possibly interacting with carbohydrate receptors on immune cells, thereby influencing immune functions (Seifert and Watzl 2007). An increase in immunoglobulin level has been observed by adding prebiotics such as pectin oligosaccharides (POS) to the diet. A significant increase in IgG and IgA is of critical significance in the fight against infections. An increase in IgA level is crucial in preventing bacterial and viral adhesion to the gut wall (Sori et al. 2022).

3.3 Digestive Health Benefits and Growth Enhancement

Healthy gut microbes are crucial for the overall health of an animal. Prebiotics are important for the improvement of digestion and nutrient absorption in the GIT. Prebiotics like FOS have a great influence on the activities of enzymes like amylase and protease. FOS-supplemented diet increased the length of villi and alleviates the diarrhea caused by poor nutrient absorption (associated with large crypts and shortened villi) (Xu et al. 2003).

The absorption of macro minerals and trace minerals increases with the increase of FOS in the diet (Wolf et al. 1998). Iron, calcium, and magnesium absorption is stimulated by indigestible oligosaccharides due to an increase in fermentation activity and a change in gut cell morphology (Scholz-Ahrens et al. 2001). Microbial fermentation of prebiotics results in the production of SCFA, and it lowers the pH of the gut. Lower pH increases calcium absorption in the colon indirectly. Introduction of polysaccharides such as chito-oligosaccharides (COS), chitosan (CTS), and chitin (CT) in aquatic diet changes the morphology of the gut and

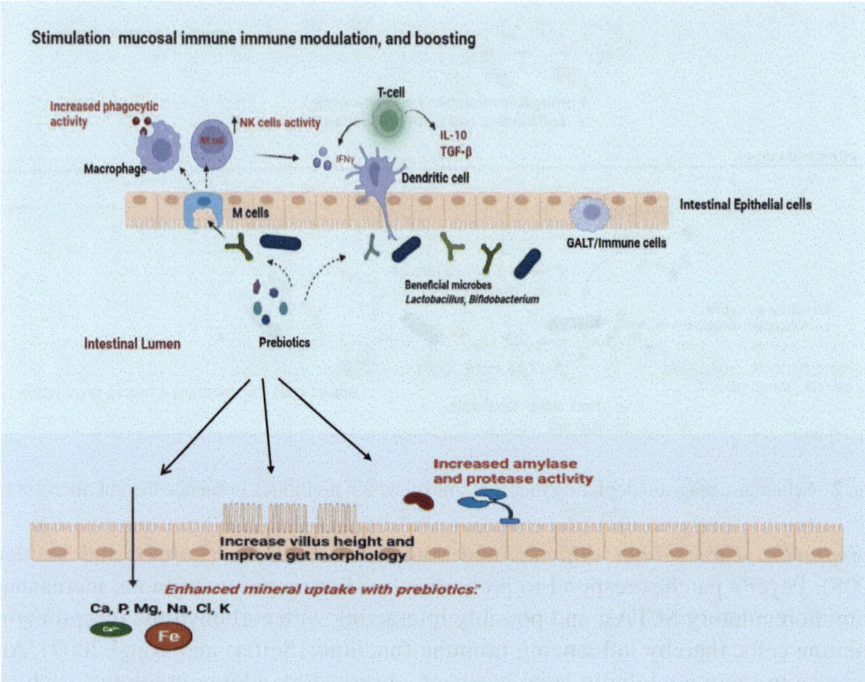

Fig. 3 Mechanisms of action of prebiotics in modulating immunity, enhancing nutrient absorption, and improving gut health

improves growth performance (Mohan et al. 2023). Diet of lambs that is supplemented with MOS and beta-glucan (BG) results in an increase in average daily gain, total dry matter intake, and growth rate. Feed conversion ratio (FCR) is improved in chicks receiving MOS and BG treatment (Shoukry et al. 2023). The figure depicts the mechanisms of action of prebiotics in modulating immunity, enhancing nutrient absorption, and improving gut health (Fig. 3).

4 Synbiotic Applications in Animal Health

Generally, synbiotics are of two types: (a) *Complementary synbiotics,* which is a probiotic in combination with a prebiotic that functions independently, and (b) *Synergistic synbiotics,* where the prebiotic is specifically designed to support the growth of the co-administered probiotic strain (Swanson et al. 2020). The synergistic action of synbiotics helps suppress harmful bacteria while promoting beneficial microbes in the host (Fazelnia et al. 2021). Synbiotics are currently recognized as a valuable strategy to support animal health by helping in the prevention and reduction of diseases (Anadón et al. 2016). Research shows that synbiotics can positively influence the gut microbiota, particularly by modulating the Firmicutes/Bacteroidetes ratio and inhibiting the harmful bacteria through mechanisms such as direct antagonism, competitive exclusion, and microbiota restoration. An increase in the Firmicutes/Bacteroidetes ratio is often associated with improved energy harvest from the diet, as Firmicutes are efficient at fermenting dietary fibers and producing SCFAs, which provide energy to the host (Sutoyo et al. 2020). However, an excessively high ratio has also been linked to obesity and metabolic disorders in some pigs and humans (Cui et al. 2013). Synbiotics also contribute to gut health by maintaining intestinal pH and supporting the integrity of the intestinal barrier. Additionally, synbiotics can help the body in combating multidrug-resistant microorganisms (Li et al. 2020; Spinler et al. 2017).

4.1 Use of Synbiotics in Ruminants

In cattle, the use of synbiotics increases the ruminal microbial synthesis, improves the intensity of fermentation, and synthesis of SCFAS. This increases the productivity of animals by reducing the effect of evolving digestive diseases and also by limiting the impact of low-quality feed (Trukhachev et al. 2022). In neonatal buffalo calves, mortality due to diarrhea is reduced by using a combination of probiotics and prebiotics because it decreases the number of pathogenic bacteria and promotes the number of gut beneficial bacteria (Sharma et al. 2024). It also improves the feed conversion efficiency and average daily weight gain by enhancing the absorption of nutrients.

In ruminants, rumen fermentation produces methane, which, in addition to decreasing the feed efficiency, also pollutes the environment, leading to global warming. Yeast (*S. cerevisiae*) supplementation is suggested to reduce CH_4 production by favoring propionate formation, which consumes metabolic hydrogen, and this, in turn, inhibits methanogenesis (Mutsvangwa et al. 1992; Palangi and Lackner 2022).

4.2 Use of Synbiotics in Aquaculture

The administration of synbiotics, either as dietary supplements or through external bathing, improves growth rate, feed efficacy, nutrient digestibility, and the immune system in marine organisms (Dehaghani et al. 2015). They release bioactive compounds and extracellular bacterial enzymes that help in the activation of precursors of digestive enzymes that enhance nutrient absorption and the efficacy of feed utilization. Synbiotics enhance immunity by releasing special proteins called cytokines from intestinal epithelial cells that help in the activity of B cells, T cells, and dendritic cells, and stimulate lipopolysaccharides to trigger tumor necrosis factors, leading to phagocytosis, nitric oxide production, and respiratory burst activity.

Synbiotics influence lysozyme activity, which plays a vital role as a key bactericidal enzyme of the innate immune system, providing a defense mechanism against numerous pathogens in fish (Hoseinifar et al. 2018; Lindsay 1986).

4.3 Use of Synbiotics in Poultry

Newly hatched chicks often experience a 48–72-hour delay in feed due to their residual yolk sac, which can decrease their growth rate. To overcome this issue, in ovo injections of beneficial bacteria, prebiotics, and synbiotics into the air sac at around 12 days of incubation are practiced, which can protect the chick from harmful pathogens and enhance their growth and digestion (Dunislawska et al. 2017). Additionally, the in-ovo administration of both prebiotics and synbiotics increases the activities of pancreatic enzymes like amylase, lipase, and hydrolase (Pruszynska-Oszmalek et al. 2015). This increased activity enhances the digestion of feed in the newly hatched chick.

In broiler chicken, synbiotics also protect from the harmful effects of heat shock during the summer season due to their positive effects on intestinal morphology and expression of heat shock protein 70 (HSP 70) (Hu et al. 2022). Feeding of chicken with synbiotics beneficially increases the growth of *Lactobacillus* species and *Bifidobacterium* species and decreases *Clostridium* and *Escherichia coli* species in intestinal contents (Markowiak et al. 2019).

In layer, a diet supplemented with synbiotics has positive effects on growth rate and egg production. It also reduces the *Salmonella* count in the gut and limits the adverse effects of salmonellosis regardless of vaccination. Birds supplemented with both synbiotics and vaccination indicate better immunity due to a high level of bile IgA (Luoma et al. 2017). Synbiotics enhance egg quality internally by reducing cholesterol and saturated fatty acid contents and increasing the unsaturated fatty acid, omega-6, and polyunsaturated fatty acid contents, including linoleic acid and alpha linolenic acid, without affecting the yolk color, albumin, and external egg quality (Tang et al. 2015).

In turkeys, ochratoxin-contaminated feed decreases body weight, causes immunosuppression, liver damage, reduces enzymatic activity, and decreases the overall performance of birds. Synbiotics minimize the effects of toxins by detoxification, adsorption of toxins, or microbial metabolism of toxins (McCormick 2013).

4.4 Use of Synbiotics in Swine

In preterm pigs, the use of synbiotics containing *Bifidobacterium breve*, GOS, FOS, and L-glutamine enhances the immune and brain development process (Andersen et al. 2019). Weaning is a challenging period in pigs due to immature immune and intestinal function, which reduces the performance of pigs. During this period, the chances of enterotoxigenic *E. coli* are increased, which can cause enteric disease in pigs (Wellock et al. 2008). Newly weaned pigs are unable to digest a plant-based diet due to an immature GIT, a different composition of plant-based diet from milk, and the presence of anti-nutritional factors. A plant-based diet containing xylan can damage the intestinal epithelial layer, resulting in increased viscosity of digesta and decreased nutrient digestibility. Use of synbiotics containing xylanase can hydrolyse the xylan, reduce the viscosity of digesta, and increase the nutrient digestibility (Moita and Kim 2022). The overall performance of newly weaned pigs is increased by reducing diarrhea and oxidative stress caused by *Escherichia coli* (Duarte et al. 2020).

Heat stress is also a major problem in pig farming because they either lack or have non-functional sweat glands. Heat stress disturbs the normal microbiota, damaging the gut barrier and allowing the pathogenic bacteria to enter the bloodstream, which causes endotoxemia in pigs (Gabler et al. 2018). This reduces the feed intake, growth rate, and carcass weight. Use of synbiotics balances the normal gut microbiota and promotes the gut barrier, which increases the feed intake, growth rate, and performance of pigs (Ringseis and Eder 2022).

4.5 Use of Synbiotics in Companion Animals

The dog's gut microbiota is highly complex and varies across different parts of the gut. Some bacterial strains, such as lactic acid bacteria and *Enterococcus faecium,* are most commonly used in dogs affected with infectious diarrhea, as they can trigger a strong immune response (Rinkinen et al. 2003). On the other hand, species like *Lactobacillus* and *Bifidobacteria* are referred to because they have more immune-balancing effects (Schmitz and Suchodolski 2016).

In beagle dogs, synbiotics containing porcine plasma have a good effect on gut health, enhancing the beneficial microbiota and reducing the incidence of diarrhea. It increases short-chain fatty acids concentration and reduces branched-chain fatty acids, and also has positive effects on the immune system (Montserrat-Malagarriga et al. 2024).

Horses are hind-gut fermenters; their cecum and colon have a diverse range of microbiota, including bacteria, protozoa, and fungi. This intestinal microbiota has a vital effect on the health and performance of horses. Synbiotics produce various types of substances that are effective against pathogenic microbes. These include lactic acid, acetic acid, formic acid, hydrogen peroxide, and ammonia (Lagounova 2022). In this way, they play a major role in immune modulation, antimicrobial production, competitive exclusion of pathogens, and inactivation of bacterial toxins, and exert a positive effect on gut health (Schoster et al. 2014).

In cats, the administration of synbiotics enhances the beneficial microbiota count and improves the fecal score in cats affected with chronic diarrhea. Beneficial effects of synbiotics include stimulation of the host immune system, restoration of the intestinal barrier, secretion of intestinal antimicrobial agents, and competitive exclusion of pathogens (Whittemore et al. 2018). In rabbits, synbiotics supplementation containing (*Saccharomyces cerevisiae* yeast and *Moringa oleifera* leaf extract) increases the beneficial gut bacteria like *Lactobacillus* and reduces the pathogenic bacteria such as *Salmonella* and coliforms that have negative effects on intestinal and cecal development. It also increases body weight gain and better feed efficiency (Hashem et al. 2021).

4.6 Innovative Synbiotics Formulations

Lactococcus lactis G01 is a bacterium isolated from curd that has strong probiotic potential to digest gluten. Researchers coated it in sodium alginate beads using banana powder as a prebiotic and aloe vera as an adsorbent to enhance its efficacy and protection from stomach acids, bile salts, and antibiotics (Anagha et al. 2023). Encapsulation of eight *Lactobacillus* strains and one *Lactococcus* strain is done by sodium alginate coating to protect these from stomach acids, bile salts, and pancreatic enzymes. To further enhance its survival during storage, probiotic coating is placed in skimmed milk with sucrose and ascorbic acid before drying (Goderska and Agudo Pena 2024). *Pediococcus acidilactici* is used as a probiotic that shows antimicrobial activity against *S. gallinarum*, a major poultry pathogen. When this probiotic and prebiotic phthalyl inulin is coated in alginate/chitosan microcapsules to protect them in the body, they show strong antimicrobial activity against *S. gallinarum* (Cui et al. 2018).

5 Challenges, Opportunities, and Future Directions in Prebiotic Applications

Currently, animal production is facing numerous challenges, including global-warming–induced heat stress and increased costs of animal rearing and transportation. Furthermore, the industry is dealing with growing demands for antibiotic-free meat, complications during the weaning period, and contamination of stored animal feeds by mycotoxins. These issues can lead to significant animal health problems, which can result in acute physiological disorders, the decline in overall productivity of animals, and subsequently pose risks to consumers' health. While multidrug antibiotic resistance continues to be an alarming threat in treating diseases, applications of natural bioactive compounds from microorganisms, plants, and algae offer promising alternatives for disease prevention, immune modulation, and tolerance enhancement in animals (Wang et al. 2022; Patterson et al. 2022). Global market forecasts an annual compound growth rate (CAGR) of 14.9% for prebiotics in the food and feed industry from 2022 to 2030. This growth is mainly driven by rising awareness of the potential benefits of functional prebiotic foods and increased demand for high-nutrition, low-calorie processed foods. The ban of antibiotic growth promoters in livestock and the recognized positive effects of probiotics in animals have further contributed to this trend (Khattak and Galgano 2023). However, challenges such as limited scalability and the inadequacy of insights in their mode of action remain barriers to their widespread utilization. Moreover, bioactive ingredients in prebiotics often face challenges such as low oral bioavailability primarily due to their interaction with different pH levels, mucus layers, gut microflora, and digestive enzymes (Rashidinejad et al. 2022). This can be improved by various encapsulation techniques like nanohydrogel, nanoparticles, and nanoemulsions to facilitate proper drug release at pre-determined locations. In this way, the targeted delivery system of prebiotics is improved by focusing on the structure and its release mechanism (Zhang et al. 2025). Commonly explored emerging technologies for the processing of food products are high-intensity ultrasound, pulsed electric field, pulsed light technology, supercritical carbon dioxide technology, cold plasma technology, and microwave processing (Qayum et al. 2023; Silva et al. 2019; Martins et al. 2019). These technologies can improve prebiotic formulations by incorporating high concentrations of bioactive compounds such as citric acid, ascorbic acid, anthocyanins, polyphenols, flavonoids, and antioxidants, thereby boosting their functional properties and overall efficacy (Balthazar et al. 2022).

Some scientists now propose that fermentable fiber, vitamins, and fatty acids should also be considered prebiotics as they can significantly influence the gut microbiota (Slavin 2013; Lockyer and Stanner 2019; Gibson et al. 2017). Prebiotics are now studied for their effects beyond the gut, such as skin health (helps in maintaining a healthy skin microbiome) as well as oral, urogenital, and colorectal health (Cunningham et al. 2021).

Recently, research has highlighted the positive health effects of bioactive compounds from certain macro- and micro-algae (Patel et al. 2021). Eco-friendly extraction technologies such as supercritical fluid, microwave-assisted, enzyme-assisted, and pressurized-liquid methods have emerged as efficient, cost-effective solutions for obtaining algal bioactive compounds. Microalgal biomass contains a diverse array of compounds, including polysaccharides, proteins, vitamins, pigments, polyunsaturated fatty acids (PUFAs), and minerals. Some of these compounds demonstrated the prebiotic potential by promoting beneficial probiotics in the host gut (Camacho et al. 2019). Scaling up of prebiotic compounds from algae remains a major challenge, mainly due to high costs of production, a smaller market size, and competition with cheaper fossil-based and microbial alternatives. Additional barriers include stringent regulations for quality and safety standards, environmental concerns, and technical limitations in biomass harvesting and stability, limiting their commercial availability and viability (Xiong et al. 2018; Khatoon and Pal 2015). Similarly, emerging research underscores the future potential of tailored prebiotics and enzyme-assisted strategies in animal nutrition. A polyphenol-rich prebiotic fiber blend (TF1) has shown distinct benefits in canine gut health by enhancing saccharolytic and postbiotic metabolism while reducing inflammatory and putrefactive processes (Fritsch et al. 2022). Similarly, *Thermobacillus xylanilyticus* feruloyl esterase (Tx-Est1), in processing prebiotic feruloylated xylo-oligosaccharides, efficiently releases bioactive ferulic acid from plant biomass at elevated temperatures (Garbelotti et al. 2023). These insights support the development of next-generation, eco-friendly prebiotics and enzyme systems to optimize gut health and valorize agricultural waste in sustainable animal nutrition solutions. Relatively simple culture-based assays are commonly used to quantify the effects of prebiotics on bacterial growth (Huebner et al. 2007). More advanced gut-simulation models, combined with modern molecular techniques, provide a more comprehensive analysis (Rastall 2010). Analytical tools like chromatography (e.g., GC, HPLC, and high-performance anion-exchange chromatography) and spectroscopic methods are critical for quantifying and characterizing prebiotics in complex food matrices (Mercier and Campargue 2012). Although highly detailed techniques like electron microscopy and enzymatic digestion paired with mass spectrometry show promise for analyzing novel polymeric prebiotics, they are less feasible for routine quality control (Zeng et al. 2018).

6 Conclusion

Eco-friendly alternatives to the therapeutic use of antimicrobials are continuously being explored, with the principal purpose of reducing AMR while enhancing sustainable animal health and productivity. Among these, the strategic use of prebiotics and probiotics in animals has offered several benefits, including elevated health status, increased disease resistance, improved growth performance and body composition, reduced malformations, and better gut morphology and microbial balance. Rather than relying on antibiotics to inhibit pathogenic microbes, the application of prebiotics and probiotics is a cost-effective and environmentally sustainable approach to prevent pathogens' colonization and combat AMR-related challenges. Synbiotics have been proven to have a positive influence on growth parameters by enhancing fat metabolism, stimulating growth through modulation of gastrointestinal physiological functions, and altering the morphology of the intestinal epithelium. To enhance the efficacy, stability, and viability of synbiotics, different strategies such as microencapsulation and environmental stress adaptation should be employed. Microencapsulation can increase the protection of probiotics by embedding them within a prebiotic-enriched matrix, allowing the synbiotic to maintain high viability and effectively deliver its potential health benefits. Furthermore, adapting probiotic strains to environmental stressors may be a promising strategy to positively alter the biosynthesis of protective cell wall components, thereby increasing their resilience. To achieve optimal benefits from prebiotic supplementation, it is essential to standardize factors such as the type of prebiotic, inclusion level, chemical structure, method of administration, and animal-specific characteristics. Furthermore, advancements in molecular microbiology and a deeper understanding of host–microbiome interactions will pave the way for precision prebiotic interventions tailored to species-specific health and production goals.

References

Alexander C, Cross TL, Devendran S, Neumer F, Theis S, Ridlon JM, Suchodolski JS, de Godoy MRC, Swanson KS (2018) Effects of prebiotic inulin-type fructans on blood metabolite and hormone concentrations and faecal microbiota and metabolites in overweight dogs. Br J Nutr 120(6):711–720. https://doi.org/10.1017/s0007114518001952

Anadón A, Martínez-Larrañaga MR, Ares I, Martínez MA (2016) Probiotics: safety and toxicity considerations. In: Nutraceuticals. Elsevier, pp 777–798

Anagha K, Sreejit V, Preetha R (2023) Probiotic with gluten reduction property and its encapsulation in synbiotic aloe vera gel-alginate capsules with banana powder as prebiotic. J Food Sci Technol 60(3):1125–1135. https://doi.org/10.1007/s13197-022-05639-2

Andersen AD, Nguyen DN, Langhorn L, Renes IB, van Elburg RM, Hartog A, Tims S, van de Looij Y, Sangild PT, Thymann T (2019) Synbiotics combined with glutamine stimulate brain development and the immune system in preterm pigs. J Nutr 149(1):36–45. https://doi.org/10.1093/jn/nxy243

Balthazar CF, Guimarães JF, Coutinho NM, Pimentel TC, Ranadheera CS, Santillo A, Albenzio M, Cruz AG, Sant'Ana AS (2022) The future of functional food: emerging technologies application on prebiotics, probiotics and postbiotics. Compr Rev Food Sci Food Saf 21(3):2560–2586. https://doi.org/10.1111/1541-4337.12962

Bautil A, Verspreet J, Buyse J, Goos P, Bedford MR, Courtin CM (2019) Age-related arabinoxylan hydrolysis and fermentation in the gastrointestinal tract of broilers fed wheat-based diets. Poult Sci 98(10):4606–4621. https://doi.org/10.3382/ps/pez159

Bednarczyk M, Stadnicka K, Kozłowska I, Abiuso C, Tavaniello S, Dankowiakowska A, Sławińska A, Maiorano G (2016) Influence of different prebiotics and mode of their administration on broiler chicken performance. Animal 10(8):1271–1279. https://doi.org/10.1017/s1751731116000173

Bhatia L, Sharma A, Bachheti RK, Chandel AK (2019) Lignocellulose derived functional oligosaccharides: production, properties, and health benefits. Prep Biochem Biotechnol 49(8):744–758. https://doi.org/10.1080/10826068.2019.1608446

Bindels LB, Delzenne NM, Cani PD, Walter J (2015) Towards a more comprehensive concept for prebiotics. Nat Rev Gastroenterol Hepatol 12(5):303–310. https://doi.org/10.1038/nrgastro.2015.47

Bornet FR, Brouns F, Tashiro Y, Duvillier V (2002) Nutritional aspects of short-chain Fructooligosaccharides: natural occurrence, chemistry, physiology and health implications. Dig Liver Dis 34(Suppl 2):S111–S120. https://doi.org/10.1016/s1590-8658(02)80177-3

Camacho F, Macedo A, Malcata F (2019) Potential industrial applications and commercialization of microalgae in the functional food and feed industries: a short review. Mar Drugs 17(6). https://doi.org/10.3390/md17060312

Capetti CCM, Vacilotto MM, Dabul ANG, Sepulchro AGV, Pellegrini VOA, Polikarpov I (2021) Recent advances in the enzymatic production and applications of xylooligosaccharides. World J Microbiol Biotechnol 37(10):169. https://doi.org/10.1007/s11274-021-03139-7

Choi H, Kim SW (2023) Characterization of β-Glucans from cereal and microbial sources and their roles in feeds for intestinal health and growth of nursery pigs. Animals (Basel) 13(13). https://doi.org/10.3390/ani13132236

Ciecierska A, Drywień ME, Hamulka J, Sadkowski T (2019) Nutraceutical functions of beta-glucans in human nutrition. Rocz Panstw Zakl Hig 70(4):315–324. https://doi.org/10.32394/rpzh.2019.0082

Clavijo V, Flórez MJV (2018) The gastrointestinal microbiome and its association with the control of pathogens in broiler chicken production: a review. Poult Sci 97(3):1006–1021. https://doi.org/10.3382/ps/pex359

Corino C, Di Giancamillo A, Modina SC, Rossi R (2021) Prebiotic effects of seaweed polysaccharides in pigs. Animals (Basel) 11(6). https://doi.org/10.3390/ani11061573

Cui C, Shen CJ, Jia G, Wang KN (2013) Effect of dietary Bacillus subtilis on proportion of Bacteroidetes and Firmicutes in swine intestine and lipid metabolism. Genet Mol Res 12(2):1766–1776. https://doi.org/10.4238/2013.May.23.1

Cui LH, Yan CG, Li HS, Kim WS, Hong L, Kang SK, Choi YJ, Cho CS (2018) A new method of producing a natural antibacterial peptide by encapsulated probiotics internalized with inulin nanoparticles as prebiotics. J Microbiol Biotechnol 28(4):510–519. https://doi.org/10.4014/jmb.1712.12008

Cunningham M, Azcarate-Peril MA, Barnard A, Benoit V, Grimaldi R, Guyonnet D, Holscher HD, Hunter K, Manurung S, Obis D, Petrova MI, Steinert RE, Swanson KS, van Sinderen D, Vulevic J, Gibson GR (2021) Shaping the future of probiotics and prebiotics. Trends Microbiol 29(8):667–685. https://doi.org/10.1016/j.tim.2021.01.003

Da Silva RA, Arenas NE, Luiza VL, Bermudez JAZ, Clarke SE (2023) Regulations on the use of antibiotics in livestock production in South America: a comparative literature analysis. Antibiotics (Basel) 12(8). https://doi.org/10.3390/antibiotics12081303

Dang G, Li J, Yin C, Wang W, Zhang K, Zhong R, Chen L, Zhang H, Schroyen M (2025) Deciphering pectin: a comprehensive overview of its origins, processing, and promising utility. ACS Omega 10(1):1–15. https://doi.org/10.1021/acsomega.4c01843

Dankowiakowska A, Kozłowska I, Bednarczyk M (2013) Probiotics, prebiotics and snybiotics in Poultry–mode of action, limitation, and achievements. J Cent Eur Agric 14(1):467–478

Davani-Davari D, Negahdaripour M, Karimzadeh I, Seifan M, Mohkam M, Masoumi S, Berenjian A, Ghasemi Y (2019) Prebiotics: definition, types, sources, mechanisms, and clinical applications. Foods 8(3):92. https://doi.org/10.3390/foods8030092

Dehaghani PG, Baboli MJ, Moghadam AT, Ziaei-Nejad S, Pourfarhadi M (2015) Effect of synbiotic dietary supplementation on survival, growth performance, and digestive enzyme activities of common carp (Cyprinus carpio) fingerlings. Czech J Anim Sci 60 (5):224–232

Duarte ME, Tyus J, Kim SW (2020) Synbiotic effects of enzyme and probiotics on intestinal health and growth of newly weaned pigs challenged with Enterotoxigenic F18(+) Escherichia coli. Front Vet Sci 7:573. https://doi.org/10.3389/fvets.2020.00573

Dunislawska A, Slawinska A, Stadnicka K, Bednarczyk M, Gulewicz P, Jozefiak D, Siwek M (2017) Synbiotics for broiler chickens-in vitro design and evaluation of the influence on host and selected microbiota populations following in Ovo delivery. PLoS One 12(1):e0168587. https://doi.org/10.1371/journal.pone.0168587

Elkomy HS, Koshich II, Mahmoud SF, Abo-Samaha MI (2023) Use of lactulose as a prebiotic in laying hens: its effect on growth, egg production, egg quality, blood biochemistry, digestive enzymes, gene expression and intestinal morphology. BMC Vet Res 19(1):207. https://doi.org/10.1186/s12917-023-03741-x

Fazelnia K, Fakhraei J, Yarahmadi HM, Amini K (2021) Dietary supplementation of potential probiotics Bacillus subtilis, Bacillus licheniformis, and Saccharomyces cerevisiae and synbiotic improves growth performance and immune responses by modulation in intestinal system in broiler chicks challenged with Salmonella Typhimurium. Probiotics Antimicrob Proteins 13(4):1081–1092. https://doi.org/10.1007/s12602-020-09737-5

Fischer C, Kleinschmidt T (2018) Synthesis of galactooligosaccharides in Milk and whey: a review. Compr Rev Food Sci Food Saf 17(3):678–697. https://doi.org/10.1111/1541-4337.12344

Fishbein L, Kaplan M, Gough M (1988) Fructooligosaccharides: a review. Vet Hum Toxicol 30(2):104–107

Fritsch DA, Jackson MI, Wernimont SM, Feld GK, Badri DV, Brejda JJ, Cochrane CY, Gross KL (2022) Adding a polyphenol-rich fiber bundle to food impacts the gastrointestinal microbiome and metabolome in dogs. Front Vet Sci 9:1039032. https://doi.org/10.3389/fvets.2022.1039032

Gabler NK, Koltes D, Schaumberger S, Murugesan GR, Reisinger N (2018) Diurnal heat stress reduces pig intestinal integrity and increases endotoxin translocation. Transl Anim Sci 2(1):1–10. https://doi.org/10.1093/tas/txx003

Gaggìa F, Mattarelli P, Biavati B (2010) Probiotics and prebiotics in animal feeding for safe food production. Int J Food Microbiol 141(Suppl 1):S15–S28. https://doi.org/10.1016/j.ijfoodmicro.2010.02.031

Garbelotti CV, Bulmer GS, Ward RJ, van Munster JM (2023) The feruloyl esterase from Thermobacillus xylanilyticus shows broad specificity for processing pre-biotic feruloylated xylooligosaccharides at high temperatures. Food Chem 405(Pt B):134939. https://doi.org/10.1016/j.foodchem.2022.134939

Gibson GR, Hutkins R, Sanders ME, Prescott SL, Reimer RA, Salminen SJ, Scott K, Stanton C, Swanson KS, Cani PD, Verbeke K, Reid G (2017) Expert consensus document: the International Scientific Association for Probiotics and Prebiotics (ISAPP) consensus statement on the definition and scope of prebiotics. Nat Rev Gastroenterol Hepatol 14(8):491–502. https://doi.org/10.1038/nrgastro.2017.75

Gibson GR, Roberfroid MB (1995) Dietary modulation of the human colonic microbiota: introducing the concept of prebiotics. J Nutr 125(6):1401–1412. https://doi.org/10.1093/jn/125.6.1401

Gobinath D, Prapulla SG (2014) Permeabilized probiotic Lactobacillus plantarum as a source of β- galactosidase for the synthesis of prebiotic galactooligosaccharides. Biotechnol Lett 36(1):153–157. https://doi.org/10.1007/s10529-013-1345-9

Goderska K, Agudo Pena S (2024) An in vitro gastrointestinal model to evaluate the tolerance of encapsulated Lactobacillus and Lactococcus strains with synbiotic containing lactobionic acid via lyophilization technique to harsh gastric conditions during storage time. Eur J Pharm Biopharm 197:114147. https://doi.org/10.1016/j.ejpb.2023.11.012

Gu J, Mao B, Cui S, Liu X, Zhang H, Zhao J, Chen W (2019) Metagenomic insights into the effects of Fructooligosaccharides (FOS) on the composition of luminal and mucosal microbiota in C57BL/6J mice, Especially the Bifidobacterium Composition. Nutrients 11(10):2431. https://doi.org/10.3390/nu11102431

Halas V, Nochta I (2012) Mannan oligosaccharides in nursery pig nutrition and their potential mode of action. Animals 2(2):261–274. https://doi.org/10.3390/ani2020261

Hashem NM, Hosny NS, El-Desoky NI, Shehata MG (2021) Effect of nanoencapsulated alginate-synbiotic on Gut microflora balance, immunity, and growth performance of growing rabbits. Polymers (Basel) 13(23). https://doi.org/10.3390/polym13234191

Hesta M, Debraekeleer J, Janssens G, De Wilde R (2006) And cat nutrition: a review. In: Trends in dietary carbohydrates research, p 179

Hoseinifar SH, Sun YZ, Wang A, Zhou Z (2018) Probiotics as means of diseases control in aquaculture, a review of current knowledge and future perspectives. Front Microbiol 9:2429. https://doi.org/10.3389/fmicb.2018.02429

Hu JY, Mohammed AA, Murugesan GR, Cheng HW (2022) Effect of a synbiotic supplement as an antibiotic alternative on broiler skeletal, physiological, and oxidative parameters under heat stress. Poult Sci 101(4):101769. https://doi.org/10.1016/j.psj.2022.101769

Huang M, Bai J, Buccato DG, Zhang J, He Y, Zhu Y, Yang Z, Xiao X, Daglia M (2024) Cereal-derived water-unextractable arabinoxylans: structure feature, effects on baking products and human health. Foods 13(15). https://doi.org/10.3390/foods13152369

Huebner J, Wehling R, Hutkins R (2007) Functional activity of commercial prebiotics. Int Dairy J 17(7):770–775

Jennings SAV, Clavel T (2024) Synthetic communities of Gut microbes for basic research and translational approaches in animal health and nutrition. Annu Rev Anim Biosci 12:283–300. https://doi.org/10.1146/annurev-animal-021022-025552

Karakan T, Ozkul C, Küpeli Akkol E, Bilici S, Sobarzo-Sánchez E, Capasso R (2021) Gut-brain-microbiota axis: antibiotics and functional gastrointestinal disorders. Nutrients 13(2):389. https://doi.org/10.3390/nu13020389

Khatoon N, Pal R (2015) Microalgae in biotechnological application: a commercial approach. In: Plant biology and biotechnology: Volume II: plant genomics and biotechnology, pp 27–47

Khattak F, Galgano S (2023) Role of pre, pro and Synbiotics in reducing zoonotic pathogen abundance, lowering antimicrobial resistance and improving food safety: old concepts with a new perspective. In: Pre and probiotics for poultry Gut health. CABI GB, pp 66–84
Lagounova M (2022) The use of dietary probiotics in horses: microbial activity during antibiotic treatment and bacterial characterization. University of Guelph
Le Blay G, Michel C, Blottière HM, Cherbut C (1999) Prolonged intake of fructo-oligosaccharides induces a short-term elevation of lactic acid-producing bacteria and a persistent increase in cecal butyrate in rats. J Nutr 129(12):2231–2235. https://doi.org/10.1093/jn/129.12.2231
Li C, Niu Z, Zou M, Liu S, Wang M, Gu X, Lu H, Tian H, Jha R (2020) Probiotics, prebiotics, and synbiotics regulate the intestinal microbiota differentially and restore the relative abundance of specific gut microorganisms. J Dairy Sci 103(7):5816–5829. https://doi.org/10.3168/jds.2019-18003
Li J, Peng C, Mao A, Zhong M, Hu Z (2024) An overview of microbial enzymatic approaches for pectin degradation. Int J Biol Macromol 254(Pt 1):127804. https://doi.org/10.1016/j.ijbiomac.2023.127804
Lian Z, Zhang Q, Xu Y, Zhou X, Jiang K (2022) Biorefinery cascade processing for converting corncob to Xylooligosaccharides and glucose by maleic acid pretreatment. Appl Biochem Biotechnol 194(10):4946–4958. https://doi.org/10.1007/s12010-022-03985-7
Lindsay GJ (1986) The significance of chitinolytic enzymes and lysozyme in rainbow trout (Salmo gairdneri) defence. Aquaculture 51(3–4):169–173
Lockyer S, Stanner S (2019) Prebiotics–an added benefit of some fibre types. Nutr Bull 44(1):74–91
Lomax AR, Calder PC (2008) Prebiotics, immune function, infection and inflammation: a review of the evidence. Br J Nutr 101(5):633–658. https://doi.org/10.1017/s0007114508055608
Luoma A, Markazi A, Shanmugasundaram R, Murugesan GR, Mohnl M, Selvaraj R (2017) Effect of synbiotic supplementation on layer production and cecal salmonella load during a Salmonella challenge. Poult Sci 96(12):4208–4216. https://doi.org/10.3382/ps/pex251
Markowiak P, Śliżewska K (2018a) The role of probiotics, prebiotics and synbiotics in animal nutrition. Gut Pathog 10:21. https://doi.org/10.1186/s13099-018-0250-0
Markowiak P, Śliżewska K (2018b) The role of probiotics, prebiotics and synbiotics in animal nutrition. Gut Pathogens 10(1). https://doi.org/10.1186/s13099-018-0250-0
Markowiak P, Śliżewska K, Nowak A, Chlebicz A, Żbikowski A, Pawłowski K, Szeleszczuk P (2019) Probiotic microorganisms detoxify ochratoxin A in both a chicken liver cell line and chickens. J Sci Food Agric 99(9):4309–4318. https://doi.org/10.1002/jsfa.9664
Martins CPC, Cavalcanti RN, Couto SM, Moraes J, Esmerino EA, Silva MC, Raices RSL, Gut JAW, Ramaswamy HS, Tadini CC, Cruz AG (2019) Microwave processing: current background and effects on the physicochemical and microbiological aspects of dairy products. Compr Rev Food Sci Food Saf 18(1):67–83. https://doi.org/10.1111/1541-4337.12409
McCormick SP (2013) Microbial detoxification of mycotoxins. J Chem Ecol 39(7):907–918. https://doi.org/10.1007/s10886-013-0321-0
Mercier G, Campargue C (2012) Partially hydrolyzed guar gum characterization and sensitive quantification in food matrices by high performance anion exchange chromatography with pulsed amperometric detection--validation using accuracy profile. J Chromatogr A 1262:180–187. https://doi.org/10.1016/j.chroma.2012.09.027
Mohan K, Rajan DK, Ganesan AR, Divya D, Johansen J, Zhang S (2023) Chitin, chitosan and chitooligosaccharides as potential growth promoters and immunostimulants in aquaculture: a comprehensive review. Int J Biol Macromol 251:126285. https://doi.org/10.1016/j.ijbiomac.2023.126285
Moita VHC, Kim SW (2022) Nutritional and functional roles of Phytase and Xylanase enhancing the intestinal health and growth of nursery pigs and broiler chickens. Animals (Basel) 12(23). https://doi.org/10.3390/ani12233322
Montserrat-Malagarriga M, Castillejos L, Salas-Mani A, Torre C, Martín-Orúe SM (2024) Use of different Synbiotic strategies to improve gut health in dogs. Animals (Basel) 14(23). https://doi.org/10.3390/ani14233366

Mutsvangwa T, Edwards I, Topps J, Paterson G (1992) The effect of dietary inclusion of yeast culture (Yea-Sacc) on patterns of rumen fermentation, food intake and growth of intensively fed bulls. Anim Sci 55(1):35–40

Newman AM, Arshad M (2020) The role of probiotics, prebiotics and Synbiotics in combating multidrug-resistant organisms. Clin Ther 42(9):1637–1648. https://doi.org/10.1016/j.clinthera.2020.06.011

Nguyen TT, Nguyen PT, Pham MN, Razafindralambo H, Hoang QK, Nguyen HT (2022) Synbiotics: a new route of self-production and applications to human and animal health. Probiotics Antimicrob Proteins 14(5):980–993. https://doi.org/10.1007/s12602-022-09960-2

Palangi V, Lackner M (2022) Management of enteric methane emissions in ruminants using feed additives: a review. Animals (Basel) 12(24). https://doi.org/10.3390/ani12243452

Pandey AK, Kumar P, Saxena M (2019) Feed additives in animal health. In: Nutraceuticals in veterinary medicine, pp 345–362

Patel AK, Singhania RR, Awasthi MK, Varjani S, Bhatia SK, Tsai ML, Hsieh SL, Chen CW, Dong CD (2021) Emerging prospects of macro- and microalgae as prebiotic. Microb Cell Factories 20(1):112. https://doi.org/10.1186/s12934-021-01601-7

Patel S, Goyal A (2012) The current trends and future perspectives of prebiotics research: a review. 3 Biotech 2(2):115–125. https://doi.org/10.1007/s13205-012-0044-x

Pattanaik AK, Kumar S, Rashmi H (2022) Probiotics, prebiotics, eubiotics and synbiotics for human and animal health and mitigation of antimicrobial resistance. Indian J Compar Microbiol Immunol Infect Dis 43(1):90–97

Patterson JA, Burkholder KM (2003) Application of prebiotics and probiotics in poultry production. Poult Sci 82(4):627–631. https://doi.org/10.1093/ps/82.4.627

Patterson R, Rogiewicz A, Kiarie EG, Slominski BA (2022) Yeast derivatives as a source of bioactive components in animal nutrition: a brief review. Front Vet Sci 9:1067383. https://doi.org/10.3389/fvets.2022.1067383

Pruszynska-Oszmalek E, Kolodziejski PA, Stadnicka K, Sassek M, Chalupka D, Kuston B, Nogowski L, Mackowiak P, Maiorano G, Jankowski J, Bednarczyk M (2015) In ovo injection of prebiotics and synbiotics affects the digestive potency of the pancreas in growing chickens. Poult Sci 94(8):1909–1916. https://doi.org/10.3382/ps/pev162

Qayum A, Rashid A, Liang Q, Wu Y, Cheng Y, Kang L, Liu Y, Zhou C, Hussain M, Ren X, Ashokkumar M, Ma H (2023) Ultrasonic and homogenization: An overview of the preparation of an edible protein-polysaccharide complex emulsion. Compr Rev Food Sci Food Saf 22(6):4242–4281. https://doi.org/10.1111/1541-4337.13221

Qin YQ, Wang LY, Yang XY, Xu YJ, Fan G, Fan YG, Ren JN, An Q, Li X (2023) Inulin: properties and health benefits. Food Funct 14(7):2948–2968. https://doi.org/10.1039/d2fo01096h

Rahman MRT, Fliss I, Biron E (2022) Insights in the development and uses of alternatives to antibiotic growth promoters in poultry and swine production. Antibiotics (Basel) 11(6). https://doi.org/10.3390/antibiotics11060766

Rashidinejad A, Bahrami A, Rehman A, Rezaei A, Babazadeh A, Singh H, Jafari SM (2022) Co-encapsulation of probiotics with prebiotics and their application in functional/synbiotic dairy products. Crit Rev Food Sci Nutr 62(9):2470–2494. https://doi.org/10.1080/10408398.2020.1854169

Rastall RA (2010) Functional oligosaccharides: application and manufacture. Annu Rev Food Sci Technol 1:305–339. https://doi.org/10.1146/annurev.food.080708.100746

Ringseis R, Eder K (2022) Heat stress in pigs and broilers: role of gut dysbiosis in the impairment of the gut-liver axis and restoration of these effects by probiotics, prebiotics and synbiotics. J Anim Sci Biotechnol 13(1):126. https://doi.org/10.1186/s40104-022-00783-3

Rinkinen M, Jalava K, Westermarck E, Salminen S, Ouwehand AC (2003) Interaction between probiotic lactic acid bacteria and canine enteric pathogens: a risk factor for intestinal enterococcus faecium colonization? Vet Microbiol 92(1–2):111–119. https://doi.org/10.1016/s0378-1135(02)00356-5

Roberfroid MB (2008) Prebiotics: concept, definition, criteria, methodologies, and products. In: Handbook of prebiotics. CRC Press, pp 57–86
Sabater-Molina M, Larqué E, Torrella F, Zamora S (2009) Dietary fructooligosaccharides and potential benefits on health. J Physiol Biochem 65(3):315–328. https://doi.org/10.1007/bf03180584
Salminen S, Collado MC, Endo A, Hill C, Lebeer S, Quigley EMM, Sanders ME, Shamir R, Swann JR, Szajewska H, Vinderola G (2021) The International Scientific Association of Probiotics and Prebiotics (ISAPP) consensus statement on the definition and scope of postbiotics. Nat Rev Gastroenterol Hepatol 18(9):649–667. https://doi.org/10.1038/s41575-021-00440-6
Sarkar S (2007) Potential of prebiotics as functional foods–a review. Nutr Food Sci 37(3):168–177
Schley PD, Field CJ (2002) The immune-enhancing effects of dietary fibres and prebiotics. Br J Nutr 87 Suppl 2:S221–S230. doi:https://doi.org/10.1079/bjnbjn/2002541
Schmitz S, Suchodolski J (2016) Understanding the canine intestinal microbiota and its modification by pro-, pre- and synbiotics - what is the evidence? Vet Med Sci 2(2):71–94. https://doi.org/10.1002/vms3.17
Scholz-Ahrens KE, Schaafsma G, van den Heuvel EG, Schrezenmeir J (2001) Effects of prebiotics on mineral metabolism. Am J Clin Nutr 73(2 Suppl):459s–464s. https://doi.org/10.1093/ajcn/73.2.459s
Schoster A, Weese JS, Guardabassi L (2014) Probiotic use in horses - what is the evidence for their clinical efficacy? J Vet Intern Med 28(6):1640–1652. https://doi.org/10.1111/jvim.12451
Seifert S, Watzl B (2007) Inulin and oligofructose: review of experimental data on immune modulation. J Nutr 137(11 Suppl):2563s–2567s. https://doi.org/10.1093/jn/137.11.2563S
Sharma AN, Chaudhary P, Grover CR, Kumar S, Mondal G (2024) Impact of synbiotics on growth performance and gut health in Murrah buffalo calves. Vet Res Commun 48(1):179–190. https://doi.org/10.1007/s11259-023-10194-y
Shoukry MM, El-Nomeary YAAE-F, Salman FM, Shakweer WME-S (2023) Improving the productive performance of growing lambs using prebiotic and probiotic as growth promoters. Trop Anim Health Prod 55(6). https://doi.org/10.1007/s11250-023-03752-8
Silva EK, Arruda HS, Eberlin MN, Pastore GM, Meireles MAA (2019) Effects of supercritical carbon dioxide and thermal treatment on the inulin chemical stability and functional properties of prebiotic-enriched apple juice. Food Res Int 125:108561. https://doi.org/10.1016/j.foodres.2019.108561
Sitanggang AB, Drews A, Kraume M (2016) Recent advances on prebiotic lactulose production. World J Microbiol Biotechnol 32(9):154. https://doi.org/10.1007/s11274-016-2103-7
Slavin J (2013) Fiber and prebiotics: mechanisms and health benefits. Nutrients 5(4):1417–1435. https://doi.org/10.3390/nu5041417
Śliżewska K, Nowak A, Barczyńska R, Libudzisz Z (2013) Prebiotyki-definicja, właściwości i zastosowanie w przemyśle. Żywność Nauka Technologia Jakość 1(86):5–20
Solis-Cruz B, Hernandez-Patlan D, Hargis BM, Tellez G (2019) Use of prebiotics as an alternative to antibiotic growth promoters in the poultry industry. In: Prebiotics and probiotics-potential benefits in nutrition and health. IntechOpen
Sori N, Kunnummal SP, Peddha MS, Khan M (2022) Prophylactic effect of pectic oligosaccharides against poly I: C- induced virus-like infection in BALB/c mice. J Food Biochem 46(12):e14459. https://doi.org/10.1111/jfbc.14459
Souza A, Gabardo S, Coelho RJS (2022) Galactooligosaccharides: physiological benefits, production strategies, and industrial application. J Biotechnol 359:116–129. https://doi.org/10.1016/j.jbiotec.2022.09.020
Spinler JK, Auchtung J, Brown A, Boonma P, Oezguen N, Ross CL, Luna RA, Runge J, Versalovic J, Peniche A, Dann SM, Britton RA, Haag A, Savidge TC (2017) Next-generation probiotics targeting Clostridium difficile through precursor-directed antimicrobial biosynthesis. Infect Immun 85(10). https://doi.org/10.1128/iai.00303-17
Sutoyo DA, Atmaka DR, Sidabutar L (2020) Dietary factors affecting firmicutes and bacteroidetes ratio in solving obesity problem: a literature review. Media Gizi Indones 15(2):94–109

Swanson KS, Gibson GR, Hutkins R, Reimer RA, Reid G, Verbeke K, Scott KP, Holscher HD, Azad MB, Delzenne NM, Sanders ME (2020) The International Scientific Association for Probiotics and Prebiotics (ISAPP) consensus statement on the definition and scope of synbiotics. Nat Rev Gastroenterol Hepatol 17(11):687–701. https://doi.org/10.1038/s41575-020-0344-2

Tang SG, Sieo CC, Kalavathy R, Saad WZ, Yong ST, Wong HK, Ho YW (2015) Chemical compositions of egg yolks and egg quality of laying hens fed prebiotic, probiotic, and synbiotic diets. J Food Sci 80(8):C1686–C1695. https://doi.org/10.1111/1750-3841.12947

Tawfick MM, Xie H, Zhao C, Shao P, Farag MA (2022) Inulin fructans in diet: role in gut homeostasis, immunity, health outcomes and potential therapeutics. Int J Biol Macromol 208:948–961. https://doi.org/10.1016/j.ijbiomac.2022.03.218

Tekin T, Dincer E (2023) Effect of resistant starch types as a prebiotic. Appl Microbiol Biotechnol 107(2–3):491–515. https://doi.org/10.1007/s00253-022-12325-y

Trukhachev VI, Buryakov NP, Shapovalov SO, Shvydkov AN, Buryakova MA, Khardik IV, Fathala MM, Komarova OE, Aleshin DE (2022) Impact of inclusion of multicomponent synbiotic Russian Holstein dairy cow's rations on Milk yield, rumen fermentation, and some blood biochemical parameters. Front Vet Sci 9:884177. https://doi.org/10.3389/fvets.2022.884177

Usman M, Noor M, Ahmed U, Khan A, Ahmad S, Khan A, Rehman A, Arshad U, Abdullah OM (2025) Prebiotic and probiotics supplementation used in animal nutrition have their effects on Gut health and animal performance. Indus J Biosci Res 3(2):577–585

Van Loo J, Vancraeynest D (2008) Prebiotics and animal nutrition. In: Handbook of prebiotics. CRC Press, pp 439–454

Wang K, Duan F, Sun T, Zhang Y, Lu L (2024) Galactooligosaccharides: synthesis, metabolism, bioactivities and food applications. Crit Rev Food Sci Nutr 64(18):6160–6176. https://doi.org/10.1080/10408398.2022.2164244

Wang L, Huo B, Huang L, Che L, Feng B, Lin Y, Xu S, Wu D, Fang Z (2022) Dietary supplementation with a mixture of herbal extracts during late gestation and lactation improves performance of sows and nursing piglets through regulation of maternal metabolism and transmission of antibodies. Front Vet Sci 9:1026088. https://doi.org/10.3389/fvets.2022.1026088

Wang Y (2009) Prebiotics: present and future in food science and technology. Food Res Int 42(1):8–12

Wellock IJ, Fortomaris PD, Houdijk JG, Kyriazakis I (2008) Effects of dietary protein supply, weaning age and experimental enterotoxigenic Escherichia coli infection on newly weaned pigs: health. Animal 2(6):834–842. https://doi.org/10.1017/s1751731108002048

Whittemore JC, Stokes JE, Laia NL, Price JM, Suchodolski JS (2018) Short and long-term effects of a synbiotic on clinical signs, the fecal microbiome, and metabolomic profiles in healthy research cats receiving clindamycin: a randomized, controlled trial. PeerJ 6:e5130. https://doi.org/10.7717/peerj.5130

Wolf BW, Firkins JL, Zhang X (1998) Varying dietary concentrations of fructooligosaccharides affect apparent absorption and balance of minerals in growing rats. Nutr Res 18(10):1791–1806. https://doi.org/10.1016/S0271-5317(98)00138-9

Xiong JQ, Kurade MB, Jeon BH (2018) Can microalgae remove pharmaceutical contaminants from water? Trends Biotechnol 36(1):30–44. https://doi.org/10.1016/j.tibtech.2017.09.003

Xu ZR, Hu CH, Xia MS, Zhan XA, Wang MQ (2003) Effects of dietary fructooligosaccharide on digestive enzyme activities, intestinal microflora and morphology of male broilers. Poult Sci 82(6):1030–1036. https://doi.org/10.1093/ps/82.6.1030

Zaman SA, Sarbini SR (2016) The potential of resistant starch as a prebiotic. Crit Rev Biotechnol 36(3):578–584. https://doi.org/10.3109/07388551.2014.993590

Zeng H, Chen P, Chen C, Huang C, Lin S, Zheng B, Zhang Y (2018) Structural properties and prebiotic activities of fractionated lotus seed resistant starches. Food Chem 251:33–40. https://doi.org/10.1016/j.foodchem.2018.01.057

Zhang Y, He J, Zeng H, Xu D, Li W, Wang Y (2025) Advances in prebiotic carbohydrate-based targeted delivery: overcoming gastrointestinal challenges for bioactive ingredients. Food Chem 466:142210. https://doi.org/10.1016/j.foodchem.2024.142210

Prebiotics as Gut Microbiome Modulators: Definitions, Mechanisms, and Emerging Applications

Maryam Fatima, Fiza Abbas, Samra, Hafsa Saeed, Ulfat Zahra, Sania Satti, Swaira Ashfaq, Nudrat Fatima, Muhammad Usama, and Shabana Naz

Abstract Prebiotics are non-digestible food components that selectively stimulate the growth and activity of beneficial microorganisms in the gastrointestinal tract, significantly contributing to host health. This chapter provides a comprehensive examination of prebiotics, encompassing their definition, classification, and dietary sources, as well as their biochemical mechanisms of action and associated health benefits. It delves into various types of prebiotics, such as fructans, galactooligosaccharides, resistant starches, and human milk oligosaccharides, emphasizing their unique characteristics and sources. The discussion elucidates the role of prebiotics in modulating gut barrier integrity, influencing immune system function, and enhancing digestive health outcomes. Furthermore, the chapter highlights the applications of synbiotics—combinations of prebiotics and probiotics—in animal health, emphasizing their potential to reduce antibiotic dependence and enhance productivity. The chapter concludes by outlining future research trajectories and challenges in optimizing the therapeutic utility of prebiotics for maintaining a balanced and resilient gut microbiome, advocating for further research and large-scale trials to unlock their full potential in both human and animal nutrition.

Keywords Prebiotics · Dietary Fibers · Synbiotic · Gut microbiota

1 Introduction

The term "prebiotics" was first used in 1995 by Gibson and Roberfroid to describe "non-digestible food components, which have a beneficial impact on the host by selectively stimulating the growth and proliferation of one or specific bacteria in the colon that significantly improve the host's health" (Glenn and Roberfroid 1995). At

M. Fatima · F. Abbas · Samra · H. Saeed · U. Zahra · S. Satti · S. Ashfaq · N. Fatima · M. Usama · S. Naz (✉)
Department of Zoology, Government College University, Faisalabad, Pakistan
e-mail: drshabananaz@gcuf.edu.pk

R. U. Khan, S. Naz (eds.), *Green Alternatives to Antibiotics in Animal Health*,
https://doi.org/10.1007/978-981-92-0309-3_9

that time, substances that could increase the number of bacteria, primarily Bifidobacteria and Lactobacilli, were also classified as prebiotics. In 2004, the definition was changed to "selectively fermented ingredients that specifically improve the activity and composition of gastrointestinal microflora and provide benefits to host health and well-being," which describes the conditions that exhibit beneficial effects on the host. The term "prebiotic" was later defined by the FAO Technical Meeting on Prebiotics in 2007 as "a non-viable food component that provides a health benefit on the host related with the transmission of microflora" and the Italian Ministry of Health's Guidelines on Probiotics and Prebiotics (2018) implemented this definition as well (Ministero Della Salute 2018). The items in this category are mostly composed of galacto-oligosaccharides (GOS), lactulose and short and long chain fructans (FOS and inulin). When introduced into a diet in small amounts (5–20 g/day), these substances promote the growth of lactobacilli and bifidobacteria, which are not the most prevalent intestinal microorganisms—with the exception of nursing infants (Guarino et al. 2020).

During a significant meeting of the International Scientific Association of Probiotics and Prebiotics (ISAPP) in 2017, prebiotics were in fact described as "a substrate that is selectively utilized by host microorganisms, conferring a health benefit" (Gibson et al. 2017).

1.1 Background

The high rate of morbidity and mortality in the world is caused by poor nutrition, a rise in food-borne illnesses, and the use of alcohol and tobacco. The rise in non-communicable diseases, including cancer, degenerative diseases, diabetes, coronary heart disease, gastrointestinal disorders, and chronic obesity, has sparked a renewed interest in prebiotics in recent years. Prebiotics are undigestible food ingredients, such as polysaccharide and non-digestible oligosaccharides (NDO), that can selectively promote the growth of beneficial microbes (lactic acid bacteria (LAB)) and have antagonistic activity against pathogenic microorganisms in the gastrointestinal tract (GIT), such as Salmonella spp. and Escherichia coli, to decrease their survival in the gut and improve the host's health (Scott et al. 2020)

The regulatory framework for prebiotics varies around the world. The Foods for Specified Health Use (FOSHU) Act, which was put into effect in Japan in the 1980s, marked the first documented regulation of other functional foods, including prebiotics. After the claims have been proven by science, the Japanese Ministry of Health, Labor, and Welfare certifies all foods and drinks that are known to have health benefits. After that, these foods are permitted to be sold commercially in Japan. But in Europe, the first countries to regulate functional foods were Sweden, the Netherlands, and the United Kingdom (Bamigbade et al. 2022)

1.2 *Prebiotic and Dietary Fibers*

In general, prebiotics are unique dietary fibers that act as a substrate for beneficial bacteria in the human gut. However, not all prebiotics are considered dietary fiber, as not all dietary fibers tend to exhibit prebiotic effects. Classification of a food ingredient as a prebiotic requires scientific characteristics that the ingredient must exhibit. These include (1) resistance to gastric acidity, hydrolysis by digestive enzymes, and GI absorption; (2) fermentation by intestinal microflora, which can be assessed in vitro by adding the appropriate carbohydrates to suspensions of colon content; and (3) improvement of intestinal bacterial growth that is advantageous to health and well-being (Slavin 2013).

In recent years, research on the impact of functional foods—specifically, prebiotics and probiotics—in human health has increased due to the growing understanding of the importance of gut bacteria in maintaining overall wellbeing. Functional foods are foods that provide more health advantages than just simple sustenance. These foods are prepared or supplemented with biologically active ingredients, such as vitamins, minerals, herbs, or other materials, in order to offer specific health benefits beyond their basic nutritional value (Gul et al. 2015).

Functional food is designed to improve overall health, reduce the risk of illness, or foster optimum well-being. Their focus is often on specific physiological processes or health concerns. Omega-3-enriched goods, probiotics (living bacteria that support gut health), prebiotics (fiber that stimulates probiotic growth), fortified meals (like calcium-fortified orange juice), and foods strong in antioxidants (like certain fruits and vegetables) are a few examples (Gul et al. 2015).

2 Types of Prebiotics

Prebiotics come in a variety of forms. The majority of prebiotics are composed of carbohydrates (oligosaccharides). Numerous studies have documented oligosaccharides, which are a subset of the carbohydrate family. Notably, some prebiotics are derived from fiber, therefore, not all of them are carbohydrates (Bamigbade et al. 2022). The most commonly investigated substances were oligosaccharide (20%), resistant starch (16%) and chemically synthesized fibers (15%). Inulin (13%), brain (13%), and cereal fiber (11%) in close succession (Hijova et al. 2019). In order to classify a compound as a prebiotic, there are the following requirements. (i) They must be resistant to gastric acidity and mammalian enzymes; (ii) They must be receptive to fermentation by gut bacteria; and (iii) They must enhance the viability and activity of beneficial microbes (Kaur et al. 2021). The various types of prebiotics are illustrated in Fig. 1.

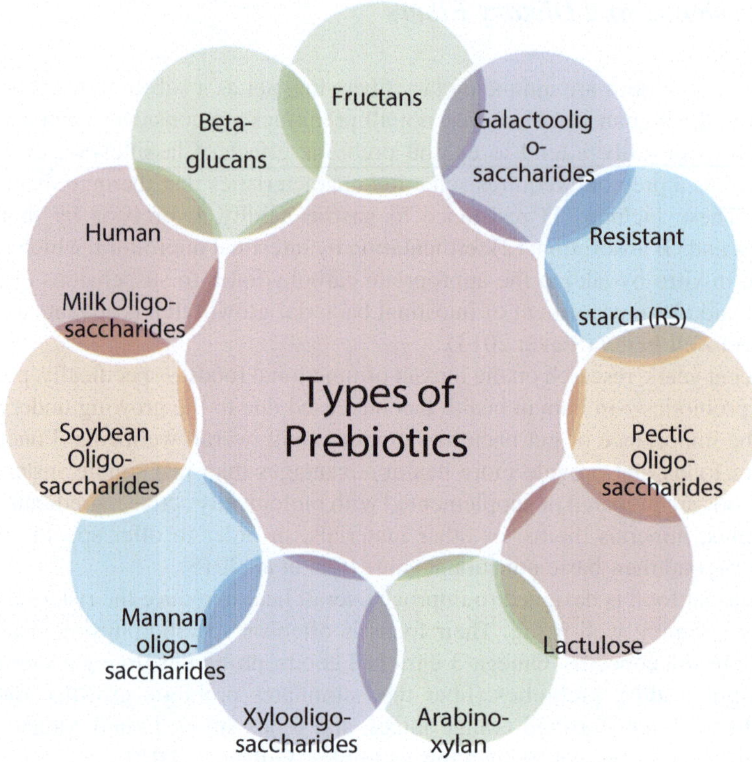

Fig. 1 Types of prebiotics

2.1 Fructans

Fructans consist of inulins and Fructooligosaccharides (FOS). Fructans are structurally composed of a linear chain of fructose joined by β (2→1) bond. Typically, they have β (2→1) linked terminal glucose units. FOS has a degree of polymerization (DP) of less than 10, whereas inulin can have a DP of up to 60. According to some earlier research, fructans have the ability to specifically promote lactic acid bacteria. However, some research conducted in recent years has indicated that one crucial factor in identifying the bacteria that can ferment fructans is their chain length. Fructans can therefore either directly or indirectly promote other bacterial species (Obayomi et al. 2024).

2.1.1 Inulin

Inulin is a polymer of fructose molecules connected by beta (2–1) glycosidic linkages. The fructans, a class of dietary fibers, include prebiotics like inulin. It is found naturally in many plants, but chicory root is one of the most abundant sources.

Significant levels of inulin can also be found in foods including Jerusalem artichokes, garlic, onions, leeks, asparagus, and bananas. It is classified as a fructans because its structure is composed of fructose units joined by a terminal glucose molecule. Longer inulin chains are referred to as polysaccharides, while shorter chains are known as oligofructans (Obayomi et al. 2024). Because of the beta configuration bonds between fructose monomers, inulin-type fructans are resistant to enzymatic hydrolysis by human salivary enzymes and small intestine digesting enzymes (Megur et al. 2022).

2.1.2 Fructooligosaccharides (FOS)

Fructooligosaccharides (FOS), a kind of prebiotic fiber, are composed of short chains of fructose molecules joined by a glucose molecule at one end (Obayomi et al. 2024). Cereal grains (barley, wheat, and oats), vegetables, fruits (artichokes, asparagus, bananas, garlic, leeks, and onions), and the blue agave plant are all sources of them (Kaur et al. 2021). They are composed of fructose units joined by beta (2–1) glycosidic linkages and glucose molecules at the end. The FOS chain may vary in length depending on the source and processing method. Like other prebiotic fibers, FOS can survive digestion in the upper gastrointestinal tract and reach the colon, where it is used as a substrate for fermentation by good bacteria (Obayomi et al. 2024).

2.2 Galactooligosaccharides (GOS)

GOS are a type of non-digestible oligosaccharide that results from lactose extension (Megur et al. 2022). They are composed of short galactose molecule chains. By promoting the proliferation of bifidobacteria, GOS helps maintain a balanced gut microbiota (Obayomi et al. 2024). They fall into two subgroups: (i) those that have extra galactose at positions C3, C4, and C6, and (ii) those that are made from lactose by enzymatic trans-glycosylation. The reaction conditions and the enzymes employed determine the product combination (Megur et al. 2022). Lactulose, an isomer of lactose, is the source of several GOSs. These GOSs made from lactulose are likewise regarded as prebiotics (Davani-Davari et al. 2019).

2.3 Resistant Starch (RS)

Resistant starch (RS) is a type of starch that is difficult for the upper gastrointestinal tract to digest. Since human pancreatic amylase cannot break down RS in the small intestine, it enters the colon and produces a substantial amount of butyrate, which may indicate that it is a prebiotic (Megur et al. 2022). It is composed of the two primary components of starch, amylose and amylopectin molecules. However, the structure of resistant starch limits its breakdown by human enzymes in the small intestine (Obayomi et al. 2024).

2.4 *Pectic Oligosaccharides (POS)*

Pectic oligosaccharides (POS) are derived from pectin, a polymer that is abundant in fruits and vegetable waste products and serves as a structural component of intra-cellular areas and plant cell walls (Megur et al. 2022). The type of fruit and its level of ripeness are two factors that can alter the structure of pectin. It is a complex polymer composed of chains of galacturonic acid molecules connected by side chains of various sugars, such as rhamnose, galactose, and arabinose (Obayomi et al. 2024). Saliva and stomach juice cannot break down pectin in humans. Moreover, pectin cannot be broken down in the small intestine by digestive enzymes such as rennet, trypsin, and pepsin. According to reports, pectin produces SCFAs during slow fermentation, which have prebiotic effects (Megur et al. 2022).

2.5 *Lactulose*

Lactulose is a manufactured non-digestible ketose disaccharide made up of fructose and galactose joined by a lactase-resistant bond. The enzyme β-galactosidase is utilized for the biocatalytic creation of lactulose, which is derived from lactose (milk sugar), which is chemically known as 4-O-β-d-galactopyranosyl-d-fructose. Since the human small intestinal mucosa lacks the enzymes necessary to break down lactulose, it passes through the large intestine unaltered. Colonic bacteria break down cellulose to produce monosaccharides, which are then converted to hydrogen, methane, and volatile fatty acids. Studies in humans have shown that lactulose has a considerably altered GM by increasing Lactobacillus, Streptococcus, and Bifidobacterium, and it has positive health effects (Megur et al. 2022).

2.6 *Arabinoxylan*

The cell walls of cereal grains, such as wheat, rye, barley, and oats, as well as some other plant sources, contain the hemicellulose polymer known as arabinoxylan. It is composed of arabinose side chains attached to xylose molecules. The basic backbone of the complex polymer arabinoxylan is composed of xylose units, to which arabinose side chains are joined. The ratio of xylose to arabinose units can vary depending on the source of the arabinoxylan. Arabinoxylan is classified as a prebiotic since it can withstand being broken down by human enzymes in the small intestine and travel all the way to the colon. It serves as a substrate for the fermentation of good bacteria such as Bifidobacterium and Lactobacillus species in the colon. This fermentation process produces short-chain fatty acids (SCFAs), such as butyrate, propionate, and acetate, which have several health benefits (Obayomi et al. 2024).

2.7 *Xylooligosaccharides (XOS)*

Oligosaccharides composed of glycosidic linkages joining xylose units are known as xylooligosaccharides (XOS). Xylooligosaccharides are groups of two to ten xylose units joined by β-(1–4) glycosidic linkages. Xylan, a hemicellulose found in plant cell walls, particularly in hardwoods and agricultural waste, is hydrolyzed to produce them. Particularly, XOS encourages the development and activity of beneficial bacteria in the gut, particularly those that are capable of fermenting xylooligosaccharides. A more diversified and balanced gut microbiota is the outcome of XOS's ability to support the growth of these bacteria by providing them with a fermentable substrate. XOS is found naturally in some plant foods, including fruits, vegetables, cereal grains, and bamboo shoots. However, enzymatic hydrolysis of xylan-rich materials such as corncobs, sugarcane bagasse, and wood chips is the most popular method of producing them. Numerous food products, including dairy products, beverages, baked goods, and dietary supplements, commonly contain functional components known as XOS (Obayomi et al. 2024).

2.8 *Mannan Oligosaccharides (MOS)*

Mannan oligosaccharides (MOS), a type of prebiotic fiber, are made from the cell walls of yeast or certain plant materials, such as yeast or certain types of seaweed. They consist of short chains of mannose molecules connected by glycosidic linkages. MOS are considered prebiotics because they survive digestion in the upper gastrointestinal tract and reach the colon undigested, where they serve as a substrate for probiotic bacterial fermentation. Mannose units joined by β-(1–4) glycosidic links form mannan. They may also contain side chains of other sugars, such as glucose or galactose, depending on their origin and processing method. The degree of polymerization (DP) in MOS can vary, ranging from short oligosaccharides to longer polysaccharides (Obayomi et al. 2024).

2.9 *Soybean Oligosaccharides (SOS)*

Soybean oligosaccharides (SOS) are oligosaccharides that are present in soybeans and include raffinose and stachyose. The gut microbiota hydrolyzes these oligosaccharides instead of the stomach or intestines' enzymes. SOS is effective for increasing the growth of the large intestine's Bifidobacteria. Because they have the same action as GOS, they are also referred to as bifidogenic. To assess the fermentation and prebiotic impact of soybean Okara on the fecal microbiota of healthy persons, an in vitro investigation was carried out. Inhibiting the growth of dangerous bacteria like Bacteroides and certain species of Clostridium, the results demonstrated an increase in the growth of Bifidobacteria and Lactobacilli. Additionally, compared to FOS, Okara's cell wall was difficult to digest, indicating the prolonged prebiotic impact (Kaur et al. 2021).

2.10 Human Milk Oligosaccharides (HMO)

HMOs are complicated, indigestible carbohydrates that were just categorized as prebiotics. Maternal breast milk has a high concentration of them (10–15 g/L). The mammary gland synthesizes the HMO chain, which ranges in length from three to fifteen carbohydrate units. HMOs are structurally made up of five monosaccharides: galactose, glucose, fucose, N-acetylglucosamine, and N-acetylneuraminic acid, also known as sialic acid. They are produced by the lactocyte's glycosyl transferases from a lactose core (galactose-β (1→4) glucose. Less than 50 HMOs, out of its wide varieties, are representatively abundant in human breast milk (Megur et al. 2022).

2.11 Beta-Glucans

Beta-glucans, a type of prebiotic fiber, are polysaccharides composed of glucose molecules joined by beta-glycosidic linkages. Some bacteria, fungi, yeasts, algae, and cereals have cell walls that contain them. New research suggests that beta-glucans have prebiotic benefits in addition to their primary role in immune system modulation. Beta-glucans are polysaccharides composed of glucose units connected by beta-glycosidic linkages. The molecular weight, degree of branching, and source of a beta-glucan can all affect its unique structure. Beta-glucans are frequently found in oats, barley, yeast (Saccharomyces cerevisiae), mushrooms (shiitake and maitake), and certain bacteria (Lactobacillus species). Beta-glucans are found in larger concentrations in the bran and germ layers of barley and oats, making them especially excellent sources of this dietary fiber. Supplements containing beta-glucan derived from mushrooms or yeast are also available. In many foods, including bread, pasta, cereal, and dietary supplements, beta-glucans are often used as beneficial components (Obayomi et al. 2024).

2.12 *Non-Carbohydrate Oligosaccharides*

While carbohydrates are more likely to fit the concept of prebiotics, some substances—like flavanols generated from cocoa—are not considered carbohydrates but are advised to be considered prebiotics. Flavanols have been shown to boost lactic acid bacteria in both in vitro and in vivo tests (Davani-Davari et al. 2019).

3 Mechanisms of Prebiotic Action

The primary mechanism of action of prebiotics involves fermentation in the colon, where prebiotics like fructo-oligosaccharides (FOS) and galacto-oligosaccharides (GOS) are fermented by good bacteria such as *Bifidobacterium* and *Lactobacillus*. Short-chain fatty acids (SCFAs), such as acetate, propionate, and butyrate, are produced during this fermentation process and are essential for a number of physiological functions, such as regulating immunological responses and preserving gut health (Guarino et al. 2020; You et al. 2022).

Prebiotics also aid in regulating the gut flora through fermentation. To preserve a balanced gut microbiome, they specifically encourage the growth of good bacteria while suppressing harmful ones. By suppressing dangerous bacteria and favoring helpful ones, this selective stimulation improves gut health in general (Guarino et al. 2020; Davani-Davari et al. 2019). By reducing the pH in the colon, the SCFAs generated during fermentation prevent harmful bacteria from growing and improve the absorption of vital minerals like calcium and magnesium. Furthermore, SCFAs have systemic effects that affect immunological and metabolic function (Guarino et al. 2020; You et al. 2022).

Prebiotics contribute to improving gut barrier function by increasing mucin secretion and enhancing tight junction integrity between epithelial cells. Preventing pathogen translocation and preserving gut health depend on a strong intestinal barrier. Furthermore, a mechanism known as cross-feeding occurs when certain good bacteria in a complex microbial ecosystem converts prebiotics into substances that act as substrates for other beneficial species. For example, Bifidobacteria fermentation products can promote the development of other advantageous species that generate butyrate, further enhancing gut health (Guarino et al. 2020; Davani-Davari et al. 2019).

In addition, the fermentation of prebiotics changes the pH levels in the gut, which can favorably shift microbial populations toward beneficial species like *Firmicutes*, which are known to produce butyrate. Overall, prebiotics work mainly through fermentation by the gut microbiota, which results in the generation of SCFA, and changes to the gut microbiota composition, improved immunological responses, improved gut barrier function, and changes to the gut environment. All of these processes work together to provide their many health advantages (You et al. 2022; Gowrishankar et al. 2021; Vuyst et al. 2004) (Fig. 2).

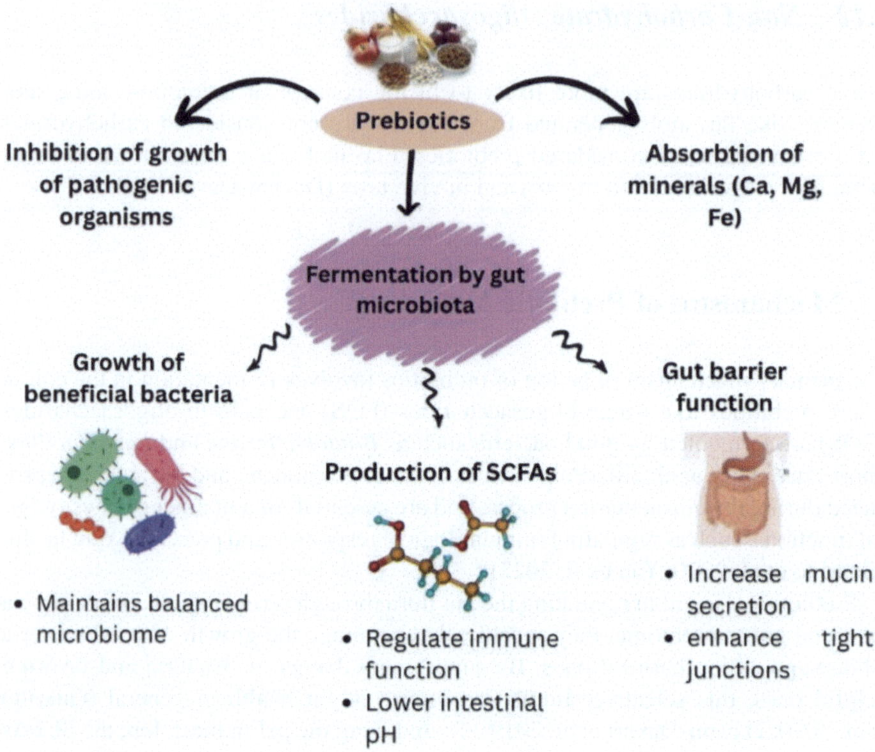

Fig. 2 Action mechanism of prebiotics

4 Synbiotic Applications in Animals

Synbiotics are a combination of probiotics and prebiotics that enhance the implantation and survival of live microbial dietary supplements in the gastrointestinal tract (GIT) by either metabolically activating or boosting the growth of the beneficial bacteria (Vuyst et al. 2004). The terms synbiotic and symbiotic are frequently used interchangeably. Symbiotic is an ecological interaction in a natural ecosystem where two species (the host and the symbiont) work together in symbiosis (Mounir et al. 2022).

The word itself was created by combining the Greek prefix "syn," which means "together," with the suffix "biotic," which means "pertaining to life" (Swanson et al. 2020). A synbiotic product beneficially affects the host in boosting the survival and implantation of live microbial dietary supplements in the gastrointestinal system by selectively stimulating the growth and/or activating the metabolism of one or a restricted number of health-promoting bacteria. Because the name "synbiotics" implies synergism, this term should be reserved for products in which the prebiotic compound(s) selectively favor the probiotic organism(s) (Pandey et al. 2015). The gut microbiota is notable for its persistence and functionality. A healthy gut community helps preserve homeostasis, which includes immunological control, and defends the host against invasive microbes. However, dietary changes, antibiotic use, aging, or infection can cause disturbances that result in a gut microbiota that can contribute to a variety of inflammatory, pathogenic, and metabolic problems, including metabolic syndrome, atopy, inflammatory bowel illnesses, and colorectal cancer (Swanson et al. 2020). In terms of avoiding GI disorders, synbiotics may be more effective than probiotics or prebiotics by themselves. Examples of synbiotics are lactobacilli and lactitol, Bifidobacteria and fructooligosaccharides, and Bifidobacteria and galactooligosaccharides (Chauhan and Chorawala 2012).

Synbiotics were developed to address potential issues with probiotic survival in the gastrointestinal tract and feature both probiotic and prebiotic qualities. Since a prebiotic's effects are primarily seen in the large intestine and a probiotic is essentially active in both the small and large intestine, the two may work in concert. Prebiotics are primarily employed as a selective medium for intestinal passage, fermentation, and probiotic strain expansion. There is evidence in the literature that probiotic microbes become more tolerant of environmental factors, such as temperature, pH, and oxygenation in an organism's intestine, as a result of using prebiotics. However, there is insufficient explanation of how an additional energy source that offers greater tolerance to these conditions works. Viable microbiological dietary supplements are produced by this mix of ingredients, and a favorable effect on the host's health is made possible by maintaining the right conditions.

When probiotics and prebiotics are taken together, the intestinal metabolic activity is modulated, the intestinal biostructure is maintained, beneficial microbiota is developed, and prospective pathogens in the gastrointestinal tract are inhibited. The inactivation of nitrosamines and carcinogens, as well as decreased amounts of unwanted metabolites, are the outcomes of using synbiotics. Short-chain fatty acids, ketones, carbon disulfides, and methyl acetates are significantly elevated when they are used, which may have a beneficial impact on the host's health. Regarding their medicinal effectiveness, synbiotics' antibacterial, anticancerogenic, and antiallergic qualities are beneficial. Additionally, they reduce diarrhea and constipation and inhibit intestinal degradation processes. As it turns out, synbiotics may be quite effective in preventing osteoporosis, lowering blood sugar and fat levels, controlling the immune system, and treating brain conditions linked to aberrant liver function (Markowiak and Śliżewska 2017).

4.1 Types of Synbiotics

There are two types of synbiotics:

4.1.1 Complementary Synbiotics

Probiotics and prebiotics (more than one of each type can be taken) make up complementary synbiotics, which function separately to provide one or more health advantages. The prebiotic has already shown selective consumption, so the study does not need to prove it.

4.1.2 Synergistic Symbiotic

A live microbe and a substrate that has been specifically used are the two parts of a synergistic synbiotic, and they are made to cooperate. The study must show a health advantage in addition to selective utilization (Gomez Quintero et al. 2022).

4.2 Animal Gut Health and Microbiota

The population of microorganisms that live in a specific environment along the gastrointestinal system is known as the gut microbiota. This community consists of bacteria, fungi, protozoa, archaea, and yeasts. Since bacteria account for more than 98% of the isolated genetic sequences found in the gut, the word and the most of the research in this field mostly apply to the bacteria found in the gastrointestinal system. Despite more than ten years of dedicated research, the gut has the most complex and prolific microbiota of any part of the body, with over 500 bacterial species found to date, many of which are still unclassified, and the overall relevance of the microbiota is still unclear.

It is believed that the enteric microbiota is important for the digestion and absorption of nutrients, the production of volatile fatty acids, amino acids, and vitamins, the preservation of intestinal mucosal integrity and gut peristalsis, and the development of the enteric immune system, which includes the organization of Peyer's patches, isolated lymphoid follicles, and the secretion of antimicrobial peptides by the epithelium. Additionally, the intestinal microbiota interacts with the brain, lungs, skin, and liver, among other organ systems, impacting their functionality in hitherto unidentified ways and pointing to the potential links between gastrointestinal dysbiosis and other medical disorders (Turner 2018). Together with the host, the gastrointestinal tract and co-existing microbiota of vertebrates have evolved to become adapted to meet the dietary requirements and preferences of each species (de Jonge et al. 2022).

4.3 Applications of Synbiotics

4.3.1 Antioxidant Properties

Various substances that have the ability to directly or indirectly counteract oxidative agents are known as antioxidants. Antioxidant characteristics and actions are thought to alleviate disorders linked to oxidative stress by preventing the negative effects of ROS and RNS. Consuming foods high in antioxidants or taking dietary supplements might boost the body's antioxidant defenses, preventing or slowing the progression of many chronic diseases. Although probiotics have long been known to be good for human health, especially because of their antioxidant qualities, research on the function of synbiotic antioxidants is still in its infancy. The mechanisms of antioxidant action are really influenced by the effects of probiotics or prebiotics alone, and particularly by the interactions of both within synbiotic preparations. In order to neutralize free radicals, each element is essential. Synbiotics' capacity to activate and translocate nuclear factors has been connected to the mechanism behind their antioxidant properties. These generate important antioxidant molecules, detoxify the generation of free radicals and singlet oxygen, and trigger the expression of the antioxidant defense enzymatic system. A number of important vitamins and regulators were also found in synbiotic dairy products, according to recent studies. These consist of antioxidants, water-soluble vitamins, and GSH, a crucial tripeptide that directly neutralizes singlet oxygen, hydroxyl radicals, and superoxide radicals chemically (Mounir et al. 2022).

4.3.2 Increase in Immunity

The intestines and other gut-related organs contain more than half of the cells that produce antibodies to combat illness. Thus, having a healthy gut flora—that is, the proper balance of beneficial bacteria and other organisms in your intestines—may reduce your risk of getting sick. So, by controlling inflammation and balancing gut flora, synbiotics strengthen the immune system.

Both humoral and cellular immune responses, both innate and adaptive, can be modulated by synbiotics. The generation of immunoglobulin-producing cells reflects the mechanism by which synbiotics alter humoral immune responses. Without triggering systemic immune responses, synbiotic-induced immune responses in the gut flora primarily boost the quantity of IgA-producing cells (Athiyyah et al. 2019).

4.3.3 Promotion of Growth

By selectively promoting the growth and/or activating the metabolism of one or a small number of health-promoting bacteria, synbiotics improve the survival and implantation of live microbial dietary supplements in the gastrointestinal tract, thereby improving host welfare. Consumers' growing health concern has resulted in a rise in the use of functional meals and drinks. Many retailers now provide functional food products that have significant health benefits. The demand for these items is also increasing as a result of changes in dietary and lifestyle habits. Synbiotics improve host health by specifically fostering growth and triggering microbial metabolism in the colon (Parhi et al. 2024).

To treat gastrointestinal disorders and strengthen the host's immune system, prebiotics, probiotics, and their mixtures may be utilized as an alternative. Prebiotics are oligomers that the gut microorganisms can employ to speed up their growth and development since they cannot be broken down by digestive enzymes. By shielding the intestinal walls from infections and limiting the proliferation of microorganisms in the GI tract, they encourage the development and activity of good bacteria. For health advantages, the most popular prebiotics are carbohydrate substrates such as dietary fibers or oligosaccharides. So, by enhancing digestion and nutrient absorption and assisting in the establishment of a balanced population of beneficial bacteria in the gastrointestinal tract, synbiotics promote growth (Sharma et al. 2023).

4.3.4 Stress Reduction

An increasing number of studies have shown that the gut microbiota is important for either preventing or causing neurological disorders, such as Parkinson's disease. It has been proposed that gut and brain health could be enhanced by modifying the gut environment with either probiotics or prebiotics. A synbiotic is a combination of prebiotics and certain probiotics that work together to support beneficial intestinal bacteria. By changing the makeup of gut microflora, generating short-chain fatty acids, and lowering endotoxin concentrations, gut microbiota targeted interventions have positive effects that reduce inflammation and oxidative stress (Mehrabani et al. 2023).

4.3.5 Anti-Inflammatory

Synbiotics are used to treat chronic illnesses by influencing the immune system's capacity to control inflammation and the anti-inflammatory response. Systemic inflammation may result from the release of pro-inflammatory mediators, including TNF-β, C-reactive protein (CRP), and ILs (i.e., IL-6 and IL-1) brought on by prolonged immune system stimulation. Even though dietary fibers like synbiotics have anti-inflammatory qualities, a number of research have demonstrated a negative correlation between dietary fiber intake and the biomarkers of systemic inflammation. Research has also looked into the effects of probiotics and synbiotics on oxidative stress and inflammation biomarkers in various disease conditions. The synthesis of metabolites by the beneficial bacteria can regulate the transition of immunological markers from a pro-inflammatory to an anti-inflammatory phenotype (Verma et al. 2024).

4.3.6 Anti-Drug Resistant

Concerns concerning drug residues in meat products and the emergence of bacteria resistant to antibiotics have grown as a result of the use of antibiotics in animal agriculture. Thus, it is now crucial to look for new alternatives in antimicrobial therapy for animal production. Natural product-based treatments, such as using prebiotics, probiotics, and synbiotics to target the gut microbiota (or microbiome), are receiving special attention. Since synbiotics are a synergistic combination of prebiotics and probiotics, they might be more effective than both probiotics and prebiotics. Probiotics are live microorganisms that help the GIT's microbiota balance, prevent harmful bacteria from growing, aid in food digestion and nutrient absorption, and strengthen the immune system (Mohammed et al. 2022).

A particularly promising strategy for lowering the development of multidrug resistance appears to be dietary control for the prevention of common illnesses. By directly inhibiting drug-resistant microorganisms or indirectly by lowering the risk of infection, functional foods containing probiotics—either living or non-living biological elements—may help mitigate antimicrobial resistance. By promoting the colonization of beneficial microorganisms in the gut, improving the metabolism of short-chain fatty acids, and promoting the growth of intestinal epithelium and mucosal immunity, prebiotics and dietary fibers have been shown to enhance gut health. A healthy gut environment lowers the likelihood of infection or dysbiosis, which may be a useful tactic to lessen the side effects of many medications (Elshaghabee and Rokana 2022).

5 Prebiotics: Future Hallmark of Science and Society

Being beneficial and viable in nature, prebiotics are proving themselves to be revolutionary in every moving step of life. The least pivotal roles that these prebiotics can play are reducing toxins in the body, cardiovascular issues, and enhancing immunity by diminishing serious health problems and colon cancer, which is the most acknowledged health problem in the present era. Most importantly, these prebiotics, while weighing with multiple health concerning benefits, have antioxidant properties and can even support the body against DNA damage. Obesity, which has been contemplated as the root problem of all diseases, sparks constipation, is a major issue, then links with numerous other dilemmas, and gives rise to cancer ultimately. The following future outcomes can be observed by keenly working on prebiotics:

5.1 Amalgamation of Different Sources for Well-Being

Tailoring of microbiome is being established for the improvement of health outcomes. Personalized prebiotics are thought to be good source along with synbiotic amalgamation. The unfold spring of future conduct involves the correct usage of plants and animals for the benefit of humanity. Worth mentioning sources include plants, typically seaweeds, polyphenols from fruits and grains, etc.

Compared to probiotics alone, synbiotics, a blend of probiotics and prebiotics, have demonstrated more promise in boosting host health and intestinal function. Researchers are becoming more interested in the use of synbiotics in the treatment of diseases, especially in the non-pharmaceutical management of hypertension. Medication is the main treatment for hypertension, a chronic illness that frequently has adverse consequences like gastrointestinal distress and exhaustion. Hence, formulation of synbiotics can be a game change. For these clinical trials are needed to be performed for the formulation of multiple sources. Immunomodulatory advantages of prebiotics, probiotics and synbiotics deliberately strengthen the aqua culture species.

The combination of multiple oligosaccharides and its byproducts with different bioactive compounds is being focused on with deep concern. Plant derived polysaccharide are broken down in the intestine by enzymes that break down polysaccharides and are emanated by gut microbiota (Wang et al. 2024). Xylooligosaccarides, isomaltooligosaccarides, chitooligosaccarides, epilactose, raffinose, lactosucrose, and glucomannans are recently derived carbohydrate prebiotics (Guan et al. 2024). These newly derived sources alter the gut microbiota's makeup and metabolic outcomes enhance overall health benefits and delaying the onset of disease.

5.2 Microbiome Targeted Therapies

In the future, microbiome therapies can lead the world toward better immunity and long-term exposure to lethal diseases. Treatments that typically target the gut microbiota help people to cure from hepatic impairments. To further understand this concept of targeted therapies, researchers are plotting various tailored compounds to be tested on the gut for its proper working. To elaborate, the bacterial strains, time duration, and dosage timing must be known. A proper series of information is needed to be planned. For example, NAFLD can be treated with microbiome therapies. Hence, MTT's are being proven as a promising outcome for the diagnosis and cure of many diseases. They possess striking impacts on curing cholesterol level, blood pressure, and other life-threatening issues.

With the manipulation of prebiotics with medicinal precision, in future, various problems can be targeted. For sound voyage of health, healthcare workers can work on prebiotics strategies that support individual's health for the diseases like obesity, IBD, diabetes and even neurological sufferings. Researchers have designed various drugs potential therapeutic targets for the creation of new generation anti-diabetic medication (Kaur et al. 2018). Prebiotics can strongly support healthy microbes, which ultimately provides immunity for the cure of multiple diseases like viruses and chronic diseases.

5.3 Elaboration of Mechanism

Researchers can further work on the understanding of mechanisms by which prebiotics work, which alter the gut microbiota and its diversity of existence, microbial metabolites, and epithelial nature of the concerned organ. For example, for neurological disorders, gut–brain-axis are needed to be understood for the detailed formulation of prebiotics. This can effectively influence the gut–brain axis and generate metabolites for having powerful impacts on brain function.

5.4 Large-Scale Trials

Commercialization of prebiotics plays a vital role in the well-being of humans. Large-scale application of these prebiotics takes a major direction for health benefits. Synthetic and novel bioengineered prebiotics can be designed to target a specific microbial population. When the matter of diet comes, proper campaigning and public awareness about the health concerning items with prebiotics should be addressed. Prebiotics products in the market should be given to the public via exact validation and verification procedures. Prebiotics items in the market should be gone through a series of whole monitoring so that no chances of fake items may be entertained. Internationally, the varied regulatory frameworks and nebulous prebiotic necessities have resulted in product outcomes like pathogenic contamination, exaggerated claims, incorrect labeling, and fake trademarks for prebiotic products (Liang et al. 2024). It can take much of the burden of the economic crisis.

Hence, prebiotics being the hallmark in future science and society can mold societal infrastructure into a sounder and more sophisticated lifestyle by reducing maximum diseases and its bioengineered products in markets serve as a paradigm shift for humanity. It is our duty to make sure the usage of these prebiotics for the safety and efficacy of humans.

References

Athiyyah AF, Widjaja NA, Fitri P, Setiowati A, Darma A, Ranuh R, Sudarmo SM (2019) Effects of a multispecies synbiotic on intestinal mucosa immune responses. Iran J Microbiol 11(4):300

Bamigbade GB, Subhash AJ, Kamal-Eldin A, Nyström L, Ayyash M (2022) An updated review on prebiotics: insights on potentials of food seeds waste as source of potential prebiotics. Molecules 27(18):5947. https://doi.org/10.3390/molecules27185947
Chauhan SV, Chorawala MR (2012) Probiotics, prebiotics and synbiotics. Int J Pharma Sci Res 3(3):711
Davani-Davari D, Negahdaripour M, Karimzadeh I, Seifan M, Mohkam M, Masoumi SJ et al (2019) Prebiotics: definition, types, sources, mechanisms, and clinical applications. Foods 8(3):92
de Jonge N, Carlsen B, Christensen MH, Pertoldi C, Nielsen JL (2022) The gut microbiome of 54 mammalian species. Front Microbiol 13:886252
Elshaghabee FM, Rokana N (2022) Mitigation of antibiotic resistance using probiotics, prebiotics and synbiotics. A review. Environ Chem Lett 20(2):1295–1308
Gibson GR, Hutkins R, Sanders ME, Prescott SL, Reimer RA, Salminen SJ, Scott K, Stanton C, Swanson KS, Cani PD et al (2017) Expert consensus document: The International Scientific Association for Probiotics and Prebiotics (ISAPP) consensus statement on the definition and scope of prebiotics. Nat Rev Gastroenterol Hepatol 14:491–502
Glenn G, Roberfroid M (1995) Dietary modulation of the human colonic microbiota: Introducing the concept of prebiotics. J Nutr 125:1401–1412
Gomez Quintero DF, Kok CR, Hutkins R (2022) The future of synbiotics: rational formulation and design. Front Microbiol 13:919725
Gowrishankar S, Kamaladevi A, Pandian SK (2021) Prebiotics mechanism of action: an overview. In: Advances in probiotics. Academic Press, pp 137–148
Guan F, Fu G, Ma Y, Zhou L, Li G, Sun C, Zhang T (2024) Spirulina polysaccharide-based prebiotic foods preparations-a promising approach for modulating gut microbiota and improving health. J Funct Foods 116:106158
Guarino MPL, Altomare A, Emerenziani S, Di Rosa C, Ribolsi M, Balestrieri P, Iovino P, Rocchi G, Cicala M (2020) Mechanisms of action of prebiotics and their effects on gastro-intestinal disorders in adults. Nutrients 12(4):1037. https://doi.org/10.3390/nu12041037
Gul K, Singh AK, Jabeen R (2015) Nutraceuticals and functional foods: the foods for the future world. Crit Rev Food Sci Nutr 56(16):2617–2627. https://doi.org/10.1080/10408398.2014.903384
Hijova E, Bertkova I, Stofilova J (2019) Dietary fibre as prebiotics in nutrition. Central Eur J Public Health 27(3):251–255
Kaur P, Mittal A, Nayak SK, Vyas M, Mishra V, Khatik GL (2018) Current strategies and drug targets in the management of type 2 diabetes mellitus. Curr Drug Targets 19(15):1738–1766
Kaur AP, Bhardwaj S, Dhanjal DS, Nepovimova E, Cruz-Martins N, Kuca K et al (2021) Plant prebiotics and their role in the amelioration of diseases. Biomolecules 11(3):440
Liang D, Wu F, Zhou D, Tan B, Chen T (2024) Commercial probiotic products in public health: Current status and potential limitations. Crit Rev Food Sci Nutr 64(19):6455–6476
Markowiak P, Śliżewska K (2017) Effects of probiotics, prebiotics, and synbiotics on human health. Nutrients 9(9):1021
Megur A, Daliri EBM, Baltriukienė D, Burokas A (2022) Prebiotics as a tool for the prevention and treatment of obesity and diabetes: classification and ability to modulate the gut microbiota. Int J Mol Sci 23(11):6097
Mehrabani S, Khorvash F, Heidari Z, Tajabadi-Ebrahimi M, Amani R (2023) The effects of synbiotic supplementation on oxidative stress markers, mental status, and quality of life in patients with Parkinson's disease: a double-blind, placebo-controlled, randomized controlled trial. J Funct Foods 100:105397
Ministero Della Salute. Linee Guida su Probiotici e Prebiotici. Revisione Marzo 2018. Available online: https://sumailombardia.info/sanita-nazionale/ministero-salute/2018/05/07/linee-guida-su-probiotici-e-prebiotici-revisione-marzo-2018/
Mohammed A, Hu J, Murugesan R, Cheng HW (2022) Effects of a synbiotic as an antibiotic alternative on behavior, production performance, cecal microbial ecology, and jejunal histomorphology of broiler chickens under heat stress. Plos One 17(9):e0274179

Mounir M, Ibijbijen A, Farih K, Rabetafika HN, Razafindralambo HL (2022) Synbiotics and their antioxidant properties, mechanisms, and benefits on human and animal health: a narrative review. Biomolecules 12(10):1443

Obayomi OV, Olaniran AF, Owa SO (2024) Unveiling the role of functional foods with emphasis on prebiotics and probiotics in human health: a review. J Funct Foods 119:106337

Pandey KR, Naik SR, Vakil BV (2015) Probiotics, prebiotics and synbiotics-a review. J Food Sci Technol 52:7577–7587

Parhi P, Liu SQ, Choo WS (2024) Synbiotics: effects of prebiotics on the growth and viability of probiotics in food matrices. Bioact Carbohydr Diet Fibre:100462

Scott KP, Grimaldi R, Cunningham M, Sarbini SR, Wijeyesekera A, Tang ML, Lee JY, Yau YF, Ansell J, Theis S (2020) Developments in understanding and applying prebiotics in research and practice—an ISAPP conference paper. J Appl Microbiol 128:934–949

Sharma AN, Chaudhary P, Kumar S, Grover CR, Mondal G (2023) Effect of synbiotics on growth performance, gut health, and immunity status in pre-ruminant buffalo calves. Sci Reports 13(1):10184

Slavin J (2013) Fiber and prebiotics: mechanisms and health benefits. Nutrients 5(4):1417–1435

Swanson KS, Gibson GR, Hutkins R, Reimer RA, Reid G, Verbeke K et al (2020) The International Scientific Association for Probiotics and Prebiotics (ISAPP) consensus statement on the definition and scope of synbiotics. Nat Rev Gastroenterol Hepatol 17(11):687–701

Turner PV (2018) The role of the gut microbiota on animal model reproducibility. Anim Models Exper Med 1(2):109–115

Verma B, Ashique S, Mishra N, Kumar N, Tyagi N, Kumar S et al (2024) Role of synbiotics on modulation of inflammation. In: Synbiotics in human health: biology to drug delivery. Springer Nature Singapore, Singapore, pp 25–54

Vuyst LD, Avonts L, Makras L (2004) Probiotics, prebiotics and gut health

Wang X, Li X, Zhang L, An L, Guo L, Huang L, Gao W (2024) Recent progress in plant-derived polysaccharides with prebiotic potential for intestinal health by targeting gut microbiota: a review. Crit Rev Food Sci Nutr 64(33):12242–12271

You S, Ma Y, Yan B, Pei W, Wu Q, Ding C, Huang C (2022) The promotion mechanism of prebiotics for probiotics: A review. Front Nutr 9:1000517

Advancing Animal Nutrition with Probiotics: Targeted Benefits and Pathogen Inhibition Mechanisms

Saima Naz, Ahmad Manan Mustafa Chatha, Qudrat Ullah, Sobia Malik, Muniba Maryam, and Urwah Ishaque

Abstract Probiotics, which are live microorganisms like bacteria and yeasts, give health benefits that go beyond basic nutrition when ingested in suitable amounts. These microbes influence the gastrointestinal microbiota by restoring balance and enhancing activity, playing a critical role in maintaining gut homeostasis. However, factors such as dietary and management practices can disrupt gut microbial communities, negatively affecting the health and performance of livestock animals. Probiotics or their byproducts can also modulate immune responses by interacting with T cells, antigen-presenting cells, and other immune components, giving them immunoregulatory properties. This chapter delves into the species-specific effects of probiotics on ruminants, livestock, aquaculture species, and monogastric animals, exploring their documented implications for animal nutrition, immune health, and overall performance.

Keywords Probiotics · Antimicrobial components · Pathogens · Microbial balance · Gut Microbiota

---

S. Naz (✉) · S. Malik · M. Maryam · U. Ishaque
Department of Zoology, Government Sadiq College Women University, Bahawalpur, Punjab, Pakistan
e-mail: saima.naz@gscwu.edu.pk

A. M. M. Chatha
Department of Entomology, Faculty of Agriculture and Environment, The Islamia University of Bahawalpur, Bahawalpur, Punjab, Pakistan

Q. Ullah
Department of Theriogenology, Faculty of Veterinary and Animal Sciences, Cholistan University of Veterinary and Animal Sciences, Bahawalpur, Punjab, Pakistan

R. U. Khan, S. Naz (eds.), *Green Alternatives to Antibiotics in Animal Health*,
https://doi.org/10.1007/978-981-92-0309-3_10

1 Introduction

Probiotics are defined as "live microorganisms that, when administered in sufficient amount, provide health benefits to the host" (Reid 2016). Initially, probiotics were introduced as microbial feed supplements (Inchingolo et al. 2023). Recently, a European expert group broadened the definition to include mechanisms beyond those mediated solely by microflora (Anee et al. 2021). Werner Kollath introduced the word "probiotic" in 1953. The term is derived from the Latin word pro and the Greek word βιo, meaning "for life." Probiotics are active compounds that play important roles in promoting a variety of health qualities (Gasbarrini et al. 2016). Over time, probiotics became widely known. Fermented foods based on probiotics have gained appeal again since the early 2000s, primarily because of claims about their health advantages. Fermented foods have been linked to the prevention of obesity, lactose intolerance, gastrointestinal disorders, and irritable bowel syndrome, in addition to chronic diarrhea, allergies, dermatitis, and bacterial and viral infections, all of which are closely associated with an unhealthy lifestyle (Maftei et al. 2024).

Recent and ongoing advances in microbiome and microbiota science have opened up new research opportunities for probiotics. The unique varieties, methods, and uses investigated thus far, as well as those now under investigation, have the potential to transform our scientific understanding of probiotics' nutritional roles and impact on human health care (Fahmy 2024).

The bulk of common probiotics comes from the lactobacilli category, which has recently been divided into 25 genera. *Bacillus acidophilus*, *Lactobacillus gasseri, Lacticaseibacillus rhamnosus, Bifidobacterium longum,* and *Bifidobacterium bifidum* are a few of these species. Additionally, several strains of other bacterial species, including *Lactococcus lactis*, *Pediococcus mesenteroides, Escherichia coli*, and even some strains of specific yeasts, including *Saccharomyces cerevisiae var. boulardii*, are considered probiotics (Fijan 2014).

Probiotics work in a variety of ways to achieve their goals. Probiotics enhance an animal's ability to function and produce while also inhibiting and controlling intestinal infections. The primary probiotic mode of action includes inhibiting pathogen adherence, creating antimicrobial components such as bacteriocins and defensins, competitively rejecting harmful bacteria, enhancing barrier function, reducing luminal pH, and altering the immune system. Because probiotics block dangerous microorganisms, they improve health problems. For instance, *Lactobacillus rhamnosus* and *Lactobacillus plantarum* may inhibit *Esherichia coli* adherence in the digestive tract (Maldonado Galdeano et al. 2019).

This chapter will discuss species-specific interactions of probiotics concerning their antimicrobial activities and immune modulatory effects in animal nutrition. Through an analysis of these areas, we hope to provide a comprehensive knowledge of probiotics' function in improving animal health and probiotic yield.

2 Rationale of the Study

In both the medical agriculture fields, bacterial antimicrobial resistance has become a serious problem globally. Over time, the cattle and poultry industries' usage of antibiotics results in the development of dangerous bacterial strains that endanger food safety and public health. As such, urgent steps are necessary to develop alternative strategies to control harmful bacterial activities. Recent studies explored that probiotics have the potential to control harmful bacterial activities and can be used as conventional antibiotics alternative. Probiotics can actually colonize in the gastrointestinal tract of living organisms and cause therapeutic effects. Probiotics can actually colonize in the gastrointestinal tract of living organisms and cause therapeutic effects. There exist probiotic cultures that have significant performance improvements in living organisms, and defined cultures might sometimes offer an attractive substitute for conventional antimicrobial therapy. Because of public health concerns, the use of probiotics as a substitute for antibiotics in living organisms has become an area of interest. The function of probiotics as antibiotic alternatives in different animals will be the main topic of this chapter.

We will explore antimicrobial properties, immune modulatory effects, and Potential applications in animal nutrition. This will offer a comprehensive overview of the benefits of overcoming microbial resistance and improving health outcomes.

3 Species-Specific Effects and Benefits

Depending on the species, probiotics have different effects. Farmers give fish, ruminants, and poultry probiotic feed additives. Gram-positive bacteria make up the majority of probiotics, although they can also contain yeast, fungus, and gram--negative bacteria (Anee et al. 2021). The suppression of pathogenic processes in animals is largely dependent on the competition between probiotic strains for nutrition and adhesion sites. As biological antagonists, probiotics compete with infections for the same vital nutrients needed for colonization and growth. Because of this rivalry, dangerous bacteria are unable to obtain nutrients, which hinders their capacity to cause diseases (Collado et al. 2010). Here are some examples of the effect of probiotics on different species.

3.1 Ruminants (e.g., Cattle, Buffalo)

The world's supply of the highest quality and quantity of proteins for human consumption, livelihood subsistence, and food security is largely derived from ruminant production (Reuben et al. 2022). However, there have been several difficulties associated with the phase-out of antibiotics in animal agriculture, including poor ruminant growth, performance, and nutrient utilization, as well as pathogen colonization, dysbiosis, and food safety concerns (Cone et al. 2021). Probiotics, also known as direct-fed microbials, are live microbial strains that, when administered in sufficient quantities, boost the host's nutrition and health. They are quickly emerging as a sustainable, safe, natural, and practical substitute for antibiotics (Reuben et al. 2022).

Many studies conducted on tropical animals proved that supplementation of probiotics to the ruminant's diet enhances the growth of beneficial microbes in the rumen. Improvement in the diet and digestion results in improved nutrition, milk yield, reproductive efficiency, and feed efficiency, as well as immunopositive responses. Use of probiotics has also been revealed to reduce unfavorable effects on the environment, including mitigation of methane generation from anaerobic fermentation by tropical feedstuff (Kulkarni et al. 2022). Probiotic work by inhabiting the GIT and assisting with the indigenous natives in the alimentary tract in the digestive tract of an animal. By encouraging the growth of advantageous mucosal microflora and preventing pathogens from colonizing the mucosa, they also indirectly improve mucosal immunity (Nalla et al. 2022). Additionally, ruminant digestive systems benefit from probiotics in terms of regulating acidity, enhancing comfort, lowering methanogenesis, promoting the intestine, and the rumen development, enhancing nutrient absorption, and boosting feed conversion ratio (Abd El-Trwab et al. 2016).

The commensal microbiota in the gastrointestinal tract is essential for forming the internal architecture and assisting immunomodulatory activities in a mutually beneficial way. Numerous alien microbes start colonizing ruminants as soon as they are born. These commensal microbes create important chemicals that aid in the host's immune system development (Raabis et al. 2019). Maintaining intestinal homeostasis and offering a defense against infections depend on the host's mucosal immune system (Yan and Polk 2011). Probiotics have positive immunomodulatory effects and are essential for maintaining the immunological and fragile intestinal mucosal cells (Raabis et al. 2019).

3.2 Poultry (e.g., Chicken, Turkey)

Due in large part to worldwide trends like the elimination or restriction of antibiotic growth boosters in chicken feed, the use of probiotics in poultry production has increased significantly (Shini and Bryden 2021). Because of the established advantages of probiotics and the rising demand for antibiotic-free chicken, their use in poultry has grown gradually over time. Probiotic bacteria are commonly found in poultry and have well-established advantages. Bifidobacterium, Lactococcus, Lactobacillus, Bacillus, Streptococcus, and yeasts, including Candida, are common probiotic microorganisms used in poultry. Important factors to consider when choosing probiotic strains are their ability to withstand gastrointestinal disorders, stick to the gastrointestinal mucosa, and keep infections out (Jha et al. 2020).

This is because studies have demonstrated that probiotics can alter intestinal immunity pathways directly as well as through their interactions with the gut flora. These interactions are essential for improving feed efficiency and preserving gut health and function (Shini and Bryden 2021). Studies on popular Bacillus species have shed light on their immunomodulatory properties, especially with regard to tight junction proteins like zonulin 1 and occluding, which would be increased in expression and improve the efficiency and integrity of the intestinal barrier. Probiotic microbes may increase levels of anti-inflammatory cytokines like TGF-β and IL-10 while balancing pro- inflammatory cytokines (Krysiak et al. 2021).

Probiotic integration is likely to improve the quality of the meat and eggs as the world's demand for such products with the population increases. Mainly, the aim of poultry farming is to deliver healthy and safe products, and probiotics have proven themselves to be effective alternatives to more conventional growth promoters for enhancing product quality, besides improved animal productivity (Hussein et al. 2020).

3.3 Aquaculture (e.g., Fish, Shrimp)

Aquaculture is one of the fastest-growing industries worldwide, and Asia currently represents around 90% of global production. Nevertheless, disease outbreaks remain a challenge to aquaculture and negatively impact overall production. The control of potential pathogens through probiotics has become increasingly important (Sahu et al. 2008). It is also worth noting that probiotics for aquatic organisms are different from those used for humans and terrestrial animals (Wang et al. 2019). The rapid evolution of probiotics in aquaculture is mainly because of the negative effects of commonly used antibiotics and broad-spectrum pesticides, which eliminate not only harmful bacteria but also beneficial ones that support aquatic health (Mohsen et al. 2020).

Research on probiotics has shown various effects, including enhanced growth, improved activity of digestive enzymes, feed utilization efficiency, increased immunity through gene transcription of immunity, an increase in favorable gut microbes, and changes in favorable modification in intestinal structure, along with disease protection. These resulted in more eco-friendly conditions for aquatic environments' management (Sumon et al. 2022).

Probiotics, which are live beneficial bacteria, inactivate inside the human body and maintain the balance of internal microbes through the intake of food and water. Probiotic bacteria create bacteriocins, siderophores, lysozymes, proteases, and hydrogen peroxides, which prevent the growth of dangerous pathogens (El-Saadony et al. 2021).

There are two forms that aquatic probiotics are usually in: dry and liquid. Dry probiotics last longer and need to be rehydrated or mixed with feed before being provided to the host. Liquid probiotics, commonly used in hatcheries for eggs, are simply mixed with feed or added directly to tanks (Decamp and Moriarty 2007).

Liquid probiotics tend to yield better results because of their lower density compared to spore and dry forms (Nageswara and Babu 2006). Probiotics can lower the risk of infectious outbreaks by fostering an environment that is less conducive to the growth of dangerous bacteria.

Traditional methods for controlling hazardous metabolites in aquaculture include water exchange and biofiltration (Jahangiri and Esteban 2018). Other systems and techniques have been proposed and implemented to improve water quality and treat aquaculture wastewater, including recirculating aquaculture systems, biofloc technology, and aquaponics (Emerenciano et al. 2017; Maucieri et al. 2018).

The formation of bacteriocins, siderophores, lysozymes, proteases, hydrogen peroxide, pH variations, and the synthesis of organic acids and ammonia are some of the elements that affect the antibacterial qualities of pond probiotics. In summary, using probiotics (whether single or multiple species) as water additives with minimal water exchanges has been shown to improve water quality by reducing total viable counts (TVC) and ammonia (NH3) levels while increasing dissolved oxygen (DO) levels (Hassan et al. 2022). These collective effects of probiotics across aquaculture, poultry, and ruminants are summaried in Fig. 1.

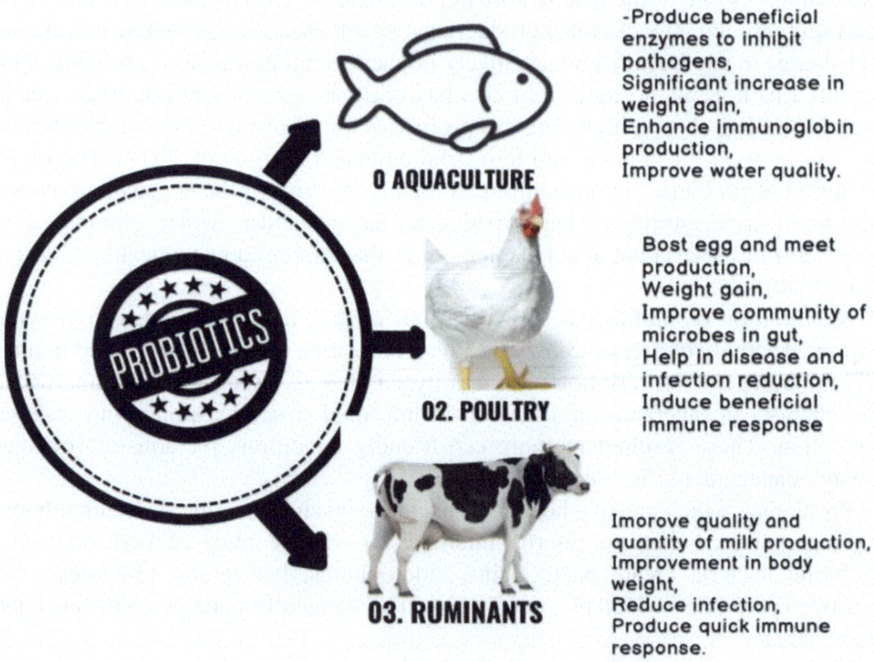

Fig. 1 Applications of probiotics in aquaculture, poultry, and ruminants

4 Mechanisms of Pathogen Inhibition

Through a number of processes that work together to support gut health and ward off infection, probiotics significantly contribute to the suppression of pathogens in animal nutrition. The major mechanism through which probiotics interfere with pathogens is by strengthening the gut's natural defense, commonly known as competitive exclusion (Gänzle et al. 2016). Germs can directly prevent the growth of pathogenic germs by producing antimicrobial compounds such as lactic acid, hydrogen peroxide, and bacteriocins (Corsetti and De Vuyst 2001). Therefore, inhibition of probiotics in animal nutrition upon pathogens involves a multi-faceted mechanism, which includes competitive exclusion, antimicrobial production, immune system modulation, restoration of the balance of microbiota, and fortification of the gut barrier. These mechanisms combine to reduce the infection risks, thereby improving health and performance in animals.

4.1 Competitive Exclusion

Probiotics are increasingly being recognized as a potential alternative to antibiotics and anti-inflammatory medications (Quigley 2011). Three different mechanisms of action can be used to describe the benefits of probiotics. Probiotics can help prevent and treat infections, chronic inflammation of the digestive tract, and possibly even the elimination of neoplastic host cells by modifying the innate and acquired immune systems (Liu et al. 2018). Probiotics can directly influence other preventing and treating infections, as well as restoring the balance of microbes in the gut (Suchodolski and Jergens 2016). Probiotics can inactivate toxins and detoxify bile salts and food components in the gut, affecting microbial and host products (Oelschlaeger 2010).

Enteropathogens must cross the intestinal mucosal barrier to start infection. Then, they penetrate the mucus layer, attach to epithelial cells, and colonize the gastrointestinal tract to cause infection (Van Zyl et al. 2020). Probiotics, which can attach to the gut, protect it from enteric infections by stopping the attachment of pathogens (Khaneghah et al. 2020) and probiotics lower the PH value in the gut, release the bactericidal peptides and proteins, or push epithelial cells to produce to the intestinal lining by blocking the surface receptors in epithelial cells or by evoking the synthesis of mucins, which is a large carbohydrate molecule that presents a protective shield along the epithelial layer (Corr et al. 2009).

4.2 Production of Antimicrobial Compounds

Probiotic strains produce antimicrobial compounds such as VFA, hydrogen peroxide, nitric oxide, and bacteriocins begin the most commonly produced (Raabis et al. 2019). AMPs, or host defense peptides (HDPs), are released by a range of organisms: commensal bacteria and even intestinal epithelial cells, providing protection against enteric pathogens. AMPs affect bacterial membranes, thereby maintaining intestinal homeostasis and modulating the composition of the microbiome and excluding enteric pathogens from colonization in the gut lumen (Mejía-Pitta et al. 2021).

Peptides known as bacteriocins have been authorized for use as antibacterial agents in the food sector. Their uses have grown over the last ten years to include food additives, cosmetics, nanotechnology, food packaging, and cancer treatment. They are receiving more interest as possible immune-modulating and clinical antimicrobials (Pereira et al. 2022; Soltani et al. 2021). Table 1 shows antimicrobial compounds produced by probiotics with their applications.

4.3 Immune Modulation

Probiotics use DNA, metabolites, and cell wall components to affect the immune system. Dead probiotics or parts of them can cause immune modulation. These are recognized by host cells via receptors, such as immunological and gastrointestinal epithelial cells (Raheem et al. 2021; Mazziotta et al. 2023). Bacteria initiate both innate and adaptive immune responses by interacting with immune cells, such as DCs, lymphocytes, epithelial cells, monocytes, and macrophages. Bacterial PAMPs interact with PRRs (TLRs, NLRs, and C-type lectin receptors) to start cell signaling (Bermudez-Brito et al. 2012; Liu et al. 2022).

Pattern recognition receptors (PRRs) allow probiotics to engage with intestinal epithelial cells and dendritic cells (DCs). By suppressing NF–κB signaling and downregulating Toll-like receptors (TLRs), this interaction influences cytokine production, initiates immune responses, including T cell differentiation, and decreases intestinal inflammation (Yang et al. 2021; Saha and Saroj 2022; Truong and Jeong 2022; Aghamohammad et al. 2022).

4.4 Modulation of Gut Microbiota

A complex ecological organism, the human gut microbiota plays a number of vital roles in the homeostatic preservation of the host's intestinal and overall health. Since many intestinal and systemic disorders have been linked to microbiota dysbiosis, there is interest in modifying it for human health (Satokari 2019). More

Table 1 Key antimicrobial compounds produced by probiotics

Compound	Mode of action	Target pathogens	Examples of producing strains	Effectiveness range	Applications	References
Bacteriocins	Disrupts cell membranes of target pathogens	Gram- positive bacteria (e.g., Listeria, Clostridium)	*Lactobacillus plantarum, Pediococcus acidilactici*	Narrow spectrum; mostly Gram-positive	Food preservation, livestock pathogen control	Bisht et al. (2024)
Bacilysin	Inhibits cell wall synthesis	Gram-positive and Gram-negative bacteria	*Bacillus subtilis*	Broad spectrum	Food preservation, pharmaceuticals , Animal feed additive	Zhang et al. (2022)
Lactocin	Disrupts cell membranes	Gram-positive bacteria	*Lactobacillus species*	Broad spectrum	Food Preservation, probiotic formulations, animal health,. Fermented beverages, Medical devices	Anumudu et al. (2024)
Pediocin PA-1	Forms pores in cell membranes	Gram- positive bacteria (e.g., Listeria)	*Pediococcus acidilactici*	Narrow spectrum	Food preservation, biopreservation, Meat products,. Dairy products,. Ready-to-eat meals, Control of Listeria monocytogenes	Khorshidianet al. (2021)
Enterocin AS-48	Disrupts cell membranes	Gram-positive and Gram-negative bacteria	*Enterococcus faecalis*	Broad spectrum	Food preservation, pharmaceutical, Dairy industry, Meat processing, Fermented foods, Probiotic supplements	Castrejón-Jiménez et al. (2024)

(continued)

Table 1 (continued)

Compound	Mode of action	Target pathogens	Examples of producing strains	Effectiveness range	Applications	References
Nisin ZP	Inhibits cell wall synthesis and forms pores	Gram- positive bacteria (e.g., Listeria)	*Lactococcus lactis*	Narrow spectrum	Food preservation, dairy products,, Meat products, Bakery products, Beverage Industry	Ozel et al. (2018)
Reutericyclin	Disrupts cell membrane	Gram- positive bacteria, some Gram-negative	*Lactobacillus reuteri*	Broad spectrum	Food preservation, Probiotic formulations, Animal feed, Fermented foods, Biocontrol in agriculture	Riesute et al. (2021)
Lactobacillin	Inhibits cell wall synthesis	Gram-positive bacteria	*Lactobacillus species*	Narrow spectrum	Food preservation, Probiotic supplements, Dairy products, Meat products, Fermented beverages	Bhattacharya et al. (2022)
Bifidocin B	Disrupts cell membranes	Gram-positive bacteria	*Bifidobacterium species*	Broad spectrum	Food preservation, Probiotic supplements, Dairy products,. Meat products, Fermented foods	Darbandi et al. (2022)
Propionicin	Inhibits cell wall synthesis	Gram-positive bacteria	*Propionibacterium species*	Broad spectrum	Bakery products, Dairy products, Meat products, Fermented foods	Huang et al. (2021)
Plantaricin	Disrupts cell membrane	Gram-positive bacteria	*Lactobacillus plantarum*	Narrow spectrum	Food preservation, probiotics, pharmaceuticals	Rocchetti et al. (2021)

Sakacin	Forms pores in cell membrane	Gram-positive bacteria (e.g., Listeria)	*Lactobacillus sakei*	Narrow spectrum	Meat preservation, biocontrol	Smaoui et al. (2024)
Leucocin	Inhibits cell wall synthesis	Gram-positive bacteria	*Leuconostoc mesenteroides*	Narrow spectrum	Dairy preservation, probiotics	Shi et al. (2016)
Acidocin	Disrupts cell membrane	Gram-positive bacteria	*Lactobacillus acidophilus*	Narrow spectrum	Dairy products, probiotic supplements	Liu et al. (2024)
Salivaricin	Inhibits protein synthesis	Gram-positive bacteria	*Streptococcus salivarius*	Narrow spectrum	Oral health, probiotics, pharmaceutical	Huang (2023)
Helveticin	Disrupts cell membrane	Gram-positive and Gram-negative bacteria	*Lactobacillus helveticus*	Broad spectrum	Dairy products, fermented foods, biocontrol	Ahansaz et al. (2023)
Garvicin	Forms pores in cell membrane	Gram-positive bacteria	*Lactococcus garvieae*	Narrow spectrum	Fish farming, aquaculture, food safety	Chen et al. (2024)
Weissellicin	Inhibits cell wall synthesis	Gram-positive bacteria	*Weissella cibaria*	Narrow spectrum	Fermented foods, probiotics	Singh et al. (2024)
Coagulin	Disrupts cell membrane	Gram-positive bacteria	*Bacillus coagulans*	Narrow spectrum	Probiotic supplements, food preservation	Zhao et al. (2023)
Paracin	Inhibits DNA replication	Gram-positive bacteria	*Lactobacillus paraplantarum*	Narrow spectrum	Dairy preservation, probiotics, fermented foods	Pei et al. (2020)

importantly, the native microbiota itself also poses a variable factor due to its probable interference with the colonization and functionality of the strains administered probiotically. On average, orally consumed strains like Lactobacillus and Bifidobacterium are only temporary colonizers, often becoming undetectable in feces even in a matter of days or weeks after administration (Torp et al. 2022).

Gut microbiota and fecal microbiota transplantation research emphasize the role of commensal species in gut health, opening the door to next-generation probiotics and live biotherapeutics that focus on core species underrepresented in dysbiosis (Hiippala et al. 2018). Probiotics also have a lot of potential in modifying gut microbiota and promoting beneficial bacteria, which is worth further study (Markowiak and Śliżewska 2017).

5 Limitations of the Study

While probiotics are mostly acclaimed for their potential in enhancing animal performance and health, some negative impacts may be realized while feeding the animals. For example, enterococci possess transmissible drug-resistant genes, and some other strains, like Bacillus cereus, can produce enterotoxins that might cause risk to the host.

Other significant problems with probiotic use include the lack of information regarding safe dosing and potential interactions between the probiotics and host cells. The other important piece of information is the limited understanding of how probiotics interact with host cells and how safe and effective their dosages are. The use of probiotics is further complicated by variability in their efficacy due to host genetics, environmental conditions, and diet. Storage and shelf-life stability are also critical factors that influence the viability and effectiveness of probiotic formulations. To overcome these limitations, more comprehensive research is needed to identify host-specific stains established standardized dosing guidelines, and access the long-term safety of probiotic use under varying conditions. Advances in molecular techniques and data-driven approaches would provide valuable insight into mitigating these challenges.

6 Recommendations

Probiotics have been demonstrated to serve as a valuable alternative to growth-promoting antibiotics by enhancing animal health, immunity, and performance. However, further research is needed to explore their full potential. The mechanisms underlying probiotic host interactions can be clarified by new and sophisticated molecular approaches, including transcriptome, proteome, and metabolomics, opening the door to more focused and efficient uses.

To improve the diagnostic, therapeutic, and production-related advantages of probiotics, future studies should concentrate on pinpointing the precise gene expression and metabolic pathways that they affect. For example, metabolic assay can help probiotic strains to specific health outcomes such as improves immunity or diseased resistance. Additionally, tailored probiotic formulations could address specific challenges in animal production systems, including stress management, disease prevention, and nutrient optimization.

To maximize the benefits of probiotics, interdisciplinary studies that integrate the molecular biology, microbiology, and animal nutrition are essential. These studies can ensure that probiotics are not only effective but also safe and sustainable for diverse livestock aquaculture system.

7 Conclusions

As the global population continues to rise, the demand for sustainable solutions in meat and fish production is becoming increasingly crucial. Probiotics provide promising evidence to meet this demand while reducing the reliance on harmful antibiotics in farm industries. By improving animal health performance and immunity, probiotics contribute to sustainable and ethical food production practices. This chapter highlights the current state-of-knowledge on probiotics, including the production methods, mechanism of action, and applications across diverse species. Probiotics have shown great potential as natural growth promoters for both nutritional and therapeutic probiotics. However, further research and innovation are needed to overcome existing challenges and fully realize their potential as an integral component of animal production systems.

References

Abd El-Trwab MM, Youssef II, Bakr HA, Fthenakis GC, Giadinis ND (2016) Role of probiotics in nutrition and health of small ruminants. Pol J Vet Sci 19(4)

Aghamohammad S, Sepehr A, Miri ST, Najafi S, Pourshafie MR, Rohani M (2022) The role of combining probiotics in preventing and controlling inflammation: a focus on the anti-inflammatory and immunomodulatory effects of probiotics in an in vitro model of IBD. Can J Gastroenterol Hepatol 2022(1):2045572

Ahansaz N, Tarrah A, Pakroo S, Corich V, Giacomini A (2023) Lactic acid bacteria in dairy foods: prime sources of antimicrobial compounds. Fermentation 9(11):964

Anee IJ, Alam S, Begum RA, Shahjahan RM, Khandaker AM (2021) The role of probiotics on animal health and nutrition. J Basic Appl Zool 82:1–16

Anumudu CK, Miri T, Onyeaka H (2024) Multifunctional applications of lactic acid bacteria: enhancing safety quality and nutritional value in foods and fermented beverages. foods 13(23):3714. https://doi.org/10.3390/foods1323371

Bhattacharya D, Nanda PK, Pateiro M, Lorenzo JM, Dhar P, Das AK (2022) Lactic acid bacteria and bacteriocins: novel biotechnological approach for biopreservation of meat and meat products microorganisms 10(10):2058. https://doi.org/10.3390/microorganisms10102058

Bermudez-Brito M, Plaza-Díaz J, Muñoz-Quezada S, Gómez-Llorente C, Gil A (2012) Probiotic mechanisms of action. Ann Nutr Metab 61(2):160–174

Bisht V, Das B, Hussain A et al (2024) Understanding of probiotic origin antimicrobial peptides: a sustainable approach ensuring food safety. npj Sci Food 8:67

Castrejón-Jiménez NS, Castrejón-Jiménez IA, Rojas-Campos TO, Chavarría-Hernández N, García-Pérez BE, Hernández-González JC, Ray S, Kumar P, Mandal M (2024) Antimicrobial peptides from lactic acid bacteria. diversity biosynthesis and applications classification of bacteriocins from lactic acid bacteria. Their Mode of Action Springer Nature Singapore, 33–65

Chen X, Liu H, Liu S, Mao J (2024) Impact of bacteriocins on multidrug-resistant bacteria and their application in aquaculture disease prevention and control. Rev Aquac

Collado MC, Gueimonde M, Salminen S (2010) Probiotics in adhesion of pathogens: mechanisms of action. In: Bioactive foods in promoting health. Academic Press, pp 353–370

Cone JW, Márquez O, Salem AZ (2021) Influence of microbial probiotics on ruminant health and nutrition: sources, mode of action and implications

Corr SC, Hill C, Gahan CG (2009) Understanding the mechanisms by which probiotics inhibit gastrointestinal pathogens. Adv Food Nutr Res 56:1–15

Corsetti A, De Vuyst L (2001) Lactic acid bacteria in sourdough fermentation. Food Res Int 34(6):479–487

Darbandi A, Asadi A, Mahdizade Ari M, Ohadi E, Talebi M, Halaj Zadeh M, Kakanj M (2022) Bacteriocins: Properties and potential use as antimicrobials. J Clin Lab Anal 36(1). https://doi.org/10.1002/jcla.v36.110.1002/jcla.24093

Decamp O, Moriarty D (2007) Aquaculture species profit from probiotics. Feed Mix 15(1):20

El-Saadony MT, Alagawany M, Patra AK et al (2021) The functionality of probiotics in aquaculture: an overview. Fish Shellfish Immunol 117:36–52

Emerenciano MGC, Martínez-Córdova LR, Martínez-Porchas M, Miranda-Baeza A (2017) Biofloc technology (BFT): a tool for water quality management in aquaculture. Water Qual 5:92–109

Fahmy AM (2024) Microbiome and human health: recent insights and future directions. Public Health 1:100002

Fijan S (2014) Microorganisms with claimed probiotic properties: an overview of recent literature. Int J Environ Res Public Health 11(5):4745–4767

Gänzle MG, Russo P, Lücke FK (2016) Lactic acid bacteria in fermented foods: metabolism and impact on food safety. In: Lactic acid bacteria: microbiological and functional aspects. Springer, pp 313–351

Gasbarrini G, Bonvicini F, Gramenzi A (2016) Probiotics history. J Clin Gastroenterol 50:S116–S119. https://doi.org/10.1097/MCG.0000000000000697

Hassan MA, Fathallah MA, Elzoghby MA, Salem MG, Helmy MS (2022) Influence of probiotics on water quality in intensified Litopenaeus vannamei ponds under minimum- water exchange. AMB Express 12(1):22

Hiippala K, Jouhten H, Ronkainen A, Hartikainen A, Kainulainen V, Jalanka J, Satokari R (2018) The potential of gut commensals in reinforcing intestinal barrier function and alleviating inflammation. Nutrients 10(8):988

Huang RCS (2023) Streptococcus salivarius secretes novel lantibiotics with dual antibiofilm and immunomodulation bioactivities (Master's thesis, University of Toronto (Canada))

Huang F, Teng K, Liu Y, Cao Y, Wang T, Ma C, Zhong J (2021) Bacteriocins: potential for human health oxidative medicine and cellular longevity. https://doi.org/10.1155/2021/5518825

Hussein EO, Ahmed SH, Abudabos AM, Suliman GM, Abd El-Hack ME, Swelum AA, Alowaimer AN (2020) Ameliorative effects of antibiotic-, probiotic-and phytobiotic-supplemented diets on the performance, intestinal health, carcass traits, and meat quality of Clostridium perfringens-infected broilers. Animals 10(4):669

Inchingolo F, Inchingolo AM, Malcangi G, De Leonardis N, Sardano R, Pezzolla C et al (2023) The benefits of probiotics on oral health: systematic review of the literature. Pharmaceuticals 16(9):1313

Jahangiri L, Esteban MÁ (2018) Administration of probiotics in the water in finfish aquaculture systems: A Review Fishes 3(3):33. https://doi.org/10.3390/fishes3030033

Jha R, Das R, Oak S, Mishra P (2020) Probiotics (direct-fed microbials) in poultry nutrition and their effects on nutrient utilization, growth and laying performance, and gut health: a systematic review. Animals 10(10):1863

Khaneghah AM, Abhari K, Eş I, Soares MB, Oliveira RB, Hosseini H et al (2020) Interactions between probiotics and pathogenic microorganisms in hosts and foods: a review. Trends Food Sci Technol 95:205–218

Krysiak K, Konkol D, Korczyński M (2021) Overview of the use of probiotics in poultry production. Animals 11(6):1620

Kulkarni NA, Chethan HS, Srivastava R, Gabbur AB (2022) Role of probiotics in ruminant nutrition as natural modulators of health and productivity of animals in tropical countries: an overview. Trop Anim Health Prod 54(2):110

Liu Y, Tran DQ, Rhoads JM (2018) Probiotics in disease prevention and treatment. J Clin Pharmacol 58:S164–S179

Liu J, Zhang H, Su Y, Zhang B (2022) Application and prospect of targeting innate immune sensors in the treatment of autoimmune diseases. Cell Biosci 12(1):68

Liu Y, Nawazish H, Farid MS, Abdul Qadoos K, Habiba UE, Muzamil M et al (2024) Health-promoting effects of Lactobacillus acidophilus and its technological applications in fermented food products and beverages. Fermentation 10(8):380

Maftei NM, Raileanu CR, Balta AA, Ambrose L, Boev M, Marin DB, Lisa EL (2024) The potential impact of probiotics on human health: an update on their health- promoting properties. Microorganisms 12(2):234

Maldonado Galdeano C, Cazorla SI, Lemme Dumit JM, Vélez E, Perdigón G (2019) Beneficial effects of probiotic consumption on the immune system. Ann Nutr Metab 74:115–124

Markowiak P, Śliżewska K (2017) Effects of probiotics, prebiotics, and synbiotics on human health. Nutrients 9(9):1021

Maucieri C, Nicoletto C, Junge R, Schmautz Z, Sambo P, Borin M (2018) Hydroponic systems and water management in aquaponics: a review. Ital J Agron 13(1):1–11

Mazziotta C, Tognon M, Martini F, Torreggiani E, Rotondo JC (2023) Probiotics mechanism of action on immune cells and beneficial effects on human health. Cells 12(1):184

Mejía-Pitta A, Broset E, de la Fuente-Nunez C (2021) Probiotic engineering strategies for the heterologous production of antimicrobial peptides. Adv Drug Deliv Rev 176:113863

Mohsen S, Dickinson JA, Somayaji R (2020) Update on the adverse effects of antimicrobial therapies in community practice. Can Fam Physician 66(9):651–659. PMID: 32933978; PMCID: PMC7491661

Nageswara PV, Babu DE (2006) Probiotics as an alternative therapy to minimize or avoid antibiotics use in aquaculture. Fishing Chimes 26(1):112–114

Nalla K, Manda NK, Dhillon HS, Kanade SR, Rokana N, Hess M, Puniya AK (2022) Impact of probiotics on dairy production efficiency. Front Microbiol 13:805963

Oelschlaeger TA (2010) Mechanisms of probiotic actions–a review. Int J Med Microbiol 300(1):57–62

Özel B, Şimşek Ö, Akçelik M, Saris PE (2018) Innovative approaches to nisin production Applied Microbiology and Biotechnology 102(15):6299–6307. https://doi.org/10.1007/s00253-018-9098-y

Pei J, Jin W, Abd El-Aty AM, Baranenko DA, Gou X, Zhang H et al (2020) Isolation, purification, and structural identification of a new bacteriocin made by Lactobacillus plantarum found in conventional kombucha. Food Control 110:106923

Pereira WA, Mendonça CMN, Urquiza AV, Marteinsson VÞ, LeBlanc JG, Cotter PD et al (2022) Use of probiotic bacteria and bacteriocins as an alternative to antibiotics in aquaculture. Microorganisms 10(9):1705
Quigley EM (2011) Therapies aimed at the gut microbiota and inflammation: antibiotics, prebiotics, probiotics, synbiotics, anti-inflammatory therapies. Gastroenterol Clin 40(1):207–222
Raabis S, Li W, Cersosimo L (2019) Effects and immune responses of probiotic treatment in ruminants. Vet Immunol Immunopathol 208:58–66
Raheem A, Liang L, Zhang G, Cui S (2021) Modulatory effects of probiotics during pathogenic infections with emphasis on immune regulation. Front Immunol 12:616713
Reid G (2016) Probiotics: definition, scope and mechanisms of action. Best Pract Res Clin Gastroenterol 30(1):17–25
Reuben RC, Elghandour MM, Alqaisi O, Cone JW, Márquez O, Salem AZ (2022) Influence of microbial probiotics on ruminant health and nutrition: sources, mode of action and implications. J Sci Food Agric 102(4):1319–1340
Rocchetti MT, Russo P, Capozzi V, Drider D, Spano G, Fiocco D (2021) Bioprospecting antimicrobials from Lactiplantibacillus plantarum: key factors underlying its probiotic action. Int J Mol Sci 22(21):12076
Riesute R, Salomskiene J, Moreno DS, Gustiene S (2021) Effect of yeasts on food quality and safety and possibilities of their inhibition. Trends Food Sci Technol 1081. https://doi.org/1010.1016/j.tifs.2020.11.022
Saha UB, Saroj SD (2022) Lactic acid bacteria: prominent player in the fight against human pathogens. Expert Rev Anti-Infect Ther 20(11):1435–1453
Sahu MK, Swarnakumar NS, Sivakumar K, Thangaradjou T, Kannan L (2008) Probiotics in aquaculture: importance and future perspectives. Indian J Microbiol 48:299–308
Satokari R (2019) Modulation of gut microbiota for health by current and next-generation probiotics. Nutrients 11(8):1921
Shi F, Wang Y, Li Y, Wang X (2016) Mode of action of leucocin K7 produced by Leuconostoc mesenteroides K7 against listeria monocytogenes and its potential in milk preservation. Biotechnol Lett 38:1551–1557
Shini S, Bryden WL (2021) Probiotics and gut health: linking gut homeostasis and poultry productivity. Anim Prod Sci 62(12):1090–1112
Singh JK, Devi PB, Reddy GB, Jaiswal AK, Kavitake D, Shetty PH (2024) Biosynthesis, classification, properties, and applications of Weissella bacteriocins. Front Microbiol 15:1406904
Smaoui S, Echegaray N, Kumar M, Chaari M, D'Amore T, Shariati MA et al (2024) Beyond conventional meat preservation: saddling the control of bacteriocin and lactic acid bacteria for clean label and functional meat products. Appl Biochem Biotechnol 196(6):3604–3635
Soltani S, Hammami R, Cotter PD, Rebuffat S, Said LB, Gaudreau H et al (2021) Bacteriocins as a new generation of antimicrobials: toxicity aspects and regulations. FEMS Microbiol Rev 45(1):fuaa039
Suchodolski JS, Jergens AE (2016) Recent advances and understanding of using probiotic- based interventions to restore homeostasis of the microbiome for the prevention/therapy of bacterial diseases. In: Virulence mechanisms of bacterial pathogens, pp 823–841
Sumon TA, Hussain MA, Sumon MAA, Jang WJ, Abellan FG, Sharifuzzaman SM et al (2022) Functionality and prophylactic role of probiotics in shellfish aquaculture. Aquacult Reports 25:101220
Torp AM, Bahl MI, Boisen A, Licht TR (2022) Optimizing oral delivery of next generation probiotics. Trends Food Sci Technol 119:101–109
Truong VL, Jeong WS (2022) Antioxidant and anti-inflammatory roles of tea polyphenols in inflammatory bowel diseases. Food Sci Human Wellness 11(3):502–511
Van Zyl WF, Deane SM, Dicks LM (2020) Molecular insights into probiotic mechanisms of action employed against intestinal pathogenic bacteria. Gut Microbes 12(1):1831339
Wang A, Ran C, Wang Y et al (2019) Use of probiotics in aquaculture of China—a review of the past decade. Fish Shellfish Immunol 86:734–755

Yan F, Polk DB (2011) Probiotics and immune health Current Opinion in Gastroenterology 27(6):496–501. https://doi.org/10.1097/MOG.0b013e32834baa4d

Yang ZJ, Wang BY, Wang TT, Wang FF, Guo YX, Hua RX et al (2021) Functions of dendritic cells and its association with intestinal diseases. Cells 10(3):583

Zhang T, Zhang Y, Li L, Jiang X, Chen Z, Zhao F, Yi Y (2022) Biosynthesis and Production of Class II Bacteriocins of Food-Associated Lactic Acid Bacteria Fermentation 8(5):217. https://doi.org/10.3390/fermentation8050217

Zhao N, Yu T, Yan F (2023) Probiotic role and application of thermophilic Bacillus as novel food materials. Trends Food Sci Technol 138:1–15

Probiotics in Animal Nutrition

Shahid Nazir, Ameer Hamza, Maooz Abdullah, Muhammad Farooq, Aman Ullah Khan, and Umair Ahsan

Abstract The irrational use of antibiotic growth promoters in animal nutrition has contributed to the development of antimicrobial resistance (AMR), creating an urgent need for alternative strategies. One promising approach to combat AMR is to enhance the animal body's natural defenses against infections through the use of probiotics. In recent years, the use of probiotics has gained significant attention as an alternative to antibiotics, not only for the prevention of enteric infections but also for enhancing gut health and nutrient absorption. Probiotics are live microorganisms that, when administered in adequate quantities, confer a health benefit to the host by beneficially modulating the gut microbiota and strengthening the immune responses. They are available in various forms for use as feed additives, supporting efficient feed digestion and enhancing nutrient availability for improved growth performance. However, the careful selection of probiotic strains is essential to achieve optimal health and production benefits. This chapter explores probiotic strains, their mechanisms of action, species-specific applications as growth promot-

S. Nazir · A. U. Khan (✉)
Institute of Microbiology, University of Veterinary and Animal Sciences, Lahore, Pakistan
e-mail: 2025-mphil-1002@uvas.edu.pk; amanullah.khan@uvas.edu.pk

A. Hamza
University of Veterinary and Animal Sciences, Lahore, Jhang, Pakistan

Department of Plant and Animal Production, Burdur Vocational School of Food, Agriculture and Livestock, Burdur Mehmet Akif Ersoy University, Burdur, Turkey

M. Abdullah
University of Veterinary and Animal Sciences, Lahore, Jhang, Pakistan

M. Farooq
Department of Clinical Sciences, University of Veterinary and Animal Sciences, Lahore, Jhang, Pakistan
e-mail: muhammad.farooq@uvas.edu.pk

U. Ahsan
Department of Plant and Animal Production, Burdur Vocational School of Food, Agriculture and Livestock, Burdur Mehmet Akif Ersoy University, Burdur, Turkey

R. U. Khan, S. Naz (eds.), *Green Alternatives to Antibiotics in Animal Health*,
https://doi.org/10.1007/978-981-92-0309-3_11

ers, and prospects as sustainable alternatives to antibiotics as growth promoters in animal nutrition.

Keywords Probiotics · Animal nutrition · Gut health · Microbiota · Immune modulation · AMR · Antibiotic alternatives

1 Introduction

The global human population is increasing rapidly and is projected to exceed 9.8 billion by 2050, intensifying food security challenges worldwide. Livestock farming and aquaculture are rapidly growing agribusiness sectors to meet the rising food demands and to support the livelihoods of the increasing global population. Feed supplements have emerged as a promising strategy to improve the health and enhance the productivity of food-producing animals, ensuring the production of large quantities of livestock, fish, and their by-products. The gastrointestinal tract (GIT) of domestic animals contains complex and dense microbial communities, including viruses, bacteria, fungi, protozoa, and archaea. Among these, bacteria are the most prevalent group with over 500 different species, e.g., *Lactobacillus* spp., *Bifidobacterium* spp., *Clostridium* spp., *Eubacterium* spp., and *Bacteroides* spp. (Maki et al. 2019). Over the past three decades, scientific studies have made substantial contributions to our current understanding of microbial ecosystems, particularly their functional diversity, composition, and critical roles in improving animal health and nutrition. Gut microbiota is responsible for the digestion and fermentation of plant polymers that possess prime importance in herbivores. They are also required for vitamin synthesis, maintenance of gut mucosal integrity and peristalsis, detoxification of potentially harmful compounds, immune stimulation, and inhibition of pathogen colonization (Skonieczna-Żydecka et al. 2020). Various environmental and management factors are believed to influence the composition and functional properties of gut microbiota in livestock and other animal species. These may include feeding practices, diet composition, and farm management (Chaucheyras-Durand and Durand 2010).

Previously, growth-promoting antibiotics like ionophores were extensively used in animal production to enhance productivity, animal health, and welfare. However, owing to the potential of antimicrobial resistance (AMR) and the persistence of chemical residues in animal- derived products have restricted their use (Liu et al. 2020b; Han et al. 2020). There is also a significant risk of the transmission of multidrug-resistant bacteria between humans and animals, suggesting a serious public health concern (Iramiot et al. 2020). Additionally, subtherapeutic doses of antibiotics negatively impact beneficial gut microbes, leading to disruptions in the delicate balance of the gut microbial ecosystem (Blajman et al. 2015). This imbalance can compromise animal health and productivity, emphasizing the importance of microbiota-friendly alternatives in animal nutrition. Therefore, alternative strategies based on more natural products, such as probiotics, have been developed to ensure safe, eco-friendly, and sustainable animal nutrition.

The term "probiotic" comes from two Greek words ("pro" and "bios") and it means "for life". Probiotics are defined as "live microbial supplements that can colonize and beneficially modulate the gut microbiota when administered in adequate amounts, thereby supporting the host's health and physiological functions" (Gibson et al. 2017; Alayande et al. 2020). They offer numerous benefits such as providing protection against physiological stress, modulation of gut microbiota, improvement of intestinal epithelial barrier integrity, and stimulation of antioxidant capacity and the immune system (Abd El-Hack et al. 2020). However, the disadvantages of the probiotics application includes inconsistency in efficiency across different individuals, species, and production systems. Moreover, some probiotic strains demonstrate lower tolerance to feed manufacturing temperatures and chlorine levels in drinking water, and their efficiency is mediated by the type of substrate in the GIT of the host, which can compromise their viability and effectiveness (Melara et al. 2022). Their primary mechanisms of action include: (a) preventing adhesion of pathogens, (b) production of antimicrobial compounds such as bacteriocins and defensins, (c) competitive exclusion of harmful microbes, (d) strengthening the intestinal barrier, (e) reduction of gut pH, and (f) regulating the immune system (van Zyl et al. 2020; Kuebutornye et al. 2020).

Nowadays, commercial animal supplements commonly contain blends of lactic acid bacteria (LAB) species, in conjunction with enzymes or prebiotics (Coman et al. 2012). These products are available in liquid form for mixing with drinking water or as dry in-feed supplements. They are typically administered to animals under periods of stress, such as transportation, climate fluctuations, transition periods, weaning, or following antibiotic treatments (Moretti et al. 2023). Probiotics, when incorporated into food and feed products, are subjected to stressors like heat, acidity, osmotic stress, and oxidation during processing, transportation, and storage (Gao et al. 2022). Their efficacy can be preserved through intrinsic resistance, adaptive responses during production, or adequate protection by proper dehydration and storage conditions (Santivarangkna et al. 2008). Matrices serve as substrates for probiotic microorganisms, providing essential nutrients and serving as delivery vehicles. Applications of protective molecules and encapsulation technologies provide opportunities for tailoring feed additives to enhance probiotic stability and performance (Kiepś and Dembczyński 2022). Effective integration of probiotics into these matrices requires careful consideration of multiple factors.

This chapter encompasses the different strains and forms of probiotics employed in animal nutrition and highlights their implications as reliable and sustainable alternatives to antibiotics. It aims to elucidate the underlying mechanisms by which probiotics exert their beneficial effects, including the inhibition of pathogenic microorganisms, modulation of the gut, and stimulation of the immune system. Species-specific impacts of probiotics across various animal species, including livestock, poultry, aquaculture, and companion animals, are discussed in detail, offering insights into their targeted application. Additionally, it intends to address the practical challenges, technological advancements, and regulatory considerations encountered in the use of probiotics. By presenting detailed future directions in research and innovation, this chapter provides researchers, veterinarians, and policymakers with a better understanding of how probiotics can be effectively utilized in animal nutrition as green alternatives to antibiotic growth promoters.

2 Types of Probiotics Used in Animal Nutrition

Probiotics include various strains of beneficial bacteria, each having unique characteristics and functions. There are numerous microecological interactions between gut microflora and the intestinal environment; therefore, the effectiveness of probiotics largely depends on the microbial species used. The common types of probiotics used in animal nutrition, along with their general functions, are presented in Fig. 1.

2.1 *Lactobacillus*

Lactobacilli are gram-positive, facultatively anaerobic, non-spore-forming bacteria that ferment glucose into lactose. They are widely used in feed and biotechnology industries as probiotics, dairy starters, vaccine carriers, and silage inoculants

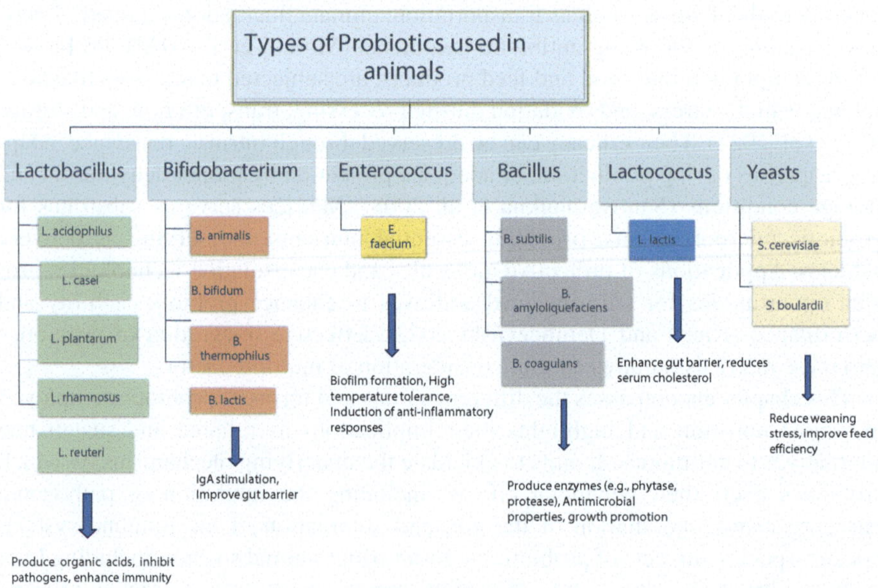

Fig. 1 Types of probiotics (bacteria and yeasts) commonly used in animal nutrition and their general functions

(Kleerebezem and Vaughan 2009; Huang et al. 2018). *Lactobacillus* spp. populate the intestines (Turroni et al. 2014), where these bacteria produce organic acids (Ridwan et al. 2008), antimicrobial peptides (Cotter et al. 2013), and H2O2 (Pridmore et al. 2008) that inhibit the growth of pathogens like *Salmonella typhimurium*, *Staphylococcus aureus*, and *E. coli* (Denev 2006). Pre-clinical studies show that *L. rhamnosus* GG can reverse intestinal dysbiosis, regulate diarrhea, and reduce chronic inflammation associated with cancer development (Banna et al. 2017). It also exhibits anticancer properties by promoting tumor cell apoptosis and reducing metastasis (Huang et al. 2022). *L. acidophilus*, *L. reuteri*, and *L. salivarius* contribute to immune system regulation in animals, particularly in poultry, mainly by inducing the production of cytokines (IFN-γ, IL-12, and IL-18) and by promoting antibody and cell-mediated immune responses (Brisbin et al. 2011). In swine, *L. plantarum* ZJ316 improves growth and carcass quality by inhibiting opportunistic pathogens and increasing villus height (Suo et al. 2012; Valeriano et al. 2017). *L. crispatus, L. vaginalis,* and *L. gasseri* protect against infections in the urogenital tract (De Gregorio et al. 2014).

2.2 *Bifidobacterium*

Bifidobacteria are gram-positive, non-motile, branched anaerobic bacteria found in the gastrointestinal tract, vagina, and mouth of mammals, including humans. All *Bifidobacteria* possess strong resistance to GI conditions, such as tolerance to acids, bile, and enzymes (bile salt hydrolase). *B. animalis* subspecies *lactis* BB-12 improves gastrointestinal health and enhances immune function by inhibiting pathogens and enhancing barrier function (Jungersen et al. 2014). Of all the strains, *B. animalis* subspecies *lactis* HY8002 induces the most IgA secretions, has the highest adhesion to Caco-2 cells, and the maximum cell survival in GIT conditions (Kim et al. 2022).

2.3 *Enterococcus*

Enterococci possess a stronger ability to adhere to intestinal mucosa and other cells compared with other probiotics. In the intestine, *E. faecium* creates a biofilm that serves as a protective agent against pathogenic bacteria, thus promoting intestinal health (Ma QingWen et al. 2018).

E. faecium enhances animal performance by hydrolyzing carbohydrates into lactic, acetic, and propionic acids, thus improving feed conversion and nutrient utilization. It also promotes immunity by triggering anti-inflammatory responses (Liu et al. 2023). Compared to other probiotics, *E. faecium* is more tolerant of high temperatures (Amaral et al. 2017) and, therefore, can be used as a feed additive for livestock.

2.4 *Bacillus*

Bacilli are rod-shaped, aerobic, or facultatively anaerobic, Gram-positive bacteria belonging to the phylum Bacillota. *Bacillus* species can form spores, which makes them more resistant to environmental stresses such as heat, cold, radiation, desiccation, and disinfectants (Cutting 2011). These species are considered safe alternatives to antibiotics in animal nutrition. *B. subtilis, B. amyloliquefaciens, B. velezensis*, and *B. licheniformis* also have antimicrobial and antifungal properties. These strains produce bacteriocins and subtilosin, which exhibit antagonistic effects against pathogenic microorganisms (Luise et al. 2022). In pigs, *B. subtilis* Natto reduces diarrhea and mortality, while improving weight gain and meat quality (Sheng et al. 2016). In cattle, it enhances growth, milk production, and digestibility (Sun et al. 2013). However, *Bacillus* species also pose a safety concern due to toxin production that is unevenly distributed across the genus (Additives and Feed 2014). Selecting *B. cereus* strains for animal production is inadvisable, but sequencing their genome and conducting bioinformatic analysis could help identify enterotoxins and cereulide synthetase (Carroll et al. 2019).

2.5 *Lactococcus*

Lactococcus lactis has been used for centuries in fermented foods and acts as a probiotic. It resists severe gastrointestinal conditions and produces lactic acid to maintain an acidic environment that prevents the growth of pathogenic bacteria. It also strengthens intestinal epithelial cells by enhancing gut barrier integrity and preventing harmful compounds from entering the bloodstream (Kondrotiene et al. 2024). *L. lactis* subsp. *Lactis* BIONCL17752 strain possesses a significant probiotic potential in poultry due to its high adaptability to GIT conditions and absence of antibiotic resistance genes. It enhances the growth and health of broiler chickens without any adverse effects on blood serum markers and organ histopathology. It also lowers serum cholesterol levels (Navale et al. 2024). *L. lactis* MA5 is a promising autochthonous probiotic for hybrid catfish, boosting nutrient digestibility and disease resistance (Huang et al. 2024).

2.6 *Yeasts*

The most extensively used yeast as a feed additive in livestock is *Saccharomyces cerevisiae.* Other yeast species used as probiotics are *Kluyveromyces marxianus, Candida utilis,* and *Saccharomyces boulardii.* In piglets, *S. cerevisiae* suppresses pathogenic bacteria, maintains intestinal flora, and improves immune function. Supplementing diets with *S. cerevisiae* fermentation broth enhances daily gain and feed intake, while decreasing feed-to-gain ratios. This mitigates weaning stress in piglets (Bin et al. 2019; Wang et al. 2016a). *S. boulardii* inhibits pathogenic bacteria in intestinal tracts, increases IgA antibody secretion, suppresses inflammation, reduces diarrhea and mortality, increases food intake, and improves meat quality in pigs. It also regulates lymphocyte count, increases IgA antibody secretion, and reduces ETEC localization (Yan and Kim 2011).

3 Forms of Probiotics

Probiotics are available in several different *forms*, each tailored to ensure the *viability, stability, and ease of delivery* of beneficial microorganisms.

3.1 Live Cultures

Live cultures are the most common form of probiotics used in animal nutrition. These viable microorganisms are administered directly to animals through feed or water. Live cultures are effective in colonizing the gut and providing immediate benefits. However, they are sensitive to environmental conditions, such as heat and moisture, which can reduce their viability during storage and feed processing. Freeze drying and microencapsulation methods enhance bacterial survival and ensure controlled release within the digestive system (Gaggìa et al. 2010).

3.1.1 Spore-Forming and Non-Spore-Forming Probiotics

Spore-forming probiotics, such as *Bacillus* species, are highly resistant to extreme conditions, including heat, pH, and mechanical stress. These probiotics remain dormant in spore form until they reach the gut, where they germinate and become active. This makes them ideal for use in pelleted feeds and other processed feeds (Cutting 2011; Khalid et al. 2022). Non-spore bacteria, such as *Lactobacillus* and *Bifidobacterium*, are more sensitive to environmental conditions and require proper handling to maintain viability. They improve gut health and immune function, but may not survive as well as spore-formers during processing and storage (Arsène et al. 2021).

3.1.2 Allochthonous and Autochthonous Probiotics

Allochthonous probiotics are strains that are not naturally present in the host's gastrointestinal tract. They are often derived from other species or environments. Their ability to colonize the host's GIT may be limited compared to autochthonous strains

(Jatoba and Jesus 2022). Autochthonous probiotics are naturally present in the host's GIT. They are isolated from the same species or closely related species. Autochthonous probiotics are generally better adapted to their host's environment, which can enhance their colonization ability and effectiveness.

3.2 *Synbiotics*

Synbiotics are a combination of probiotics and prebiotics, which are non-digestible food ingredients that promote the growth of beneficial bacteria in the gut. The synergy between probiotics and prebiotics enhances the survival and activity of probiotic strains, leading to improved gut health and performance in animals. Common prebiotics used in synbiotics include fructo-oligosaccharides (FOS) and mannan-oligosaccharides (MOS) (Gibson and Roberfroid 1995).

4 Selection Criteria for Probiotic Strains

Probiotics are selected based on their non-pathogenic nature (Seferoğlu and Kırkan 2022), tolerance to low pH, and high bile concentrations (Shewale et al. 2014), and their ability to adhere to the gastrointestinal mucosa. They should withstand manufacturing, transportation, and storage conditions while maintaining viability and integrity. Indigenous strains with minimal nutritional requirements, such as low amino acid and vitamin dependency, are preferred (Vanbelle et al. 1990). The selection of new probiotic organisms involves identifying strains or even genera of microorganisms that demonstrate the most beneficial or specific effects. This assessment primarily focuses on the safety and the benefit-to-risk ratio linked to each specific probiotic strain (Kaur et al. 2021). Spore-forming bacteria like *Bacillus* are particularly valued for their resistance against environmental conditions such as heat, cold, radiation, desiccation, and disinfectants, making them ideal for feed processing and storage (Cutting 2011).

5 Mechanisms of Action

Probiotics exert their effects in animals through complex mechanisms. These mechanisms primarily include pathogen inhibition, gut microbiota regulation, and nutrient absorption through digestive enzyme production.

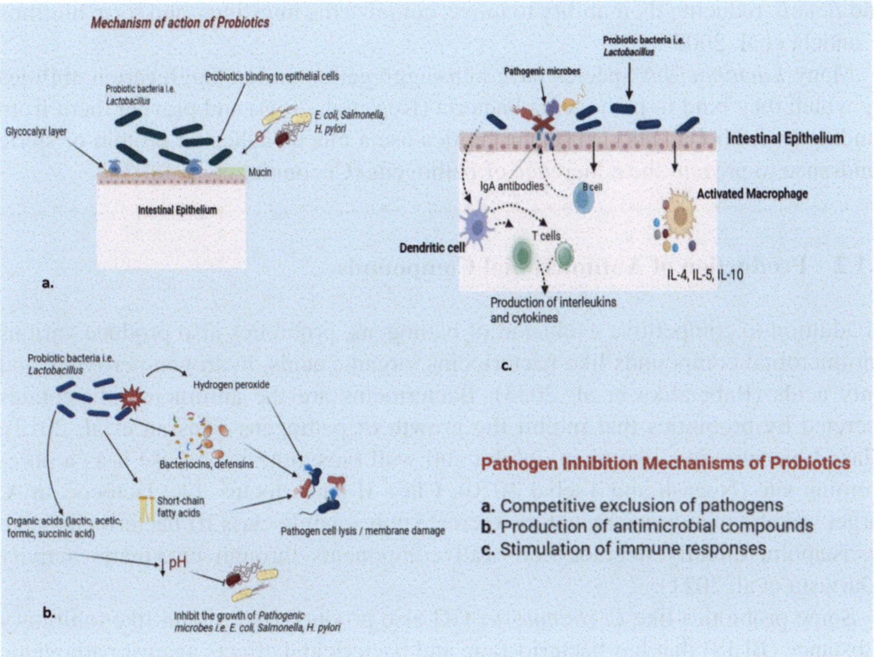

Fig. 2 Mechanism of action of probiotics

5.1 Pathogen Inhibition Mechanisms

The schematic diagram depicts the different pathogen inhibition mechanisms of probiotics (Fig. 2).

5.1.1 Competitive Exclusion of Pathogens

Competitive exclusion is a process by which probiotics prevent pathogens from colonizing the GIT (Bermudez-Brito et al. 2012). The *adhesion* to the mucosa of the GIT is an essential property of a probiotic (Kos et al. 2003). Glycolipid and glycoprotein-rich glycocalyx layers protect the epithelial cells of the intestines. Probiotics such as lactic acid bacteria possess surface proteins that enhance their hydrophobicity and adhesion to epithelial cells, depleting the available binding sites for pathogenic bacteria (Radziwill-Bienkowska et al. 2017).

Lactobacillus and *Bifidobacteria* inhibit various pathogens, such as *Helicobacter pylori, L. monocytogenes, and Rotavirus,* from binding to receptors in the gastrointestinal tract (Servin 2004). *L. acidophilus, L. plantarum, B. longum*, and *B. lactis* inhibit the flagellar motility of enteropathogenic bacteria, including *S. cholerasuis* and *E. coli,* reducing their ability to move, colonize the intestines, and form biofilms (Candela et al. 2008).

Many *Lactobacillus* species have auto-aggregation and co-aggregation abilities by which they bind to pathogenic bacteria (Kos et al. 2003) and prevent them from binding to epithelial cells. Some probiotics use a mucus-adhesion protein or steric hindrance to prevent the adherence of pathogens (Coconnier et al. 1993).

5.1.2 Production of Antimicrobial Compounds

In addition to competitive exclusion of pathogens, probiotics also produce various antimicrobial compounds like bacteriocins, organic acids, hydrogen peroxide, and fatty acids (Rabetafika et al. 2023). Bacteriocins are the antimicrobial peptides secreted by probiotics that inhibit the growth of pathogens (Hassan et al. 2012). Class I bacteriocins, like nisin, inhibit cell wall biosynthesis and use it as a pore-forming site (Negash and Tsehai 2020). Class II bacteriocins, like lactococcin A, target specific membrane receptors to create pores, while class III bacteriocins, like lysostaphin, directly degrade cell wall components through enzymatic activity (Darvishi et al. 2021).

Some probiotics like *L. rhamnosus* GG also produce bacteriocin-like inhibitory substance (BLIS) that has bacteriostatic and bactericidal effects against pathogenic bacteria (Cleveland et al. 2001). Probiotics such as *Lactobacillus, Lactococcus,* and *Bifidobacterium* ferment carbohydrates, producing lactic, acetic, formic, and succinic acid. Some probiotics, particularly *Lactobacillus* species, produce hydrogen peroxide as a byproduct of their metabolism, which slows down the growth of pathogenic bacteria and eventually causes death (Hertzberger et al. 2014).

5.1.3 Stimulation of Immune Responses

Probiotics enhance non-specific immune responses by promoting phagocytic activity through macrophage activation. Activated macrophages produce cytotoxic molecules and secrete immunoregulatory cytokines, initiating the destruction of pathogens (Mishra et al. 2008). Cytokines are proteins released by immune cells to regulate immune response and inflammation. They play a crucial role in cell activation, growth, and differentiation. Probiotics such as *Streptococcus thermophilus* ST285 and lactic acid bacteria induce the release of anti-inflammatory cytokines such as IL-4, IL-5, and IL-10, suppressing pro-inflammatory cytokines secretion such as IL-1, IL-2, IL-6, IL-12, IL-18, gamma interferon, and TNF-α (Dargahi et al. 2020).

5.2 *Gut Microbiota Modulation*

5.2.1 Balancing Beneficial and Harmful Bacteria

The gut microflora in animals is an intricate ecosystem of microorganisms, primarily bacteria, residing in the gastrointestinal tract. These bacteria form a complex microbial ecosystem in the digestive tract, with competition and symbiosis (OhashiI and Ushida 2009), which maintains intestinal homeostasis by performing metabolic, immunologic, and protective functions (Guarner and Malagelada 2003; Hill and Round 2021). Probiotic bacteria work in various ways to prevent enteropathogenic bacteria from colonizing the tract. They induce competitive exclusion (Bermudez-Brito et al. 2012) by competing with pathogenic microorganisms for nutrients and binding sites, and enhancing the colonization of beneficial bacteria (Monteagudo-Mera et al. 2019). Probiotics also prevent pathogens by the production of bacteriocins (Hassan et al. 2012), phagocytosis of pathogens (Mishra et al. 2008), stimulation of cytokines, and the stimulation of antibody secretion (Dargahi et al. 2020).

5.2.2 Enhancing Gut Barrier Function

Probiotics can restore the gut barrier by enhancing tight junction signaling and regulating intestinal epithelial cells' apoptosis and proliferation. They also enhance the expression of genes and proteins involved in tight junction signaling, thereby protecting against intestinal inflammation (Al-Sadi et al. 2021). Probiotics such as *Lactobacillus*, *Bifidobacterium*, and *A. muciniphila* elicit mucin expression and mucus secretion from goblet cells, promoting nutrient uptake and preventing bacterial penetration (Paone and Cani 2020).

5.3 *Nutrient Utilization and Digestive Enzymes Production*

Probiotics improve animal digestion, nutrient absorption, and enzyme activity in the gut microbiota. They also enhance the digestibility of essential amino acids and increase chicken body weight (Zhang and Kim 2014) and improve calcium bioaccumulation (Chawla et al. 2013). They can increase food digestibility in ruminants. Mixture of *L. acidophilus* NP51 and *Propionibacterium freudenreichii* NP24 improves the digestion of neutral detergent fiber and crude protein in cows (Boyd et al. 2011). In monogastric animals, probiotics affect mineral absorption by changing the gut's pH level, degrading cholesterol, and producing digestive enzymes (Yoo and Kim 2016).

6 Species-Specific Effects and Benefits

6.1 Ruminants

6.1.1 Improved Fiber Digestion and Milk Production

Feed additives such as probiotic bacteria, yeast culture, and fibrinolytic enzymes have been utilized extensively in recent years to enhance ruminal fermentation and, in turn, animal performance, especially in growing animals (Mousa et al. 2022). A ruminant's ability to digest more fiber in the rumen and improve nutritional digestibility can be achieved by including enzyme-based products having fibrinolytic qualities in the diet (El-Garhy et al. 2020; Mousa et al. 2022).

6.1.2 Reduction of Methane Emission and Ruminal Acidosis

One of the primary causes of CH4 emissions is agriculture. Compared to preindustrial times, methane concentrations in the atmosphere have doubled and grown quickly (SUN et al. 2023). Because of ruminal fermentation in ruminants, animal production accounts for about 37% of world CH4 emissions and 14.5% of global anthropogenic greenhouse gas emissions within the agricultural sector (Opio et al. 2013). CH4 has a substantially shorter atmospheric lifespan than CO2, ranging from 8 to 11 years; however, CH4 is more than 25 times more effective than CO2 in terms of trapping atmospheric heat (Mendonça et al. 2017). Probiotics significantly improve ruminal fermentation, feed digestibility, and rumen pH stabilization. These microorganisms can efficiently colonize the rumen or remain there for a brief period to change the enzymatic activity and metabolic efficiency (Sun et al. 2023).

C. utilis and *S. cerevisiae* are two of the most common and beneficial yeasts. They can tolerate the rumen environment and can continue their metabolism at a pH of 3.0. They can reduce methanogens' efficiency by utilizing hydrogen. The predominant fibrinolytic bacterium in the rumen is *Ruminococcus flavefaciens* (El-Zaiat et al. 2024). Its addition to the diets may increase digestion and fermentation in the rumen, and it can also decrease the emission of methane in sheep (Hassan et al. 2020). In low-starch diets, the addition of *Propionibacterium P63* and *L. rhamnosus 32* enhanced rumen pH and reduced CH4 synthesis (Philippeau et al. 2017). Overall, probiotics are a viable method to lower methane emissions in ruminants by modifying rumen fermentation pathways and encouraging a healthy microbial balance. Reducing greenhouse gas emissions in livestock systems improves feed efficiency while simultaneously promoting environmental sustainability.

6.1.3 General Uses of Probiotics in Ruminants

Studies have reported that *Bacillus* has probiotic properties and helps in improving the microbial load, digestion, and immunity of animals (Jia et al. 2018; Hashem et al. 2021; Jiang et al. 2020). *B. subtilis* can change the fermentation pattern in the calf's rumen, and it also enhances the growth and performance of the animals (Hashem et al. 2021). Yeast supplementation also improves ruminant nutrition in several ways, such as regulating rumen pH, raising volatile fatty acids, decreasing ammonia content, and improving nutrient digestibility (Wang et al. 2016b). *S. cerevisiae* is the most widely used yeast strain, which can improve weight gain, feed efficiency, alter the composition of the microbial environment, and boost nutrient digestibility (Stella et al. 2007).

6.2 *Poultry*

6.2.1 Enhanced Growth Performance and Feed Conversion Ratio

The poultry industry's reputation in the agri-food sector has improved in recent years (Krysiak et al. 2021). Since chicken feed makes up over 70% of overall production costs, its assessment is essential to the long-term financial success of the business (Agawane and Lonkar 2004). According to estimates, by the year 2050, the growing global population is expected to drive a 60% increase in agricultural output and consumption over current levels (Alexandratos and Bruinsma 2012). As a result, there will be a greater need for meat, dairy, eggs, fruits, and vegetables (Ponnampalam et al. 2019). Therefore, it is important to improve feed efficiency while minimizing costs (Agawane and Lonkar 2004). Probiotic supplementation in poultry promotes gut health by boosting the microbial population's optimal equilibrium in digestive enzymes and nutrient absorption, inhibiting the growth of infections, lowering mortality, and speeding up growth (Mahmood et al. 2014). The intestinal populations of beneficial bacteria in birds are kept in balance by probiotics, which are crucial for their effective feed conversion, growth, and production (Ivanov 2003).

Probiotics may promote growth by selectively regulating insulin signals in various tissues and by increasing the generation of SCFAs in chickens. In tissues, these SCFAs serve as a source of energy. In particular, butyrate has demonstrated beneficial effects by upregulating insulin receptors and by selectively transferring resources from the liver and adipose tissues toward the muscles (Mahajan et al. 1999). By directly promoting epithelial cell proliferation, probiotics induce SCFAs synthesis, which supports intestinal health and integrity and serves as an epigenetic regulator of the expression of several genes, leading to enhanced growth and improved general health (Mátis et al. 2015).

Several studies concluded that probiotics appear to have a good impact on poultry's physiological characteristics and productivity. According to a study, adding phytase or a probiotic with phytase in chicken diets continuously improve feed intake, weight gain, and feed conversion ratio (Derakhshan et al. 2023). Hussein et al (Hussein et al. 2020) found that the addition of probiotics and photobiotics to broiler feed, either separately or in combination, enhances the intestinal health and performance of the chicks. Additionally, including probiotics in laying hens' diet increased both the quantity and quality of their eggs (Chung et al. 2015).

6.2.2 Feed Conversion Ratio

Commercial probiotics can be used as an alternative method to boost chicken nutrition and are among the biological ways to improve feed consumption effectiveness (Al-Otaibi et al. 2023). Probiotics aid in the synthesis of vitamins and the metabolism of minerals, including biotin and vitamins B1, B2, B12, and K, all of which are necessary for healthy chicken growth and metabolism (Dhama et al. 2008; Dhama and Singh 2010). Increased feed consumption and a notable improvement in broiler FCR following probiotic treatment were reported by Landy and Kavyani (2013), Soomro et al. (2019). Probiotics enhance the function of FCR through a variety of mechanisms, including changing the intestinal flora, promoting the growth of gram-positive bacteria and anaerobic microorganisms that generate hydrogen peroxide and lactic acid (Aziz Mousavi et al. 2018).

6.2.3 Reduction of Common Diseases

To improve animal welfare and increase productivity, the poultry industry has increased the usage of certain probiotic cultures (Dhama and Singh 2010; Hajati and Rezaei 2010). Probiotics can help eliminate dangerous microorganisms from hens' gastrointestinal tracts (Song et al. 2014). Probiotic therapy greatly improves birds' gut immunity and is particularly effective against coccidian parasites and enteric bacterial illnesses. Adding probiotics to feed could effectively reduce oocysts and prevent necrotic enteritis, as well as *E. acervulina* and *E. tenella* infections (Lee et al. 2007). It was found that probiotic supplements in chicken inhibited several pathogens, including *Salmonella enteritidis*, *E. Coli*, *Staphylococcus aureus*, *Clostridium perfringens*, *Listeria monocytogenes*, *Campylobacter jejuni*, and *Candida albicans* (Dhama and Singh 2010). In commercial turkey brooding facilities, probiotics may significantly lower *Salmonella* colonization and idiopathic diarrhea in turkeys and broilers (Higgins et al. 2011; Aliyi 2023). Probiotics may also help to increase antibody levels in chicks against viral diseases, including ND and IBD (Boirivant and Strober 2007). Additionally, probiotic-induced immune system stimulation showed increased immunoglobulin production, increased activity of macrophages and lymphocytes, and increased γ-interferon production (Yang et al. 2009).

6.3 *Aquaculture*

6.3.1 Growth Performance

Aquaculture has made a substantial contribution to the animal protein production industry in almost every country around the globe during the past few decades, supplying about half of the seafood consumed globally. Beneficial probiotic bacteria are essential for maintaining the healthy growth and functioning of aquatic animals because they provide a source of vitamins, nutrients, and digestive enzymes that improve feed utilization, nutrient absorption, and growth performance (Aragona et al. 2018; Nath et al. 2019). The improved growth performance of cultured fish after probiotic diet feeding may be due to improved food quality and feed efficiency, which in turn raises the fish's appetite (Al-Dohail et al. 2009). Asian sea bass Lates calcarifer fed probiotics (*B. licheniformis* and *B. subtilis*) as nutritional supplements grew noticeably better than those given the basal diet (Adorian et al. 2019). Another study discovered that dietary supplementation of *B. aerius* strain B81e had a significant impact on *Pangasius bocourti* weight increase, specific growth rate, and feed utilization efficiency (Yanbo and Zirong 2006). So, probiotics greatly improve aquaculture growth performance by boosting immune response, nutrient absorption, and general health, which raises productivity.

6.3.2 Enhanced Water Quality and Disease Resistance

Studies have demonstrated that probiotics improve the habitat for fish by lowering the growth of dangerous phytoplankton and pathogenic bacteria and by bioremediating organic waste in rearing water (Zorriehzahra et al. 2016; Elsabagh et al. 2018). The addition of *Bacillus* to the shrimp hatchling culture in the green water system resulted in improved water temperature, pH, dissolved oxygen, ammonia, and hydrogen sulfide (Rahman et al. 2021). Compared to Gram-negative bacteria, *Bacillus* species are more effective in converting organic materials, which benefits aquaculture's water quality (Gobi et al. 2018). Probiotics, prebiotics, therapeutic herbs, and immunostimulants have been gaining interest in aquaculture as functional feed additives (Dawood et al. 2018). Probiotics can strengthen immunity and protect several fish species, including *Labeo rohita, Oreochromis niloticus, Epinephelus bruneus, Oncorhynchus mykiss*, and *Cyprinus carpio,* from a variety of infections (Rehman et al. 2020). In *Oreochromis niloticus*, food supplementation of *B. amyloliquefaciens* enhances innate immunity and disease resistance with notable increases in kidney IL-1 and TNF-alpha mRNA levels (Reda and Selim 2015). In conclusion, by enhancing gut health, regulating immune responses, and inhibiting harmful bacteria, probiotics greatly increase disease resistance in aquaculture and support healthier and more sustainable aquatic farming methods.

6.4 Companion Animals (Dogs and Cats)

For the improvement of gut health, probiotics are usually used to support the intestinal health of companion animals. A probiotic containing *L. acidophilus* (strain DSM13241) was found to improve the gut flora and stool consistency in dogs with non-specific dietary sensitivities (Pascher et al. 2008). Dogs with diarrhea fed probiotics containing *Lactobacillus* and *Bifidobacterium* strains showed improvements in microbiota structure, diversity, and regulation of microbial functional pathways (Xu et al. 2019). Puppies with GIT illnesses receiving dietary *L. rhamnosus* MP01 or *L. plantarum* MP02 isolated from canine milk, exhibited a positive clinical response (Fernández et al. 2019).

The feces of healthy cats contain a considerable number of lactic acid-producing bacteria, including *Bifidobacteria*, which could be used to choose probiotics (Handl et al. 2011). Oral administration of fermented products from *B. licheniformis* decreased *C. perfringens* in the stools of cats with chronic diarrhea and alleviated the diarrhea (Lee et al. 2022). In conclusion, the administration of single and multi-strain probiotics may lessen the need for antibiotics, which are known to have negative side effects and contribute to antibiotic resistance, and provide an alternate approach to the management and prevention of GI problems in dogs and cats.

6.5 Swine

6.5.1 Improved Gut Health and Weight Gain

Probiotics can be administered to pigs at various periods of their development (Liu et al. 2020a). *Lactobacillus* species are thought to be one of the most promising substitutes among the investigated probiotics, making them a safe option to replace AGPs in pigs (Dowarah et al. 2017). Maternal probiotic treatment has been shown to improve the digestive health of piglets not only during the lactation phase but also after weaning (Galli et al. 2024). Probiotics have the potential to enhance growth performance by improving intestinal shape and barrier function (Wang et al. 2020), increasing enzyme activity (Deng et al. 2020), producing more short-chain volatile fatty acids, and reducing diarrhea (Tian et al. 2020). *L. plantarum* has been identified as one of the most effective feed probiotics, positively impacting development and gastrointestinal health in weaned piglets (Mappley et al. 2013). Probiotics feeding to sows throughout pregnancy and lactation may thus influence the immune system and intestinal health of the offspring, even at the nursery stage (Wang et al. 2021). Among the alternative and natural feed additives, probiotics are superior to antibiotics since they are safe to use over an extended time and don't have any serious adverse effects (Virdi et al. 2019). According to Meng et al. (2010) dietary administration of probiotics (*C. butyricum*-endospore complex and *B. subtilis*-endospore) might successfully minimize feed gain and enhance average daily gain.

6.5.2 Reduced Incidence of Diarrhea

It is well known that pre-weaning piglet diarrhea is a serious issue in the swine industry. Increased piglet mortality, watery stools, cramping in the abdomen, shivering, reduced feed intake, growth retardation, and malabsorption of water and electrolytes into the gut are all consequences of diarrhea (Jacobson 2022). To enhance intestinal health and lower the prevalence of diarrhea in newborn piglets, multiple approaches have been used, including the use of antibiotics, different preventive measures, and feed additive applications. To lessen drug residues and antibiotic resistance in the swine production chain, probiotics have been used as an alternative feed additive (Barba-Vidal et al. 2019). Gram-positive obligate anaerobic *C. butyricum* has been recognized for its probiotic potential, which includes modifying gut microbiota, enhancing intestinal barrier functions, or providing defense against harmful bacteria, all of which promote growth and lower the incidence of diarrhea in piglets (Tang 2024).

7 Probiotics as Alternatives to Antibiotics

7.1 Role of Probiotics in Reducing Antibiotic Dependence

The discovery of antibiotics undoubtedly had a role in lowering morbidity and mortality. However, the development of antibiotic resistance is a challenging issue in humans and animals alike. Probiotics may lessen overall drug dependence by reducing the number of antibiotic-induced adverse effects caused by disruption of the gut microbiota (Evans et al. 2016). Probiotics have several proven health-promoting benefits, including improving host immunity, reestablishing homeostasis, encouraging a healthy gut microbiota, and inhibiting possible infections, all of which may help lower antibiotic dose or dependence. Therefore, using probiotics and antibiotics together may reduce selection pressure and lower the chance that antibiotic resistance will emerge and spread (Yang et al. 2024).

7.2 Positive Effects of Probiotics in Treating Bacterial Infection

Research by dos Santos et al. (2011) showed that the probiotic *L. delbrueckii* protected mice against *L. monocytogenes*. Their findings indicated that this probiotic stimulated the synthesis of TNF-α and IFN-γ, cytokines that help activate macrophages by increasing their nitric oxide secretion to eradicate intracellular bacteria. It has been shown that a combination of probiotic strains (*Enterococcus*, *Pediococcus*, *Lactobacillus*, and *Bifidobacterium*) effectively reduces the colonization of *C. jejuni* in the ceca of broiler chickens (Ghareeb et al. 2012). Oral administration of *Lactobacillus*-based probiotic culture (FM-B11) in *S. enteritidis*-challenged considerably limits the colonization of *S. enteritidis* (Higgins et al. 2008). *S. typhimurium*-challenged pigs fed diets supplemented with *L. plantarum* CJLP56 showed enhanced growth and decreased fecal pathogen shedding, possibly through improved immune responses, gut microbiota modulation, and changes to the structure and function of the intestinal epithelium (Gebru et al. 2010). In mouse experiments, the administration of the probiotic *L. casei* (CRL-431) decreased the severity of *S. typhimurium* infection (Castillo et al. 2011).

8 Limitations, Current Challenges, and Future Directions in Probiotics Applications

Although probiotics have certain benefits, they have some drawbacks and limitations in comparison with antibiotics, including their susceptibility to harsh stressors, limiting their survival rate and, therefore, the ability to colonize the gut (Rabetafika et al. 2023). Hung and colleagues (Hung et al. 2012) discovered that using the probiotic *B. coagulans* in the diet decreased the average daily feed intake by 8% during the broiler grower-finisher phase (days 22–42). Similarly, Amerah et al. (2013) observed a 2% decrease in feed intake with the use of a commercial probiotic containing three strains of *B. subtilis*. The ultimate body weight did not increase when *E. faecium* and *L. amylovorus* were fed to pigs (Ross et al. 2010). Szabo et al. (Szabó et al. 2009) discovered that in pigs experimentally infected with *Salmonella enterica serovar Typhimurium* DT104, the clinical symptoms were not improved by the administration of *E. faecium* NCIMB 10415.

The global market for probiotics in animal feed is estimated to reach USD 7.3 billion by 2026. Factors driving this growth are ongoing innovations in the animal feed industry, rising consumer awareness of probiotic and prebiotic benefits, increasing concerns over zoonotic diseases, and growing disposable income among major companies (Moretti et al. 2023). Advanced biotechnological methods in animal health and nutrition are significantly promoting animal growth, health, and productivity. Especially, the application of direct-fed microbials (probiotics), including both genetically modified and natural strains, has minimized many challenges in global livestock production (Bhogoju and Nahashon 2022). Probiotics commonly utilized in animal nutrition are generally considered harmless. However, potential risks associated with probiotics use include the transmission of ARGs, contagions from probiotic strains, and the production of enterotoxins or emetic toxins. Most research usually deals with the efficacy of probiotics with a limited focus on the safety, primarily discussing *Lactobacillus* and *Bifidobacterium* (Shanahan 2012). Therefore, further studies are needed to ensure the safety of a broader range of probiotic strains in animal nutrition.

Recent studies have increasingly promoted the use of probiotics in both preventive and therapeutic strategies within future veterinary health services (Melara et al. 2022). This highlights the importance of advancing research efforts toward improving the ability of probiotics to beneficially alter the intestinal microbiota, leading to increased overall intestinal health and biological function in the host. In the face of rising health challenges, probiotics may play a critical role in strengthening host resistance against pathogens inhabiting the gut environment, in addition to external microbial threats. Furthermore, the emerging evidence suggests that probiotics may hold therapeutic value in treating some newly arising diseases in animals, making them a pivotal component in sustainable animal healthcare (Cunningham et al. 2021). With advancements in bioinformatics tools, more emphasis is being placed on enhancing the reproducibility of microbiome studies and minimizing biases that may arise during the collection and processing of samples, sequencing, and data analysis. Any of these could potentially obscure important biological patterns, including individual variations (Sinha et al. 2017; Casals-Pascual et al. 2020). A new set of omics technologies, such as genomics, metagenomics, transcriptomics, proteomics, and metabolomics, is now developing into fully integrated meta technologies for better analysis and interpretation through machine learning and artificial intelligence (Bokulich et al. 2018; Lloyd-Price et al. 2019).

Antibiotic-producing bacteria with significant positive effects can now be quickly and thoroughly characterized using next-generation sequencing and liquid chromatography mass spectrometry analysis (Ling et al. 2015). Emerging evidence indicates their potential therapeutic value in managing a range of chronic conditions (Abouelela and Helmy 2024). Further research into specific gene expression pathways, particularly those identified through metabolomics, is needed to explore the complex diagnostic and therapeutic applications of probiotics. Focusing on targeted probiotic applications may provide insights into addressing various disease-related concerns in both humans and animals. Raising awareness and enhancing understanding among policymakers and consumers regarding the benefits of probiotics is critical for advancing the probiotic feed industry.

9 Conclusion

Probiotics have emerged as a promising and effective substitute to the irrational use of antibiotic growth promoters in animal production. They have contributed significant beneficial effects to animal health and productivity primarily through the modulation of host microbiota, exerting antimicrobial effects, improving antioxidant activity, stimulating immune function, and positively affecting the intestinal morphology. The administration of probiotics by the oral route has improved the growth rate, feed conversion efficiency, nutrient absorption, and utilization by promoting gut microflora balance and enhancing intestinal health. Furthermore, they play a central role in reducing the incidence of diarrheal syndromes and bacterial infections in both ruminants and non-ruminants. However, their efficacy is influenced by numerous factors, including specific strain used, dose, health status of the host, age, species, diet, physiological status, and the production category. So, appropriate care is essential while selecting candidate probiotic strains. A better understanding of the structure and functional properties of gut microbiota, as well as the intricate relationships between gut microbiome and host cells, is fundamental to boost and advance probiotic research. Emerging omics technologies and modern molecular methods, including quorum sensing, PCR, fluorescent in situ hybridization (FISH), and genome-wide association studies (GWAS), are powerful tools to better characterize and understand the effects of probiotics on the balance of the GIT microflora. Implementation of these techniques will help the researchers to better understand how probiotics modify the composition, function, and stability of the GIT microbiota, and gene exchange or transfer either horizontally or vertically, resulting in more targeted and effective applications in animal health.

References

Abd El-Hack ME, El-Saadony MT, Shafi ME, Qattan SYA, Batiha GE, Khafaga AF, Abdel-Moneim AE, Alagawany M (2020) Probiotics in poultry feed: a comprehensive review. J Anim Physiol Anim Nutr (Berl) 104(6):1835–1850. https://doi.org/10.1111/jpn.13454

Abouelela ME, Helmy YA (2024) Next-generation probiotics as novel therapeutics for improving human health: current trends and future perspectives. Microorganisms 12(3). https://doi.org/10.3390/microorganisms12030430

Additives EP, Feed PSA (2014) Guidance on the assessment of the toxigenic potential of Bacillus species used in animal nutrition. EFSA J 12(5):3665

Adorian TJ, Jamali H, Farsani HG, Darvishi P, Hasanpour S, Bagheri T, Roozbehfar R (2019) Effects of probiotic bacteria Bacillus on growth performance, digestive enzyme activity, and hematological parameters of Asian Sea Bass, Lates calcarifer (Bloch). Probiot Antimicrob Proteins 11(1):248–255. https://doi.org/10.1007/s12602-018-9393-z

Agawane SB, Lonkar PS (2004) Effect of probiotic containing Saccharomyces boulardii on experimental ochratoxicosis in broilers: hematobiochemical studies. J Vet Sci 5(4):359–367

Alayande KA, Aiyegoro OA, Ateba CN (2020) Probiotics in animal husbandry: applicability and associated risk factors. Sustainability 12(3):1087

Al-Dohail MA, Hashim R, Aliyu-Paiko M (2009) Effects of the probiotic, Lactobacillus acidophilus, on the growth performance, haematology parameters and immunoglobulin concentration in African Catfish (Clarias gariepinus, Burchell 1822) fingerling. Aquac Res 40(14):1642–1652

Alexandratos N, Bruinsma J (2012) World agriculture towards 2030/2050: the 2012 revision

Aliyi M (2023) Review on application of probiotics in poultry farm. Int J Probiot Diet 3:58–65

Al-Otaibi AM, Abd El-Hack ME, Dmour SM, Alsowayeh N, Khafaga AF, Ashour EA, Nour-Eldeen MA, Świątkiewicz S (2023) A narrative review on the beneficial impacts of probiotics on poultry: an updated knowledge. Ann Anim Sci 23(2):405–418

Al-Sadi R, Nighot P, Nighot M, Haque M, Rawat M, Ma TY (2021) Lactobacillus acidophilus induces a strain-specific and toll-like receptor 2–dependent enhancement of intestinal epithelial tight junction barrier and protection against intestinal inflammation. Am J Pathol 191(5):872–884

Amaral DMF, Silva LF, Casarotti SN, Nascimento LCS, Penna ALB (2017) Enterococcus faecium and Enterococcus durans isolated from cheese: survival in the presence of medications under simulated gastrointestinal conditions and adhesion properties. J Dairy Sci 100(2):933–949. https://doi.org/10.3168/jds.2016-11513

Amerah A, Quiles A, Medel P, Sánchez J, Lehtinen M, Gracia M (2013) Effect of pelleting temperature and probiotic supplementation on growth performance and immune function of broilers fed maize/soy-based diets. Anim Feed Sci Technol 180(1–4):55–63

Aragona M, Lauriano ER, Pergolizzi S, Faggio C (2018) Opuntia ficus-indica (L.) Miller as a source of bioactivity compounds for health and nutrition. Nat Prod Res 32(17):2037–2049. https://doi.org/10.1080/14786419.2017.1365073

Arsène MMJ, Davares AKL, Andreevna SL, Vladimirovich EA, Carime BZ, Marouf R, Khelifi I (2021) The use of probiotics in animal feeding for safe production and as potential alternatives to antibiotics. Vet World 14(2):319–328. https://doi.org/10.14202/vetworld.2021.319-328

Aziz Mousavi SMA, Mahmoodzadeh Hosseini H, Mirhosseini SA (2018) A review of dietary probiotics in poultry. J Appl Biotechnol Reports 5(2):48–54

Banna GL, Torino F, Marletta F, Santagati M, Salemi R, Cannarozzo E, Falzone L, Ferrau F, Libra M (2017) Lactobacillus rhamnosus GG: an overview to explore the rationale of its use in cancer. Front Pharmacol 8:603. https://doi.org/10.3389/fphar.2017.00603

Barba-Vidal E, Martín-Orúe SM, Castillejos L (2019) Practical aspects of the use of probiotics in pig production: a review. Livest Sci 223:84–96

Bermudez-Brito M, Plaza-Díaz J, Muñoz-Quezada S, Gómez-Llorente C, Gil A (2012) Probiotic mechanisms of action. Ann Nutr Metab 61(2):160–174

Bhogoju S, Nahashon S (2022) Recent advances in probiotic application in animal health and nutrition: a review. Agriculture 12(2):304

Bin D, Shuang X, Lu L, Xi C (2019) Research progress on application of Saccharomyces cerevisiae in animal production. Feed Res 7:114–116

Blajman JE, Zbrun MV, Astesana DM, Berisvil AP, Romero Scharpen A, Fusari ML, Soto LP, Signorini ML, Rosmini MR, Frizzo LS (2015) Probiotics in broilers' rearing: a strategy for intensive production models. Rev Argent Microbiol 47(4):360–367. https://doi.org/10.1016/j.ram.2015.08.002

Boirivant M, Strober W (2007) The mechanism of action of probiotics. Curr Opin Gastroenterol 23(6):679–692. https://doi.org/10.1097/MOG.0b013e3282f0cffc

Bokulich NA, Kaehler BD, Rideout JR, Dillon M, Bolyen E, Knight R, Huttley GA, Gregory Caporaso J (2018) Optimizing taxonomic classification of marker-gene amplicon sequences with QIIME 2's q2-feature-classifier plugin. Microbiome 6(1):90. https://doi.org/10.1186/s40168-018-0470-z

Boyd J, West J, Bernard J (2011) Effects of the addition of direct-fed microbials and glycerol to the diet of lactating dairy cows on milk yield and apparent efficiency of yield. J Dairy Sci 94(9):4616–4622

Brisbin JT, Gong J, Orouji S, Esufali J, Mallick AI, Parvizi P, Shewen PE, Sharif S (2011) Oral treatment of chickens with lactobacilli influences elicitation of immune responses. Clin Vaccine Immunol 18(9):1447–1455. https://doi.org/10.1128/CVI.05100-11

Candela M, Perna F, Carnevali P, Vitali B, Ciati R, Gionchetti P, Rizzello F, Campieri M, Brigidi P (2008) Interaction of probiotic Lactobacillus and Bifidobacterium strains with human intestinal epithelial cells: adhesion properties, competition against enteropathogens and modulation of IL-8 production. Int J Food Microbiol 125(3):286–292. https://doi.org/10.1016/j.ijfoodmicro.2008.04.012

Carroll LM, Wiedmann M, Mukherjee M, Nicholas DC, Mingle LA, Dumas NB, Cole JA, Kovac J (2019) Characterization of emetic and diarrheal Bacillus cereus strains from a 2016 foodborne outbreak using whole-genome sequencing: addressing the microbiological, epidemiological, and bioinformatic challenges. Front Microbiol 10:144. https://doi.org/10.3389/fmicb.2019.00144

Casals-Pascual C, González A, Vázquez-Baeza Y, Song SJ, Jiang L, Knight R (2020) Microbial diversity in clinical microbiome studies: sample size and statistical power considerations. Gastroenterology 158(6):1524–1528. https://doi.org/10.1053/j.gastro.2019.11.305

Castillo NA, Perdigón G, de Moreno de Leblanc A (2011) Oral administration of a probiotic Lactobacillus modulates cytokine production and TLR expression improving the immune

response against Salmonella enterica serovar Typhimurium infection in mice. BMC Microbiol 11:177. https://doi.org/10.1186/1471-2180-11-177
Chaucheyras-Durand F, Durand H (2010) Probiotics in animal nutrition and health. Benef Microb 1(1):3–9. https://doi.org/10.3920/bm2008.1002
Chawla S, Katoch S, Sharma K, Sharma V (2013) Biological response of broiler supplemented with varying dose of direct fed microbial. Vet World 6(8):521
Chung SH, Lee J, Kong C (2015) Effects of multi strain probiotics on egg production and quality in laying hens fed diets containing food waste product. Int J Poult Sci 14(1):19
Cleveland J, Montville TJ, Nes IF, Chikindas ML (2001) Bacteriocins: safe, natural antimicrobials for food preservation. Int J Food Microbiol 71(1):1–20
Coconnier M, Chauviere G, Servin AL (1993) Adhering heat-killed human Lactobacillus acidophilus strain LB, inhibits the process of pathogenically of diarrhoeagenic bacteria in culture human intestinal cells. J Diarrhoeal Dis Res 11:235–242
Coman MM, Cecchini C, Verdenelli MC, Silvi S, Orpianesi C, Cresci A (2012) Functional foods as carriers for SYNBIO®, a probiotic bacteria combination. Int J Food Microbiol 157(3):346–352. https://doi.org/10.1016/j.ijfoodmicro.2012.06.003
Cotter PD, Ross RP, Hill C (2013) Bacteriocins — a viable alternative to antibiotics? Nat Rev Microbiol 11(2):95–105. https://doi.org/10.1038/nrmicro2937
Cunningham M, Azcarate-Peril MA, Barnard A, Benoit V, Grimaldi R, Guyonnet D, Holscher HD, Hunter K, Manurung S, Obis D, Petrova MI, Steinert RE, Swanson KS, van Sinderen D, Vulevic J, Gibson GR (2021) Shaping the future of probiotics and prebiotics. Trends Microbiol 29(8):667–685. https://doi.org/10.1016/j.tim.2021.01.003
Cutting SM (2011) Bacillus probiotics. Food Microbiol 28(2):214–220. https://doi.org/10.1016/j.fm.2010.03.007
Dargahi N, Matsoukas J, Apostolopoulos V (2020) Streptococcus thermophilus ST285 alters pro-inflammatory to anti-inflammatory cytokine secretion against multiple sclerosis peptide in mice. Brain Sci 10(2):126
Darvishi N, Fard NA, Sadrnia M (2021) Genomic and proteomic comparisons of bacteriocins in probiotic species Lactobacillus and Bifidobacterium and inhibitory ability of Escherichia coli MG 1655. Biotechnol Reports 31:e00654
Dawood MA, Koshio S, Esteban MÁ (2018) Beneficial roles of feed additives as immunostimulants in aquaculture: a review. Rev Aquac 10(4):950–974
De Gregorio PR, Tomas MSJ, Terraf MCL, Nader-Macias MEF (2014) In vitro and in vivo effects of beneficial vaginal Lactobacilli on pathogens responsible for urogenital tract infections. J Med Microbiol 63(Pt 5):685–696. https://doi.org/10.1099/jmm.0.069401-0
Denev S (2006) Role of Lactobacilli in gastrointestinal ecosystem. Bulgarian J Agr Sci
Deng B, Wu J, Li X, Zhang C, Men X, Xu Z (2020) Effects of Bacillus subtilis on growth performance, serum parameters, digestive enzyme, intestinal morphology, and colonic microbiota in piglets. AMB Express 10(1):212. https://doi.org/10.1186/s13568-020-01150-z
Derakhshan M, Ghasemian SO, Gholami-Ahangaran M (2023) The effects of probiotic and phytase on growth performance, biochemical parameters and antioxidant capacity in broiler chickens. Vet Med Sci 9(2):860–866
Dhama K, Singh S (2010) Probiotics improving poultry health and production: an overview. Poultry Punch 26(3):41
Dhama K, Mahendran M, Tomar S, Chauhan R (2008) Beneficial effects of probiotics and prebiotics in livestock and poultry: the current perspectives. Intas Polivet 9(1):1–12
dos Santos LM, Santos MM, de Souza Silva HP, Arantes RM, Nicoli JR, Vieira LQ (2011) Monoassociation with probiotic Lactobacillus delbrueckii UFV-H2b20 stimulates the immune system and protects germfree mice against Listeria monocytogenes infection. Med Microbiol Immunol 200(1):29–38. https://doi.org/10.1007/s00430-010-0170-1
Dowarah R, Verma AK, Agarwal N (2017) The use of lactobacillus as an alternative of antibiotic growth promoters in pigs: a review. Anim Nutr 3(1):1–6. https://doi.org/10.1016/j.aninu.2016.11.002

El-Garhy G, Abd El-Mola A, Azzaz H, Mousa G (2020) Influence of using pectinase enzymes in the ration on nutrient digestibility, blood chemistry, milk composition and economics of lactating buffaloes. J Anim Health Prod 8(4):158–167

Elsabagh M, Mohamed R, Moustafa EM, Hamza A, Farrag F, Decamp O, Dawood MA, Eltholth M (2018) Assessing the impact of Bacillus strains mixture probiotic on water quality, growth performance, blood profile and intestinal morphology of Nile tilapia. Oreochromis niloticus Aquacult Nutr 24(6):1613–1622

El-Zaiat HM, Al-Marzooqi W, Al-Kharousi K (2024) Exploring rumen fermentation and microbial populations in Dhofari goats fed a chitosan-added diet. Anim Biotechnol 35(1):2337748

Evans M, Salewski RP, Christman MC, Girard SA, Tompkins TA (2016) Effectiveness of lactobacillus helveticus and Lactobacillus rhamnosus for the management of antibiotic-associated diarrhoea in healthy adults: a randomised, double-blind, placebo-controlled trial. Br J Nutr 116(1):94–103. https://doi.org/10.1017/s0007114516001665

Fernández L, Martínez R, Pérez M, Arroyo R, Rodríguez JM (2019) Characterization of Lactobacillus rhamnosus MP01 and Lactobacillus plantarum MP02 and assessment of their potential for the prevention of gastrointestinal infections in an experimental canine model. Front Microbiol 10:1117

Gaggìa F, Mattarelli P, Biavati B (2010) Probiotics and prebiotics in animal feeding for safe food production. Int J Food Microbiol 141(Suppl 1):S15–S28. https://doi.org/10.1016/j.ijfoodmicro.2010.02.031

Galli GM, Andretta I, Levesque C, Stefanello T, Carvalho CL, Perez Pelencia JY, Bueno Martins G, Souza de Lima Cony B, Romeiro de Oliveira C, Franceschi CH, Kipper M (2024) Using probiotics to improve nutrient digestibility and gut-health of weaned pigs: a comparison of maternal and nursery supplementation strategies. Front Vet Sci 11:1356455. https://doi.org/10.3389/fvets.2024.1356455

Gao X, Kong J, Zhu H, Mao B, Cui S, Zhao J (2022) Lactobacillus, Bifidobacterium and Lactococcus response to environmental stress: mechanisms and application of cross-protection to improve resistance against freeze-drying. J Appl Microbiol 132(2):802–821. https://doi.org/10.1111/jam.15251

Gebru E, Lee J, Son J, Yang S, Shin S, Kim B, Kim M, Park S (2010) Effect of probiotic-, bacteriophage-, or organic acid-supplemented feeds or fermented soybean meal on the growth performance, acute-phase response, and bacterial shedding of grower pigs challenged with Salmonella enterica serotype Typhimurium. J Anim Sci 88(12):3880–3886

Ghareeb K, Awad WA, Mohnl M, Porta R, Biarnés M, Böhm J, Schatzmayr G (2012) Evaluating the efficacy of an avian-specific probiotic to reduce the colonization of Campylobacter jejuni in broiler chickens. Poult Sci 91(8):1825–1832. https://doi.org/10.3382/ps.2012-02168

Gibson GR, Roberfroid MB (1995) Dietary modulation of the human colonic microbiota: introducing the concept of prebiotics. J Nutr 125(6):1401–1412. https://doi.org/10.1093/jn/125.6.1401

Gibson GR, Hutkins R, Sanders ME, Prescott SL, Reimer RA, Salminen SJ, Scott K, Stanton C, Swanson KS, Cani PD, Verbeke K, Reid G (2017) Expert consensus document: the International Scientific Association for Probiotics and Prebiotics (ISAPP) consensus statement on the definition and scope of prebiotics. Nat Rev Gastroenterol Hepatol 14(8):491–502. https://doi.org/10.1038/nrgastro.2017.75

Gobi N, Vaseeharan B, Chen JC, Rekha R, Vijayakumar S, Anjugam M, Iswarya A (2018) Dietary supplementation of probiotic Bacillus licheniformis Dahb1 improves growth performance, mucus and serum immune parameters, antioxidant enzyme activity as well as resistance against Aeromonas hydrophila in tilapia Oreochromis mossambicus. Fish Shellfish Immunol 74:501–508. https://doi.org/10.1016/j.fsi.2017.12.066

Guarner F, Malagelada JR (2003) Gut flora in health and disease. Lancet 361(9356):512–519. https://doi.org/10.1016/s0140-6736(03)12489-0

Hajati H, Rezaei M (2010) The application of prebiotics in poultry production. Int J Poult Sci 9(3):298–304

Han T, Zhang Q, Liu N, Wang J, Li Y, Huang X, Liu J, Wang J, Qu Z, Qi K (2020) Changes in antibiotic resistance of Escherichia coli during the broiler feeding cycle. Poult Sci 99(12):6983–6989. https://doi.org/10.1016/j.psj.2020.06.068

Handl S, Dowd SE, Garcia-Mazcorro JF, Steiner JM, Suchodolski JS (2011) Massive parallel 16S rRNA gene pyrosequencing reveals highly diverse fecal bacterial and fungal communities in healthy dogs and cats. FEMS Microbiol Ecol 76(2):301–310. https://doi.org/10.1111/j.1574-6941.2011.01058.x

Hashem NM, Hosny NS, El-Desoky NI, Shehata MG (2021) Effect of nanoencapsulated alginate-synbiotic on gut microflora balance, immunity, and growth performance of growing rabbits. Polymers (Basel) 13(23). https://doi.org/10.3390/polym13234191

Hassan M, Kjos M, Nes I, Diep D, Lotfipour F (2012) Natural antimicrobial peptides from bacteria: characteristics and potential applications to fight against antibiotic resistance. J Appl Microbiol 113(4):723–736

Hassan A, Gado H, Anele UY, Berasain MAM, Salem AZM (2020) Influence of dietary probiotic inclusion on growth performance, nutrient utilization, ruminal fermentation activities and methane production in growing lambs. Anim Biotechnol 31(4):365–372. https://doi.org/10.1080/10495398.2019.1604380

Hertzberger R, Arents J, Dekker HL, Pridmore RD, Gysler C, Kleerebezem M, de Mattos MJ (2014) H(2)O(2) production in species of the Lactobacillus acidophilus group: a central role for a novel NADH-dependent flavin reductase. Appl Environ Microbiol 80(7):2229–2239. https://doi.org/10.1128/aem.04272-13

Higgins SE, Higgins JP, Wolfenden AD, Henderson SN, Torres-Rodriguez A, Tellez G, Hargis B (2008) Evaluation of a Lactobacillus-based probiotic culture for the reduction of Salmonella enteritidis in neonatal broiler chicks. Poult Sci 87(1):27–31. https://doi.org/10.3382/ps.2007-00210

Higgins SE, Wolfenden AD, Tellez G, Hargis BM, Porter TE (2011) Transcriptional profiling of cecal gene expression in probiotic- and Salmonella-challenged neonatal chicks. Poult Sci 90(4):901–913. https://doi.org/10.3382/ps.2010-00907

Hill JH, Round JL (2021) SnapShot: microbiota effects on host physiology. Cell 184(10):2796–2796. e2791

Huang CH, Li SW, Huang L, Watanabe K (2018) Identification and classification for the Lactobacillus casei group. Front Microbiol 9:1974. https://doi.org/10.3389/fmicb.2018.01974

Huang R, Wu F, Zhou Q, Wei W, Yue J, Xiao B, Luo Z (2022) Lactobacillus and intestinal diseases: mechanisms of action and clinical applications. Microbiol Res 260:127019. https://doi.org/10.1016/j.micres.2022.127019

Huang J, Jordan HR, Older CE, Griffin MJ, Allen PJ, Wise DJ, Goodman PM, Reifers JG, Yamamoto FY (2024) Lactococcus lactis MA5 is a potential autochthonous probiotic for nutrient digestibility enhancement and bacterial pathogen inhibition in hybrid catfish (Ictalurus punctatus × I. furcatus). J Fish Dis 47(10):e13997. https://doi.org/10.1111/jfd.13997

Hung AT, Lin S-Y, Yang T-Y, Chou C-K, Liu H-C, Lu J-J, Wang B, Chen S-Y, Lien T-F (2012) Effects of Bacillus coagulans ATCC 7050 on growth performance, intestinal morphology, and microflora composition in broiler chickens. Anim Prod Sci 52(9):874–879

Hussein EOS, Ahmed SH, Abudabos AM, Aljumaah MR, Alkhlulaifi MM, Nassan MA, Suliman GM, Naiel MAE, Swelum AA (2020) Effect of antibiotic, phytobiotic and probiotic supplementation on growth, blood indices and intestine health in broiler chicks challenged with Clostridium perfringens. Animals (Basel) 10(3). https://doi.org/10.3390/ani10030507

Iramiot JS, Kajumbula H, Bazira J, Kansiime C, Asiimwe BB (2020) Antimicrobial resistance at the human-animal interface in the pastoralist communities of Kasese District, South Western Uganda Sci Rep 10 (1):14737. doi:https://doi.org/10.1038/s41598-020-70517-w

Ivanov I (2003) A balancing act--Optimising the gut microflora-Achieve and maintain the optimal balance with antibiotics, probiotics and organic acids. Poultry International 42(7):33–37

Jacobson M (2022) On the infectious causes of neonatal piglet Diarrhoea-a review. Vet Sci 9(8). https://doi.org/10.3390/vetsci9080422

Jatoba A, Jesus GFA (2022) Autochthonous and allochthonous lactic acid bacteria: action on the hematological and intestinal microbiota for two species of Astyanax genus. An Acad Bras Cienc 94(suppl 4):e20201611. https://doi.org/10.1590/0001-3765202220201611

Jia P, Cui K, Ma T, Wan F, Wang W, Yang D, Wang Y, Guo B, Zhao L, Diao Q (2018) Influence of dietary supplementation with Bacillus licheniformis and Saccharomyces cerevisiae as alternatives to monensin on growth performance, antioxidant, immunity, ruminal fermentation and microbial diversity of fattening lambs. Sci Rep 8(1):16712. https://doi.org/10.1038/s41598-018-35081-4

Jiang B, Wang T, Zhou Y, Li F (2020) Effects of enzyme + bacteria treatment on growth performance, rumen bacterial diversity, KEGG pathways, and the CAZy spectrum of Tan sheep. Bioengineered 11(1):1221–1232. https://doi.org/10.1080/21655979.2020.1837459

Jungersen M, Wind A, Johansen E, Christensen JE, Stuer-Lauridsen B, Eskesen D (2014) The science behind the probiotic strain Bifidobacterium animalis subsp. lactis BB-12((R)). Microorganisms 2(2):92–110. https://doi.org/10.3390/microorganisms2020092

Kaur S, Kaur R, Rani N, Sharma S, Joshi M (2021) Sources and selection criteria of probiotics. In: Advances in probiotics for sustainable food and medicine, pp 27–43

Khalid A, Khalid F, Mahreen N, Hussain SM, Shahzad MM, Khan S, Wang Z (2022) Effect of spore- forming probiotics on the poultry production: a review. Food Sci Anim Resour 42(6):968–980. https://doi.org/10.5851/kosfa.2022.e41

Kiepś J, Dembczyński R (2022) Current trends in the production of probiotic formulations. Foods 11(15). https://doi.org/10.3390/foods11152330

Kim JY, Bang SJ, Kim JY, Choi EJ, Heo K, Shim JJ, Lee JL (2022) The probiotic strain Bifidobacterium animalis ssp. lactis HY8002 potentially improves the mucosal integrity of an altered intestinal microbial environment. Front Microbiol 13:817591. https://doi.org/10.3389/fmicb.2022.817591

Kleerebezem M, Vaughan EE (2009) Probiotic and gut Lactobacilli and bifidobacteria: molecular approaches to study diversity and activity. Ann Rev Microbiol 63:269–290. https://doi.org/10.1146/annurev.micro.091208.073341

Kondrotiene K, Zavistanaviciute P, Aksomaitiene J, Novoslavskij A, Malakauskas M (2024) Lactococcus lactis in dairy fermentation—health-promoting and probiotic properties. Fermentation 10(1):16

Kos B, Šušković J, Vuković S, Šimpraga M, Frece J, Matošić S (2003) Adhesion and aggregation ability of probiotic strain Lactobacillus acidophilus M92. J Appl Microbiol 94(6):981–987. https://doi.org/10.1046/j.1365-2672.2003.01915.x

Krysiak K, Konkol D, Korczyński M (2021) Overview of the use of probiotics in poultry production. Animals (Basel) 11(6). https://doi.org/10.3390/ani11061620

Kuebutornye FKA, Abarike ED, Lu Y, Hlordzi V, Sakyi ME, Afriyie G, Wang Z, Li Y, Xie CX (2020) Mechanisms and the role of probiotic Bacillus in mitigating fish pathogens in aquaculture. Fish Physiol Biochem 46(3):819–841. https://doi.org/10.1007/s10695-019-00754-y

Landy N, Kavyani A (2013) Effects of using a multi-strain probiotic on performance, immune responses and cecal microflora composition in broiler chickens reared under cyclic heat stress condition

Lee SH, Lillehoj HS, Dalloul RA, Park DW, Hong YH, Lin JJ (2007) Influence of Pediococcus-based probiotic on coccidiosis in broiler chickens. Poult Sci 86(1):63–66. https://doi.org/10.1093/ps/86.1.63

Lee TW, Chao TY, Chang HW, Cheng YH, Wu CH, Chang YC (2022) The effects of bacillus licheniformis- fermented products on the microbiota and clinical presentation of cats with chronic diarrhea. Animals (Basel) 12(17). https://doi.org/10.3390/ani12172187

Ling LL, Schneider T, Peoples AJ, Spoering AL, Engels I, Conlon BP, Mueller A, Schäberle TF, Hughes DE, Epstein S, Jones M, Lazarides L, Steadman VA, Cohen DR, Felix CR, Fetterman KA, Millett WP, Nitti AG, Zullo AM, Chen C, Lewis K (2015) A new antibiotic kills pathogens without detectable resistance. Nature 517(7535):455–459. https://doi.org/10.1038/nature14098

Liu H, Wang S, Zhang D, Wang J, Zhang W, Wang Y, Ji H (2020a) Effects of dietary supplementation with Pediococcus acidilactici ZPA017 on reproductive performance, fecal microbial flora and serum indices in sows during late gestation and lactation. Asian Australas J Anim Sci 33(1):120–126. https://doi.org/10.5713/ajas.18.0764

Liu YL, Yan T, Li XY, Duan YL, Yang X, Yang XJ (2020b) Effects of Bacillus subtilis and antibiotic growth promoters on the growth performance, intestinal function and gut microbiota of pullets from 0 to 6 weeks. Animal 14(8):1619–1628. https://doi.org/10.1017/s1751731120000191

Liu ZL, Chen YJ, Meng QL, Zhang X, Wang XL (2023) Progress in the application of enterococcus faecium in animal husbandry. Front Cell Infect Microbiol 13:1168189. https://doi.org/10.3389/fcimb.2023.1168189

Lloyd-Price J, Arze C, Ananthakrishnan AN, Schirmer M, Avila-Pacheco J, Poon TW, Andrews E, Ajami NJ, Bonham KS, Brislawn CJ, Casero D, Courtney H, Gonzalez A, Graeber TG, Hall AB, Lake K, Landers CJ, Mallick H, Plichta DR, Prasad M, Rahnavard G, Sauk J, Shungin D, Vázquez-Baeza Y, White RA 3rd, Braun J, Denson LA, Jansson JK, Knight R, Kugathasan S, McGovern DPB, Petrosino JF, Stappenbeck TS, Winter HS, Clish CB, Franzosa EA, Vlamakis H, Xavier RJ, Huttenhower C (2019) Multi-omics of the gut microbial ecosystem in inflammatory bowel diseases. Nature 569(7758):655–662. https://doi.org/10.1038/s41586-019-1237-9

Luise D, Bosi P, Raff L, Amatucci L, Virdis S, Trevisi P (2022) Bacillus spp. probiotic strains as a potential tool for limiting the use of antibiotics, and improving the growth and health of pigs and chickens. Front Microbiol 13. https://doi.org/10.3389/fmicb.2022.801827

Ma QingWen MQ, Wang XinLing WX, Wang YuJing WY, Huang AiXiang HA (2018) Preliminary screening and isolation and identification of Lactic acid bacteria in goat dairy cake. China Dairy Ind 46(3):8–12

Mahajan P, Sahqo J, Panda P (1999) Effects of probiotic feeding and seasons on the growth performance and carcass quality of broilers

Mahmood K, Rahman S, Hussain I, Abbas R, Khaliq T, Arif J, Mahmood F (2014) Non-antibiotic strategies for the control of necrotic enteritis in poultry. Worlds Poult Sci J 70(4):865–879

Maki JJ, Klima CL, Sylte MJ, Looft T (2019) The microbial pecking order: utilization of intestinal microbiota for poultry health. Microorganisms 7(10). https://doi.org/10.3390/microorganisms7100376

Mappley LJ, Tchórzewska MA, Nunez A, Woodward MJ, Bramley PM, La Ragione RM (2013) Oral treatment of chickens with lactobacillus reuteri LM1 reduces Brachyspira pilosicoli-induced pathology. J Med Microbiol 62(Pt 2):287–296. https://doi.org/10.1099/jmm.0.051862-0

Mátis G, Kulcsár A, Turowski V, Fébel H, Neogrády Z, Huber K (2015) Effects of oral butyrate application on insulin signaling in various tissues of chickens. Domest Anim Endocrinol 50:26–31. https://doi.org/10.1016/j.domaniend.2014.07.004

Melara EG, Avellaneda MC, Valdivié M, García-Hernández Y, Aroche R, Martínez Y (2022) Probiotics: Symbiotic relationship with the animal host. Animals (Basel) 12(6). https://doi.org/10.3390/ani12060719

Mendonça R, Müller RA, Clow D, Verpoorter C, Raymond P, Tranvik LJ, Sobek S (2017) Organic carbon burial in global lakes and reservoirs. Nat Commun 8(1):1694. https://doi.org/10.1038/s41467-017-01789-6

Meng QW, Yan L, Ao X, Zhou TX, Wang JP, Lee JH, Kim IH (2010) Influence of probiotics in different energy and nutrient density diets on growth performance, nutrient digestibility, meat quality, and blood characteristics in growing-finishing pigs. J Anim Sci 88(10):3320–3326. https://doi.org/10.2527/jas.2009-2308

Mishra VK, Mohammad G, Jha A, Mohammad G (2008) Immunomodulation and anticancer potentials of yogurt probiotic. EXCLI J 7(1):177–184

Monteagudo-Mera A, Rastall RA, Gibson GR, Charalampopoulos D, Chatzifragkou A (2019) Adhesion mechanisms mediated by probiotics and prebiotics and their potential impact on human health. Appl Microbiol Biotechnol 103:6463–6472

Moretti AF, Brizuela NS, Bravo-Ferrada BM, Tymczyszyn EE, Golowczyc MA (2023) Current applications and future trends of dehydrated lactic acid bacteria for incorporation in animal feed products. Fermentation 9(8):742

Mousa G, Allak M, Hassan O (2022) Influence of fibrolytic enzymes supplementation on lactation performance of Ossimi ewes. Adv Anim Vet Sci 10(1):27–34

Nath S, Matozzo V, Bhandari D, Faggio C (2019) Growth and liver histology of Channa punctatus exposed to a common biofertilizer. Nat Prod Res 33(11):1591–1598. https://doi.org/10.1080/14786419.2018.1428586

Navale VD, Yadav R, Khilari A, Dharne M, Shanmugam D, Vamkudoth KR (2024) Dietary supplementation of Lactococcus lactis subsp. lactis BIONCL17752 on growth performance, and gut microbiota of broiler chickens. Probiot Antimicrob Proteins. https://doi.org/10.1007/s12602-024-10313-4

Negash AW, Tsehai BA (2020) current applications of Bacteriocin. Int J Microbiol 2020(1):4374891. https://doi.org/10.1155/2020/4374891

OhashiI Y, Ushida K (2009) Health-beneficial effects of probiotics: its mode of action. Anim Sci J 80(4):361–371. https://doi.org/10.1111/j.1740-0929.2009.00645.x

Opio C, Gerber P, Mottet A, Falcucci A, Tempio G, MacLeod M, Vellinga T, Henderson B, Steinfeld H (2013) Greenhouse gas emissions from ruminant supply chains–a global life cycle assessment

Paone P, Cani PD (2020) Mucus barrier, mucins and gut microbiota: the expected slimy partners? Gut 69(12):2232–2243

Pascher M, Hellweg P, Khol-Parisini A, Zentek J (2008) Effects of a probiotic Lactobacillus acidophilus strain on feed tolerance in dogs with non-specific dietary sensitivity. Arch Anim Nutr 62(2):107–116. https://doi.org/10.1080/17450390801892583

Philippeau C, Lettat A, Martin C, Silberberg M, Morgavi DP, Ferlay A, Berger C, Nozière P (2017) Effects of bacterial direct-fed microbials on ruminal characteristics, methane emission, and milk fatty acid composition in cows fed high- or low-starch diets. J Dairy Sci 100(4):2637–2650. https://doi.org/10.3168/jds.2016-11663

Ponnampalam E, Bekhit A, Bruce H, Scollan N, Muchenje V, Silva P, Jacobs J (2019) Production strategies and processing systems of meat: current status and future outlook for innovation–a global perspective. In: Sustainable meat production and processing. Elsevier, pp 17–44

Pridmore RD, Pittet AC, Praplan F, Cavadini C (2008) Hydrogen peroxide production by actoL-bacillus johnsonii NCC 533 and its role in anti-salmonella activity. FEMS Microbiol Lett 283(2):210–215. https://doi.org/10.1111/j.1574-6968.2008.01176.x

Rabetafika HN, Razafindralambo A, Ebenso B, Razafindralambo HL (2023) Probiotics as antibiotic alternatives for human and animal applications. Encyclopedia 3(2):561–581

Radziwill-Bienkowska JM, Robert V, Drabot K, Chain F, Cherbuy C, Langella P, Thomas M, Bardowski JK, Mercier-Bonin M, Kowalczyk M (2017) Contribution of plasmid-encoded peptidase S8 (PrtP) to adhesion and transit in the gut of Lactococcus lactis IBB477 strain. Appl Microbiol Biotechnol 101:5709–5721

Rahman A, Shefat SHT, Chowdhury MA, Khan SU (2021) Effects of probiotic bacillus on growth performance, immune response and disease resistance in aquaculture. J Aquacult Res Dev 12(4):634

Reda RM, Selim KM (2015) Evaluation of Bacillus amyloliquefaciens on the growth performance, intestinal morphology, hematology and body composition of Nile tilapia, Oreochromis niloticus. Aquac Int 23:203–217

Rehman A, Arif M, Sajjad N, Al-Ghadi MQ, Alagawany M, Abd El-Hack ME, Alhimaidi AR, Elnesr SS, Almutairi BO, Amran RA, Hussein EOS, Swelum AA (2020) Dietary effect of probiotics and prebiotics on broiler performance, carcass, and immunity. Poult Sci 99(12):6946–6953. https://doi.org/10.1016/j.psj.2020.09.043

Ridwan BU, Koning CJ, Besselink MG, Timmerman HM, Brouwer EC, Verhoef J, Gooszen HG, Akkermans LM (2008) Antimicrobial activity of a multispecies probiotic (Ecologic 641)

against pathogens isolated from infected pancreatic necrosis. Lett Appl Microbiol 46(1):61–67. https://doi.org/10.1111/j.1472-765X.2007.02260.x
Ross GR, Gusils C, Oliszewski R, de Holgado SC, González SN (2010) Effects of probiotic administration in swine. J Biosci Bioeng 109(6):545–549. https://doi.org/10.1016/j.jbiosc.2009.11.007
Santivarangkna C, Kulozik U, Foerst P (2008) Inactivation mechanisms of lactic acid starter cultures preserved by drying processes. J Appl Microbiol 105(1):1–13. https://doi.org/10.1111/j.1365-2672.2008.03744.x
Seferoğlu Y, Kırkan Ş (2022) Roles of probiotics in animal health. Anim Health Prod Hygiene 11(1):40–46. https://doi.org/10.53913/aduveterinary.1060132
Servin AL (2004) Antagonistic activities of lactobacilli and bifidobacteria against microbial pathogens. FEMS Microbiol Rev 28(4):405–440. https://doi.org/10.1016/j.femsre.2004.01.003
Shanahan F (2012) A commentary on the safety of probiotics. Gastroenterol Clin N Am 41(4):869–876. https://doi.org/10.1016/j.gtc.2012.08.006
Sheng QK, Zhou KF, Hu HM, Zhao HB, Zhang Y, Ying W (2016) Effect of Bacillus subtilis Natto on meat quality and Skatole content in TOPIGS pigs. Asian Australas J Anim Sci 29(5):716–721. https://doi.org/10.5713/ajas.15.0478
Shewale RN, Sawale P, Khedkar C, Singh A (2014) Selection criteria for probiotics: a review. Int J Probiot Prebiot 9:17–22
Sinha R, Abu-Ali G, Vogtmann E, Fodor AA, Ren B, Amir A, Schwager E, Crabtree J, Ma S, Abnet CC, Knight R, White O, Huttenhower C (2017) Assessment of variation in microbial community amplicon sequencing by the Microbiome Quality Control (MBQC) project consortium. Nat Biotechnol 35(11):1077–1086. https://doi.org/10.1038/nbt.3981
Skonieczna-Żydecka K, Jakubczyk K, Maciejewska-Markiewicz D, Janda K, Kaźmierczak-Siedlecka K, Kaczmarczyk M, Łoniewski I, Marlicz W (2020) Gut biofactory-neurocompetent metabolites within the gastrointestinal tract. A scoping review. Nutrients 12(11). https://doi.org/10.3390/nu12113369
Song J, Xiao K, Ke YL, Jiao LF, Hu CH, Diao QY, Shi B, Zou XT (2014) Effect of a probiotic mixture on intestinal microflora, morphology, and barrier integrity of broilers subjected to heat stress. Poult Sci 93(3):581–588. https://doi.org/10.3382/ps.2013-03455
Soomro RN, Abd El-Hack ME, Shah SS, Taha AE, Alagawany M, Swelum AA, Hussein EOS, Ba-Aawdh HA, Saadeldin I, El-Edel MA, Tufarelli V (2019) Impact of restricting feed and probiotic supplementation on growth performance, mortality and carcass traits of meat-type quails. Anim Sci J 90(10):1388–1395. https://doi.org/10.1111/asj.13290
Stella A, Paratte R, Valnegri L, Cigalino G, Soncini G, Chevaux E, Dell'Orto V, Savoini G (2007) Effect of administration of live Saccharomyces cerevisiae on milk production, milk composition, blood metabolites, and faecal flora in early lactating dairy goats. Small Rumin Res 67(1):7–13
Sun P, Wang JQ, Deng LF (2013) Effects of Bacillus subtilis natto on milk production, rumen fermentation and ruminal microbiome of dairy cows. Animal 7(2):216–222. https://doi.org/10.1017/S1751731112001188
Sun J, Zhao G, Li MM (2023) Using nutritional strategies to mitigate ruminal methane emissions from ruminants. Frontiers of Agricultural Science and Engineering 10(3):390–402
Suo C, Yin Y, Wang X, Lou X, Song D, Wang X, Gu Q (2012) Effects of Lactobacillus plantarumZJ316 on pig growth and pork quality. BMC Vet Res 8(1):89. https://doi.org/10.1186/1746-6148-8-89
Szabó I, Wieler LH, Tedin K, Scharek-Tedin L, Taras D, Hensel A, Appel B, Nöckler K (2009) Influence of a probiotic strain of Enterococcus faecium on Salmonella enterica serovar Typhimurium DT104 infection in a porcine animal infection model. Appl Environ Microbiol 75(9):2621–2628. https://doi.org/10.1128/aem.01515-08
Tang X (2024) Probiotic roles of Clostridium butyricum in piglets: considering aspects of intestinal barrier function. Animals (Basel) 14(7). https://doi.org/10.3390/ani14071069

Tian Z, Wang X, Duan Y, Zhao Y, Zhang W, Azad MAK, Wang Z, Blachier F, Kong X (2020) Dietary supplementation with Bacillus subtilis promotes growth and gut health of weaned piglets. Front Vet Sci 7:600772. https://doi.org/10.3389/fvets.2020.600772

Turroni F, Ventura M, Buttó LF, Duranti S, O'Toole PW, Motherway MO, van Sinderen D (2014) Molecular dialogue between the human gut microbiota and the host: a Lactobacillus and Bifidobacterium perspective. Cell Mol Life Sci 71(2):183–203. https://doi.org/10.1007/s00018-013-1318-0

Valeriano VDV, Balolong MP, Kang DK (2017) Probiotic roles of Lactobacillus sp. in swine: insights from gut microbiota. J Appl Microbiol 122(3):554–567. https://doi.org/10.1111/jam.13364

van Zyl WF, Deane SM, Dicks LMT (2020) Molecular insights into probiotic mechanisms of action employed against intestinal pathogenic bacteria. Gut Microbes 12(1):1831339. https://doi.org/10.1080/19490976.2020.1831339

Vanbelle M, Teller E, Focant M (1990) Probiotics in animal nutrition: a review. Arch für Tierernaehr 40(7):543–567. https://doi.org/10.1080/17450399009428406

Virdi V, Palaci J, Laukens B, Ryckaert S, Cox E, Vanderbeke E, Depicker A, Callewaert N (2019) Yeast-secreted, dried and food-admixed monomeric IgA prevents gastrointestinal infection in a piglet model. Nat Biotechnol 37(5):527–530. https://doi.org/10.1038/s41587-019-0070-x

Wang B, Wang S, Guo C, Huan H, Gao Y, Bai X, Yang J, Zhou L (2016a) Effects of Saccharomyces cerevisiae fermentation broth on growth performance, small intestine development and immune function of small intestinal mucosa of weaned piglets. Chin J Anim Nutr 28:4014–4022

Wang Z, He Z, Beauchemin KA, Tang S, Zhou C, Han X, Wang M, Kang J, Odongo NE, Tan Z (2016b) Comparison of two live Bacillus species as feed additives for improving in vitro fermentation of cereal straws. Anim Sci J 87(1):27–36. https://doi.org/10.1111/asj.12346

Wang M, Wu H, Lu L, Jiang L, Yu Q (2020) Lactobacillus reuteri promotes intestinal development and regulates mucosal immune function in newborn piglets. Front Vet Sci 7:42. https://doi.org/10.3389/fvets.2020.00042

Wang K, Hu C, Tang W, Azad MAK, Zhu Q, He Q, Kong X (2021) The enhancement of intestinal immunity in offspring piglets by maternal probiotic or synbiotic supplementation is associated with the alteration of gut microbiota. Front Nutr 8:686053. https://doi.org/10.3389/fnut.2021.686053

Xu H, Zhao F, Hou Q, Huang W, Liu Y, Zhang H, Sun Z (2019) Metagenomic analysis revealed beneficial effects of probiotics in improving the composition and function of the gut microbiota in dogs with diarrhoea. Food Funct 10(5):2618–2629. https://doi.org/10.1039/c9fo00087a

Yan L, Kim I (2011) Effect of dietary grape pomace fermented by Saccharomyces boulardii on the growth performance, nutrient digestibility and meat quality in finishing pigs. Asian- Australas J Anim Sci 24(12):1763–1770

Yanbo W, Zirong X (2006) Effect of probiotics for common carp (Cyprinus carpio) based on growth performance and digestive enzyme activities. Anim Feed Sci Technol 127(3–4):283–292

Yang Y, Iji P, Choct M (2009) Dietary modulation of gut microflora in broiler chickens: a review of the role of six kinds of alternatives to in-feed antibiotics. Worlds Poult Sci J 65(1):97–114

Yang S, Qiao J, Zhang M, Kwok L-Y, Matijašić BB, Zhang H, Zhang W (2024) Prevention and treatment of antibiotics-associated adverse effects through the use of probiotics: a review. J Adv Res

Yoo JY, Kim SS (2016) Probiotics and prebiotics: present status and future perspectives on metabolic disorders. Nutrients 8(3):173

Zhang Z, Kim I (2014) Effects of multistrain probiotics on growth performance, apparent ileal nutrient digestibility, blood characteristics, cecal microbial shedding, and excreta odor contents in broilers. Poult Sci 93(2):364–370

Zorriehzahra MJ, Delshad ST, Adel M, Tiwari R, Karthik K, Dhama K, Lazado CC (2016) Probiotics as beneficial microbes in aquaculture: an update on their multiple modes of action: a review. Vet Q 36(4):228–241. https://doi.org/10.1080/01652176.2016.1172132

Probiotics in Animal Nutrition: Mechanisms, Species-Specific Benefits, and Gut Health Modulation

Muhammad Usama, Nudrat Fatima, Swaira Ashfaq, Sania Satti, Fizza Abbas, Maryam Fatima, Samra, Hafsa Saeed, Ulfat Zahra, and Shabana Naz

Abstract Live microorganisms, known as probiotics, offer significant health benefits when consumed in appropriate amounts, especially in promoting gut health and modulating the gastrointestinal microbiome. These beneficial bacteria and yeasts exert their effects through mechanisms such as competitive inhibition, adherence to intestinal cells, and the production of antimicrobial substances like bacteriocins, hydrogen peroxide, and organic acids. These actions help prevent the growth of harmful pathogens and strengthen the immune system's defense against infections. Probiotics are increasingly used in animal nutrition as a substitute for antibiotics, offering advantages such as enhanced digestion, immune response, and overall well-being across various species, including ruminants, poultry, swine, aquaculture species, and companion animals. The use of probiotics dates back to ancient practices involving fermented foods, with contemporary research expanding their potential applications, focusing on the identification of specific probiotic strains for optimized performance. Consequently, probiotics present a sustainable approach to boosting livestock productivity, reducing reliance on antibiotics, and improving food safety.

Keywords Probiotics · Gut health · Gastrointestinal microbiome · Antimicrobial substances · Animal nutrition · Antibiotics · Livestock productivity · Food safety · Immune system · Fermented foods

---

M. Usama · N. Fatima · S. Ashfaq · S. Satti · F. Abbas · M. Fatima · Samra · H. Saeed · U. Zahra · S. Naz (✉)
Department of Zoology, Government College University, Faisalabad, Pakistan
e-mail: drshabananaz@gcuf.edu.pk

R. U. Khan, S. Naz (eds.), *Green Alternatives to Antibiotics in Animal Health*,
https://doi.org/10.1007/978-981-92-0309-3_12

1 Introduction

Probiotics are live microorganisms that, when consumed in sufficient quantities, confer health benefits to the host, particularly through their positive influence on gut health. These microorganisms, primarily bacteria and yeasts, play a critical role in maintaining a balanced intestinal microbiota, which is essential for overall health (He et al. 2021; Lin et al. 2020). The main route of probiotics is via the oral administration of food or supplements, allowing them to directly interact with the gastrointestinal system. Once ingested, probiotics colonize the gut, where they outcompete harmful pathogens for resources like space, nutrients, and adhesion sites (Hutkins et al. 2022). This phenomenon, known as competitive exclusion, is one of the core mechanisms by which probiotics exert their beneficial effects. They also adhere to the gut epithelium using various surface proteins, carbohydrates, and lipids, preventing pathogens from establishing infections (Collado et al. 2007; Azcarate-Peril et al. 2011). Furthermore, probiotics produce antimicrobial compounds, such as bacteriocins, hydrogen peroxide, and organic acids, which lower the gut's pH and disrupt the metabolic activities of pathogens (Pérez-Díaz et al. 2020; Chen et al. 2020).

In addition to their competitive actions, probiotics influence the immune system by stimulating immune cells like macrophages, dendritic cells, and T cells, which enhances the host's ability to fight infections and maintain homeostasis (Ouwehand et al. 2002; Vinderola et al. 2020). Probiotics also play a pivotal role in modulating the gut environment, including altering redox potential and strengthening the gut barrier, which helps prevent pathogen translocation across the epithelial layer (Turner 2009; Gänzle 2020). The consumption of probiotics offers a natural alternative to antibiotics, especially in livestock nutrition, where their use has become integral in improving digestion, boosting immune function, and reducing pathogen load, thus supporting animal health and productivity (Farag and Abdel-Alim 2020; Chaucheyras-Durand and Durand 2010).

The history of probiotics dates back to ancient civilizations, where fermented foods such as yogurt and kefir were consumed for their purported health benefits. The term "probiotic," derived from the Latin word *pro vita* (for life), reflects the long-standing relationship between beneficial microorganisms and both human and animal health (Vinderola et al. 2022; Gogineni et al. 2013). Modern probiotic research, particularly in animal nutrition, has focused on identifying specific strains that can enhance gut health, improve nutrient absorption, and reduce the risk of gastrointestinal diseases (Santacroce et al. 2019). As the global demand for sustainable, antibiotic-free livestock production continues to rise, probiotics offer an environmentally friendly alternative, helping to ensure the safety and quality of animal-derived products.

Probiotics have shown species-specific benefits across a variety of animals. In ruminants such as cattle and sheep, probiotics like *Saccharomyces cerevisiae* and *Lactic acid bacteria* (LAB) enhance digestion, boost milk production, and support immune function (Chaucheyras-Durand and Durand 2010). In poultry, probiotics such as *Bacillus subtilis* and *Lactobacillus plantarum* improve gut health, increase egg production, and reduce pathogen load (Lin et al. 2018). Swine benefit from probiotics like *Bacillus amyloliquefaciens* and *Saccharomyces cerevisiae*, which support growth, reduce weaning stress, and lower antibiotic use (Bottalico et al. 2020). Similarly, in aquaculture, probiotics such as *Lactobacillus plantarum* and *Pediococcus acidilactici* enhance immune function, growth, and disease resistance, providing a sustainable solution for improving fish and shrimp health (Gänzle 2020). For companion animals, probiotics like *Lactobacillus acidophilus* and *Bifidobacterium animalis* support digestion, boost immune function, and promote overall well-being (Patel et al. 2023).

Probiotics function through multiple mechanisms to inhibit the growth of pathogens and promote gut health. The production of antimicrobial compounds, including bacteriocins, hydrogen peroxide, and organic acids, helps create an environment that is hostile to harmful microorganisms (Gudiña et al. 2010; Martin et al. 2019). Additionally, probiotics modulate the gut's pH, redox potential, and nutrient availability, thereby further limiting the survival and proliferation of pathogens (Marco et al. 2006; Tzortzis et al. 2022). The ability of probiotics to produce anti- adhesive compounds, such as exopolysaccharides and biosurfactants, also plays a critical role in preventing the adhesion of pathogens to the gut epithelium (Fanning et al. 2012; Gudiña et al. 2010). These combined mechanisms highlight the diverse ways in which probiotics can support gut health, making them a valuable tool for both human and animal welfare (Miller et al. 2024).

As research advances, the identification of novel probiotic strains and the optimization of their mechanisms of action hold great promise for improving human and animal health, offering an eco- friendly alternative to traditional antibiotics, and enhancing the sustainability of livestock and aquaculture industries (Sánchez et al. 2021; Xu et al. 2023).

1.1 Defining and Concept of Probiotics

Probiotics are live microorganisms that, when administered in adequate amounts, confer health benefits to the host animal. These microorganisms, which include various species of bacteria and yeast, work by modulating the balance and activity of the gastrointestinal microbiota, thereby contributing to gut health and overall well-being. The most commonly used probiotic strains are lactic acid bacteria, such as *Lactobacillus* and *Bifidobacterium*, and yeast species like *Saccharomyces* (Chaucheyras-Durand and Durand 2010). Probiotics can be delivered in the form of feed additives or as fermented products, with the primary goal being to enhance gut homeostasis improve nutrient absorption, and reduce pathogenic infections (Santacroce et al. 2019).

1.2 History and Evolution of Probiotics in Animal Nutrition

The use of probiotics traces back to ancient civilizations, where fermented foods were consumed for their perceived health benefits. The term "probiotic," derived from the Latin word *pro vita* (for life), reflects the long-standing association between beneficial microorganisms and human and animal health. Early examples of fermented foods, such as yogurt, kefir, and cheese, have been linked to the discovery and development of probiotics. The modern understanding of probiotics evolved with the work of scientists like Metchnikoff, Escherich, and Tissier, who first recognized the health-promoting effects of lactic acid bacteria in the late nineteenth and early twentieth centuries. Over time, this understanding expanded to include their use in animal nutrition, particularly as a replacement for antibiotic growth promoters, which were banned in 2006 (Gogineni et al. 2013). Today, probiotics are a cornerstone of livestock nutrition, with ongoing research aimed at improving their efficacy and expanding their use in diverse animal species.

1.3 Importance of Probiotics in Animal Nutrition

The importance of probiotics in animal nutrition cannot be overstated, as they contribute to several critical aspects of livestock health and productivity. First, probiotics enhance gut health by maintaining a balanced microbiota, which supports nutrient absorption and prevents the overgrowth of harmful pathogens. By improving intestinal health, probiotics can help reduce the incidence of diseases caused by gastrointestinal pathogens such as *Salmonella*, *Campylobacter*, *Listeria*, and *Yersinia*. Moreover, probiotics have been shown to enhance immune function, promoting better disease resistance and recovery in animals. Probiotics also play a role in improving animal performance. In ruminants, for example, they help optimize fermentation in the rumen, leading to better digestion and increased feed efficiency. In monogastric animals such as poultry, pigs, and fish, probiotics can improve growth rates and reduce the need for antibiotics, making them an environmentally friendly and sustainable option for modern animal husbandry (Shen et al. 2022; Chaucheyras-Durand and Durand 2010). As the global demand for animal products continues to rise, the application of probiotics in animal nutrition will remain a vital tool in ensuring both the health of animals and the safety of the food supply.

2 Species-Specific Effects and Benefits of Probiotics

2.1 Ruminants (Cattle, Sheep, Goats)

Probiotics like *Saccharomyces cerevisiae*, lactic acid bacteria (LAB), *Aspergillus oryzae*, and *Bacillus* spp. offer several benefits:

- *Improved digestion*: Enhance fiber digestion and rumen fermentation.
- *Increased milk production*: Help in better nutrient utilization, leading to higher milk yields.
- *Enhanced immune function*: Probiotics support overall immune health, reducing the likelihood of infections and diseases.

Strains:

- *Saccharomyces cerevisiae*
- Lactic acid bacteria (LAB)
- *Aspergillus oryzae*
- *Bacillus* spp.

2.2 Poultry (Chickens, Turkeys)

Probiotics such as *Bacillus subtilis*, *Lactobacillus plantarum*, *Pediococcus acidilactici*, and *Clostridium butyricum* provide:

- *Improved gut health*: Enhance gut flora balance and digestion.
- *Increased egg production*: Support better feed conversion, leading to more eggs.
- *Reduced pathogen load*: Probiotics help control harmful pathogens like *Salmonella*, reducing disease incidence.
- *Enhanced growth performance*: Boost the growth rate and overall health of poultry.

Strains:

- *Bacillus subtilis*
- *Lactobacillus* spp.
- *Pediococcus acidilactici*
- *Clostridium butyricum*

2.3 Swine (Pigs)

For swine, probiotics like *Bacillus amyloliquefaciens*, *Saccharomyces cerevisiae*, and *Lactobacillus rhamnosus* offer:

- *Improved growth and feed conversion*: Boost nutrient digestion and weight gain.
- *Gut health support*: Help balance gut microbiota, improving digestion and immune responses.
- *Reduced weaning stress*: Probiotics ease the transition from milk to solid food, preventing diarrhea and other digestive issues.
- *Reduced antibiotic usage*: By enhancing immune function, probiotics reduce the need for antibiotics.

Strains
- *Bacillus amyloliquefaciens*
- *Saccharomyces cerevisiae*
- *Lactobacillus rhamnosus*

2.4 Aquaculture (Fish, Shrimp)

Aquatic species benefit from probiotics like *Bacillus subtilis*, *Lactobacillus plantarum*, and *Pediococcus acidilactici*:

- *Enhanced immune function*: Strengthen disease resistance, especially against *Vibrio* infections.
- *Improved growth rates*: Promote better digestion and feed utilization.
- *Better water quality*: Reduce harmful pathogens, ensuring a healthy aquaculture environment.
- *Disease resistance*: Improve disease resistance in both fish and shrimp, reducing the need for antibiotics.

Strains
- *Bacillus subtilis*
- *Lactobacillus plantarum*
- *Saccharomyces cerevisiae*
- *Bacillus pumilus*

2.5 Companion Animals (Dogs, Cats)

Probiotics like *Lactobacillus acidophilus*, *Bifidobacterium animalis*, and *Enterococcus faecium* provide:

- *Digestive health*: Support gastrointestinal health and improve nutrient absorption.
- *Immune system regulation*: Enhance immune responses, helping to prevent infections and inflammation.
- *Balanced gut microbiota*: Improve the balance of gut flora, reducing the risk of gastrointestinal disorders.

- *Overall well-being*: Enhance the overall health of companion animals by supporting digestion and immunity.

Strains:
- *Lactobacillus acidophilus*
- *Bifidobacterium animalis*
- *Enterococcus faecium*

3 Competitive Exclusion

Competitive exclusion is a mechanism by which probiotics outcompete pathogenic microorganisms for space, nutrients, and adhesion sites in the gut. This process prevents the colonization and proliferation of pathogens, thereby promoting a healthy gut microbiota. Emerging research highlights the role of advanced genomic tools in identifying specific probiotic strains with superior competitive exclusion capabilities (Hutkins et al. 2022).

3.1 Adhesion to the Gut Epithelium

Probiotics adhere to the gut epithelium, preventing pathogens from colonizing and causing infections. This adhesion is mediated by various adhesins, such as proteins, carbohydrates, and lipids. The adhesins on the surface of probiotics interact with specific receptors on the gut epithelial cells, facilitating stable attachment and hindering pathogen colonization. Studies have shown that the adhesion of probiotics, such as *Lactobacillus* species, is a critical factor in their ability to exert beneficial effects (Collado et al. 2007). For example, *Lactobacillus rhamnosus* GG has demonstrated high adhesion efficiency, making it a model probiotic in adhesion studies (Azcarate-Peril et al. 2011) (Table 1).

Table 1 Comparison of adhesion efficiencies of probiotic strains

Probiotic strain	Adhesion efficiency (%)	Pathogens inhibited	References
Lactobacillus rhamnosus GG	85%	*E. coli*, *Salmonella* spp.	Azcarate-Peril et al. (2011)
Bifidobacterium breve	72%	*Helicobacter pylori*	Fanning et al. (2012)
Lactobacillus plantarum	65%	*Clostridium difficile*	Lin et al. (2018)

3.2 Production of Antimicrobial Compounds

Probiotics produce antimicrobial compounds, such as bacteriocins, hydrogen peroxide, and organic acids, which inhibit the growth of pathogens. These compounds are synthesized through mechanisms like fermentation and enzymatic reactions, creating a hostile environment for pathogenic microorganisms (Pérez-Díaz et al. 2020). Recent advancements have identified novel bacteriocins, such as plantaricin, with enhanced pathogen inhibition potential (Chen et al. 2020).

3.3 Competition for Nutrients

Probiotics compete with pathogens for essential nutrients, such as carbohydrates, proteins, and vitamins, limiting the resources available for pathogen survival and proliferation. For instance, *Lactobacillus plantarum* has been shown to outcompete *Escherichia coli* in nutrient-rich environments (Lin et al. 2018). A recent study highlights how probiotics metabolize oligosaccharides more efficiently, depriving pathogens of critical energy sources (Tzortzis et al. 2022).

3.4 Modulation of the Gut Environment

Probiotics modulate the gut environment by altering factors such as pH, redox potential, and temperature, making it less conducive for pathogenic growth. The production of lactic acid by probiotics like *Lactobacillus acidophilus* lowers the pH, creating an unfavorable condition for many pathogens (Marco et al. 2006) (Table 2).

4 Production of Antimicrobial Compounds

Probiotics synthesize various antimicrobial compounds that inhibit pathogen growth and activity through diverse mechanisms.

Table 2 Key environmental modulations by probiotics

Factor	Impact of probiotics	Examples
pH	Lowered, inhibiting alkaline bacteria	*Lactobacillus acidophilus*
Redox potential	Balanced to disrupt pathogen survival	Reactive oxygen species
Nutrient availability	Restricted for pathogens	Carbohydrate competition

4.1 Bacteriocin Production

Bacteriocins are proteinaceous compounds produced by probiotics that inhibit the growth of closely related bacterial species. These compounds, such as nisin and pediocin, function by disrupting bacterial membranes or inhibiting cell wall synthesis (Cotter et al. 2005). Recent research suggests that engineered probiotics can produce tailored bacteriocins with broader spectra of activity (Sánchez et al. 2021).

4.2 Hydrogen Peroxide Production

Hydrogen peroxide is a potent antimicrobial agent produced by probiotics through enzymatic reactions. It creates oxidative stress for pathogens, impairing their cellular functions and viability (Martin et al. 2019). Modern studies have focused on enhancing hydrogen peroxide production through genetic modifications (Xu et al. 2023).

4.3 Organic Acid Production

Probiotics produce organic acids, including lactic acid and acetic acid, during fermentation. These acids lower the pH of the gut environment, inhibiting the growth of pH-sensitive pathogens such as *Clostridium difficile* (Gänzle 2020) (Table 3).

5 Modulation of the Immune System

Probiotics influence the immune system, enhancing its ability to fight infections and maintain homeostasis.

Table 3 Antimicrobial compounds produced by probiotics

Compound type	Example compound	Mechanism of action	Pathogen inhibited	References
Bacteriocin	Nisin	Disrupts bacterial membranes	*Listeria monocytogenes*	Sánchez et al. (2021)
Hydrogen peroxide	Reactive oxygen species	Oxidative stress to pathogens	*Candida albicans*	Martin et al. (2019)
Organic acids	Lactic acid, acetic acid	Lowers pH, inhibits growth	*Clostridium difficile*	Gänzle (2020)

Table 4 Cytokine modulation by probiotics

Cytokine	Effect of modulation	Probiotic strain involved	References
IL-10	Anti-inflammatory	*Lactobacillus casei Shirota*	Ng et al. (2009)
TNF-α	Reduced inflammation	*Bifidobacterium longum*	Vinderola et al. (2020)
IFN-γ	Enhanced antiviral defense	*Lactobacillus rhamnosus*	Ouwehand et al. (2002)

5.1 Stimulation of Immune Cells

Probiotics activate immune cells such as macrophages, dendritic cells, and T cells, stimulating the production of cytokines and other immune molecules. This immune activation enhances the host's ability to combat infections and reduce disease severity (Ouwehand et al. 2002). Recent studies suggest that probiotics enhance the maturation of dendritic cells, improving antigen presentation (Vinderola et al. 2020) (Table 4).

6 Production of Anti-adhesive Compounds

Probiotics produce anti-adhesive compounds that prevent pathogens from adhering to the gut epithelium, reducing their ability to establish infections.

6.1 Production of Exopolysaccharides

Probiotics produce exopolysaccharides—complex carbohydrates that act as a physical barrier preventing pathogen adhesion. For instance, *Bifidobacterium breve* produces exopolysaccharides that inhibit the adhesion of *Helicobacter pylori* (Fanning et al. 2012). Recent research has identified exopolysaccharides with biofilm-disrupting properties (Patel et al. 2023).

6.2 Production of Biosurfactants

Biosurfactants are surface-active molecules synthesized by probiotics, disrupting pathogen adhesion to epithelial cells. *Lactobacillus rhamnosus* is known to produce biosurfactants that inhibit the adhesion of *Salmonella enterica* (Gudiña et al. 2010) (Table 5).

Table 5 Comparison of anti-adhesive compounds

Compound type	Mechanism	Pathogens targeted	References
Exopolysaccharides	Physical barrier	*Helicobacter pylori*	Fanning et al. (2012)
Biosurfactants	Disrupts adhesion	*Salmonella enterica*	Gudiña et al. (2010)
Mucins	Enhances mucosal barrier	*Campylobacter jejuni*	Lindén et al. (2008)

7 Alteration of the Gut Environment

Probiotics alter the gut environment to create unfavorable conditions for pathogen growth.

7.1 *Modulation of Gut pH*

Probiotics produce organic acids, such as lactic acid and acetic acid, during fermentation, which lower the gut pH. This acidic environment inhibits the growth of pH-sensitive pathogens, including *Salmonella* and *Clostridium difficile* (Marco et al. 2006).

7.2 *Reduction of Redox Potential*

By producing metabolites that influence oxidation–reduction reactions, probiotics can balance redox potential in the gut, disrupting the metabolic activities of pathogens that rely on specific redox conditions (Gänzle 2020).

7.3 *Enhancement of Gut Barrier Function*

Probiotics can strengthen the gut barrier by enhancing tight junction protein expression, reducing permeability, and preventing pathogen translocation across the epithelial layer (Turner 2009) (Table 6).

8 Conclusion

Probiotics serve as essential agents in fostering gut health and overall vitality in both humans and animals. Their multifaceted mechanisms—ranging from competitive exclusion and antimicrobial secretion to immune modulation—play a crucial role in maintaining a balanced intestinal microbiota and mitigating the proliferation of harmful pathogens. In animal husbandry, probiotics present a sustainable,

Table 6 Environmental modifications by probiotics

Environmental factor	Probiotic action	Impact on pathogens	References
pH	Acidification	Inhibits *Salmonella* spp.	Marco et al. (2006)
Redox potential	Balancing redox reactions	Disrupts *C. difficile*	Gänzle (2020)
Gut barrier function	Enhanced tight junctions	Blocks pathogen translocation	Turner (2009)

antibiotic-alternative solution, enhancing digestive efficiency, fortifying immune defenses, and improving overall productivity. The increasing incorporation of probiotics across various species, including ruminants, poultry, swine, aquaculture, and companion animals, highlights their transformative potential for optimizing livestock management and ensuring food safety. With ongoing advancements in probiotic research, the identification of novel strains promises to further refine their therapeutic efficacy and contribute to the development of eco-friendly, resilient farming systems.

9 Limitations of the Study

While this chapter highlights the various beneficial mechanisms of probiotics, several limitations must be acknowledged:

1. *Strain-Specific Variability*: The effects of probiotics are highly strain-specific. Generalizing results from one strain to another may lead to inaccuracies.
2. *Short-Term Studies*: Many studies on probiotics are short-term, making it difficult to assess long-term impacts on health and disease prevention.
3. *Limited Understanding of Mechanisms*: Although several mechanisms have been proposed, the exact molecular pathways through which probiotics exert their effects remain partially understood.
4. *Host-Specific Responses*: Probiotic efficacy can vary significantly between individuals due to differences in gut microbiota composition, genetics, and lifestyle factors.
5. *Regulatory Challenges*: The lack of standardized guidelines for probiotic formulations and dosages poses challenges for both research and clinical application.
6. *Safety Concerns*: While probiotics are generally considered safe, there is a potential risk of infections or adverse effects, particularly in immunocompromised individuals.

Future research should focus on addressing these limitations by conducting long-term, multi-strain studies and developing a deeper understanding of probiotic–host interactions.

10 Strategic Recommendations for Advancing Probiotic Research and Application

1. Enhancing Strain-Specific Efficacy: Future research should focus on improving the efficacy of specific probiotic strains by investigating their mechanisms of action and how they can be tailored to the needs of different animal species. This includes exploring strain-specific properties like antimicrobial production, adhesion ability, and immune modulation to optimize their use in animal nutrition.
2. Long-Term Impact Assessment: While short-term studies have demonstrated the benefits of probiotics in animal health, it is essential to conduct long-term trials to evaluate their lasting effects, including any potential side effects or changes in gut microbiota composition over time. This will help ensure the safe and sustainable use of probiotics.
3. Standardization of Dosage and Application Protocols: The development of standardized protocols for the use of probiotics in animal husbandry is crucial. Research should focus on determining optimal dosages, frequency of administration, and methods of delivery across various species, ensuring consistency and effectiveness in probiotic supplementation.
4. Exploration of Novel Probiotic Sources: Researchers should investigate new sources of probiotics, particularly non-traditional or underexplored microorganisms, to expand the range of beneficial strains available. By exploring diverse ecosystems, including marine and extreme environments, novel probiotics with unique properties could be discovered.
5. Integration with Sustainable Agricultural Practices: Probiotics should be integrated into broader sustainable agricultural practices. This includes combining probiotics with organic farming, precision livestock farming, and other environmentally friendly practices to reduce reliance on antibiotics and enhance the overall health and productivity of livestock.
6. Cross-Species Health Benefits: Probiotics should be further explored for their potential dual benefits for both animal and human health, particularly in areas like gut microbiota modulation and disease resistance. Collaborative research between human and veterinary medicine could lead to more integrated probiotic solutions for both animals and humans.
7. Impact of Probiotics on Gut Microbiome Diversity: Research should explore how probiotics can contribute to maintaining or enhancing microbiome diversity in animal guts. This will help in understanding how probiotics interact with native microbial populations and influence overall health, especially in preventing dysbiosis and promoting resilience to pathogens.
8. Probiotics as Part of Holistic Animal Health Programs: Probiotics should be considered as an integral part of holistic animal health management programs, alongside nutrition, vaccinations, and environmental health. This comprehensive approach will promote sustainable practices in livestock production, leading to healthier animals and fewer diseases.

References

Azcarate-Peril MA et al (2011) Lactobacillus rhamnosus GG adhesion efficiency. J Appl Microbiol 110:1421–1428

Bottalico L, Castellaneta F, Charitos IA (2020) From hydrotherapy to the discovery of the gut microbiota: the historical gastrointestinal health concept. Pharmacophore 11(2):82–90

Chaucheyras-Durand F, Durand H (2010) Probiotics in animal nutrition and health. Benef Microbes 1:3–9

Chen J et al (2020) Plantaricin production by probiotics and its potential for pathogen inhibition. Microorganisms 8:1614

Collado MC et al (2007) Adhesion of Lactobacillus species to the intestinal epithelium and its role in the probiotic effects. Microb Ecol Health Dis 19:58–67

Cotter PD et al (2005) Bacteriocins: novel antimicrobial compounds from lactic acid bacteria. Food Res Int 38:1127–1141

Fanning S et al (2012) Exopolysaccharides production by Bifidobacterium breve and its role in pathogen inhibition. Food Control 23:123–130

Farag EF, Abdel-Alim GA (2020) The effect of Bacillus subtilis on the bacterial content in rabbits caeci. J Appl Vet Sci 5(4):10–16

Gänzle MG (2020) Organic acids in probiotics: their role in pathogen inhibition. Food Res Int 137:109453

Gogineni VK, Morrow LE, Gregory PJ, Malesker MA (2013) Probiotics: history and evolution. J Anc Dis Prev Rem 1:1–7

Gudiña EJ et al (2010) Biosurfactants produced by Lactobacillus rhamnosus and their impact on Salmonella adhesion. J Appl Microbiol 109:1203–1211

He J, Wu C, Zhang Y (2021) Probiotics in companion animals: a review of their effects on health. J Anim Sci 99:171–182. https://doi.org/10.1093/jas/skab221

Hutkins RW et al (2022) Advanced genomic tools in probiotic research. Food Res Int 159:105107

Lin X et al (2018) Probiotic competition for nutrients in the gut and its effect on pathogen inhibition. Microorganisms 6:47

Lin X et al (2020) Probiotic characteristics of Lactobacillus plantarum AR113 and its molecular mechanism of antioxidant. LWT 126:109278

Lindén SK, Florin TH, McGuckin MA (2008) Mucin dynamics in intestinal bacterial infection. PLOS ONE 3(12):e3952

Marco ML et al (2006) Probiotic strains and modulation of the gut environment. Int Dairy J 16:854–859

Martin R et al (2019) Hydrogen peroxide production by probiotics as an antimicrobial mechanism. J Microb Biotechnol 12:1147–1156

Miller EA et al (2024) Survey of antimicrobial and probiotic use practices in wildlife rehabilitation in the United States. PLoS One 19:e0308261

Ng QS et al (2009) Anti-inflammatory cytokine modulation by Lactobacillus casei Shirota. J Clin Immunol 29:551–560

Ouwehand A et al (2002) Effects of probiotics on immune response in the gut. Clin Nutr 21:222–229

Patel S et al (2023) Exopolysaccharides with biofilm-disrupting properties in probiotics. Bioorg Chem 123:105688

Pérez-Díaz JI et al (2020) Antimicrobial production by probiotics and its role in gut health. J Food Sci 85:2742–2753

Sánchez B et al (2021) Engineered probiotics producing tailored bacteriocins. Nat Biotechnol 39:1067–1074

Santacroce L, Charitos IA, Bottalico L (2019) A successful history: probiotics and their potential as antimicrobials. Expert Rev Anti-Infect Ther 17:635–646

Shen J et al (2022) Tibetan pig-derived probiotic Lactobacillus amylovorus SLZX20-1 improved intestinal function via producing enzymes and regulating intestinal microflora. Front Nutr 9:846991

Turner JR (2009) Intestinal mucosal barrier function in health and disease. Nat Rev Immunol 9:779–790

Tzortzis G et al (2022) Probiotics and oligosaccharide metabolism in the gut. Food Res Int 151:110737

Vinderola G et al (2022) The concept of postbiotics. Foods 11:1077

Vinderola G et al (2020) Probiotic modulation of dendritic cell maturation. Clin Immunol 215:108461

Xu R et al (2023) Genetic modification for enhanced hydrogen peroxide production by probiotics. Appl Microbiol Biotechnol 107:2121–2133

Probiotics, Prebiotics, and Synbiotics: Enhancing Gut Health and Productivity in Animal Production

Razia Kausar, Rabia Bashir, Tayyba Batool, Ghulam Murtaza, and Zahida Nasreen

Abstract Probiotics and prebiotics perform a vital role in improving the health, nutrition, and productivity of animals. Probiotics are live microorganisms that confer nutritional and health benefits to the host by improving gut microbes, improving digestion, and boosting the immune system. The influences of probiotics are species-specific and perform a definitive function in the host, for example, the *Lactobacillus, Bifidobacterium*, and *Saccharomyces* strains are responsible for the modulation of intestinal microbes and inhibit pathogens. Prebiotics, on the other hand, are non-digestible dietary fibers that stimulate the growth and activity of beneficial bacteria selectively. The combined application of both these products is known as synbiotics, which has gained attention for its synergistic effects on growth performance, feed efficiency, and disease resistance. These functional feed supplements are also beneficial as a replacement for the antibiotics growth promoters (AGPs), along with the increasing demand for sustainable and antibiotic-free livestock production. This chapter explores the applications, sources, mechanisms, and benefits of probiotics and prebiotics in animal nutrition, emphasizing their role in balancing gut health, increasing immune responses, and developing sustainable animal production. Future research is needed to focus on strain-specific effects, dosage optimization, and long-term effects on animal species in livestock and aquaculture.

R. Kausar (✉)
Department of Anatomy, Faculty of Veterinary Science, University of Agriculture, Faisalabad, Pakistan
e-mail: razia.kausar@uaf.edu.pk

R. Bashir · T. Batool
Department of Zoology, Wildlife and Fisheries, University of Agriculture, Faisalabad, Pakistan

G. Murtaza
Department of Anatomy, Faculty of Veterinary Science, University of Agriculture, Faisalabad, Pakistan

Faculty of Veterinary and Animal Sciences, Gomal University, Dera Ismail Khan, Pakistan

Z. Nasreen
Department of Zoology, University of Mianwali, Mianwali, Pakistan

R. U. Khan, S. Naz (eds.), *Green Alternatives to Antibiotics in Animal Health*,
https://doi.org/10.1007/978-981-92-0309-3_13

Keywords Probiotics · Gut microbes · Prebiotics · Synbitics · Growth promoters · Feed supplements · Animal nutrition

1 Introduction

It is predicted that there will be nine billion people on the planet by 2050. The need for food derived from plants and animals is inextricably linked to the ongoing expansion of the human population. Because of this, scientists are searching for ways to enhance food production with low costs and adhere to strict quality and safety regulations. The production of higher-quality food is impacted by the types of feed additives utilized (Markowiak and Katarzyna 2018).

Animal intestinal diseases like Salmonella and Yersinia are an obvious cause of contamination of food and zoonosis. Animal production is thus inextricably linked to consumer nutrition and health. As a result, new techniques for animal breeding are used with the goal of improving the safety and quality of meat while also taking preservation of the environment and animal welfare into consideration (Kumar et al. 2023). Strict requirements must be met by both livestock feed and nutritional supplements without raising the price of animal breeding at the same time. Probiotics are "live strains of strictly selected microorganisms which when administered in adequate amounts, confer a health benefit on the host." The International Scientific Association for Probiotics and Prebiotics (ISAPP) maintained the definition in 2013 (Hill et al. 2014).

Prebiotics are utilized as supplements to feed in addition to probiotics. In 1995, the prebiotic idea was originally proposed. The fact that they may now be included in other groups, such as animal nutrition, rather than only human food, is another significant factor. On the other hand, standards have been upheld regarding the specific processes of microbiota regulation and the existence of proven positive impacts on host (Gibson et al. 2017). A mixture of both, probiotics and prebiotics that have beneficial effects for living organisms and also increases their survival is termed as synbiotic (Cencic and Chingwaru 2010). Enhancing the survival of probiotic bacteria in the gastrointestinal system is the primary objective of that kind of combination. Synbiotics were developed to address potential challenges in probiotic survival in the gastrointestinal system and possess both prebiotic and probiotic qualities (Rioux et al. 2005). Probiotics have a positive effect on intestinal balance and provide a barrier of defense for the digestive system. Conversely, prebiotics provide probiotic bacteria with nourishment and energy (Roberfroid 2007).

2 Applications of Probiotics in Animals Nutrition

Probiotics, which are classified as non-pathogenic bacteria, are now often employed as feed additions for farmed animals. By maintaining a healthy gastrointestinal environment and enhancing its function, their usage aims to increase production performance and avoid diseases (Mountzouris et al. 2009). Furthermore, probiotics improve the rumen's microbial ecosystem as described in Fig. 1 (Musa et al. 2009), feed conversion rate (Whitley et al. 2009), and digestibility and absorption of nutrients (Abd El-Ghani 2004). These improvements improve the animals' production performance. According to Guillot (2003), probiotics may potentially have an antagonistic effect on pathogenic microbes by blocking their attachment sites.

By promoting fiber breakdown in vitro (Rodrigues et al. 2008), in situ (Krueger et al. 2008), and in vivo (Salem et al. 2007), scientists have recently shown that adding fiber-degrading enzymes to the feeds of dairy cows and livestock in feedlots can enhance feed utilization and animal performance. Enzyme feeding frequently results in higher feed intake, which may be partially owing to the diet's enhanced palatability as a result of the sugars produced during the hydrolysis of pre-ingestive fiber. To decrease gut fill and increase feed intake, post-ingestive activities of enzymes, such as enhanced digestive rate is required (Midhun et al. 2019).

By raising the overall numbers of ruminal microbes, directly fed enzymes can also enhance the microbial colonization, which will accelerate the pace at which fiber is broken down and the synthesis of proteins by rumen bacteria (Nsereko et al.

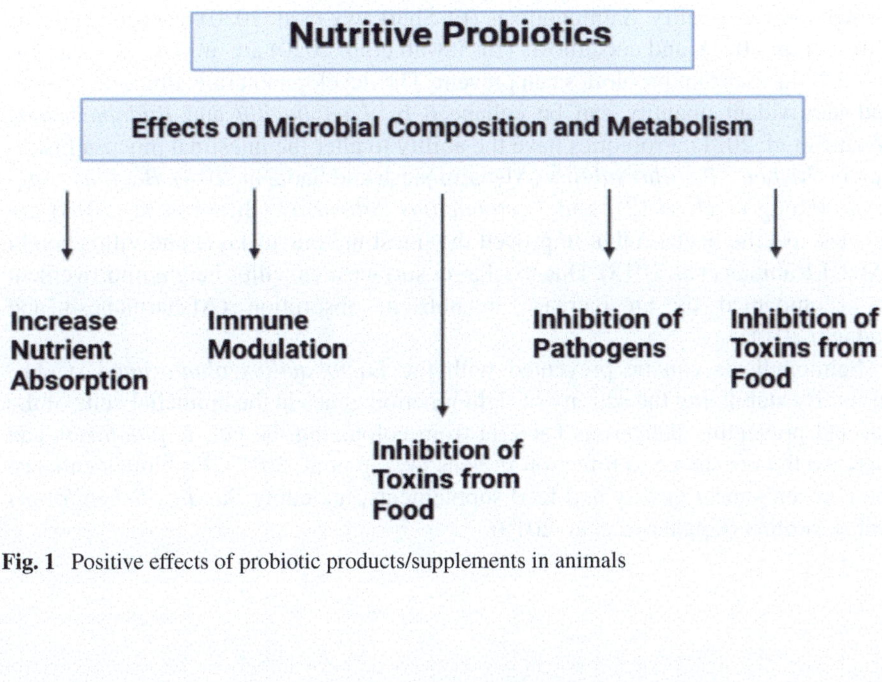

Fig. 1 Positive effects of probiotic products/supplements in animals

2002). Due to economic and environmental concerns, the introduction of exogenous enzymes as supplements in chicken's diets has attracted a lot of interest. Although the effectiveness of a number of commercial enzyme products has been clearly demonstrated, their route of action is still somewhat unclear. Furthermore, the development of the immunological organs was hastened by the addition of enzymes (Kalmendal and Tauson 2012).

Probiotics are thought to be extremely beneficial for controlling fish infections because they can increase nutritional competitiveness and the development of antimicrobial chemicals. The use of microorganisms as probiotics and an alternative to antibiotics is necessary in aquaculture due to ongoing infectious issues (Irianto and Austin 2002). According to a trials' findings, probiotics used as antibiotics can enhance fish development (Kumar et al. 2006). Aquaculture uses a wide variety of microbial strains as antibiotics. *Enterococcus*, *Bicdobacterium*, *Bacillus*, *Saccharomyces*, *Vibrio*, *Lactobacillus*, and *B. subtilis* are among the major antibiotics used in aquaculture. Fish feed containing *Bacillus* species is applied to enhance feed utilization, growth, and bacterial infection control (Asghar et al. 2022).

3 Species-Specific Effects

3.1 Role of Probiotics in Poultry

According to several studies (Abd El-Hack et al. 2020), probiotics can increase the growth rates of poultry. Salmonellosis (El-Sharkawy et al. 2020), necrotic enteritis (Anee et al. 2021), and coccidiosis (El-Sawah et al. 2020) are among the gastrointestinal illnesses that probiotics can prevent. The development rate, immune system, and antioxidant quantity can be enhanced by *Lactobacillu* and *Bifdobacterium* (Zhang et al. 2021). Probiotics have the ability to alter the intestinal mucosa histology in chickens. *Bacillus subtilis* (Afsharmanesh and Sadaghi 2014), *Bacillus coagulans* (Hung et al. 2012), and *Lactobacillus salivarius* (Biloni et al. 2013) are species-specific bacteria that improved the intestinal mucus layer and villus height (Abdel-Rahman et al. 2013). Due to a larger surface area, villus height improvement is accompanied by an increase in nutrient absorption (Afsharmanesh and Sadaghi 2014).

Salmonellosis can be prevented with the *Lactobacillus plantarum* LTC-113 strain. By stabilizing the activity of tight junction genes in the epithelial cells of the gut and preventing dangerous bacteria from colonizing the gut, *L. plantarum* can increase the resistance to infection in hens (Wang et al. 2018). Probiotics enhance the chicken's meat quality and feed supplements, including *Bacillus licheniformis* and *B. subtilis* (Upadhaya et al. 2019).

3.2 *Role of Probiotics in Ruminants*

Lactic acid bacteria (Weimer 2015) and *Saccharomyces cerevisiae* (Elghandour et al. 2020) are examples of probiotics that are commonly used for ruminants and provide substantial health benefits. Ruminants, when given the probiotics as feed supplements shows the gain in their body weight. For example, a probiotic combination of *Enterococcus fecium*, *Lactobacillus reuteri*, and *L. alimentarius* improved the goat's normal bodyweight by 9% in just 2 months (Apas et al. 2010). Probiotics can help dairy cattle produce more milk. Probiotic microbes like *Bifidobacterium bifidum* can prevent milk allergy reactions, whereas *Bacillus subtilis* can enhance milk production (Ma et al. 2020).

3.3 *Role of Probiotics in Aquaculture*

A wide variety of probiotics are used by farmers in fish farming, and *Bacillus subtilis* is common among all (Olmos et al. 2020). According to Kuebutornye et al. (2020), *Bacillus* probiotics by themselves can reduce the number of harmful microbes in fish, including *Aeromonas*, *Vibrio*, *Pseudomonas,* and *Streptococcus*. *Lactobacillus* spp. (Giri et al. 2013) is also utilized as probiotics in aquaculture. Aquatic animals' growth rates, consumption of feed, and digestive enzyme functions can all be enhanced by probiotics. Probiotics effectively aid in aquatic animals' nutritional digestion by producing extracellular enzymes such as lipase, protease, and carbohydrase (Anee et al. 2021).

The development rate and larval metamorphosis of white leg prawns can be accelerated by feeding them the supplements of probiotics as well as beneficial microorganisms, including gram-positive cocci, Bifidobacterium, and Photosynthetic Bacteria. Additionally, it reduces the quantity of *vibrio* infections and encumbers partial molting during development (Wang et al. 2020). Fish immune responses can be improved by probiotics. In addition, the immunity is strengthened by *Bacillus pumilus* and *B. licheniformis* in Nile tilapia (Ramesh et al. 2015).

4 Benefits/Advantages of Probiotics

All animals and humans' digestive systems and overall health are significantly influenced by intestinal probiotics, especially bacteria (Fuller 1992). The selected type, the consumed amount, the frequency and length of exposure and the most important individual's physiological state are some of the variables that will affect the probiotic's benefits (Kopp-Hoolihan 2001). The following are the few advantages of using probiotics practically:

(i) *Farm Animal Growth Promotion*: Probiotic bacteria break down hydrocarbons, which implies the meal is being broken down into its fundamental elements. This permits nearly complete absorption via the digestive tract. Probiotics significantly improve quick cellular development and growth and general nutrition (Mohammed et al. 2021). For example, in young pigs, *Bifidobacteria* and *Lactobacillus* decreased mortality and enhanced weight gain. Additionally, compared to unsupplemented piglets, supplemented piglets with *Bacillus coagulans* performed more effectively, had reduced mortality, enhanced body growth, and feed conversion (Ezema 2013).

(ii) *Defense Against Intestinal Infections*: Probiotics cleanse the digestive tract as they attach themselves to the intestinal walls behind the layer of sludge and pull out the accumulated debris, and after that, this waste is normally flushed out. Probiotics can help to prevent and even to cure fungal and yeast infections (Oyetayo et al. 2003). By producing antimicrobial molecules (bacteriocins and organic acids), competing with pathogens for nutrients, stimulating intestinal immune responses, and adhering to the intestinal mucosa to prevent attachment, probiotic strains mixed in feed may shield from abdominal pathogens through different potential mechanisms (Ezema 2013).

(iii) *Reduction of Lactose Intolerance*: Most people lose their capacity to digest lactose between the ages of 10 and 20. This results in a reduction in the intake of milk products, which eliminates a high-quality source of calcium and protein. Ingested lactose is hydrolytically digested by *Lactoba acidophilus* and *L. bifidus*. In persons who are lactose intolerant, consuming milk products containing live *Lactobacillus* is therefore more tolerated and may even reduce malabsorption (Fuller 1992).

(iv) *Constipation Relief*: Probiotics rapidly alleviate constipation and restore regular bowel motions. You can take *Lactobacillus* both while and after taking antibiotics. This aids in reducing diarrhea brought on by antibiotics, which kill both "good" and "bad" bacteria in the intestinal tract indiscriminately (Fuller 1992).

(v) *Bioavailability and Nutrient Synthesis*: Certain amino acids, including lysine from particular strains of *Lactobacillus plantarum*, are produced by probiotic bacteria and are subsequently directly absorbed. They generate B vitamins, which are biological catalysts in food metabolism and aid in stress reduction. These include folic acid, niacin, riboflavin, and pantothenic acid (Ezema 2013).

5 Mechanisms of Action of Probiotics

Through the synthesis of antagonistic chemicals or competitive elimination (competition for nutritional attachment sites), probiotic bacteria can suppress infections. An aquatic ecosystem's water quality is enhanced by probiotic bacteria, which directly absorb or break down organic waste. In addition to their potential use in animal nutrition by enhancing feed digestibility and utilization, the diverse

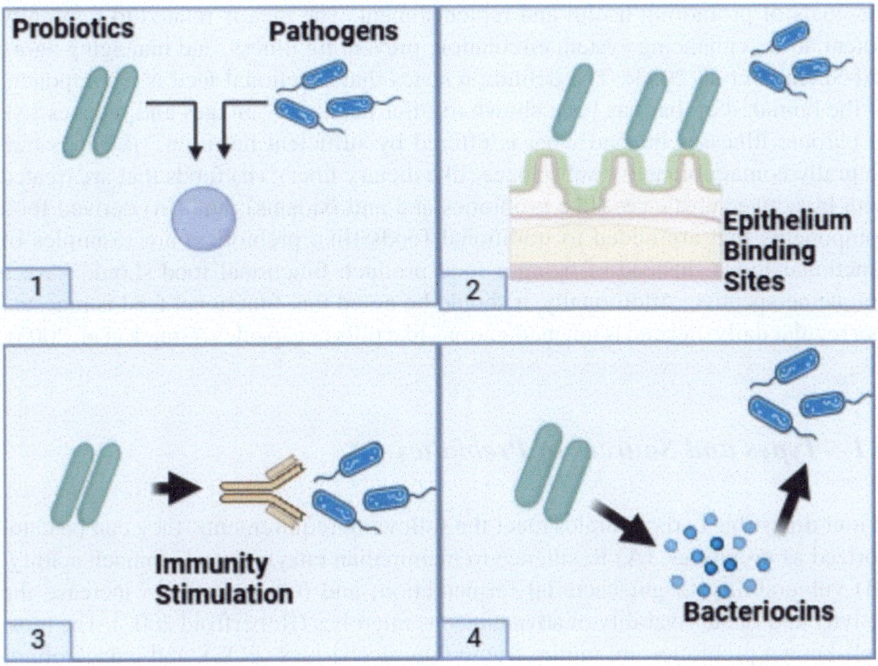

Fig. 2 Mechanism of pathogen inhibition. Probiotics tend to be less likely to be utilized by pathogens and compete with them for vital nutrients (**a**). By decreasing the amount of ground surface accessible for pathogenic colonization, they bind to adhesion sites and lessen pathogen reliance (**b**). Probiotic-induced immune cell signaling results in cytokine release and pathogen death (**c**). Finally, probiotics that directly release bacteriocins from bacteria (**d**) (Khoder et al. 2016)

exoenzymes produced by beneficial microbial cultures, including lipase, amylase, and protease, aid in the breakdown of the unconsumed feed and feces in the pond (Vijayaram and Kannan 2018).

The application of probiotics is one of the main tenets of all microbial therapies to increase output. Probiotics work by inhibiting a pathogen by producing compounds that resemble bacteriocin, competing for attachment sites, competing for nutrients as described in Fig. 2 (especially iron in marine microbes), changing the pathogens' enzymatic activity, stimulating the immune system, and providing nutritional advantages like better feed digestibility and utilization (Kesarcodi-Watson et al. 2008).

6 Prebiotics

Herbal remedies and food have been combined to create a variety of distinctive traditional functional meals. In certain nations, traditionally herbal items are widely utilized as medicine in everyday meals, functional feed, and dietary supplements for the goals of promoting health and replenishment. The idea is related to immunity potentiation, enhancing system circulation, preventing illness, and managing aging (Al-Sherajia et al. 2013). The definition states that functional feed is a component of the human diet that has been shown to offer health advantages and reduces risk of chronic illnesses beyond what is offered by sufficient nutrition. (i) Foods that naturally contain bioactive substances (like dietary fiber); (ii) foods that are treated with bioactive substances (like probiotics and antioxidants); and (iii) derived food components that are added to traditional foods (like prebiotics) are examples of functional foods. Instead of being a food product, functional food should have a unique perspective. Additionally, it should be noted that functional feed is ingested as a regular daily diet and is not medications like pills or capsules (Grajek et al. 2005).

6.1 Types and Sources of Prebiotics

If non-digestible carbohydrates meet the following requirements, they can be categorized as prebiotics. (A) Resilience to mammalian enzymes and stomach acidity; (B) vulnerability to gut bacterial fermentation; and (C) capacity to increase the activity and/or survivability of advantageous microbes (Roberfroid 2002). The most well-known prebiotics are inulin, fructooligosaccharides (FOS), and galactooligosaccharides (GOS). GOS are made of chains of galactose monomers and are indigestible. They are generated from lactose, which is found naturally in mammalian milk. Soluble dietary fibers include inulin and inulin-type fructans (Roberfroid 2005). Furthermore, the same beneficial benefits as inulin-type fructans may be obtained by adjusting the transfer time via the stomach through dietary fiber that contains a variety of non-starch polysaccharides, including cellulose, dextrins, pectins, beta-glucans, waxes, and lignin (Napolitano et al. 2009). Table 1 summarizes a number of prebiotics along with their sources.

Beneficial metabolites are released when host bacteria selectively break down prebiotics. Prebiotics have acquired awareness for their physiological and systemic effects and are no longer only bifidobacteria and lactobacilli enhancers (Ashaolu

Table 1 List of sources and types of prebiotics

Sources	Type	References
Cow's milk	Galactooligosaccharides	Alander et al. (2001)
Asparagus, garlic, onion, wheat, banana, barley, and rye	Fructooligosaccharides	Sangeetha et al. (2005)
Honey, sugarcane juice	Isomaltulose	Lina et al. (2002)
Water-soluble glucans	Cyclodextrins	Singh et al. (2002)
Soybean	Oligosaccharide	Mussatto and Mancilha (2007)
Wheat	Arabinoxylooligosaccharides	Eeckhaut et al. (2008)
Potato starch	Enzyme-resistant dextrin	Barczynska et al. (2012)

2020). Prebiotics may thus not always contain carbohydrates and can be administered to other unusual "substrates" in the body than the stomach, such as the skin, where the probiotics can colonize.

Prebiotics' immune-boosting effects are linked to probiotic stimulation and the production of their metabolites, including short-chain and branched-chain fatty acids. These effects can also originate from other rare microbial taxa, such as roseburia, eubacteria, or faecalibacteria (Guimaraes et al. 2020). Lactulose, an isomer of lactose, is thought to be an excellent prebiotic because it can inhibit gut bacteria like Bacteroides and Enterobacteriaceae to grow while *Bifidobacterium*, *Lactobacillus,* and *Streptococcus* are promoted (Peng et al. 2020). The indigestible carbohydrates from wheat and barley in the colon can also function as prebiotics (Moshfegh et al. 2018).

7 Applications of Synbiotics

As compared to the action of the probiotic or prebiotic separately, a suitable cocktail of both ingredients as a single mixture should provide a better effect. The specific combination of a probiotic and a prebiotic is likely to be linked to the beneficial effect of synbiotics (Scavuzzi et al. 2014). The majority of research conducted in the last 10 years on probiotics and prebiotics suggested that they can enhance animals' immune systems and growth performance. (Akhter et al. 2015). The most widely utilized prebiotics and probiotic microbes in animal feeding were covered in earlier sections. Synbiotic products that combine bacteria from the *Bifdobacterium* or *Lactobacillus* genera appear to be the most commonly used types. Examples of synbiotic formulations used for cattle nutrition that are available in the market are shown in Table 2.

By selectively improving the growth and/or activating the metabolism of one or a number of health-promoting bacteria, a synbiotic product aids the host in improving the ability to survive in the digestive system. The term "synbiotics," which

Table 2 List of commercial synbiotic used in livestock nutrition (Markowiak and Katarzyna 2018)

Brand name	Probiotics	Prebiotics	Subject
Synbiotic poultry (Vetafarm)	*Lactobacillus*: *acidophilus, plantarum, rhamnosus, brevis, Bifdobacterium*: *Streptococcus thermophilus*	Inulin	Poultry
Biomin®	*Enterococcus faecium*	Fructo-oligosacharides	Poultry, pigs
PoultryStar®	*Bifdobacterium animalis, enterococcus faecium, lactobacillus*: *Salivarius, Pediococcus acidilactici,*	Inulin	Poultry
DigestAid™	*Saccharomyces*: *Cerevisiae, boulardii*	Mannano-oligosacharides	Horses

suggests synergism, should only be used for goods when the probiotic organism or organisms are preferentially favored by the prebiotic compounds (Cencic and Chingwaru 2010).

Synbiotics were created to help probiotics overcome potential survival challenges. Observations demonstrating the increased survival of probiotic bacteria throughout their route through the upper gastrointestinal system seem to support the use of synbiotics. Maintaining intestinal balance and a healthy body is facilitated by probiotics' more effective colon implantation as well as the growth-promoting effects of ubiquitous bacteria (Pandey et al. 2015). It is believed that the most important probiotics engaged in synbiotic having antioxidant activities are strains of *Lactobacillus* and *Bifidobacterium*. It has been discovered that synbiotics that combine *L. casei* and inulin are effective compounds that shield the body from the harmful effects of free radicals. Synbiotics may increase the activity of certain antioxidant enzymes and the antioxidant capacity of blood plasma.

8 Conclusion

The world's population is growing steadily. There was an urgent need for some safe yet efficient ways to produce fish and meat to feed this constantly expanding population. In order to address this issue with food production and replace the damaging use of antibiotics in agricultural sectors, probiotics are an essential requirement. However, it is not possible to ensure that any particular probiotic will be safe when used with common strains. Unwanted characteristics of certain probiotics might include virulence factors, transmissible antibiotic resistance, and releasing undesired, harmful biochemical compounds. The different types of probiotics must be selected for each species in a certain habitat in order to avoid any causations. Each probiotic has different effects and responses in various organisms. Therefore, it is necessary to determine the favorable conditions for the probiotics to live, colonize, develop, and affect the hosts in a given environment.

References

Abd El-Ghani AA (2004) Influence of diet supplementation with yeast culture (*Saccharomyces cerevisiae*) on performance of Zaraibi goats. Small Rumin Res 52(3):223–229

Abd El-Hack ME, El-Saadony MT, Shafi ME, Qattan SY, Batiha GE, Khafaga AF, Abdel-Moneim AM, Alagawany M (2020) Probiotics in poultry feed: a comprehensive review. J Anim Physiol Anim Nutr 104(6):1835–1850

Abdel-Rahman HA, Shawky SM, Ouda H, Nafeaa AA, Orabi SH (2013) Effect of two probiotics and bioflavonoids supplementation to the broilers diet and drinking water on the growth performance and hepatic antioxidant parameters. Glob Vet 10(6):734–741

Afsharmanesh M, Sadaghi B (2014) Effects of dietary alternatives (probiotic, green tea powder, and Kombucha tea) as antimicrobial growth promoters on growth, ileal nutrient digestibility, blood parmeters, and immune response of broiler chickens. Comp Clin Pathol 23:717–724

Akhter N, Wu B, Memon AM, Mohsin M (2015) Probiotics and prebiotics associated with aquaculture: a review. Fish Shellfish Immunol 45(2):733–741

Alander M, Matto J, Kneifel W, Johansson M, Kogler B, Crittenden R (2001) Effect of galacto-oligosaccharide supplementation on human fecal micro flora and on survival and persistence of *Bifidobacterium lactis* Bb-12 in the gastrointestinal tract. Int Dairy J 11(10):817–825

Al-Sherajia SH, Amin I, Mohd YM, Shuhaimi M, Rokiah MY, Fouad AH (2013) Prebiotics as functional foods: a review. J Funct Foods 5(4):1542–1553. ISSN 1756-4646

Anee IJ, Alam S, Begum RA, Shahjahan RM, Khandaker AM (2021) The role of probiotics on animal health and nutrition. J Basic Appl Zool 82(52):1–6

Apas AL, Dupraz J, Ross R, González SN, Arena ME (2010) Probiotic administration effect on fecal mutagenicity and microflora in the goat's gut. J Biosci Bioeng 110(5):537–540. https://doi.org/10.1016/j.jbiosc.2010.06.005

Asghar K, Rasool F, Parveen S, Akmal M, Anjum KM, Mahmood S (2022) Effect of probiotics on growth and health status of *Labeo rohita*. Pak J Agric Sci 59(1):135

Ashaolu TJ (2020) Immune boosting functional foods and their mechanisms: a critical evaluation of probiotics and prebiotics. Biomed Pharmacother 130(20):1–11

Barczynska R, Slizewska K, Jochym K, Kapusniak J, Libudzisz Z (2012) The tartaric acid- modified enzymeresistant dextrin from potato starch as potential prebiotic. J Funct Foods 4(4):954–962

Biloni A, Quintana CF, Menconi A, Kallapura G, Latorre J, Pixley C, Layton S, Dalmagro M, Hernandez-Velasco X, Wolfenden A, Hargis BM (2013) Evaluation of effects of EarlyBird associated with FloraMax-B11 on *salmonella enteritidis*, intestinal morphology, and performance of broiler chickens. Poult Sci 92(9):2337–2346

Cencic A, Chingwaru W (2010) The role of functional foods, nutraceuticals, and food supplements in intestinal health. J Nutr 2(6):611–625

Eeckhaut V, Van IF, Dewulf J, Pasmans F, Haesebrouck F, Ducatelle R, Courtin CM, Delcour JA, Broekaert WF (2008) Arabinoxylooligosaccharides from wheat bran inhibit Salmonella colonization in broiler chickens. Poult Sci 87(11):2329–2334

Elghandour MM, Tan ZL, Abu Hafsa SH, Adegbeye MJ, Greiner R, Ugbogu EA, Cedillo Monroy J, Salem AZ (2020) *Saccharomyces cerevisiae* as a probiotic feed additive to non and pseudo-ruminant feeding: a review. J Appl Microbiol 128(3):658–674. https://doi.org/10.1111/jam.14416

El-Sawah AA, Aboelhadid SM, El-Nahass EN, Helal HE, Korany AM, El-Ashram S (2020) Efficacy of probiotic Enterococcus faecium in combination with diclazuril against coccidiosis in experimentally infected broilers. J Appl Microbiol 129(4):1020–1028

El-Sharkawy H, Tahoun A, Rizk AM, Suzuki T, Elmonir W, Nassef E, Shukry M, Germoush MO, Farrag F, Bin-Jumah M, Mahmoud AM (2020) Evaluation of Bifidobacteria and Lactobacillus probiotics as alternative therapy for *Salmonella typhimurium* infection in broiler chickens. J Anim 10(6):1023

Ezema C (2013) Probiotics in animal production: a review. J Vet Med Anim Health 5(11):308–316. ISSN 2141-2529

Fuller R (1992) History and development of probiotics. In: Probiotics. Springer, Dordrecht, pp 1–8. https://doi.org/10.1007/978-94-011-2364-8_1

Gibson GR, Hutkins R, Sanders ME, Prescott SL, Reimer RA, Salminen SJ, Scott K, Stanton C, Swanson KS, Cani PD, Verbeke K (2017) Expert consensus document: The International Scientific Association for Probiotics and Prebiotics (ISAPP) consensus statement on the definition and scope of prebiotics. Nat Rev Gastroenterol Hepatol 14(8):491–502

Giri SS, Sukumaran V, Oviya M (2013) Potential probiotic *Lactobacillus plantarum* VSG3 improves the growth, immunity, and disease resistance of tropical freshwater fish, *Labeo rohita*. Fish Shellfish Immunol 34(2):660–666. https://doi.org/10.1016/j.fsi.2012.12.008

Grajek W, Olejnik A, Sip A (2005) Probiotics, prebiotics and antioxidants as functional foods: a review. Acta Biochim Pol 52(3):665–671

Guillot JF (2003) Probiotic feed additives. J Vet Pharmacol Ther 26(1):52–55

Guimaraes JT, Balthazar CF, Silva R, Rocha RS, Graca JS, Esmerino EA (2020) Impact of probiotics and prebiotics on food texture. Curr Opin Food Sci 33:38–44. https://doi.org/10.1016/j.cofs.2019.12.002

Hill C, Guarner F, Reid G, Gibson GR, Merenstein DJ, Pot B, Sanders ME (2014) Expert consensus document: The International Scientific Association for Probiotics and Prebiotics consensus statement on the scope and appropriate use of the term probiotic. Nat Rev Gastroenterol Hepatol 11(8):506–514

Hung AT, Lin SY, Yang TY, Chou CK, Liu HC, Lu JJ, Wang B, Chen SY, Lien TF (2012) Effects of *Bacillus coagulans* ATCC 7050 on growth performance, intestinal morphology, and microflora composition in broiler chickens. Anim Prod Sci 52(9):874–879

Irianto A, Austin B (2002) Use of probiotics to control furunculosis in rainbow trout, *Oncorhynchus mykiss* (Walbaum). J Fish Dis 25(6):333–342

Kalmendal R, Tauson R (2012) Effects of a xylanase and protease, individually or in combination, and an ionophore coccidiostat on performance, nutrient utilization, and intestinal morphology in broiler chickens fed a wheat-soybean meal-based diet. Poult Sci 91(6):1387–1393

Kesarcodi-Watson A, Kaspar H, Lategan MJ, Gibson L (2008) Probiotics in aquaculture: the need, principles and mechanisms of action and screening processes. Aquaculture 274(1):1–14

Khoder G, Al-Menhali AA, Al-Yassir F, Karam SM (2016) Potential role of probiotics in the management of gastric ulcer. Exp Ther Med 12(1):3–17

Kopp-Hoolihan L (2001) Prophylactic and therapeutic uses of probiotics: a review. J Am Diet Assoc 101(2):229–241

Krueger NA, Adesogan AT, Staples CR, Krueger WK, Kim SC, Littell RC, Sollenberger LE (2008) Effect of method of applying fibrolytic enzymes or ammonia to Bermudagrass hay on feed intake, digestion, and growth of beef steers. J Anim Sci 86(4):882–889

Kuebutornye FK, Abarike ED, Lu Y, Hlordzi V, Sakyi ME, Afriyie G, Wang Z, Li Y, Xie CX (2020) Mechanisms and the role of probiotic Bacillus in mitigating fish pathogens in aquaculture. Fish Physiol Biochem 46:819–841. https://doi.org/10.1007/s10695-019-00754-y

Kumar R, Mukherjee SC, Prasad KP, Pal AK (2006) Evaluation of *Bacillus subtilis* as a probiotic to Indian major carp *Labeo rohita* (Ham.). Aquac Res 37(12):1215–1221

Kumar P, Abubakar AA, Verma AK, Umaraw P, Adewale Ahmed M, Mehta N, Sazili AQ (2023) New insights in improving sustainability in meat production: opportunities and challenges. Crit Rev Food Sci Nutr 63(33):11830–11858

Lina BAR, Jonker D, Kozianowsky G (2002) Isomaltulose (Palatinose): a review of biological and toxicological studies. Food Chem Toxicol 40(10):1375–1381

Ma ZZ, Cheng YY, Wang SQ, Ge JZ, Shi HP, Kou JC (2020) Positive effects of dietary supplementation of three probiotics on milk yield, milk composition and intestinal flora in Sannan dairy goats varied in kind of probiotics. J Anim Physiol Anim Nutr 104(1):44–55. https://doi.org/10.1111/jpn.13226

Markowiak P, Katarzyna S (2018) The role of probiotics, prebiotics and synbiotics in animal nutrition. Gut Pathog 10(1):1–20

Midhun SJ, Neethu S, Arun D, Vysakh A, Divya L, Radhakrishnan EK, Jyothis M (2019) Dietary supplementation of *Bacillus licheniformis* HGA8B improves growth parameters, enzymatic profile and gene expression of *Oreochromis niloticus*. Aquaculture 505:289–296. https://doi.org/10.1016/j.aquaculture.2019.02.064

Mohammed AA, Zaki RS, Negm EA, Mahmoud MA, Cheng HW (2021) Effects of dietary supplementation of a probiotic (*Bacillus subtilis*) on bone mass and meat quality of broiler chickens. Poult Sci 100(3):100906. https://doi.org/10.1016/j.psj.2020.11.073

Moshfegh AJ, Friday JE, Goldman JP, Ahuja JKC (2018) Presence of inulin and oligofructose in the diets of Americans. J Nutr 129(7):1407–1411. https://doi.org/10.1093/jn/129.7.1407

Mountzouris KC, Balaskas C, Xanthakos I, Tzivinikou A, Fegeros K (2009) Effects of a multi-species probiotic on biomarkers of competitive exclusion efficacy in broilers challenged with *Salmonella enteritidis*. Br Poult Sci 50(4):467–478

Musa HH, Wu SL, Zhu CH, Seri HI, Zhu GQ (2009) The potential benefits of probiotics in animal production and health. J Anim Vet Adv 8(2):313–321

Mussatto SI, Mancilha IM (2007) Non-digestible oligosaccharides: a review. Carbohydr Polym 68(3):587–597

Napolitano A, Costabile A, Martin-Pelaez S, Vitaglione P, Klinder A, Gibson GR, Fogliano V (2009) Potential prebiotic activity of oligosaccharides obtained by enzymatic conversion of durum wheat insoluble dietary fiber into soluble dietary fiber. Nutr Metab Cardiovasc Dis 19(4):283–290

Nsereko VL, Beauchemin KA, Morgavi DP, Rode LM, Furtado AF, McAllister TA, Iwaasa AD, Yang WZ, Wang Y (2002) Effect of a fibrolytic enzyme preparation from *Trichoderma longibrachiatum* on the rumen microbial population of dairy cows. Can J Microbiol 48(1):14–20

Olmos J, Acosta M, Mendoza G, Pitones V (2020) *Bacillus subtilis*, an ideal probiotic bacterium to shrimp and fish aquaculture that increase feed digestibility, prevent microbial diseases, and avoid water pollution. Arch Microbiol 202:427–435. https://doi.org/10.1007/s00203-019-01757-2

Oyetayo VO, Adetuyi FC, Akinyosoye FA (2003) Safety and protective effect of *Lactobacillus acidophilus* and *Lactobacillus casei* used as probiotic agent in vivo. Afr J Biotechnol 2(11):448–452

Pandey KR, Suresh RN, Babu VV (2015) Probiotics, prebiotics and synbiotics- a review. J Food Sci Technol 52(12):7577–7587

Peng M, Tabashsum Z, Anderson M, Truong A, Ashley K, Joselyn P, Ahlam A, Jacob B, Shaik O, Debabrata B (2020) Effectiveness of probiotics, prebiotics, and prebiotic-like components in common functional foods: comprehensive reviews in food science and food safety. Compr Rev Food Sci Food Saf 19(4):1908–1933. https://doi.org/10.1111/1541-4337.12565

Ramesh D, Vinothkanna A, Rai AK, Vignesh VS (2015) Isolation of potential probiotic *Bacillus* spp. and assessment of their subcellular components to induce immune responses in *Labeo rohita* against *Aeromonas hydrophila*. Fish Shellfish Immunol 45(2):268–276. https://doi.org/10.1016/j.fsi.2015.04.018

Rioux KP, Madsen KL, Fedorak RN (2005) The role of enteric microfora in infammatory bowel disease: human and animal studies with probiotics and prebiotics. Clin J Gastroenterol 34(3):465–482

Roberfroid MB (2002) Functional food concept and its application to prebiotics. Dig Liver Dis 34(2):105–110

Roberfroid MB (2005) Introducing inulin-type fructans. Br J Nutr 93(1):13–25

Roberfroid MB (2007) Prebiotics: the concept revisited1. J Nutr 137(3):830–837

Rodrigues MA, Pinto P, Bezerra RM, Dias AA, Guedes CV, Cardoso VM, Cone JW, Ferreira LM, Colaco J, Sequeira CA (2008) Effect of enzyme extracts isolated from white-rot fungi on chemical composition and in vitro digestibility of wheat straw. Anim Feed Sci Technol 141(3–4):326–338

Salem AF, El-Adawy M, Gado H, Khalil M (2007) Feed intake, nutrient digestibility and animal growth performance in Sheep and Goats fed wheat straw. J Dairy Sci 90:107

Sangeetha PT, Ramesh MN, Prapulla SG (2005) Recent trends in the microbial production, analysis and application of fructooligosaccharides. Trends Food Sci Technol 16(10):442–457

Scavuzzi BM, Henrique FC, Miglioranza LHS (2014) Impact of prebiotics, probiotics and synbiotics on components of the metabolic syndrome. Ann Nutr Disord Ther 1(2):1–13

Singh M, Sharma R, Banerjee VC (2002) Biotechnological applications of cyclodextrins. Biotechnol Adv 20(5–6):341–359

Upadhaya SD, Rudeaux F, Kim IH (2019) Efficacy of dietary *Bacillus subtilis* and *Bacillus licheniformis* supplementation continuously in pullet and lay period on egg production, excreta microflora, and egg quality of Hyline-Brown birds. Poult Sci 98(10):4722–4728. https://doi.org/10.3382/ps/pez184

Vijayaram S, Kannan S (2018) Probiotics: the marvelous factor and health benefits. Biomed Biotechnol Res J 2(1):1–8

Wang L, Li L, Lv Y, Chen Q, Feng J, Zhao X (2018) *Lactobacillus plantarum* restores intestinal permeability disrupted by *Salmonella* infection in newly-hatched chicks. Sci Rep 8(1):2229

Wang R, Guo Z, Tang Y, Kuang J, Duan Y, Lin H, Jiang S, Shu H, Huang J (2020) Effects on development and microbial community of shrimp *Litopenaeus vannamei* larvae with probiotics treatment. AMB Express 10:1–4. https://doi.org/10.1186/s13568-020-01041-3

Weimer PJ (2015) Redundancy, resilience, and host specificity of the ruminal microbiota: implications for engineering improved ruminal fermentations. Front Microbiol 6:296. https://doi.org/10.3389/fmicb.2015.00296

Whitley NC, Cazac D, Rude BJ, Jackson-O'Brien D, Parveen S (2009) Use of commercial probiotics supplement in meat goats. J Anim Sci 87(3):723–728

Zhang L, Zhang R, Jia H, Zhu Z, Li H, Ma Y (2021) Supplementation of probiotics in water beneficial growth performance, carcass traits, immune function, and antioxidant capacity in broiler chickens. Open Life Sci 16(1):311–322

Other Non-antibiotic Approaches

From Pathogen to Protection: Modern Vaccination Technologies for Reducing Antibiotic Dependence in Animals

Saima Naz, Ahmad Manan Mustafa Chatha, Sobia Malik, Bushra Irfan, Urwah Ishaque, and Qudrat Ullah

Abstract The global rise in antimicrobial resistance (AMR) is closely tied to the widespread use of antibiotics in animal agriculture. This chapter presents vaccination as a scientifically robust, scalable, and sustainable strategy to reduce antibiotic reliance in livestock and aquaculture. It critically examines traditional and next-generation vaccine technologies, including mRNA, DNA, nanoparticle-based, and DIVA-compatible platforms across species such as cattle, poultry, swine, and fish. Emphasis is placed on immunological mechanisms, species-specific applications, and the integration of vaccination into biosecurity and herd health frameworks. Barriers to adoption ranging from economic constraints to cold chain limitations are analyzed alongside global and regional policy efforts. Case studies from South Asia and Southeast Asia illustrate real-world successes and challenges. The chapter underscores the need for coordinated One Health approaches that couple innovation with inclusive policy, veterinary extension, and surveillance systems to position vaccination as a cornerstone of AMR mitigation and sustainable animal health management.

Keywords Antimicrobial resistance (AMR) · Veterinary vaccines · Immunoprophylaxis · Livestock health · mRNA vaccines · One Health

---

S. Naz (✉) · S. Malik · B. Irfan · U. Ishaque
Department of Zoology, Government Sadiq College Women University, Bahawalpur, Punjab, Pakistan
e-mail: saima.naz@gscwu.edu.pk

A. M. M. Chatha
Department of Entomology, Faculty of Agriculture and Environment, The Islamia University of Bahawalpur, Bahawalpur, Punjab, Pakistan

Q. Ullah
Department of Theriogenology, Faculty of Veterinary and Animal Sciences, Cholistan University of Veterinary and Animal Sciences, Bahawalpur, Punjab, Pakistan

R. U. Khan, S. Naz (eds.), *Green Alternatives to Antibiotics in Animal Health*,
https://doi.org/10.1007/978-981-92-0309-3_14

1 Introduction

Antibiotics have been widely used in animal agriculture for decades, serving therapeutic, prophylactic, and growth-promoting purposes (Estany-Gestal et al. 2024). High-density livestock systems, particularly in poultry, swine, and cattle, rely heavily on sub- therapeutic antibiotic use to maintain productivity (Speksnijder et al. 2024).

As a result, approximately 73% of global antimicrobial consumption occurs in animal production (Gustafson and Bowen 1997). This widespread use contributes significantly to antimicrobial resistance (AMR), especially in zoonotic pathogens such as Salmonella, Campylobacter, and E. coli, which can transmit to humans via food, direct contact, or environmental exposure (Nollkaemper 2025). Regulatory efforts in regions like Sweden and the EU to restrict growth-promoting antibiotics have shown mixed results, curbing AMR but also increasing disease outbreaks due to reduced herd immunity (Barrett et al. 2021).

With new antibiotic development slowing, the urgency for effective alternatives has grown. Although options like bacteriophages, probiotics, and phytochemicals offer some benefits, vaccines stand out as the most targeted, preventive solution (Mulchandani et al. 2024). Unlike antibiotics, vaccines do not leave residues, do not drive resistance, and can reduce infection risks before they occur (Cheng et al. 2014). Global health authorities, including WHO, FAO, and WOAH, advocate vaccination within the One Health framework, emphasizing the interconnectedness of human, animal, and environmental health (Sazykin et al. 2021). AMR is now among the top global health threats, with the WHO estimating 700,000 annual deaths, projected to rise to 10 million by 2050 without urgent action (Sepordeh et al. 2024). Advancements in biotechnology have enabled novel vaccine platforms, such as DNA, mRNA, vector-based, and subunit vaccines, offering scalable, species--specific tools to combat resistant pathogens (Seo et al. 2024). Yet, adoption remains uneven, particularly in low- and middle-income countries (LMICs), where infrastructure, access, and policy limitations persist (Zhang et al. 2024). This chapter explores the role of vaccines as a sustainable, science-based alternative to antibiotics in animal agriculture. It discusses immune mechanisms, vaccine technologies, implementation challenges, and integrated strategies to reduce antibiotic reliance and strengthen One Health resilience (Alhassan et al. 2025).

2 Rationale of the Study

Antimicrobial resistance (AMR) is escalating globally, driven largely by excessive antibiotic use in livestock and aquaculture. Vaccination offers a preventive, species--specific solution that reduces disease incidence and antibiotic reliance without contributing to resistance or leaving harmful residues (Zheng et al. 2025). Despite advances in vaccine technology, such as DNA, mRNA, and nanoparticle platforms, uptake remains limited due to economic, infrastructural, and educational barriers, especially in low-resource settings (Ates et al. 2020). This chapter explores vaccination's role within One Health frameworks, emphasizing its integration into animal health systems as a sustainable, science-driven strategy to combat AMR and support safer, more resilient food production.

3 Immune System and Vaccine Mechanisms

Animal immunity comprises two key components: the innate immune system, which provides immediate but non-specific defense (e.g., skin, mucosal barriers), and the adaptive immune system, which generates targeted, long-lasting protection via B and T lymphocytes (Callaway et al. 2021). Vaccines exploit the adaptive system by introducing antigens, either weakened, killed, or subunit forms of pathogens, to safely stimulate an immune response without causing disease (Alayande et al. 2020).

Upon vaccination, B cells generate antibodies against extracellular pathogens (humoral immunity), while T cells attack infected host cells (cell-mediated immunity), especially crucial for intracellular infections like Brucella (Caneschi et al. 2023). Critically, memory cells formed after vaccination allow for rapid and potent responses upon re-exposure, reducing disease severity and the need for antibiotics (Alayande et al. 2020).

3.1 Species-Specific Examples

- *Poultry*: Newcastle Disease vaccines elicit strong antibody responses, reducing mortality and antibiotic use while preventing zoonotic spillover (Guha et al. 2025).
- *Cattle*: Brucellosis and BVD vaccines stimulate both humoral and cellular immunity, enhancing herd health and reducing therapeutic interventions (Rame Hau et al. 2025).
- *Aquaculture*: Oral and injectable vaccines for Vibrio and Streptococcus species reduce bacterial load and antibiotic contamination in aquatic environments (Thakur et al. 2025).

3.2 Designing Effective Vaccines

Vaccine efficacy depends on the pathogen type, route of infection, and host immune profile. For instance, respiratory pathogens like NDV require mucosal immunity, while gut infections in fish are better addressed via oral or immersion vaccines (Gul et al. 2025). Additionally, species differences such as the simpler adaptive immunity in fish necessitate customized vaccine formulations (Din et al. 2025).

Modern vaccine design aligns immunological understanding with delivery methods, ensuring effective disease prevention without contributing to AMR. This targeted, preventive approach supports the One Health model by safeguarding animal, human, and environmental health (Guo et al. 2025).

4 Strategic Role of Vaccination in Reducing Antibiotic Use

Vaccination is a cornerstone of modern animal health management—not only preventing disease but also significantly reducing the need for antibiotics (Sazykin et al. 2021). Unlike metaphylactic antibiotic use, which treats whole groups based on assumed risk, prophylactic vaccination builds herd immunity without promoting antimicrobial resistance (Hoelzer et al. 2018).

4.1 Prevention of Secondary Bacterial Infections

Vaccination against primary viral pathogens also reduces the incidence of secondary bacterial infections, a major cause of antibiotic overuse. For example, infectious bronchitis in poultry increases susceptibility to E. coli, while swine influenza predisposes pigs to Mycoplasma infections (Saad and Ahmed 2018). Preventing these viral triggers indirectly curbs antibiotic use (Garrido-Mantilla et al. 2025).

4.2 Evidence of Impact

Empirical studies and national programs show significant antibiotic reductions linked to vaccination. Immunizing poultry against Mycoplasma gallisepticum and cattle against clostridial infections reduces disease incidence and the need for antibiotics, often by over 30% (Thakur et al. 2025).

4.3 Biosecurity Integration

Vaccines are most effective when paired with comprehensive herd health strategies—including strict hygiene, biosecurity protocols, and proper nutrition. Clean environments and immune- supportive diets improve vaccine outcomes by reducing pathogen exposure and boosting immune responses (Katz et al. 2024).

4.4 One Health Perspective

From a One Health standpoint, reduced antibiotic use lowers the environmental burden of resistant microbes and drug residues. Manure containing antibiotics or resistant bacteria can contaminate water sources, impacting ecosystems and public health (Reddy et al. 2022). Vaccines help prevent this by reducing disease and the need for antimicrobial treatments.

4.5 Policy Alignment

Vaccination also aligns with international AMR policies. WHO, FAO, and WOAH emphasize its role in preserving antimicrobial efficacy and safeguarding food safety (Cabello and Godfrey 2016). Vaccines help prevent this by reducing disease and the need for antimicrobial treatments (Rodrigues da Costa and Diana 2022).

5 Innovations in Veterinary Vaccine Development

Recent advances in biotechnology have revolutionized veterinary vaccine design, making them safer, more effective, and better suited to global disease challenges and antimicrobial resistance (AMR) (Thomas et al. 2022) (Table 1). These innovations enhance immune targeting, enable species-specific applications, and offer scalable production, especially valuable for low-resource regions and emergent zoonoses.

6 Species-Specific Vaccination Strategies

Vaccination protocols must be tailored to species-specific immune profiles, disease burdens, and production systems. Below is an overview of targeted strategies across key livestock and aquaculture sectors.

6.1 Cattle (Bovine)

- *Brucellosis*: Mass vaccination with Brucella abortus S19 or RB51 is critical to controlling zoonotic transmission. Programs in Pakistan, India, and Latin America integrate these into national eradication efforts with FAO and WHO support (Dadar et al. 2021).

Table 1 Innovations in veterinary vaccine design, technological advancements, implementation challenges, and field readiness

Sr#	Technology	Scientific principle	Key advantages	Major challenges	Current applications/ Development stage	References
1	DIVA Vaccines	Marker/deletion-based; distinguish vaccinated from infected	Supports surveillance and trade	Needs diagnostics; regulatory hurdles	Used in FMD, CSF, AI	Sun et al. (2024)
2	mRNA Vaccines	Encodes antigens via lipid nanoparticles	Broad viral protection	Complex formulation	Used in pandemics	Arevalo et al. (2022)
3	Nanovaccines	Polymer/nanoparticle delivery	Stimulates cell and antibody responses	Stability, targeting	Tested in cancer, malaria	Aarestrup (2015)
4	Reverse Vaccinology	Genome-guided antigen discovery	Efficient recombinant development	In silico limits	Preclinical use	Abbas et al. (2022)
5	Viral Vectored Vaccines	Use viruses to deliver antigens	Strong immunity	Mutation risk	Used for Ebola, SARS-CoV-2	Ahmed et al. (2025)
6	Plant-Based Vaccines	Engineered edible plants	Cost-effective, oral delivery	Regulatory issues	Regulatory issues	Aida et al. (2021)
7	Plant-Based Vaccines	Non-infectious antigen mimic	High safety, strong immunity	Complex manufacturing	HBV, papillomavirus	Akintan et al. (2024)
8	DNA Vaccines	Gene-encoded antigens	Long-lasting immunity	Low uptake	FMD, rabies	Al Amin et al. (2020)
9	Peptide Vaccines	Synthetic antigenic peptides	Specific and safe	Need adjuvants	Autoimmune trials	La Guidara et al. (2024)
10	Phage Display Vaccines	Display antigens on bacteriophages	Strong T cell activation	Contamination risk	Diagnostics, therapy	Ragothaman and Yoo (2023)
11	AI Epitope Mapping	AI to identify targets	Faster, precise design	Bias, data privacy	Tool for vaccine R&D.	Amer and Amer (2024)
12	saRNA Vaccines	Self-replicating RNA vectors	High antigen yield	Mutation risk	COVID-19 trials	Alaoui Mdarhri et al. (2022)
13	Thermostable Vaccines	Encapsulated, heat- resistant	No cold chain needed	Limited antigens	Use for NDV	Okechukwu et al. (2023)

14	Needle-Free Vaccines	Mucosal, transdermal delivery	Non-invasive, self-administered	Stability, uptake	Outbreak response	Alaoui Mdarhri et al. (2022)
15	CRISPR-Guided Vaccines	Gene editing for antigen selection	Custom vaccines	Off-target effects	Early preclinical	Yirgu et al. (2023)
16	Exosome Vaccines	Cell-derived vesicle carriers	Efficient delivery	Scale-up issues	Cancer trials	Alayande et al. (2020)
17	Artificial Immune Systems	Algorithmic immune simulation	Design optimization	Black-box behavior	Modeling only	Ahmad and Narayanan (2011)
18	Gene Editing	Modifies antigen genes	Targeted expression	Mutation risks	Research stage	Albarella et al. (2025)
19	Biosynthetic Vaccines	Synthetic glycosylation systems	Safe and targeted	Pathway engineering	Bacterial diseases	Pan et al. (2016)
20	Hybrid Vaccines	Human-animal immunobridging	Cross-species protection	Ethical concerns	Experimental	Aarestrup (1999)
21	Nanoparticle Immunotherapy	Nanocarrier-mediated stimulation	High bioavailability	Rapid clearance	Cancer models	Li et al. (2023)
22	Live Attenuated Vaccines	Weakened pathogens	Strong immunity	Reversion risk	Veterinary standard	Ahmed et al. (2025)
23	Bacterial Ghost	Non-viable bacterial shells	Envelope antigen exposure	Stability	Inflammation, tumors	Jalava et al. (2002)
24	Adjuvants	Enhance immune responses	Boost weak antigens	Toxicity, approval	Routine use	Cabello and Godfrey 2016)
25	Liposome Vaccines	Encapsulated antigen delivery	Efficient delivery	Leakage, cost	Immunity booster	La Guidara et al. (2024)
26	APC Targeting	Targets antigen-presenting cells	High specificity	Regulatory barriers	Nanovaccine delivery	Melgoza-González et al. (2023)
27	Oral Vaccines	Encapsulated in feed forms	Easy administration	Regulatory concerns	Aquaculture, poultry	Radhakrishnan et al. (2023)

(continued)

Table 1 (continued)

Sr#	Technology	Scientific principle	Key advantages	Major challenges	Current applications/ Development stage	References
28	Thermorespo nsive Hydrogels	Gel-based delivery	Localized release	Slow activation	Drug delivery	Alaoui Mdarhri et al. (2022)
29	Dendritic Cell Vaccines	Patient-derived immune cells	Safe and targeted	Complex processing	Glioblastoma therapy	Albarella et al. (2025)
30	Recombinant Proteins Vaccines	Expressed antigen proteins	Scalable, cost- effective	Production limits	Pandemic use	Pollet et al. (2021)

- *Bovine Viral Diarrhea (BVD)*: Vaccines mitigate immunosuppression and secondary infections. Both modified live and inactivated forms are widely used in herd health programs (Mohsin et al. 2019).
- *Foot and Mouth Disease (FMD)*: Controlled via polyvalent inactivated vaccines, especially in endemic zones. Logistics remain challenging due to serotype variation and cold chain needs (Windsor et al. 2020).

6.2 Poultry (Avian Species)

- *Newcastle Disease (ND)*: A major threat worldwide. Live and inactivated vaccines are administered via spray, water, or injection, enabling mass immunization (Lepczyński et al. 2024; Nhung et al. 2017).
- *Infectious Bronchitis Virus (IBV)*: Frequent antigenic shifts require multivalent or strain-specific vaccines, updated regularly to maintain efficacy (Abbas et al. 2022).
- *Avian Influenza (H5, H9 subtypes)*: Endemic regions rely on inactivated and recombinant vectored vaccines, supported by DIVA strategies for trade and surveillance (Rafique et al. 2023).
- *Coccidiosis (Protozoal Disease)*: Live attenuated vaccines replace antibiotics in broiler production, improving gut health and aligning with antibiotic-free market demands (Peek and Landman 2011).

6.3 Swine (Porcine)

- *PRRS*: Causes reproductive failure and respiratory disease. MLVs dominate despite challenges from viral diversity and limited cross-protection (Trevisi et al. 2022).
- *Swine Influenza Virus (SIV)*: Zoonotic and mutable. Trivalent vaccines are common; antigenic drift necessitates regular updates (Haile et al. 2025).
- *Mycoplasma Hypermania*: Widespread vaccination reduces chronic respiratory disease and cuts antibiotic usage (Nagy et al. 2025).
- *Classical Swine Fever (CSF)*: Managed using live attenuated vaccines like the C-strain, especially in Asia and Latin America (Zhang et al. 2025).

6.4 Aquaculture Species

- *Viral Hemorrhagic Septicemia (VHS)*: DNA vaccines show promise in preclinical trials for salmonids and carp (Hasani et al. 2025).

- *Bacterial Kidney Disease (BKD)*: Hard to control due to intracellular persistence; recombinant protein vaccines under study (Ahmed et al. 2025).
- *Aeromonas & Vibrio spp.*: Injectable and oral vaccines lower bacterial load and antibiotic use in finfish (Miryala and Swain 2025).
- *Francisella & Yersinia ruckeri*: Emerging threats in tilapia and salmonids. Subunit and mucosal delivery strategies are in development (Wangkahart et al. 2025).

6.5 Regional Highlights: South and Southeast Asia

- *Pakistan & India*: Backyard poultry vaccinated with thermostable ND vaccines (e.g., I-2 strain) via eye drop or drinking water over 80% efficacy in rural settings (Wang et al. 2025).
- *Bangladesh*: Community-led ND vaccination has halved poultry mortality and improved women's household income (Bessell et al. 2023).
- *Southeast Asia*: Oral/immersion vaccines used in export-oriented aquaculture to meet EU/US residue-free import standards (Nhung et al. 2016).

7 Barriers to Vaccine Adoption

Despite vaccines being a proven tool to reduce antibiotic use, several systemic barriers continue to hinder widespread adoption, especially in low- and middle-income countries (LMICs).

7.1 Economic and Infrastructure Limitations

High upfront vaccine costs, especially without subsidies or insurance, discourage smallholder farmers. Vaccines also require cold chain logistics and trained personnel, both lacking in rural regions (Barrett et al. 2021). Many areas lack stable electricity and transport, leading to vaccine spoilage, reduced effectiveness, and farmer distrust (Coyne et al. 2019).

7.2 Knowledge Gaps and Sociocultural Resistance

Limited understanding of vaccine function, combined with misinformation, creates resistance. In some regions, farmers believe vaccines can cause infertility or reduced productivity, while others prioritize antibiotics for their immediate results (Cuong et al. 2018). Mistrust is amplified by low literacy, poor veterinary extension, and inadequate outreach strategies.

7.3 Technological and Pathogen-Specific Challenges

Pathogen evolution presents a constant challenge. Rapid mutation in viruses like Avian Influenza and PRRS often leads to vaccine mismatch, especially in regions lacking diagnostic infrastructure (Hughes et al. 2021). Meanwhile, advanced vaccines like those using mRNA, CRISPR, or GMOs face skepticism rooted in ethical, religious, or food safety concerns (Ashok et al. 2017).

7.4 Regulatory and Policy Barriers

Vaccine adoption is often slowed by fragmented regulations, delayed approvals, and inconsistent quality control. Many countries lack harmonized frameworks, making it difficult to ensure vaccine safety, efficacy, and international trade compliance. The absence of DIVA- supportive policies also hampers the adoption of modern recombinant vaccines (Cuong et al. 2018).

7.5 Environmental and Political Instability

Vaccination programs are disrupted by conflict, livestock migration, and extreme weather events. Nomadic herders along borders (e.g., Pakistan–Afghanistan) are difficult to reach consistently. Climate-related stressors like floods and heatwaves also reduce immune responses and complicate vaccine storage and delivery (Lhermie et al. 2017).

8 Strengthening Vaccine Systems Through Policy, Extension, and Global Support

Widespread vaccination in animal agriculture requires more than scientific innovation it demands alignment with national policy, veterinary extension systems, and international cooperation. Integrated support structures are essential to maximize vaccination's role in reducing antibiotic dependence and combating AMR.

8.1 National Frameworks and Public–Private Partnerships

Countries like Pakistan and India have embedded vaccination into their livestock development strategies, linking disease control to food security and trade compliance. Programs like India's NADCP aim for large-scale FMD immunization, while Pakistan's Livestock Vision 2025 promotes preventive health through structured immunization (Dadar et al. 2021). Public–private partnerships (PPPs) play a pivotal role in delivering these goals local manufacturers, such as Indian Immunologicals Ltd., collaborate with governments to produce and distribute vaccines efficiently (Parlapani and Boziaris 2025).

8.2 Extension Services and Field-Level Implementation

Veterinary extension networks ensure vaccines reach farmers and are correctly administered. In many LMICs, para-veterinarians and Community-Based Animal Health Workers (CAHWs) support vaccination efforts, particularly in hard-to-reach pastoral areas (Clark et al. 2012). Training programs now include handling novel platforms like DIVA and thermostable vaccines, bridging scientific advances with real-world deployment.

8.3 International Collaboration and Emergency Readiness

Global bodies such as WHO, FAO, and WOAH support countries via technical guidance, emergency vaccine banks, and funding for research and cold-chain infrastructure (Farrukh et al. 2025). Programs like the FAO-WHO-OIE Tripartite Alliance and donor initiatives from USAID, GALVmed, and the Gates Foundation have expanded vaccine access for neglected species and in high-risk regions (Horvat and Kovačević 2025).

8.4 Surveillance, Digital Tools, and Compliance Incentives

Modern vaccine programs rely on digital platforms—QR-coded certificates, mobile vaccination apps, and e-health records improve traceability and outbreak response (Sanders et al. 2020). DIVA-compatible vaccines support surveillance and trade continuity, while zoning and compartmentalization policies allow safe exports even during regional disease outbreaks (Farrukh et al. 2025). Some governments now incentivize compliance with tax breaks or market access for vaccinated herds.

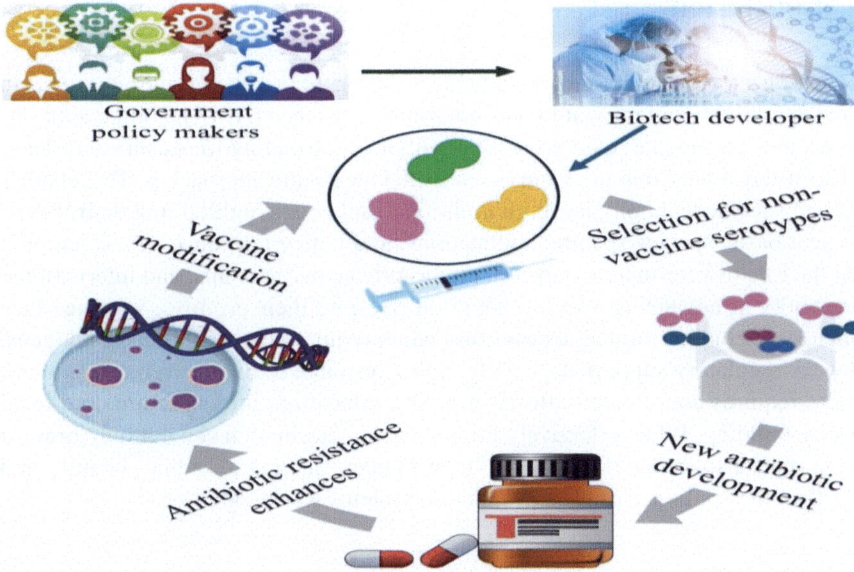

Fig. 1 Integrated framework for vaccine-based AMR reduction

8.5 Gaps and the Way Forward

Despite momentum, gaps persist: weak regulatory systems, underinvestment in minor species, and exclusion of women and marginalized groups from decision-making remain challenges (Lambrou et al. 2021). Inter-agency data sharing is also limited. Strengthening these areas is essential for building resilient, equitable vaccine systems aligned with One Health principles (Parlapani and Boziaris 2025) (Fig. 1).

9 Limitations of the Study

This chapter is based solely on secondary literature and lacks primary field data, which may limit insights into real-world vaccine adoption challenges such as hesitancy, infrastructure gaps, or cultural barriers. Much of the referenced data comes from high-income countries, potentially limiting applicability to low-resource settings. Additionally, emerging technologies like mRNA and CRISPR-based vaccines are still evolving, with limited field validation in veterinary contexts. Pathogen diversity, species-specific immune responses, and differing regulatory landscapes also constrain cross-regional generalizations. These limitations highlight the need for localized research, implementation trials, and policy coordination to ensure that vaccination strategies are both context-appropriate and globally scalable.

10 Conclusion

Vaccination is a strategic and sustainable solution for reducing antibiotic use in animal agriculture and mitigating antimicrobial resistance (AMR). Unlike antibiotics, vaccines provide targeted protection without contributing to resistance or leaving harmful residues, making them essential to One Health approaches. This chapter reviewed vaccine technologies, both traditional and novel, highlighting their immunological basis, species-specific applications, and policy relevance. It also emphasized the role of veterinary extension, public-private partnerships, and international frameworks in advancing vaccine adoption. Despite their promise, vaccines face challenges including limited access, cold chain requirements, evolving pathogens, and weak regulatory support, especially in low-resource settings. Overcoming these barriers requires coordinated efforts in policy, education, infrastructure, and local capacity-building. When effectively implemented, vaccination can not only prevent disease but also reduce antibiotic dependence, safeguard public health, and strengthen the resilience of food production systems worldwide.

References

Aarestrup FM (1999) Association between the consumption of antimicrobial agents in animal husbandry and the occurrence of resistant bacteria among food animals. Int J Antimicrob Agents 12(4):279–285

Aarestrup FM (2015) The livestock reservoir for antimicrobial resistance: a personal view on changing patterns of risks, effects of interventions and the way forward. Philos Trans R Soc B: Biol Sci 370(1670):20140085

Abbas G, Yu J, Li G (2022) Novel and alternative therapeutic strategies for controlling avian viral infectious diseases: focus on infectious bronchitis and avian influenza. Front Vet Sci 9:933274

Ahmad W, Narayanan A (2011) Principles and methods of artificial immune system vaccination of learning systems. Artificial immune systems: 10th international conference, ICARIS 2011, Cambridge, UK, July 18-21, 2011. Proceedings 10, Springer

Ahmed I, Ishtiyaq S, Sayed SF (2025) An overview on understanding the major bacterial fish diseases in freshwater salmonids. Front Aquacult 4:1515831

Aida V, Pliasas VC, Neasham PJ, North JF, McWhorter KL, Glover SR, Kyriakis CS (2021) Novel vaccine technologies in veterinary medicine: a herald to human medicine vaccines. Front Vet Sci 8:654289

Akintan O, Gebremedhin KG, Uyeh DD (2024) Animal feed formulation— connecting technologies to build a resilient and sustainable system. Animals 14(10):1497

Al Amin M, Hoque MN, Siddiki AZ, Saha S, Kamal MM (2020) Antimicrobial resistance situation in animal health of Bangladesh. Vet World 13(12):2713

Alaoui Mdarhri H, Benmessaoud R, Yacoubi H, Seffar L, Assimi HG, Hamam M, Boussettine R, Filali-Ansari N, Lahlou FA, Diawara I (2022) Alternatives therapeutic approaches to conventional antibiotics: advantages, limitations and potential application in medicine. Antibiotics 11(12):1826

Alayande KA, Aiyegoro OA, Ateba CN (2020) Probiotics in animal husbandry: applicability and associated risk factors. Sustainability 12(3):1087

Albarella D, Dall'Ara P, Rossi L, Turin L (2025) Bacteriophage therapy in freshwater and saltwater aquaculture species. Microorganisms 13(4):831

Alhassan MY, Kabara MK, Ahmad AA, Abdulsalam J, Habib HI (2025) Revisiting antibiotic stewardship: veterinary contributions to combating antimicrobial resistance globally. Bull Natl Res Cent 49(1):25

Amer MM, Amer AM (2024) Artificial intelligence: current and future role in veterinary and public medicine. Egypt J Vet Sci:1–12

Arevalo CP, Bolton MJ, Le Sage V, Ye N, Furey C, Muramatsu H, Alameh M-G, Pardi N, Drapeau EM, Parkhouse K (2022) A multivalent nucleoside-modified mRNA vaccine against all known influenza virus subtypes. Science 378(6622):899–904

Ashok A, Brison M, LeTallec Y (2017) Improving cold chain systems: challenges and solutions. Vaccine 35(17):2217–2223

Ates HC, Roberts JA, Lipman J, Cass AE, Urban GA, Dincer C (2020) On-site therapeutic drug monitoring. Trends Biotechnol 38(11):1262–1277

Barrett JR, Innes GK, Johnson KA, Lhermie G, Ivanek R, Greiner Safi A, Lansing D (2021) Consumer perceptions of antimicrobial use in animal husbandry: a scoping review. PLoS One 16(12):e0261010

Bessell PR, Salmon G, Schnier C, Tjasink K, Al-Riyami L, Peters A (2023) A high level estimation of the net economic benefits to small-scale livestock producers arising from animal health product distribution initiatives. Front Vet Sci 10:1171989

Cabello FC, Godfrey HP (2016) Even therapeutic antimicrobial use in animal husbandry may generate environmental hazards to human health. Environ Microbiol 18(2):311–313

Callaway TR, Lillehoj H, Chuanchuen R, Gay CG (2021) Alternatives to antibiotics: a symposium on the challenges and solutions for animal health and production. MDPI

Caneschi A, Bardhi A, Barbarossa A, Zaghini A (2023) The use of antibiotics and antimicrobial resistance in veterinary medicine, a complex phenomenon: a narrative review. Antibiotics 12(3):487

Cheng G, Hao H, Xie S, Wang X, Dai M, Huang L, Yuan Z (2014) Antibiotic alternatives: the substitution of antibiotics in animal husbandry? Front Microbiol 5:217

La Guidara C, Adamo R, Sala C, Micoli F (2024) Vaccines and monoclonal antibodies as alternative strategies to antibiotics to fight antimicrobial resistance. Int J Mol Sci 25(10):5487. https://doi.org/10.3390/ijms25105487

Clark S, Daly R, Jordan E, Lee J, Mathew A, Ebner P (2012) Extension education symposium: the future of biosecurity and antimicrobial use in livestock production in the United States and the role of extension. J Anim Sci 90(8):2861–2872

Coyne LA, Latham SM, Dawson S, Donald IJ, Pearson RB, Smith RF, Williams NJ, Pinchbeck GL (2019) Exploring perspectives on antimicrobial use in livestock: a mixed-methods study of UK pig farmers. Front Vet Sci 6:257

Cuong NV, Padungtod P, Thwaites G, Carrique-Mas JJ (2018) Antimicrobial usage in animal production: a review of the literature with a focus on low-and middle-income countries. Antibiotics 7(3):75

Dadar M, Tiwari R, Sharun K, Dhama K (2021) Importance of brucellosis control programs of livestock on the improvement of one health. Vet Q 41(1):137–151

Din MU, Liu X, Jiang H, Ahmad S, Lai X, Wang X (2025) Advancing vaccine development in genomic era: a paradigm shift in vaccine discovery. Prog Biomed Eng 7(2)

Estany-Gestal A, Salgado-Barreira A, Vazquez-Lago JM (2024) Antibiotic use and antimicrobial resistance: a global public health crisis. Antibiotics 13(9):900

Farrukh M, Munawar A, Nawaz Z, Hussain N, Hafeez AB, Szweda P (2025) Antibiotic resistance and preventive strategies in foodborne pathogenic bacteria: a comprehensive review. Food Sci Biotechnol:1–29

Garrido-Mantilla J, Sanhueza J, Alvarez J, Pittman JS, Davies P, Torremorell M, Culhane MR (2025) Reduction of influenza A virus prevalence in pigs at weaning after using custom-made influenza vaccines in the breeding herds of an integrated swine farm system. Viruses 17(2):240

Guha S, Misra S, Ghosh B (2025) Assessment of the antibiotic resistome pattern of bacteria isolated from poultry faeces. J Stress Physiol Biochem 21(2):156–166

Gul I, Hassan A, Chikan NA, Haq E, Ganai NA, Careem MFA, Shabir N (2025) Vetinformatics in vaccine design for the control of animal diseases. In: Bioinformatics in veterinary science: vetinformatics. Springer, pp 387–414

Guo X, Xiong J, Liu H, Wang H, Cheng Q, Tang F, Zeng Z, Zhang H, Yin H (2025) Targeted screening, distribution and sources of antibiotics in drinking water sources and its risk assessment in Sichuan Province, China. Environ Manage:1–13

Gustafson RH, Bowen R (1997) Antibiotic use in animal agriculture. J Appl Microbiol 83(5):531–541

Haile B, Liu C, Carrai M, Go YY, Yip CK, Tet NL, Yeung YGA, Luk KY, Havas KA, Ivanek R (2025) Characterization of biosecurity practices and viral infections on pig farms in Hong Kong. Prev Vet Med:106569

Hasani S, Nofouzi K, Meshkiniy S, Enferadi A, Jalilzadeh N, Ravvaz A (2025) A comprehensive review of herbal and synthetic therapeutics in managing viral diseases in fish health. J Hellenic Vet Med Soc 76(1):8805–8818

Hoelzer K, Bielke L, Blake DP, Cox E, Cutting SM, Devriendt B, Erlacher-Vindel E, Goossens E, Karaca K, Lemiere S (2018) Vaccines as alternatives to antibiotics for food producing animals. Part 2: new approaches and potential solutions. Vet Res 49:1–15

Horvat O, Kovačević Z (2025) Human and veterinary medicine collaboration: synergistic approach to address antimicrobial resistance through the lens of planetary health. Antibiotics 14(1):38

Hughes A, Roe E, Hocknell S (2021) Food supply chains and the antimicrobial resistance challenge: on the framing, accomplishments and limitations of corporate responsibility. Environ Plann A: Econ Space 53(6):1373–1390

Jalava K, Hensel A, Szostak M, Resch S, Lubitz W (2002) Bacterial ghosts as vaccine candidates for veterinary applications. J Control Release 85(1–3):17–25

Katz SE, Banerjee R, O'Leary ST, Campbell JD, Ardura MI, Bryant KA, Caserta MT, Frenck RW, Gerber JS, John CC (2024) Use of antibiotics in animal agriculture: implications for pediatrics: technical report. Pediatrics 154(4)

Lambrou AS, Innes GK, O'Sullivan L, Luitel H, Bhattarai RK, Basnet HB, Heaney CD (2021) Policy implications for awareness gaps in antimicrobial resistance (AMR) and antimicrobial use among commercial Nepalese poultry producers. Glob Health Res Policy 6:1–9

Lepczyński A, Herosimczyk A, Bucław M, Adaszyńska-Skwirzyńska M (2024) Antibiotics in avian care and husbandry-status and alternative antimicrobials. Phys Sci Rev 9(2):701–753

Lhermie G, Gröhn YT, Raboisson D (2017) Addressing antimicrobial resistance: an overview of priority actions to prevent suboptimal antimicrobial use in food-animal production. Front Microbiol 7:2114

Li Y, Li X, Yi J, Cao Y, Qin Z, Zhong Z, Yang W (2023) Nanoparticle-mediated STING activation for cancer immunotherapy. Adv Healthc Mater 12(19):2300260

Melgoza-González EA, Bustamante-Córdova L, Hernández J (2023) Recent advances in antigen targeting to antigen-presenting cells in veterinary medicine. Front Immunol 14:1080238

Miryala KR, Swain B (2025) Advances and challenges in Aeromonas hydrophila vaccine development: immunological insights and future perspectives. Vaccines 13(2):202

Mohsin M, Van Boeckel TP, Saleemi MK, Umair M, Naseem MN, He C, Khan A, Laxminarayan R (2019) Excessive use of medically important antimicrobials in food animals in Pakistan: a five-year surveillance survey. Glob Health Action 12(sup1):1697541

Mulchandani R, Zhao C, Tiseo K, Pires J, Van Boeckel TP (2024) Predictive mapping of antimicrobial resistance for Escherichia coli, Salmonella, and Campylobacter in food- producing animals, Europe, 2000–2021. Emerg Infect Dis 30(1):96

Nagy EZ, Földi D, Madzig F, Wehmann E, Orosz A, Kempf A, Buza L, Mátyus J, Búza L, Grózner D (2025) Cross-sectional study of Mycoplasma hyopharyngis, Mycoplasma hyopneumoniae, Mycoplasma hyorhinis and Mycoplasma hyosynoviae in the tonsils of fattening pigs from Central-Eastern Europe. Porcine Health Manage 11(1):11

Nhung NT, Cuong NV, Thwaites G, Carrique-Mas J (2016) Antimicrobial usage and antimicrobial resistance in animal production in Southeast Asia: a review. Antibiotics 5(4):37

Nhung NT, Chansiripornchai N, Carrique-Mas JJ (2017) Antimicrobial resistance in bacterial poultry pathogens: a review. Front Vet Sci 4:126

Nollkaemper A (2025) The quadripartite: building alliances between the WHO, the FAO, WOAH and UNEP. In: Unity in diversity: perspectives on the Law of International Organizations. Brill Nijhoff, pp 157–175

Okechukwu D, Nnamani P, Attama A, Onugwu A, Okore V, Onuigbo E (2023) Thermostability and immunostimulatory activity of cationic liposome-based newcastle disease virus vaccine. Sci View J 4:308–317

Pan C, Sun P, Liu B, Liang H, Peng Z, Dong Y, Wang D, Liu X, Wang B, Zeng M (2016) Biosynthesis of conjugate vaccines using an O-linked glycosylation system. MBio 7(2). https://doi.org/10.1128/mbio.00443-00416

Parlapani FF, Boziaris IS (2025) Fish, antibiotic resistance, and food safety at the global level. In: Handbook of public health nutrition: international, national, and regional perspectives. Springer, pp 1–21

Peek H, Landman W (2011) Coccidiosis in poultry: anticoccidial products, vaccines and other prevention strategies. Vet Q 31(3):143–161

Pollet J, Chen W-H, Strych U (2021) Recombinant protein vaccines, a proven approach against coronavirus pandemics. Adv Drug Deliv Rev 170:71–82

Radhakrishnan A, Vaseeharan B, Ramasamy P, Jeyachandran S (2023) Oral vaccination for sustainable disease prevention in aquaculture—an encapsulation approach. Aquac Int 31(2):867–891

Rafique S, Rashid F, Mushtaq S, Ali A, Li M, Luo S, Xie L, Xie Z (2023) Global review of the H5N8 avian influenza virus subtype. Front Microbiol 14:1200681

Ragothaman M, Yoo SY (2023) Engineered phage-based cancer vaccines: current advances and future directions. Vaccine 11(5):919

Rame Hau EE, Sharma M, Khanal BKS, Sly PD, Mikkelsen D, Clark N, Soares Magalhães RJ (2025) The prevalence and risk factors associated with the presence of antibiotic residues in milk from peri-urban dairy cattle farms in Kathmandu, Nepal. Antibiotics 14(1):98

Reddy S, Barathe P, Kaur K, Anand U, Shriram V, Kumar V (2022) Antimicrobial resistance and medicinal plant products as potential alternatives to antibiotics in animal husbandry. In: Antimicrobial resistance: underlying mechanisms and therapeutic approaches, pp 357–384

Rodrigues da Costa M, Diana A (2022) A systematic review on the link between animal welfare and antimicrobial use in captive animals. Animals 12(8):1025

Saad M, Ahmed MBM (2018) Necessary usage of antibiotics in animals. In: Antibiotic use in animals, vol 9, Issue 51000. IntechOpen, Janeza Trdine, pp 9–23

Sanders P, Vanderhaeghen W, Fertner M, Fuchs K, Obritzhauser W, Agunos A, Carson C, Borck Høg B, Dalhoff Andersen V, Chauvin C (2020) Monitoring of farm-level antimicrobial use to guide stewardship: overview of existing systems and analysis of key components and processes. Front Vet Sci 7:540

Sazykin I, Khmelevtsova L, Seliverstova EY, Sazykina M (2021) Effect of antibiotics used in animal husbandry on the distribution of bacterial drug resistance. Appl Biochem Microbiol 57:20–30

Seo J, Kloprogge F, Smith AM, Karu K, Ciric L (2024) Antibiotic residues in UK foods: exploring the exposure pathways and associated health risks. Toxics 12(3):174

Sepordeh S, Jafari AM, Bazzaz S, Abbasi A, Aslani R, Houshmandi S, Rad AH (2024) Postbiotic as novel alternative agent or adjuvant for the common antibiotic utilized in the food industry. Curr Pharm Biotechnol 25(10):1245–1263

Speksnijder DC, Page SW, Wagenaar JA, Prescott JF (2024) Antimicrobial stewardship in food-producing animals. In: Antimicrobial therapy in veterinary medicine, pp 459–485

Sun Z, Hong W, Xue C, Dong N (2024) A comprehensive review of antibiotic resistance gene contamination in agriculture: challenges and AI-driven solutions. Sci Total Environ 953:175971

Thakur R, Singh A, Dhanwar R, Kadam S, Waghmare U, Lodha T, Lopes BS, Prakash O (2025) Global perspectives on residual antibiotics: environmental challenges and trends. Discov Sustain 6(1):232

Thomas S, Abraham A, Rodríguez-Mallon A, Unajak S, Bannantine JP (2022) Challenges in veterinary vaccine development. In: Vaccine design: methods and protocols, Volume 2. Vaccines for veterinary diseases, pp 3–34
Trevisi P, Amatucci L, Ruggeri R, Romanelli C, Sandri G, Luise D, Canali M, Bosi P (2022) Pattern of antibiotic consumption in two Italian production chains differing by the endemic status for porcine reproductive and respiratory syndrome. Front Vet Sci 9:840716
Wang D, Tan M, Nov P (2025) Southeast Asia burden and trend of gastrointestinal tract cancers from 1990 to 2021 and its prediction to 2050: findings from the global burden of disease study 2021. Int J Color Dis 40(1):1–16
Wangkahart E, Lee P-T, Chong C-M (2025) Antimicrobial resistance (AMR) in farmed aquatic organisms. In: Antimicrobial resistance in aquaculture and aquatic environments. Springer, pp 65–89
Windsor P, Khounsy S, Earp F, MacPhillamy I, Young J, Bush R (2020) Managing welfare and antimicrobial-resistance issues in treating foot-and-mouth disease lesions: a new therapeutic approach. Vet Med: Res Rep 11:99–107
Yirgu A, Mekonnen Y, Eyado A, Staropoli A, Vinale F (2023) Biological activities of Viscum tuberculatum aqueous leaf extract. Pharm Biol 61(1):30–36
Zhang T, Nickerson R, Zhang W, Peng X, Shang Y, Zhou Y, Luo Q, Wen G, Cheng Z (2024) The impacts of animal agriculture on one health—bacterial zoonosis, antimicrobial resistance, and beyond. One Health 18:100748
Zhang L, Liang D, Tian Y, Liang J, Li X, Liu C, Liang J, Luo TR, Li X (2025) Classical swine fever virus envelope glycoproteins erns, E1, and E2 activate IL-10-STAT1- MX1/OAS1 antiviral pathway via replacing classical IFNα/β. Biomolecules 15(2):200
Zheng S, Li Y, Chen C, Wang N, Yang F (2025) Solutions to the dilemma of antibiotics use in livestock and poultry farming: regulation policy and alternatives. Toxics 13(5):348

Vaccines: Green Alternatives for Animal Health and Antimicrobial Reduction

Muhammad Waqar, Muhammad Nauman, Dian Tria Fatmila, and Qudrat Ullah

Abstract The rise of antimicrobial resistance (AMR) has escalated the quest for viable alternatives to antibiotics for animal health. Immunization has evolved as an effective and ecologically sound intervention to avoid infectious diseases, diminish the consumption of antibacterials, and work toward the general One Health objectives across the world. This chapter describes the essential role of vaccination in preventing major livestock and poultry diseases such as Newcastle disease, avian influenza, and foot-and-mouth disease, among others, that highly affect animal productivity as well as public health. It gives an insightful discussion on the types and mechanisms of vaccines as well as recent advances in technology, such as recombinant DNA vaccines, mRNA vaccines, thermostable formulations, and universal vaccines, targeting variable pathogens. Coinciding challenges discussed in the chapter include antigenic variation, cold chain constraints, and socioeconomic barriers to the uptake of calves among smallholder farmers. Improving the delivery of vaccines through measures such as mobile clinics, digital tools, and community-based animal health workers is elaborated here with the success stories from Africa and Asia. The priority is given to the role of interdisciplinary cooperation, policy frameworks, and global initiatives in increasing access to vaccines on equal terms and long-term sustainability. Bringing scientific innovation and practical models together can ensure that vaccination can be seen as a keystone in decreasing antibiotic dependence and increasing food security and animal well-being, coupled with environmental health.

Keywords Animal health · Antibiotic alternatives · Antimicrobial resistance (AMR) · Disease prevention · One Health · Vaccination · Vaccine development

M. Waqar (✉) · Q. A. Fatmila · Q. Ullah
Food Technology and Innovation Research Center of Excellence, School of Agricultural Technology and Food Industry, Walailak University, Nakhon Si Thammarat, Thailand
e-mail: diantriafatmila@usu.ac.id

M. Nauman
Department of Biology, Faculty of Science and Technology, Nantes University, Nantes, France

R. U. Khan, S. Naz (eds.), *Green Alternatives to Antibiotics in Animal Health*,
https://doi.org/10.1007/978-981-92-0309-3_15

1 Introduction

Globally, antibiotics have been an integral part of livestock and poultry production systems mainly because they double up in the treatment of bacterial infections and growth performance enhancement. The extensive use of antimicrobial agents has contributed largely to better productivity, decreased mortality, and improved food safety in intensive farming systems (Ravikumar et al. 2022). This extensive use of antibiotics has stirred major worries about the incidence and spread of antimicrobial resistance (AMR) as a major threat to human and animal health. In many countries, especially in resource-limited settings, antibiotics are frequently applied in prophylactic or even as growth promoters, without adequate veterinary control (Donadeu et al. 2019). This wildcat utilization expedites selection pressure on the microbial communities, and the microbial populations end up evolving resistant strains that can spread across the biotic species and ecosystems. The poultry industry has been specifically mentioned as one of the biggest sources of AMR because of its large-scale basis of operations and recurrent use of antibiotics (Abdallah et al. 2023).

The emergence of AMR at the global scale is now known to be one of the most important public health issues of the twenty-first century. Resilient pathogens from animals can easily be transmitted to humans through direct entry, eating contaminated food products, or being exposed to the environment (Helmy et al. 2023). AMR threatens the effective prevention and curing of an increasing variety of infections due to bacteria, parasites, viruses, and fungi (Mehand et al. 2018).

Campylobacter jejuni, *Salmonella* spp., and *Escherichia coli* resistant to antibiotics have been proven to be increasing in their prevalence at commercial farms and markets, therefore requiring urgent intervention measures that would eliminate reliance on antibiotic usage (Rabie and Amin Girh 2020; Taha-Abdelaziz et al. 2018a). The economic impact is also enormous, with rising costs of treatments, long illness duration, and increased mortality rate for both producers and consumers all over the world. With the rising issues about AMR and the reduced effectiveness of traditional antibiotics, there is an urgent need for sustainable alternatives for maintaining the health and productivity of animals while reducing the risk of developing resistance. Among the many approaches being carried out, including probiotics, prebiotics, phytogenics, and bacteriophages, vaccination is a very promising and eco-friendly strategy (Kulkarni et al. 2022).

Vaccines provide an aggressive approach that precludes outbreaks of diseases, hence eliminating the majority of the need for curative antibiotic action. Additionally, vaccination correlates quite strongly with the One Health concept, which refers to the interconnection of human, animal, and environmental health (Donadeu et al. 2019). Through control of disease outbreaks in animal populations, vaccines help boost food security, stabilize the economy, and ensure public health protection. Stimulation of the immune system by vaccination also makes it recognize and neutralize a particular pathogen, making it effective in the prevention of diseases. In poultry and livestock, effective vaccination programs can decrease the prevalence of infectious diseases, for example, Newcastle disease, avian influenza, infectious bursal disease, foot-and-mouth disease, etc. (Spackman and Pantin-Jackwood 2014; Swayne et al. 2014). Through population control of these diseases, vaccines aid in minimizing the requirement for antibiotic treatments, especially those that are used for secondary bacterial infection after a viral attack. For instance, Newcastle disease vaccination in Tanzania has registered significant improvements in flocks' survival and productivity, thus reducing the need for antimicrobial interventions (Knueppel et al. 2010). In addition, the advancements in vaccine technologies (including recombinant DNA vaccine, virus-like particles (VLPs), and mRNA-based platforms are already making it possible to create more effective, more stable, and broadly protective vaccines that can be used in various field conditions (Nurzijah et al. 2022; Ravikumar et al. 2022). Such innovations make vaccination as the primary tool of combating AMR more attainable and sustainable.

This chapter will attempt to give an overview of how vaccination strategies can be considered a viable and green option compared to antibiotics for supporting the animal's health. It will cover the scientific background, technological aspect, implementation barriers, and future of vaccine implementation in livestock and poultry systems.

2 Understanding Disease Prevention Through Vaccination

The basic principles of vaccination, types of vaccines that can be used in animal health, their mechanisms of action, and the ways they help to prevent diseases especially the reduction of usage of antibiotics will be discussed in this part.

Definition and Principles of Vaccination
Vaccination as a form of medical intervention is meant to induce an immunity response in an individual that will enable him/her to recognize and fight certain pathogens. In veterinary practice, vaccines play a key role in disease control in livestock and poultry, thus improving the welfare and augmenting the production of animals and food safety (Ravikumar et al. 2022). The idea behind vaccination revolves around the fact of immunological memory. If a strong enough animal is subjected to a vaccine normally comprising weakened or inactivated pathogen, or its parts the corresponding immune system responds with the formation of memory cells. Such cells stay in the body and can quickly react against exposure to the actual pathogen in the future, creating disease prevention or decreasing the severity of the sickness (Spackman and Pantin-Jackwood 2014).

Classes of Vaccines on Animal Health
Different types of vaccines are being used currently in animal health, which have their respective merits and demerits (Table 1).

Mechanism of Action: How Vaccines Stimulate Immunity
Upon injection of the vaccine, an antigenic material is introduced into the body of the animal. This substance is identified by antigen-presenting cells (APCs), for

Table 1 Types of vaccines (Nurzijah et al. 2022; Ravikumar et al. 2022; Spackman and Pantin-Jackwood 2014)

Vaccine type	Description	Advantages	Limitations
Live attenuated	A weakened form of the pathogen that does not cause disease	Strong, long-lasting immunity	Risk of reversion to virulence; not safe for all species
Inactivated	Killed pathogens	Safer than live vaccines	Requires adjuvants; may need multiple doses
Subunit	Uses only parts of the pathogen (e.g., proteins)	High specificity; safer	May have lower immunogenicity
Recombinant	Genetically engineered vaccines using vector technology	Can target multiple antigens	Complex production process
mRNA vaccines	Deliver mRNA- encoding antigenic proteins	Rapid development; no risk of infection	Requires cold storage; new technology
DNA vaccines	Plasmids encoding antigenic proteins	Long-term immunity; stable	Low immune system in some animals
Virus-like particles (VLPs)	Mimics virus structure without genetic material	Safe, highly immunogenic	Expensive to produce

example, dendritic cells and macrophages. These cells process the antigen and then present it to T-helper cells, which in turn activate B-cells to produce antibodies as well as cytotoxic T-cells, which kill infected cells (Gupta et al. 2014).

The Two Major Arms of the Immune Response That Are Triggered by Vaccines Include

- *Humoral Immunity*: B-cell-formed antibodies can neutralize pathogens before they get to invade host cells.
- *Cell-Mediated Immunity*: T-cells kill already infected cells and modulate the immune response.

Successful vaccines must procure both responses to optimize protection. For instance, live attenuated vaccines tend to develop strong cell-mediated immunity, while the subunit vaccines depend on humoral immunity to a greater extent. Adjuvants are very important in enhancing the body's immune response by increasing innate immunity and enhancing antigen presentation. Some of the common adjuvants have been oil emulsions, aluminum salts, and relatively new nanoparticle-based formulations (O'Hagan and Fox 2015).

Herd Immunity in Animal Populations

Herd immunity can refer to the situation when an adequate percentage of a population gets protected from disease through vaccination or prior infection and thereby indirectly protects those not immunized against the disease. The attainment of herd immunity becomes an integral part of livestock farming in controlling outbreaks of contagious diseases like Newcastle disease, avian influenza, and foot-and-mouth disease (Knueppel et al. 2010). For instance, massive vaccination efforts against Newcastle disease in Tanzania have contributed to a great extent to reducing incidence and mortality cases in backyard poultry. This has resulted in higher survival rates of the flocks and reduced the need for antibiotics in secondary bacterial infections (Knueppel et al. 2010). Herd immunity thresholds are varied when it comes to the disease and transmission dynamics. When dealing with highly contagious diseases such as foot-and-mouth disease, the vaccine coverage must be more than 70–80% to avoid continued transmission within herds (Swayne et al. 2014).

Significance of the Vaccine Efficacy and Duration of Protection
Vaccine efficacy is given as the possibility of a vaccine offering protection against clinical disease under the best conditions; effectiveness refers to reality. Time of protection is the other critical aspect, particularly in commercial livestock, where long-term immunity decreases the frequent need for vaccination.

Several Factors Influence Vaccine Efficacy

- *Antigenic variation:* Some viruses, like avian influenza virus and Newcastle disease virus, undergo antigenic drift and shift, implicating regular modification of the vaccine strains (Wang et al. 2016).
- *Route of administration*: Intranasal, oral, and in-ovo routes are more convenient but could be different in immunogenicity as compared with injectable vaccines (Williams and Hopkins 2011).
- *Immune status of the animal:* Young animals could have maternal antibody interference, thus vaccination needs to be carried out very carefully at the right time (Chase 2022).
- *Storage and handling:* Poor storage can diminish vaccine potency, particularly in situations with live vaccines and mRNA-based formulations (Uddin and Roni 2021).

New methods of designing vaccines include reverse vaccinology and computational modeling (e.g., COBRA, Computationally Optimized Broadly Reactive Antigen) for creating vaccines with greater cross-protection and sustained immunity (Giles and Ross 2011).

Knowledge of the way vaccines operate and their applications in animal populations is critical in leveraging the potential of using alternatives to antibiotics. The vaccines that trigger strong and persistent immune responses not only save the lives of individual animals but also contribute to public health and the environment due to a decline in antibiotic use.

3 The Diseases of Livestock and Poultry That Are Controlled by the Use of Vaccines

This section deals with priority infectious diseases of livestock and poultry that have already entered the scope of vaccination programs. Some of them are Newcastle disease (ND), avian influenza (AI), infectious bursal disease (IBD), foot-and-mouth

Table 2 The diseases and types of vaccines

Disease	Causative agent	Host species	Vaccine type	Key references
Newcastle disease (ND)	APMV-1	Chickens, turkeys	Live attenuated, inactivated, recombinant	Knueppel et al. (2010)
Avian influenza (AI)	Influenza A virus	Chickens, ducks	Inactivated, recombinant, mRNA	Kilany et al. (2016)
Infectious bursal disease (IBD)	IBDV	Young chickens	Live, immune complex, recombinant	Rashid et al. (2019)
Marek's disease (MD)	Herpesvirus	Chickens	Serotype 1, 2, 3 vaccines	Hein et al. (2021)
Infectious bronchitis (IB)	Coronavirus	Chickens	Live, inactivated, recombinant	Cavanagh (2007)
Fowl pox	Avipoxvirus	Chickens, turkeys	Live, recombinant	Hein et al. (2021)
Foot-and-mouth disease (FMD)	Aphthovirus	Cattle, pigs, sheep	Inactivated, synthetic, DNA	FAO (2015)
Brucellosis	Brucella spp.	Cattle, goats, sheep	Live attenuated, killed	Khurana et al. (2021)
Rabies	Rabies virus	Dogs, livestock	Inactivated, oral	Mehand et al. (2018)
Peste des Petits ruminants (PPR)	Morbillivirus	Sheep, goats	Live attenuated, thermostable	FAO (2015)

disease (FMD), brucellosis, rabies, and peste des petits ruminants (PPR). Each of these diseases is a serious economic and public health problem, which is highly transmissible, zoonotic, or both. The diseases and types of vaccines used are also shown in this section, briefly in Table 2.

3.1 Newcastle Disease (ND)

Newcastle disease is one of the most economically crippling viral diseases in global poultry. The Newcastle disease virus (NDV), which belongs to the *Paramyxoviridae* family, causes respiratory, neurological, and gastrointestinal symptoms in birds with mortality rates achieving 100% in naïve flocks (Dzogbema et al. 2021). Vaccine is the most important tool for controlling ND, particularly in smallholder farming systems where biosecurity might not be well developed. Some of the commonly used live attenuated vaccines are La Sota and Hitchner B1 strains, mostly given through drinking water, spray, or intraocular routes (Senne et al. 2004; Vilela et al. 2022). Studies have demonstrated that vaccination improves flock survival by a large margin and also reduces the rectification by antibiotics for secondary infections (Knueppel et al. 2010).

3.2 Avian Influenza (AI)

Another major problem in poultry production around the world is avian influenza, which is caused by influenza A viruses (Rehman et al. 2023). Highly pathogenic avian influenza (HPAI) H5N1 and H7N9 strains are also disturbing, as zoonotic potential and causes widespread outbreaks in birds and humans (Charostad et al. 2023; Swayne et al. 2014). Managing AI has been adopted in related countries through integrated intervention strategies that include vaccination against the disease. Killed oil-emulsified vaccines that rely on inactivated H5 or H7 subtypes are the most common; new recombinant vector vaccines provide better cross-protection and less vaccine shedding (Suarez and Pantin-Jackwood 2017). The success of the vaccination is mainly based on strain matching and updating the formulation of the vaccine regularly as a result of antigenic drift and shift (Tumpey et al. 2005).

3.3 Infectious Bursal Disease (IBD)

Otherwise referred to as Gumboro disease, IBD is an acute viral infection of young chickens that is highly contagious and infective and is caused by the infectious bursal disease virus (IBDV) (Orakpoghenor et al. 2021). It leads to immunosuppression, birds become more susceptible to other diseases, and vaccine efficacy is reduced (Hoerr 2010). Live attenuated vaccines (intermediate and intermediate plus strains) are commonly used for commercial poultry to immunize poultry against classical and variant IBDVs. Vaccination timeliness is very important to prevent maternal antibody interference and provoke a proper immune response (Gowe et al. 1993). Field studies in Morocco and Bangladesh have shown that appropriate vaccination programs can decrease IBD-related morbidity and mortality (Farraye et al. 2017; Rashid et al. 2019).

3.4 Foot-and-Mouth Disease (FMD)

FMD is a highly contagious viral malady of cloven-hoofed animals like cattle, sheep, goats as well as pigs. It results in vesicular lesions at the mouth and the feet causing lameness, appetite loss, and low milk production (Donadeu et al. 2019). Vaccination using inactivated whole-virus vaccines (adjuvanted with saponin or oil emulsions) plays a major role in the control of FMD. Emergency vaccination during outbreaks can help contain and reduce repercussions that lead to mass culling. The effectiveness of the vaccine though depends on the type of serotype, dose, and duration of immunity (Marangon and Capua 2006).

3.5 Brucellosis

Brucellosis is a zoonotic disease that is caused by infection with *Brucella* species, mostly bovines (*B. abortus*), caprines (*B. melitensis*), and pigs (*B. suis*) (Dadar et al. 2021; Khurana et al. 2021). It results in abortions, infertility, and losses in trade occasioned by the restrictions (Azzam et al. 2023). Vaccination is one of the important methods that can be used in controlling brucellosis. Common vaccines used in cattle include RB51, while in small ruminants, there is Rev-1 (Khurana et al. 2021). Despite the advances, hardships exist, the inability to differentiate between vaccinated and infected animals using traditional diagnostic tests further confounds surveillance and eradication efforts (Roth and Sandbulte 2021).

3.6 Rabies

Rabies, as a disease, is mostly a disease of wildlife, but domestic animals, particularly dogs, are the main source of human infections. Vaccination of livestock and companion animals discontinues the cycle of transmission and ensures the safety of both populations, humans and animals (Mehand et al. 2018). The application of live attenuated and inactivated rabies in veterinary medicine has been effective in containing epidemics in endemic regions. Oral bait vaccines have also been developed for wildlife management (Greenwood 2014).

3.7 Peste des Petits Ruminants (PPR)

PPR is a viral disease, which is highly contagious and infects small ruminants, including sheep and goats. Morbillivirus triggers fever, pneumonia, diarrhea, and high mortality rates, especially among naïve populations (Acosta et al. 2019). A live attenuated vaccine: based on the Nigerian 75/1 strain provides lifelong protection and is widely used for control campaigns in Africa and Asia (Zhao et al. 2021). Global plans to eradicate PPR by 2030 are in progress through coordinated vaccination and surveillance programs (FAO 2015).

4 Vaccine Development Technologies and Innovations

The area of veterinary vaccine development has also changed dramatically over the past several decades due to the rapid development of molecular biology, immunology, and biotechnology. Old techniques, including live attenuated and inactivated vaccines, have been supported by modern techniques such as recombinant DNA

Table 3 Vaccine development technologies and key features

Technology	Description	Advantages	Limitations
Live attenuated	A weakened form of the pathogen	Strong, long-lasting immunity	Risk of reversion to virulence
Inactivated	Killed pathogen	Safer than live vaccines	Requires adjuvants; multiple doses
Subunit	Purified antigenic proteins	High specificity; safe	Lower immunogenicity
Recombinant vector	Pathogen gene inserted into a vector	Can induce broad immunity	Complex development process
DNA vaccines	Plasmid DNA encoding antigen	Long-term immunity; stable	Low immunogenicity in some animals
mRNA vaccines	mRNA encoding antigenic proteins	Rapid development; no infection risk	Cold storage required
Reverse vaccinology	Genome-based antigen discovery	Accelerates candidate selection	Requires genomic data
Nanoparticle	Encapsulate antigens and adjuvants	Enhanced delivery and stability	Costly to develop
DIVA technology	Designed to distinguish vaccinated from infected	Facilitates surveillance and trade	Limited to specific diseases
One health vaccines	Target zoonotic and cross-species pathogens	Broad public health impact	Regulatory complexity

technology, virus-like particles (VLPs), mRNA vaccines, and bioinformatics-based design strategies. Such innovations are opening new opportunities for producing safer and better vaccines for use in livestock and poultry that have broad protection effects. Key features of vaccine development technologies in this section are shown in Table 3.

4.1 *Tried and Tested and Modern Vaccine Development Platforms*

In the context of history, vaccinations were designed using an empirical approach employing whole pathogens either attenuated (weakened) or inactivated/ killed. Even though these traditional vaccines have been largely responsible for controlling many infectious diseases, they have their limitations like chances of reversion to virulence in live vaccines and diminished immunogenicity in the case of inactivated vaccines (Ravikumar et al. 2022).

Modern Vaccine Platforms Aim to Overcome These Challenges by Focusing on Specific Antigenic Components or Utilizing Advanced Delivery Systems

- *Subunit Vaccines:* These vaccines have purified antigens (for example, proteins or peptides) of the pathogen and not the whole organism. Subunit vaccines are moderately safer and safer, more stable, although adjuvants may be needed to promote and stimulate immune responses (Milić et al. 2017).

- *Recombinant Vector Vaccines:* These include the process of insertion of genes coding for the antigens of the pathogen into an inert viral or bacterial vector. The vector then displays the antigen in the host and triggers an immune response. For instance, the Newcastle disease virus (NDV) has been applied as a vector to produce antigens of the avian influenza virus and infectious laryngotracheitis virus, providing dual protection in chickens (Zhao et al. 2014).
- *DNA Vaccines:* These include plasmid DNA encoding antigens. After administration, the host cells take up the DNA and synthesize the antigen, presenting it to the immune system. DNA vaccines have such benefits as being easy and long-term to produce; however, their immunogenicity is highly pronounced in large animals (Shen et al. 2015).
- *mRNA Vaccines:* Recently popularized because of their human health success, the mRNA vaccines are messenger RNA strands that contain the target antigen sequence. Once it enters the cell, the mRNA is translated into protein, which sets both humoral and cellular immune responses. These vaccines can be easily developed and modified, which makes them optimal for new pathogens (Liang et al. 2017).

4.2 Reverse Vaccinology and Bioinformatics-Assisted Design (Vaccinomics)

Reverse vaccinology is a paradigm shift in vaccine development from culture-based to genome-based resources. From the genome-wide analysis of a pathogen, investigators can pinpoint possible antigenic proteins, which can be vaccine candidates without having to grow the organism in culture (Jorge and Dellagostin 2017). The bioinformatics tools now permit epitope prediction, i.e., the regions of the antigens recognized by the immune system, to be included in subunit vaccines. This technique, which goes by the name of vaccinomics, is the one that allows for the rational vaccine design to match immunity stimulus aimed at being broad and durable. Computational modeling also helps optimize antigen structure for greater stability and cross-protection benefits (Caradonna and Schmidt 2021). For example, the computational antigenic cartography helped to monitor the changes in influenza virus strains and make recommendations regarding the update of vaccines. This method measures antigenic distances between circulating and vaccine strains and advises decisions about when a strain change is needed (Wang et al. 2016).

4.3 Applications of Nanotechnology in the Formulation of Vaccines

Novel methods of vaccine delivery and efficacy have been brought by nanotechnology. Nanoparticles (NP)-based vaccines can provide encapsulation of antigens and adjuvants to protect them from degradation and direct them to the immune cells. Also, nanoparticles can mimic the size and shape of viruses to increase immunogenicity via pattern recognition receptors (Nooraei et al. 2021). PLGA nanoparticles were used for the delivery of CpG oligodeoxynucleotides and *Campylobacter jejuni* lysate in poultry, with improved mucosal immunity and a change of gut microbiota (Taha-Abdelaziz et al. 2018b). Carbon nanotubes investigated for in-ovo delivery of immunostimulatory molecules were able to improve hatchability and immune responses in broiler chickens (Gunawardana et al. 2015).

4.4 DIVA Technology: Differences Between Infected and Vaccinated Animals

During disease control programs, it is difficult to differentiate between vaccinated and naturally infected animals, particularly concerning serological surveillance. DIVA (Differentiating Infected from Vaccinated Animals) technology solves this problem by developing vaccines that lack some of the antigens contained in wild-type diseases (Mahony et al. 2024). Diagnostic tests can later be used to detect antibodies against these missing antigens to find the infected individuals (Spackman and Pantin-Jackwood 2014).

This approach is very important in foot-and-mouth disease (FMD) when vaccination is commonly used together with culling policies. DIVA-compatible vaccines enable continued trade and preserve disease-free status (Marangon and Capua 2006).

4.5 One Health Approach in the Trans-Species Development of Vaccines

The idea of One Health understands the interrelatedness of human health with the health of animals and the environment. In vaccine development, this approach promotes interdisciplinary cooperation to develop vaccines for zoonotic diseases, which would decrease the rate of antibiotic use in agriculture (Donadeu et al. 2019). Cross-species vaccines are being formulated to ensure that animals, as well as humans, are immunized from common pathogens. For instance, rabies vaccines applied in wildlife management can protect them from spilling over into domestic animals and human beings. Likewise, the universal influenza vaccines currently in development would offer protection against other species, thus diminishing the risk of the emergence of the pandemic (Nachbagauer et al. 2017).

5 The Smallholder Farmers' Vaccine Adoption Influencers

Vaccination is an important weapon in the war against diseases in the animal sphere, but for smallholder farmers in developing nations, vaccination has been sporadic throughout many low- and middle-income countries. Although the long-term effects of vaccination include increased productivity, lowered mortality, and reduced reliance on antibiotics, several socioeconomic, logistical, and cultural factors determine whether smallholder farmers will vaccinate their livestock and poultry or not. Key factors influencing vaccine adoption among smallholder farmers are shown in Table 4.

5.1 Economic Barriers

Some of the main hindrances to vaccine acceptance include cost. Most smallholder farmers live on meager incomes, and perhaps, they focus on immediate needs rather than preventive health measures. The direct cost for the procurement of vaccines plus other indirect costs like expenditures for transportation to vaccination sites or even the hiring of veterinary services can be arduous (Donadeu et al. 2019). Also, the realized return on investment in vaccination may not always be immediate, particularly in areas with sporadic incidences of disease. Depending on the situation, some farmers may think that the risk of the disease is low and, consequently, do not consider the need to spend money on vaccines. When there are no government

Table 4 Key factors influencing vaccine adoption among smallholder farmers

Factor	Description	Impact on vaccine uptake
Economic cost	High cost of vaccines and associated services	Reduces willingness to invest
Knowledge & awareness	Limited understanding of vaccine benefits	This leads to skepticism and underutilization
Accessibility	Poor infrastructure and vaccine supply chain	Limits physical access to vaccines
Cultural beliefs	Preference for traditional remedies	Encourages vaccine hesitancy
Role of CAHWs	Trusted local providers of animal health services	Enhances access and acceptance
Government & NGO support	Availability of subsidies, training, and outreach	Facilitates sustained adoption

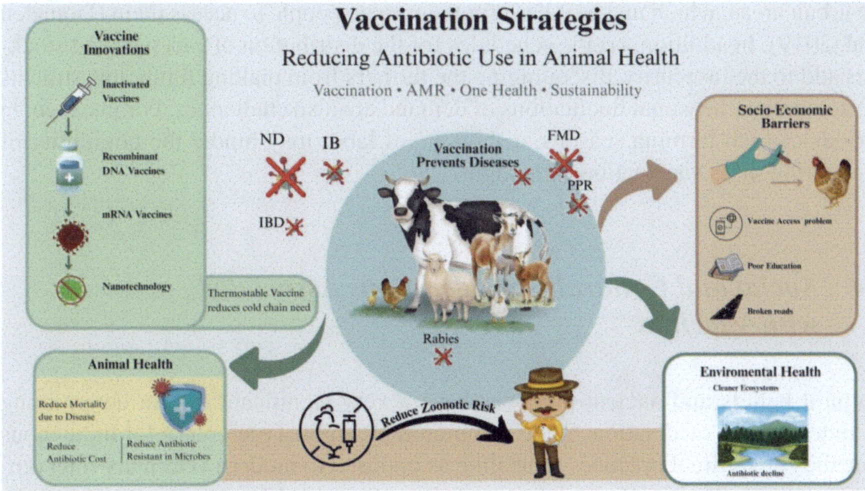

Fig. 1 Vaccination strategies

subsidies or community-based programs, this perception is even more strengthened (Fig. 1).

5.2 Knowledge Gaps and Awareness

Lack of knowledge on the value and benefits of vaccination seriously inhibits uptake. Many of the smallholder farmers have limited access to formal education or extension services that could train them in animal health management. Unfounded beliefs about vaccine allowance on disease, and reductions of animal growth are common and lead to vaccine mistrust (Etuah et al. 2020). From the studies in Uganda and Ghana, it is found that farmers who are trained regularly by extension workers from agriculture or community animal health workers (CAHWs) are more likely to practice vaccination (Nuvey et al. 2025; Serra et al. 2022). Such findings emphasize the need for targeted educational interventions made for local situations and levels of literacy (Acosta et al. 2019).

5.3 Accessibility and Supply Chain Issues

In cases where the farmers are willing to vaccinate, poor infrastructure and a weak supply chain mean that they simply can't access the vaccines. Rural areas could not afford to have cold storage units for preserving live vaccines, and this can get spoiled and less effective. There are instances where the vaccines can only be available in the urban areas, which makes it hard for the remote people to access them (Donadeu et al. 2019). In addition, erratic schedules for the distribution of vaccines and stock-outs add to the insecurity, discouraging the farmers from making future investments in vaccination. Seasonal fluctuations of demand are also challenges. For instance, in periods of high farming seasons, a shortage of labor may hinder the possibility of conducting mass vaccination campaigns.

5.4 Social and Cultural Considerations Associated with Vaccines

Cultural beliefs and ancient practices have a very significant role in determining attitudes toward vaccination. Some communities exhibit a preference for Indigenous remedies or spiritual practices of healing as opposed to modern veterinary intervention practices. Skepticism against external actors, namely, government officials or non-governmental organizations (NGOs), also prevents trust in vaccination programs (Ipara et al. 2021). Rumors or misinformation on vaccines were quickly disseminated via informal channels in other areas, causing massive opposition. For example, fears of experimentation with vaccines or making animals infertile have been reported in sections of East Africa, leading to low demand despite evidence of the benefits (Lees et al. 2024; Mutua et al. 2019).

5.5 The Role of Community Animal Health Workers (CAHWs)

CAHWs have become important mediators of formal veterinary systems and rural farming societies. Trained CAHWs can administer vaccines, educate farmers on vaccination, and also monitor disease outbreaks, enhancing access and acceptance of vaccination services (Serra et al. 2022). Research in Tanzania indicates that with CAHW-led programs, there has been an increase in the vaccination coverage of diseases such as Newcastle in backyard poultry. These programs are effective since CAHWs are usually trusted members of the community, and they can provide the services using a culturally sensitive manner (Knueppel et al. 2010). However, support in terms of training, equipment, and financial incentives needs to be provided all the time to sustain CAHW programs. Without institutional support, these initiatives might not maintain quality and consistency in the long run.

5.6 Several Case Studies Illustrate how Contextual Factors Shape Vaccine Adoption

Uganda A study done in rural districts reveals that even though there was a high level of awareness of the Newcastle vaccine, utilization was low due to money and lack of consistent supply. Farmers who belonged to cooperative groups or utilized mobile veterinary services were more likely to vaccinate their flocks (Acosta et al. 2019).

Ghana In northern Ghana, a CAHW project contributed immensely toward an increased rate of vaccinating poultry. Attributed factors behind the success were community engagement strategies, such as village meetings as well as farmer field schools, which focused on the economics of vaccination (Etuah et al. 2020).

Nigeria Despite a comparatively sophisticated system of veterinary services, the uptake of vaccines among smallholder poultry is still suboptimal. Some of the key challenges include the presence of counterfeit vaccines in informal markets and the lack of cooperation between different actors from the public and the private sectors. However, pilot projects with the use of mobile clinics have proved to be promising in enhancing access and alleviating distrust (Ipara et al. 2021).

6 Methods of Enhancing Vaccination UpTake and Delivery

Increasing vaccination and delivery in animal agriculture, particularly among smallholding farmers, is key for improved use of vaccination, food security, and reduction in the use of antibiotics in livestock farming, especially in small farms. Some of the strategies have established themselves as useful instruments in overcoming barriers, including cost, convenience, public awareness, and logistical barriers. Such are the establishment of vaccine banks, the use of public–private partnerships, and low-cost solutions such as mobile and integrated delivery systems, training programs, policy frameworks, and digital tools (Table 5).

6.1 Establishment of Vaccine Banks and Antigen Stockpiles

It is one of the promising ways of guaranteeing constant vaccine availability through the opening of vaccine banks and antigen stockpiles. Such reserves make it possible to deploy quickly in times of disease outbreak without depending only on the commercial production cycles that may be slow or disrupted through issues in the supply chain (Donadeu et al. 2019). For instance, the Global Foot-and-Mouth Disease

Table 5 Key strategies to improve vaccine uptake and delivery

Strategy	Description	Impact
Vaccine banks & stockpiles	Centralized reserves for emergency use	Ensures availability during outbreaks; stabilizes pricing
Public–private partnerships	Collaboration between governments and private firms	Increase production, improve affordability, expands reach
Mobile and integrated delivery	Vaccination via mobile clinics and CAHWs	Enhances access in remote areas; supports one health
Training programs	Farmer education and CAHW skill-building	Increases knowledge, improves technique, builds trust
Policy support	Government subsidies, regulations, national strategies	Promotes adoption, ensures quality, aligns with AMR goals
Digital tools	Apps, SMS alerts, cloud-based tracking	Improves logistics, enables real-time monitoring, enhances communication

Vaccine Bank, which is monitored by the Food and Agriculture Organization (FAO), has proved important in efforts to support emergency vaccination campaigns in the endemic regions. In the same way, regional pledges like the East African Vaccine Bank have enhanced the accessibility of Newcastle disease vaccines in countries such as Uganda and Tanzania (Acosta et al. 2019). Such banks not only create a buffer to the shortage period but also assist in stabilizing the prices and thus make the vaccines affordable to small-scale producers. What is more, they facilitate research through the provision of reference antigens for diagnostic purposes and vaccine development (Table 5).

6.2 *Public–Private Partnerships in the Production of Vaccines and Distribution*

Partnerships between public entities, private firms, and international organizations help increase the rate of vaccine coverage. The public-private partnerships (PPPs) unite the capabilities of both sectors. Public monitoring guarantees quality control and fair opportunity, but private industry lends innovation and capacity to production.

Successful PPP Models Include

- ILRI is working with the biotech companies to produce thermostable vaccines for resource-constrained environments.
- GAVI Alliance has diversified its attention from human vaccines toward veterinary vaccines, particularly in cases of zoonotic diseases.
- Local manufacturers associating with international donors to increase the production of poultry vaccines; e.g., those against avian influenza and infectious bursal disease (Etuah et al. 2020).

Such partnerships cut costs on economies of scale and ensure that vaccines find their way to distant and neglected communities.

6.3 Mobile Deliveries as Well as Integrated Delivery Systems (One Health Model)

The usual static vaccination points become inaccessible for rural people because of poor infrastructure and a lack of transport. Mobile vaccination units overcome this problem by taking the vaccines to farms and villages. These units are especially efficient when being incorporated into the general health services in the approach of One Health, where the interdependence of human, animal, and environmental health is acknowledged. In Nigeria, Newcastle disease vaccines, dewormers, and nutrition products have been delivered through mobile clinics to improve the health of the flock and engage farmers positively (Ipara et al. 2021). In East Africa, mobile teams are supported by community animal health workers (CAHWs) and carry door-to-door vaccinations and post-incidence disease surveillance that enables real-time monitoring and response (Bugeza et al. 2017). However, the One Health approach also promotes cross-sector collaboration, for example, the combination of rabies vaccination campaigns in domestic animals and animals in the wild to reduce the chances of spillover events (Ahmed et al. 2025).

6.4 Training Programs for Farmers and Animal Health Workers

Knowledge gaps are still a vital barrier to vaccine adoption. Training programs targeted at the farmers and the frontline animal health workers are important in strengthening awareness, attaining trust, and enhancing technical skills.

Major Components of Successful Training Initiatives Are

- Farmer field schools, which show the economic gains of the vaccination process.
- Hands-on workshops to CAHWs on the handling of vaccines, vaccine administration, and adverse reaction management.
- Audiovisual tools like videos and illustrated guides to cater to different literacy levels.

A project by the International Fund for Agricultural Development (IFAD) in northern Ghana trained more than 2000 farmers and CAHWs in techniques of vaccinating poultry. This increased the uptake of vaccines, which also caused the reduction of the burden of the diseases (Etuah et al. 2020).

6.5 Policy Frameworks and Government Support

Policies of the government are instrumental in creating the environment for vaccine adoption. Effective policy frameworks include:

- Subsidy or cost sharing to take the financial pressure off the farmers.
- Regulatory reforms for speeding up the approval and registration of vaccines.
- Investment in cold chain facilities to retain vaccine potency.
- National animal health strategies that emphasize vaccination within the AMR plans.

For example, in its national action plan on antimicrobial resistance, India makes a great point of increasing animal vaccination programs as one of the significant interventions. To the same extent, Kenya has also introduced legislation that requires vaccination for specific livestock diseases, supported by government-funded distribution mechanisms (Donadeu et al. 2019). In addition, strong policy support builds confidence in vaccine safety and effectiveness, setting aside misinformation and opposition from communities.

6.6 Digital Tools and Mobile Technology for Tracking and Education on Vaccines

Technology-based improvements are changing the way vaccines are monitored, delivered, and even marketed. Mobile apps, SMS alerts, and cloud-based databases are growing in use to:

- Monitor the vaccine stocks and expiration date.
- Remind farmers of the upcoming schedules of vaccination.
- Gather real-time data on disease outbreaks and vaccine coverages.
- Offer educational content through the multimedia available.

In Ethiopia, there is a digital platform called EthioVax, via which veterinary officers can monitor vaccination campaigns and report adverse events while carrying smartphones. In Bangladesh, an SMS-based system informs local poultry farmers about local vaccination drives and good practices. These technologies create transparency, inculcate accountability, and provide timely information to the farmers, which results in positive health outcomes.

7 Effect of Vaccination in Decrease of Antimicrobial Use

For several decades, livestock and poultry operations at local as well as international scales have depended on antibiotics for protection against disease and growth enhancement, particularly in extensive farming practices. However, the use and

Table 6 Key impacts of vaccination on reducing antimicrobial use

Impact area	Description	Evidence/examples
Disease prevention	Vaccines prevent infection, reducing the need for treatment	Newcastle disease vaccination in Tanzania (Knueppel et al. 2010)
Economic savings	Lower treatment costs and improved productivity	Increased broiler output in vaccinated farms (Sharma 1999)
Environmental protection	Reduces antibiotic residues in soil and water	Lower detection of resistant E. coli in organic farms (Etuah et al. 2020)
Public health benefits	Reduces zoonotic disease transmission	Rabies and Campylobacter control via vaccination (Etuah et al. 2020)

abuse of antimicrobials have resulted in the occurrence of antibiotic-resistant pathogens, which are major hindrances to both animal and human health. As a response to this, vaccination has emerged as a strong alternative option to decrease the incessant call for antibiotic treatments by avoiding infectious diseases in the first place (Table 6).

7.1 *Quantitative Evidence That Examines a Connection Between Vaccinations and Antibiotic Use Decrease*

Some studies have shown that the use of effective vaccination programs will decrease the prevalence of infectious diseases, hence limiting the need for the use of therapeutic and prophylactic antibiotics. For instance, in Tanzania, control programs using live attenuated vaccines for mass vaccination against Newcastle disease have led to an increase in the survival rate of the flock and a decrease in secondary bacterial infections, which are treated using antibiotics (Knueppel et al. 2010). In commercial poultry production, vaccination against avian influenza and infectious bronchitis has also been linked with reduced consumption of antibiotics. Research carried out in China showed that farms that had implemented broad vaccination strategies recorded a decline of 40% in the use of antibiotics on farms as compared to non-vaccinated flocks (Morgan et al. 1992). To a similar extent, the integration of compulsory vaccination procedures into poultry health plans of countries that are members of the European Union led to pronounced drops in antibiotic prescriptions, especially for respiratory and enteric diseases (Ravikumar et al. 2022). Conducted a meta-analysis on smallholder farming systems in sub-Saharan Africa, established that vaccination against the major viral diseases like Newcastle disease and infectious bursal disease reduced the antibiotic treatment by an average of 30–50% due to fewer outbreaks of the disease and improved immune status of immunized birds (Donadeu et al. 2019).

7.2 *Economic Advantages of Immunization Through the Lower Cost of Treatment in Terms of Cost Saving*

Apart from health advantages, vaccination leads to economic sustainability because overhead expenses for veterinarian services are reduced and productivity increased. Costs of treatment of sick animals using antibiotics, such as the cost of drugs, labor and possible losses in yield can be significant, especially concerning small-scale farmers. Morbidity and mortality rates in vaccinated flocks are relatively lower; this means more weight gains, better feed conversion ratios, and higher egg production. For example, a study conducted in India on Marek's disease vaccine on farms demonstrated 15% more marketable broiler output compared to controls (Sharma 1999). Additionally, vaccination diminishes the risk of culling due to diseases, which is widely practiced during outbreaks and may result in massive financial losses. Vaccines assist stability in production cycles as well as the regular income to producers since they minimize disease transmission.

7.3 *Environmental Benefits: Reduced Antibiotic Residues in Soil and Water*

Among the antibiotics used in livestock farming, many get into the surroundings in feces, runoff, and wastewater, forming part of the surrounding pollution and spreading resistant bacteria. Such residues can survive in soil and water bodies, propagating the multiplication of antibiotic-resistant genes in microbial populations (Lin et al. 2021). Vaccination can alleviate the issue by reducing the amount of antibiotics getting into the ecosystem. Those farms that practice vaccination-oriented health management strategies report reduced amounts of antibiotic residues in their waste products (Buchy et al. 2020; Van Epps and Blaney 2016). Not only is this a contribution toward environmental conservation but also in line with the calls to contain antimicrobial resistance (AMR) across the world as put forward by the World Health Organization (WHO) and the Food and Agriculture Organization (FAO) (Helmy et al. 2023). For instance, there were very low rates of detection of resistant Escherichia coli and Salmonella types in surrounding farming grounds and water sources at organic German poultry farms that depended on extensive levels of vaccination rather than antibiotics (Taha-Abdelaziz et al. 2018b).

7.4 *Co-benefits of Public Health by Decrease in Zoonotic Infections*

Various animal pathogens have zoonotic potential, which implies that they can be transferred to people. Direct risks for public health are presented by diseases such as avian influenza, rabies, brucellosis, and campylobacteriosis, especially where livestock and humans live side-by-side. Vaccination is vital in stopping the chain of transmission of these diseases. For instance, the rabies control programs through mass dog vaccination have significantly reduced cases of human rabies in several African and Asian countries (Abela-Ridder et al. 2016). On the same note, vaccination of poultry against *Campylobacter jejuni*, a major cause of bacterial gastroenteritis in human beings, has been found to reduce contamination of meat products and decrease cases of foodborne illnesses (Taha-Abdelaziz et al. 2018b). Indirectly, vaccination promotes the protection of human health and reduces the burden on healthcare systems, reducing the prevalence of zoonotic diseases in animal populations. It also reduces the human need for antibiotic use, hence reducing the spread of AMR.

8 Problems and Limitations of Animal Vaccination Programs

Despite the extensive advantages of vaccinations against diseases and decreased usage of antibiotics, there are various obstacles and limitations to the use and effectiveness of animal vaccination programs, especially in low- and middle-income countries. These are technical issues, logistical challenges, biological restrictions, regulatory bottlenecks, and socio-cultural factors (Table 7).

8.1 Technical Challenges: Vaccine Stability, Storage, and Administration

A cold chain requirement for the vaccine is one of the major technical barriers to successful vaccination. A good number of vaccines, particularly live attenuated, are temperature sensitive and should be kept at definite temperatures (usually between

Table 7 Summary table key challenges in animal vaccination programs

Challenge	Description	Impact
Vaccine stability	Sensitivity to heat and light reduces shelf life	Loss of potency; ineffective immunization
Antigenic variation	Rapid mutation of pathogens	Reduced vaccine effectiveness over time
Logistical barriers	Poor infrastructure and cold chain access	Limited reach in remote and underserved areas
Regulatory delays	Lengthy approval processes	Delayed response to disease outbreaks
Farmer mistrust	Misconceptions and cultural resistance	Low uptake despite vaccine availability
Surveillance gaps	Lack of real-time data tracking	Difficulty in evaluating program success

2 and 8 °C) to remain potent (Donadeu et al. 2019). White space or resource-limited areas are often subject to spoilage of vaccines when not refrigerated before administration. In addition, poor handling during transportation or storage can compromise vaccine efficacy, thus inducing poor immune responses in animals. For instance, Newcastle disease vaccines may become useless if subjected to heat or direct rays of the sun hence destabilizing control of disease in tropical areas (Knueppel et al. 2010). The path of administration also plays a role in terms of the performance of vaccines. Although injectable vaccines are known to be effective in providing strong immunization, they do need trained personnel and proper equipment. Practicable and aerosol-based delivery methods are more convenient and may lead to a variable or decreased immunogenicity (Spackman and Pantin-Jackwood 2014).

8.2 Biological Limitations: Antigenic Variation and Immune Evasion

Many pathogens are associated with antigenic variation, and broad protection vaccines are hard to achieve. Viruses that cause such diseases, such as avian influenza virus (AIV) and Newcastle disease virus (NDV), are undergoing high rates of mutation due to antigenic drift and shift, which call for consistent strain updates of vaccines (Wang et al. 2016). For example, the hemagglutinin (HA) protein of AIV changes fast and makes the existing vaccines ineffective against the emerging strains. This challenge is made even worse by the need for strain-matching of circulating viruses and vaccine components, which necessitates ongoing surveillance and timely actions (Tumpey et al. 2005). Some pathogens use immune evasion mechanisms to mitigate vaccine-based protection. For instance, Marek's disease virus can undergo latent infections and avoid host-mediated immune responses, thus requiring the designing of extremely potent vaccines endowed with broad cross-protection capabilities (Osterrieder et al. 2006).

8.3 Logistical Constraints: Cold Chain, Remote Area Access

Besides technical ones, logistical barriers are dramatic challenges for effective vaccination coverage. Failure to develop transport infrastructure in rural areas makes access to veterinary services and vaccine distribution centers hard. However, mobile vaccination teams are being used to mitigate this problem, but their work is sometimes handicapped by funding and workforce allowances (Acosta et al. 2019). Further, the delivery of last-mile vaccines, particularly in conflict regions/areas or geographically isolated areas, is still a significant problem. The absence of robust supply chains means that seemingly brilliant vaccination campaigns do not reach target populations. This is a particularly significant problem for smallholder farmers who depend on the informal market for selling the animals and who engage with formal veterinary systems. Another long-standing logistical problem is cold chain management. Powder outages, unreliable refrigeration units, and poor cold storage facility maintenance lead to vaccine wastage and the efficiency of campaigns (Ipara et al. 2021).

8.4 Regulatory and Approval Delays on New Vaccines

The way new animal vaccine development and licensing are accomplished is long and bureaucratic, minimizing innovation and speedy reaction to diseases on the rise. Regulatory agencies must conduct elaborate safety, efficacy, and quality tests of the vaccine before its use commercially, and this may take several years before it is sanctioned for commercial use (Roth and Sandbulte 2021). There is no clear-cut way of emergency use authorization in many countries, which delays the rollout of vaccines during outbreaks. This could be noted during the H5N1 avian influenza crisis when the delay in vaccine approval enabled the virus to spread out of control throughout poultry populations before control measures could be taken (Suarez and Pantin-Jackwood 2017). In addition, standardization of regulation among regions continues to be an obstacle, with limitations to the global supply of effective vaccines. Different registration requirements mean that a vaccine approved in one country may not be available in another when the disease burdens are similar (Marangon and Capua 2006).

8.5 Trust in the Farmer and Spread of Vaccine Misinformation

Social and cultural perceptions play a major role in adoption rates, even if vaccines are available. In other communities, misperceptions about vaccines (e.g., myths that vaccines can cause illness or infertility) result in vaccine hesitancy among livestock keepers (Donadeu et al. 2019). Non-trust in government institutions or external actors (NGOs, veterinarians) also discourages resistance against vaccination programs. There might be a perception of imposition by farmers rather than as a valuable health intervention, especially when there is a lack of cultural insensitivity to communication efforts (Acosta et al. 2019). Furthermore, anecdotal experience about adverse reactions either real or perceived can effectively destroy confidence in vaccines. Not having clear explanations and subsequent support, the bad perceptions can continue and multiply inside farming networks, further complicating uptake (Ipara et al. 2021).

8.6 *Need to Have Better Surveillance and Monitoring Systems*

Vaccination programs rely on strong systems to monitor and surveillance the diseases and monitor their vaccine coverage, disease outbreak detection, and program evaluation. However numerous countries do not have comprehensive mechanisms for gathering data hence making it challenging to measure vaccine effectiveness or to figure out those who do not receive the vaccines (Queenan et al. 2017). Some of the digital tools that can be used to enhance surveillance include mobile apps, SMS alerts, and cloud-based databases, which are promising solutions. Nevertheless, adoption is still limited by technological barriers, untimely training, and underinvestment in the digital infrastructure. Lacking accurate data, policymakers are having a hard time planning resources and even using them in an evidence-based manner for vaccine distribution and disease control tactics.

9 Future Trends and Emerging Trends in Animal Immunization

The industry of animal vaccination is rapidly changing due to the progress in biotechnology, immunology, and data science. With the rising demand for sustainable replacement of antibiotics over the globe, researchers are looking at new ways to increase vaccine efficacy, delivery, and accessibility. Some of the key emerging trends (Table 8) include universal vaccines against variable pathogens, thermostable vaccines for settings that are resource-limited, delivery via oral and aerosols, incorporation of artificial intelligence (AI) in the design of vaccines, personalized vaccines from a genomic standpoint, and global initiatives beginning to address vaccine equity.

9.1 *Universal Vaccines for Variable Pathogens*

Antigenic variation among pathogens like the avian influenza virus (AIV), Newcastle disease virus (NDV), and infectious bursal disease virus (IBDV) is one of the formidable problems in veterinary vaccinology. These viruses experience recurrent mutations through mechanisms such as antigenic drift and shift, thereby rendering the existing vaccines throughout time (Ravikumar et al. 2022). To solve this problem, scientists are creating universal vaccines that target the conserved portion of the virus protein, thus providing broad protection against different strains and serotypes. For instance, the studies on universal AIV vaccines have been based on the M2e protein and the conserved parts of the hemagglutinin stalk domain, which can exhibit little variation among strains. Correspondingly, research has focused on conserved epitopes of NDV fusion (F) and hemagglutinin-neuraminidase (HN)

Table 8 Emerging trends in animal vaccination

Trend	Description	Benefits
Universal vaccines	Target conserved pathogen regions for broad protection	Reduces the need for frequent updates; effective against multiple strains
Thermostable vaccines	Formulated to withstand temperature extremes	Eliminates cold chain dependency; improves rural access
Oral/aerosol delivery	Non-invasive administration via feed, water, or inhalation	Easy to apply; scalable for mass vaccination
AI/ML in vaccine design	Accelerates antigen discovery and formulation optimization	Speeds up development; improves precision
Personalized vaccines	Tailored to animal genetics and immune profiles	Enhances efficacy; supports selective breeding
Global vaccine equity programs	International funding and distribution networks	Increases access in low-income regions; promotes one health

proteins for cross-protective immunity. Progress in reverse vaccinology and computational modeling is increasing the pace of discovery of these conserved targets and improving the basis for a rational design of vaccines that have extended coverage.

9.2 *Thermostable Vaccines for Resource-Limited Settings*

Logistical issues inherent in cold chain sustainability still pose yet another limitation to vaccine distribution, especially in remote areas or areas with high temperatures where facilities to support refrigeration are scarce. To avoid this limitation, researchers are coming up with thermostable vaccines that can withstand higher temperatures without losing potency. Thermostable formulations are typically produced via lyophilization (freeze-drying) methods and using stabilizing excipients (e.g., sugars, amino acids, polymers). For example, mesoporous silica nanoparticles have been applied for the stabilization of E2 glycoprotein vaccines against bovine viral diarrhea virus, whereby shelf life is increased, and there is no need for constant refrigeration (Liu et al. 2023). Not only do these innovations cut costs for storage and transport but they also enhance access to vaccines in poverty-stricken countries in which power failure and lack of infrastructure inhibit the conventional vaccination alternatives (Donadeu et al. 2019).

9.3 *Methods of Vaccine Delivery Through the Means of the Oral and Aerosol*

Conventional injectable vaccines also require trained staff as well as sterilized equipment, which limits their use in mass vaccination campaigns in the smallholder farming system. As a response, scientists are investigating non-invasive delivery mechanisms such as those that are rendered orally or through aerosol vaccines, which are practically advantageous in terms of ease of administration and scalability (Han et al. 2023; Hao et al. 2022).

Oral vaccines through feed or water for poultry have been promising for poultry diseases such as Newcastle disease and coccidiosis. Studies have shown that if antigens are encapsulated in nanoparticles or if live bacterial vectors are used, then mucosal immunity is improved, as well as vaccine uptake (Acevedo-Villanueva et al. 2021). Aerosol vaccination, especially through spray or intranasal routes, is gaining popularity, as it can induce systemic as well as local immunity in the respiratory tract, into which many pathogens invade the host for the first time (Hao et al. 2022). Recombinant vaccines in the form of chitosan-coated nanoparticles demonstrated increased immunogenicity and protection efficacy against avian influenza in experimental models (Hajam et al. 2020).

9.4 AI and Machine Learning Integration in Vaccine Design

Artificial intelligence (AI) and machine learning (ML) are transforming the field of vaccine development since they allow the discovery of antigens in a shorter time, predict immune responses, and enhance the formulation of vaccines. Computational tools may help to examine gigantic amounts of genomic data for peptides, the potential epitopes, and cross-reactivity between strains to markedly accelerate the vaccine design procedure (Gupta et al. 2014). For instance, deep learning algorithms are used to model the antigenic evolution of influenza viruses, and this lets the scientists predict the upcoming modification of the strain and adjust the vaccine correspondingly. Similarly, AI-driven platforms are being utilized for optimizing adjuvant combinations and delivery systems to push vaccine performance while decreasing adverse effects (Liang et al. 2017). Such technologies also enable real-time surveillance and prediction of outbreaks, thus supporting vaccination rollout toward disease emergencies.

9.5 Customized Vaccines from Genomics and Immunogenetics

The idea of personalized medicine is now moving into veterinary vaccinology, where it is increasingly being realized that there is genetic variation in animals, and it affects the reactions of animals toward vaccines. Researchers can use breakthroughs in genomics and immunogenetics to develop regimes of breed- or even population-specific vaccination that suit their corresponding immune profiles. Researchers have been able to determine links between major histocompatibility complex (MHC) haplotypes and immunity induced by the vaccine in chickens, and have attempted selective breeding programs to have enhanced response to the vaccine (Sharma 1999). Additionally, the CRISPR/Cas9 gene-editing technology is being studied to modify the livestock's immune-related genes to enhance the efficacy of the vaccines (Atasoy et al. 2019). Biomarkers for monitoring the immune status and shaping booster schedules may also find application in personalized vaccines, guaranteeing the best protection over the entire lifespan of an animal.

9.6 Vaccine Equity Global Initiatives and Funding Mechanisms

Considering the critical role of vaccines in accomplishing global health security, various international organizations are expanding efforts to ensure vaccine equity in low and middle-income countries. Food and Agriculture Organization (FAO), World Organization for Animal Health (WOAH), and Global Alliance for Livestock Veterinary Medicines (GALVmed) are collaborating to increase access to essential animal vaccines through subsidies, technology transfer, and regional banks for vaccines (Queenan et al. 2017). Apart from that, donor-funded programs, including the Bill & Melinda Gates Foundation and CGIAR Research Programme on Livestock, are providing support for the development and delivery of low-cost vaccines for neglected diseases of smallholder farmers. Community-based delivery models, capacity building, and gender-sensitive outreach are the keystones of these programs to allow inclusive access.

10 Conclusion

Vaccination is one of the building blocks of the worldwide initiative to lessen the use of antibiotics in animal agriculture. As shown throughout this chapter, the vaccines present a sustainable and efficient solution by preventing communicable diseases before they set in, thus reducing the call to utilize therapeutic antibiotics. Apart from addressing the increasing danger of antimicrobial resistance (AMR), the vaccination approach also improves animal welfare, enhances productivity, and contributes to food security. The existing evidence highlights the significance of vaccination in the control of important diseases, including Newcastle, avian influenza, foot-and-mouth, and others that play a vital role in the health of livestock. Nevertheless, even with effective vaccines available, there are difficulties in implementing widespread adoption, especially among small-scale farmers, who are financially, logistically, and informationally challenged.

Recently, innovations in vaccine research, like thermostable formulation, universal vaccines, and new delivery approaches to vaccines, are opening a gateway for cost-effective and widely protective immunization approaches. The use of digital tools, AI-led design, and One Health practices enhances the ability of vaccines to become green substitutes for antibiotics. In the future, the focus of research should be on the provision of equitable vaccine distribution, better surveillance systems, and community-based education programs for high uptake levels of vaccines. The development of next-generation vaccines and regulation curriculums that facilitate the rapid mobilization of investment in times of outbreaks will be vital. Through inter-sector collaboration and scientific development, the global community can transform into a more sustainable model of animal health management: one that views vaccination as the key to ensuring animal and human health in the post-antibiotic age.

References

Abdallah N, Tekelioğlu BK, Kurşun K, Baylan M, Elçi Ü (2023) Vaccination and poultry (chicken) production. J Agric Food Environ Anim Sci 4(1):119–136

Abela-Ridder B, Knopf L, Martin S, Taylor L, Torres G, De Balogh K (2016) 2016: the beginning of the end of rabies? Lancet Glob Health 4(11):e780–e781

Acevedo-Villanueva K, Renu S, Gourapura R, Selvaraj R (2021) Efficacy of a nanoparticle vaccine administered in-ovo against Salmonella in broilers. PLoS One 16(4):e0247938

Acosta D, Hendrickx S, McKune S (2019) The livestock vaccine supply chain: why it matters and how it can help eradicate peste des petits Ruminants, based on findings in Karamoja, Uganda. Vaccine 37(43):6285–6290

Ahmed MM, Okesanya OJ, Othman ZK, Ibrahim AM, Adigun OA, Ukoaka BM, Abdi MI, Lucero-Prisno DE III (2025) Holistic approaches to Zoonoses: integrating public health, policy, and one health in a dynamic global context. Zoonotic Dis 5(1):5

Atasoy MO, Rohaim MA, Munir M (2019) Simultaneous deletion of virulence factors and insertion of antigens into the infectious laryngotracheitis virus using NHEJ-CRISPR/Cas9 and cre-lox system for construction of a stable vaccine vector. Vaccine 7(4):207

Azzam MM, Al-Abdullatif A, Akasha M, Alhotan R, Suliman G, Lahaye L, Santin E (2023) Effects of protected complex of biofactors and antioxidants on growth performance, serum biochemistry, meat quality, and intestinal antioxidant and immunomodulatory-related gene expressions of broiler chickens. Poult Sci 102(6):102666

Buchy P, Ascioglu S, Buisson Y, Datta S, Nissen M, Tambyah PA, Vong S (2020) Impact of vaccines on antimicrobial resistance. Int J Infect Dis 90:188–196

Bugeza J, Kankya C, Muleme J, Akandinda A, Sserugga J, Nantima N, Okori E, Odoch T (2017) Participatory evaluation of delivery of animal health care services by community animal health workers in Karamoja region of Uganda. PLoS One 12(6):e0179110

Caradonna TM, Schmidt AG (2021) Protein engineering strategies for rational immunogen design. NPJ Vaccines 6(1):154

Cavanagh D (2007) Coronavirus avian infectious bronchitis virus. Vet Res 38(2):281–297

Charostad J, Rukerd MRZ, Mahmoudvand S, Bashash D, Hashemi SMA, Nakhaie M, Zandi K (2023) A comprehensive review of highly pathogenic avian influenza (HPAI) H5N1: an imminent threat at doorstep. Travel Med Infect Dis 55:102638

Chase CC (2022) Acceptable young calf vaccination strategies—what, when, and how? Vet Clin Food Anim Pract 38(1):17–37

Dadar M, Tiwari R, Sharun K, Dhama K (2021) Importance of brucellosis control programs of livestock on the improvement of one health. Vet Q 41(1):137–151

Donadeu M, Nwankpa N, Abela-Ridder B, Dungu B (2019) Strategies to increase adoption of animal vaccines by smallholder farmers with focus on neglected diseases and marginalized populations. PLoS Negl Trop Dis 13(2):e0006989

Dzogbema KF-X, Talaki E, Batawui KB, Dao BB (2021) Review on Newcastle disease in poultry. Int J Biol Chem Sci 15(2):773–789

Etuah S, Ohene-Yankyera K, Liu Z, Mensah JO, Lan J (2020) Determinants of cost inefficiency in poultry production: evidence from small-scale broiler farms in the Ashanti region of Ghana. Trop Anim Health Prod 52:1149–1159

FAO, W (2015) Manual on global strategy for the control and eradication of Pesti des Petits Ruminants. The World Organisation for Animal Health

Farraye FA, Melmed GY, Lichtenstein GR, Kane SV (2017) ACG clinical guideline: preventive care in inflammatory bowel disease. Off J Am Coll Gastroenterol ACG 112(2):241–258

Giles BM, Ross TM (2011) A computationally optimized broadly reactive antigen (COBRA) based H5N1 VLP vaccine elicits broadly reactive antibodies in mice and ferrets. Vaccine 29(16):3043–3054

Gowe R, Fairfull R, McMillan I, Schmidt G (1993) A strategy for maintaining high fertility and hatchability in a multiple-trait egg stock selection program. Poult Sci 72(8):1433–1448

Greenwood B (2014) The contribution of vaccination to global health: past, present and future. Philos Trans R Soc B Biol Sci 369(1645):20130433

Gunawardana T, Foldvari M, Zachar T, Popowich S, Chow-Lockerbie B, Ivanova MV, Tikoo S, Kurukulasuriya S, Willson P, Gomis S (2015) Protection of neonatal broiler chickens following in ovo delivery of oligodeoxynucleotides containing CpG motifs (CpG-ODN) formulated with carbon nanotubes or liposomes. Avian Dis 59(1):31–37

Gupta SK, Deb R, Dey S, Chellappa MM (2014) Toll-like receptor-based adjuvants: enhancing the immune response to vaccines against infectious diseases of chicken. Expert Rev Vaccines 13(7):909–925

Hajam IA, Senevirathne A, Hewawaduge C, Kim J, Lee JH (2020) Intranasally administered protein coated chitosan nanoparticles encapsulating influenza H9N2 HA2 and M2e mRNA molecules elicit protective immunity against avian influenza viruses in chickens. Vet Res 51:1–17

Han S, Lee P, Choi H-J (2023) Non-invasive vaccines: challenges in formulation and vaccine adjuvants. Pharmaceutics 15(8):2114

Hao F, Bai Y, Xie X, Yuan T, Wei Y, Xiong Q, Gan Y, Zhang L, Zhang Z, Shao G (2022) Phenotypic characteristics and protective efficacy of an attenuated Mycoplasma hyopneumoniae vaccine by aerosol administration. Vaccine 40(42):6074–6083

Hein R, Koopman R, García M, Armour N, Dunn JR, Barbosa T, Martinez A (2021) Review of poultry recombinant vector vaccines. Avian Dis 65(3):438–452

Helmy YA, Taha-Abdelaziz K, Hawwas HAE-H, Ghosh S, AlKafaas SS, Moawad MM, Saied EM, Kassem II, Mawad AM (2023) Antimicrobial resistance and recent alternatives to antibiotics for the control of bacterial pathogens with an emphasis on foodborne pathogens. Antibiotics 12(2):274

Hoerr FJ (2010) Clinical aspects of immunosuppression in poultry. Avian Dis 54(1):2–15

Ipara BO, Otieno DJ, Nyikal R, Makokha NS (2021) The contribution of extensive chicken production systems and practices to Newcastle disease outbreaks in Kenya. Trop Anim Health Prod 53:1–13

Jorge S, Dellagostin OA (2017) The development of veterinary vaccines: a review of traditional methods and modern biotechnology approaches. Biotechnol Res Innov 1(1):6–13

Khurana SK, Sehrawat A, Tiwari R, Prasad M, Gulati B, Shabbir MZ, Chhabra R, Karthik K, Patel SK, Pathak M (2021) Bovine brucellosis–a comprehensive review. Vet Q 41(1):61–88

Kilany WH, Safwat M, Mohammed SM, Salim A, Fasina FO, Fasanmi OG, Shalaby AG, Dauphin G, Hassan MK, Lubroth J (2016) Protective efficacy of recombinant Turkey herpes virus (rHVT-H5) and inactivated H5N1 vaccines in commercial mulard ducks against the highly pathogenic avian influenza (HPAI) H5N1 clade 2.2. 1 virus. PLoS One 11(6):e0156747

Knueppel D, Cardona C, Msoffe P, Demment M, Kaiser L (2010) Impact of vaccination against chicken Newcastle disease on food intake and food security in rural households in Tanzania. Food Nutr Bull 31(3):436–445

Kulkarni RR, Gaghan C, Gorrell K, Sharif S, Taha-Abdelaziz K (2022) Probiotics as alternatives to antibiotics for the prevention and control of necrotic enteritis in chickens. Pathogens 11(6):692

Lees S, Alcayna-Stevens L, Bowmer A, Marchant M, Enria L, Vanderslott S (2024) Political and community logics of emergent disease vaccine deployment: anthropological Insights from DRC, Uganda and Tanzania. Anthropologica 66(1):1–28

Liang F, Lindgren G, Lin A, Thompson EA, Ols S, Röhss J, John S, Hassett K, Yuzhakov O, Bahl K (2017) Efficient targeting and activation of antigen- presenting cells in vivo after modified mRNA vaccine administration in rhesus macaques. Mol Ther 25(12):2635–2647

Lin Z, Yuan T, Zhou L, Cheng S, Qu X, Lu P, Feng Q (2021) Impact factors of the accumulation, migration and spread of antibiotic resistance in the environment. Environ Geochem Health 43:1741–1758

Liu R, Luo C, Pang Z, Zhang J, Ruan S, Wu M, Wang L, Sun T, Li N, Han L (2023) Advances of nanoparticles as drug delivery systems for disease diagnosis and treatment. Chin Chem Lett 34(2):107518

Mahony TJ, Briody TE, Ommeh SC (2024) Can the revolution in mRNA-based vaccine technologies solve the intractable health issues of current ruminant production systems? Vaccine 12(2):152

Marangon S, Capua I (2006) Control of avian influenza in Italy: from stamping out to emergency and prophylactic vaccination. Dev Biol 124:109–115

Mehand MS, Al-Shorbaji F, Millett P, Murgue B (2018) The WHO R&D blueprint: 2018 review of emerging infectious diseases requiring urgent research and development efforts. Antivir Res 159:63

Milić N, Nišavić J, Zorić A, Krnjaić D, Radojičić M, Stanojković A (2017) Overview of current advances in the development of subunit and recombinant vaccines against Newcastle disease virus. Biotechnol Anim Hus 33(1):1–11

Morgan RW, Gelb J Jr, Schreurs CS, Lütticken D, Rosenberger JK, Sondermeijer PJ (1992) Protection of chickens from Newcastle and Marek's diseases with a recombinant herpesvirus of turkeys vaccine expressing the Newcastle disease virus fusion protein. Avian Dis:858–870

Mutua E, de Haan N, Tumusiime D, Jost C, Bett B (2019) A qualitative study on gendered barriers to livestock vaccine uptake in Kenya and Uganda and their implications on rift valley fever control. Vaccine 7(3):86

Nachbagauer R, Liu W-C, Choi A, Wohlbold TJ, Atlas T, Rajendran M, Solórzano A, Berlanda-Scorza F, García-Sastre A, Palese P (2017) A universal influenza virus vaccine candidate confers protection against pandemic H1N1 infection in preclinical ferret studies. NPJ Vaccines 2(1):26

Nooraei S, Bahrulolum H, Hoseini ZS, Katalani C, Hajizade A, Easton AJ, Ahmadian G (2021) Virus-like particles: preparation, immunogenicity and their roles as nanovaccines and drug nanocarriers. J Nanobiotechnol 19:1–27

Nurzijah I, Elbohy OA, Kanyuka K, Daly JM, Dunham S (2022) Development of plant-based vaccines for prevention of avian influenza and Newcastle disease in poultry. Vaccine 10(3):478

Nuvey FS, Fink G, Hattendorf J, Haydon DT, Fokou G, Addo KK, Zinsstag J, Esse-Dibby C, Bonfoh B (2025) Effects of community action on animal vaccination uptake, antimicrobial usage, and farmers' wellbeing in Ghana: study protocol for a cluster-randomized controlled trial. One Health 20:100952

O'Hagan DT, Fox CB (2015) New generation adjuvants–from empiricism to rational design. Vaccine 33:B14–B20

Orakpoghenor O, Oladele SB, Abdu PA, Markus TP, Andamin AD, Umar BN, Esievo KAN (2021) Very virulent infectious bursal disease virus infection caused changes in cloacal temperature and clinical manifestations in pigeons (Columba livia domestica) and is transmitted to sentinel chickens. Vet Res Commun 45:335–342

Osterrieder N, Kamil JP, Schumacher D, Tischer BK, Trapp S (2006) Marek's disease virus: from miasma to model. Nat Rev Microbiol 4(4):283–294

Queenan K, Garnier J, Nielsen LR, Buttigieg S, Meneghi D d, Holmberg M, Zinsstag J, Rüegg S, Häsler B, Kock R (2017) Roadmap to a One Health agenda 2030. CABI Rev 2017:1–17

Rabie NS, Amin Girh ZM (2020) Bacterial vaccines in poultry. Bull Natl Res Cent 44:1–7

Rashid M, Rashid MI, Akbar H, Ahmad L, Hassan MA, Ashraf K, Saeed K, Gharbi M (2019) A systematic review on modelling approaches for economic losses studies caused by parasites and their associated diseases in cattle. Parasitology 146(2):129–141

Ravikumar R, Chan J, Prabakaran M (2022) Vaccines against major poultry viral diseases: strategies to improve the breadth and protective efficacy. Viruses 14(6):1195

Rehman S, Effendi MH, Witaningruma AM, Nnabuikeb UE, Bilal M, Abbas A, Abbas RZ, Hussain K (2023) Avian influenza (H5N1) virus, epidemiology and its effects on backyard poultry in Indonesia: a review. F1000Res 11:1321

Roth JA, Sandbulte MR (2021) The role of veterinary vaccines in livestock production, animal health, and public health. In: Veterinary vaccines: principles and applications. Wiley, Hoboken, pp 1–10

Senne D, King D, Kapczynski D (2004) By vaccination. Dev Biol Basel 119:165–170

Serra R, Ludgate N, Fiorillo Dowhaniuk K, McKune SL, Russo S (2022) Beyond the gender of the livestock holder: learnings from intersectional analyses of PPR vaccine value chains in Nepal, Senegal, and Uganda. Animals 12(3):241

Sharma J (1999) Introduction to poultry vaccines and immunity. Adv Vet Med 41:481–494

Shen SY, Chang WC, Yi HH, Tsai S-S, Liu HJ, Liao P-C, Chuang KP (2015) Development of a subunit vaccine containing recombinant chicken anemia virus VP1 and pigeon IFN-γ. Vet Immunol Immunopathol 167(3–4):200–204

Spackman E, Pantin-Jackwood MJ (2014) Practical aspects of vaccination of poultry against avian influenza virus. Vet J 202(3):408–415

Suarez DL, Pantin-Jackwood MJ (2017) Recombinant viral-vectored vaccines for the control of avian influenza in poultry. Vet Microbiol 206:144–151

Swayne DE, Spackman E, Pantin-Jackwood M (2014) Success factors for avian influenza vaccine use in poultry and potential impact at the wild bird–agricultural interface. EcoHealth 11:94–108

Taha-Abdelaziz K, Hodgins DC, Alkie TN, Quinteiro-Filho W, Yitbarek A, Astill J, Sharif S (2018a) Oral administration of PLGA-encapsulated CpG ODN and campylobacter jejuni lysate reduces cecal colonization by Campylobacter jejuni in chickens. Vaccine 36(3):388–394

Taha-Abdelaziz K, Yitbarek A, Alkie TN, Hodgins DC, Read LR, Weese JS, Sharif S (2018b) PLGA-encapsulated CpG ODN and campylobacter jejuni lysate modulate cecal microbiota composition in broiler chickens experimentally challenged with C. jejuni. Sci Rep 8(1):12076

Tumpey TM, García-Sastre A, Taubenberger JK, Palese P, Swayne DE, Pantin- Jackwood MJ, Schultz-Cherry S, Solórzano A, Van Rooijen N, Katz JM (2005) Pathogenicity of influenza viruses with genes from the 1918 pandemic virus: functional roles of alveolar macrophages and neutrophils in limiting virus replication and mortality in mice. J Virol 79(23):14933–14944

Uddin MN, Roni MA (2021) Challenges of storage and stability of mRNA-based COVID-19 vaccines. Vaccine 9(9):1033

Van Epps A, Blaney L (2016) Antibiotic residues in animal waste: occurrence and degradation in conventional agricultural waste management practices. Curr Pollut Rep 2:135–155

Vilela J, Rohaim MA, Munir M (2022) Avian Orthoavulavirus Type-1 as vaccine vector against respiratory viral pathogens in animal and human. Vaccine 10(2):259

Wang Y, Davidson I, Fouchier R, Spackman E (2016) Antigenic cartography of H9 avian influenza virus and its application to vaccine selection. Avian Dis 60(1s):218–225

Williams C, Hopkins B (2011) Field evaluation of the accuracy of vaccine deposition by two different commercially available in ovo injection systems. Poult Sci 90(1):223–226

Zhao W, Spatz S, Zhang Z, Wen G, Garcia M, Zsak L, Yu Q (2014) Newcastle disease virus (NDV) recombinants expressing infectious laryngotracheitis virus (ILTV) glycoproteins gB and gD protect chickens against ILTV and NDV challenges. J Virol 88(15):8397–8406

Zhao H, Njeumi F, Parida S, Benfield CT (2021) Progress towards eradication of peste des petits ruminants through vaccination. Viruses 13(1):59

Revolutionizing Animal Health: Exploring the Potential of Bacteriophages in Veterinary Medicine

Saima Naz, Ahmad Manan Mustafa Chatha, Qudrat Ullah, Sobia Malik, Saba Iqbal, and Urwah Ishaque

Abstract Veterinary practice is facing an increasing challenge due to the rise of antimicrobial-resistant (AMR) bacteria. This expanding worldwide health issue has been exacerbated by the extensive use of antibiotics in livestock. In this regard, bacteriophage viruses that selectively infect and kill bacteria present a viable substitute for treating animal bacterial illnesses. This article investigates the mechanisms, advantages, difficulties, and possible applications of bacteriophages in veterinary medicine. Bacteriophages are being favored for biotechnological drive and used to treat infectious bacteria in a variety of areas of concern. Although a few distresses over the use of phage-related solutions. Phage treatment is believed to be more effective than antibiotics in treating bacterial infections. Although, as a result of recent exposure of resistant bacteria to antibiotics, phage therapy's applications have become an unavoidable option for research. To deal with the difficulty of antimicrobial resistance, research workers are examining the use of phage therapy as a substitute and/or supplement to antibiotics act toward and stop infections both in animals and humans. In the time of growing antibiotic resistance, thoughtful and joining the power of bacteriophages could hold the key to communicate and pressing the health challenges globally. This chapter is an attempt to collect almost relevant details relating to applications and complications linked through bacteriophages.

S. Naz (✉) · S. Malik · S. Iqbal · U. Ishaque
Department of Zoology, Government Sadiq College Women University, Bahawalpur, Punjab, Pakistan
e-mail: saima.naz@gscwu.edu.pk

A. M. M. Chatha
Department of Entomology, Faculty of Agriculture and Environment, The Islamia University of Bahawalpur, Bahawalpur, Punjab, Pakistan

Q. Ullah
Department of Zoology, Government Sadiq College Women University, Bahawalpur, Punjab, Pakistan

Department of Theriogenology, Cholistan University of Veterinary and Animal Sciences, Bahawalpur, Pakistan

R. U. Khan, S. Naz (eds.), *Green Alternatives to Antibiotics in Animal Health*,
https://doi.org/10.1007/978-981-92-0309-3_16

Keywords Bacteriophages · Lytic · Lysogenic cycle · Resistance · Antibiotics · Pathogen

1 Introduction

Overview of Bacteriophages and Their Biology

Phages, also known as bacteriophages, are tiny viruses that have a diameter of 20–200 nm. For reproduction and propagation, they are totally dependent on the metabolic machinery of bacteria (Walker 2006). Two scientists, Felix d'Herelle (1917) and Fredrick Twort (1915), independently recognized the phages' significance in the environment. The production of new phages is influenced by temperature, light, available nutrition, and other environmental conditions (Jassim et al. 2017). The coat protein winds up the genome (either RNA or DNA) of the phage (Clark and March 2006). The phage uses the bacterial molecular machinery to proliferate inside the cell during the lytic cycle. The recently formed phage leads the bacteria to lyse after rapid and widespread multiplication inside the host cell, facilitating further phage–host interactions (Dadgostar 2019). The phages strike the host bacterium in the lysogenic cycle and insert their DNA directly inside the genome of the bacteria, passing the genetic information to upcoming generations through replication cycles. These phages wait for the favorable conditions (such as ultraviolet light, temperature, oxidative stress, etc.) to activate specific genes to break free from the latent lysogenic cycle and start replicating their genetic material within the bacterial chromosome (Abedon et al. 2017). "Prophage" is a viral genome incorporated in host cell nucleic acid and a non-infective form of bacteriophage; this is noticeably different from the term "provirus" for viral DNA that has been encompass into eukaryote cells, these phages are called temperate phages (Ling et al. 2022).

Lifecycles

There are two life cycles for bacteriophages: the lytic or virulent life cycle and the temperate or lysogenic life cycle. About 5000 phages have been extensively characterized using electron microscopy (Ackermann 2001). There are two types of phages: virulent and temperate phages. T4 and other virulent phages follow a lytic life cycle. While the capsid and tail stay outside, the phage genome is transferred into the bacterial cytoplasm (Rakhuba et al. 2010). When the host cell bursts, the progeny is discharged into the outside, just as most phages can do on their own. A prophage is a virus that incorporates its DNA into the bacterial chromosome during its lysogenic life cycle. Although temperate bacteriophages also include lytic genes, their lack of expression may be caused by the presence of repressors. Bacteria with integrated genomes change into the lytic form, or lysogen, after infection (Wick et al. 2005).

Instantly, next to the integration with the genome, the bacteriophage life cycle is called the eclipse period (Favrin et al. 2001). The eclipse period, as defined by Favrin et al. (2001), is the time between bacteriophage or other virus infections and the maturation of the virus within the host cell. The host's biosynthesis machinery is eventually controlled by the DNA. Within the bacterial cell, various proteins are synthesized independently. These include lysis proteins that induce the host cell to lyse and coat proteins that aid in the creation of capsids (Kutter and Sulakvelidze 2004).

By substituting bacterial DNA for the phage DNA, a durable association is formed (Verheust et al. 2010). Usually, the viral DNA is incorporated into the bacterial host's chromosome during this cycle. Later on in this life cycle, the lysogen changes into the lytic form, which causes lytic cells to become active. (Verheust et al. 2010).

Historical Context and Rediscovery

In 1915, Frederick Twort was the first to elaborate on the distinctive area of lysis connected to bacteriophages. He noticed ultramicroscopic virus particles in the vaccinia virus culture and recognized them as lytic phages, which were the contaminants of the viral culture (Fruciano and Bourne 2007). Although Felix d'Herelle discovered the cause of this phenomenon and finalized that bacterial viruses were responsible for the plaques, in 1917, at the Pasteur Institute of Paris, France, he coined the word "bacteriophage," which refers to "bacteria eater." Ehrlich clarified that the bacteriophage model was going to be the "magic bullet" that would transform the field of public health. In 1919, at the Hospital of Infants-Maladies in Paris (Ling et al. 2022), D'Herelle envisioned using phages therapeutically, and he was accredited with the first recorded clinical application of phage therapy. Phage suspensions were given systemically via oral methods and/or by injection or topically (Karn et al. 2024). D'Herelle's initial investigations encountered intense debates and condemnation for inadequate bacterial control, regardless of a small number of successful trials. Nonetheless, he carried on to advance phage therapy in the early twentieth century, treating dysentery, cholera, and the bubonic plague. Later on, many phage therapy facilities and industrial phage manufacturing facilities were established in Europe and India (Matsuzaki et al. 2005).

Importances of Phages as an Alternative to Antibiotics

The structure of the cell wall differentiates gram-negative bacteria from gram-positive bacteria, which affects how antimicrobial agents penetrate and are retained by the latter. These bacteria's outer membrane works as a permeability barrier, preventing the cell from being penetrated by a few medications and antibiotics. One of the primary causes of the intrinsic antibiotic resistance seen in the gram-negative bacteria (GNB) is this characteristic (Abedon et al. 2017). Gene mutations that result in the formation of excess efflux pumps or antibiotic-inactivating enzymes (ESBL Carbapenamase makers, extended spectrum beta-lactamase) within the bacteria can give these bacteria resistance. The resistance can be also acquired through, plasmids, transposons/integrons, and via horizontal gene transfer from other bacteria (Rakhuba et al. 2010).

With regard to GNB, research recommended that genes that (resistant) and related insertion elements transferred on plasmids are regularly found concentrated in large multi-resistance regions (MRR), especially in the Enterobacteriaceae family (Benyamini 2024). For example, ESBL and plasmids expressing carbapenemase may harbor resistance genes for other classes of antibiotics, such as fluoroquinolones and aminoglycosides. "The Big Five Carbapenemase producers" are particularly important among these acquired resistances and include: KPC (*Klebsiellapneumoniaecarbapenemase*), IMP (*Imipenemasemetallo-betalactamase*), NDM (*New Delhi metallo-betalactamase),* VIM (Verona integron-encoded *metallo-beta-lactamase*), and OXA (*Oxacillincarbapenemases*).

Classification of Bacteriophages

Phages are so numerous and varied that no one classification scheme can adequately account for all of their distinctive characteristics (Table 1). Soil, fresh water, marine ecosystems, and the human gut are all parts of the "phage-verse," also known as the universe of phage (Hanauer et al. 2017). Similarly, the forms, behaviors, and genetic sequences of phages vary. To learn about each new phage, researchers employ a variety of categorization techniques that generalize certain traits. Although individual research may still yield surprises, some of a novel phage's behaviors can be predicted by comparing it to others already reported in the phage-verse (Ramsey et al. 2020). With the growth of metagenomics research and concentrated efforts to isolate culturable phages from the environment, which are supported by research and educational initiatives globally, the rate of adding novel phage sequences to our databases keeps increasing (Fletcher et al. 2024).

Phages Potential in Veterinary

The current global rise of multi-drug-resistant (MDR) bacteria has been related to the use of antibiotics on chicken farms (Kaonga et al. 2021). Through direct contact with the environment and food chain, MDR bacteria can transfer from animals that produce food to people (Kirbis and Krizman 2015). One-Health difficulties are still a major topic in the poultry sector, pre-harvest as well as post-harvest. Before the live birds are sent to processing facilities, efforts must be made to minimize contamination and/or food-borne pathogen infection (Abd-El Wahab et al. 2023). Physical and chemical methods are the mainstay of post-harvest operations, although they are not always effective in lowering the pathogen load due to a number of issues, such as the limited temperature and concentration that may be achieved. In this regard, developing more potent substitute antimicrobials is essential to reducing the hazards to chicken production by selectively destroying infections without disrupting the microbiota and preventing food-borne pathogen contamination of the food chain (Benrabia et al. 2020). Bacteriophages usage or phage therapy as medicines may be a significant substitute for antibiotics in the chicken industry and may help manage the burden of MDR (Moye et al. 2018).

This chapter aims to explore the biology, replication mechanism, and therapeutic potential of bacteriophages, particularly in combating antibiotic resistance. It covers two life cycles of phages, which they have role in phage interaction and their application in phage therapy. Additionally, this review examines the historical context of phage discovery and their use in veterinary medicine. The scope includes ongoing research into phage ecology and the potential of phages as an alternative to antibiotics in treating bacterial infections.

Table 1 Classification of bacteriophages

Family	Member	Nucleic acid (structure)	Life cycle (Lytic/ lysogenic)	Sensitivity to external factor (temperature, pH, and salinity)	Reference
Myoviridae	T4	Linear dsDNA (NE)	Lytic	Stable at 37 °C and pH of 6–7.4	Klak et al. (2010)
	gspC	Linear dsDNA (NE)	Lytic	Sensitive to salt environment	Wilson et al. (2004)
	YP6	Linear dsDNA (NE)	Lytic	Stable at 60 °C and pH 4–10	Zhang et al. (2025)
	Wc4	Linear dsDNA (NE)	Lytic	Stable at 60 °C	Kering et al. (2020)
	MJ1	Linear dsDNA (NE)	Lytic	Stable at 80 °C	Jamal et al. (2015)
	WZ1	Linear dsDNA (NE)	Lytic	Stable at 65 °C	Jamal et al. (2015)
Siphoviridae		Linear dsDNA (NE)	Both	Stable at 4 °C and pH 3–11	Jepson and March (2004)
	T1	Linear dsDNA (NE)	Lytic	Resistant to 37 °C, sensitive to pH < 3	Faquet et al. (2005)
	TSP4	Linear dsDNA (NE)	Lytic	Survive at 65 °C	Lin et al. (2010)
		Linear dsDNA (NE)	Both	Stable at pH 11	Jepson and March (2004)
	Salmp- p7	Linear dsDNA (NE)	Lytic	Stable at 30–60 °C and pH 4–12	Chen et al. (2025)
Podoviridae	28B	Linear dsDNA (NE)	Lytic	Extremely resistant to dry environment, very stable in pH 9 (Urine)	Prigent et al. (2005) and Hoglund et al. (2002)
	CX5	Linear dsDNA (NE)	Lytic	Stable at 60 °C	Kering et al. (2020)
	P-PSG- 11	Linear dsDNA (NE)	Lysogenic	Stable at 60 °C	Kering et al. (2020)

(continued)

Table 1 (continued)

Family	Member	Nucleic acid (structure)	Life cycle (Lytic/ lysogenic)	Sensitivity to external factor (temperature, pH, and salinity)	Reference
Corticviridae	PM2	Superhelical, circular dsDNA (NE)	Lytic	NaCl and $CaCl_2$ dependent, show stability at pH 6–8	Faquet et al. (2005)
Tectiviridae	PRD1	Linear dsDNA (NE)	Lytic	Stable at 4 °C, and pH 5–8	Faquet et al. (2005) and Ackermann et al. (2004)
	AP50	Linear dsDNA (NE)	Lytic	Can be dried or lose activity at −80 °C	Ackermann et al. (2004)
Leviviridae	MS2	Linear ssRNA (NE)	Lytic	Inactivated at pH 6–8 and 5–35 °C	Feng et al. (2003)
	Qβ	Linear ssRNA (NE)	Lytic	In alkaline environment better survive than in acidic environment	Feng et al. (2003)
	PP7	Linear ssRNA (NE)	Lytic	Thermostable	
Cystoviridae	Phi6	Linear segmented RNA (LPE)	Lytic	Show stability at pH 6	
Inoviridae	M13	Circular ssDNA (E)	Lysogenic	Stable at 37 °C and pH 6–9	Tey et al. (2009)
Lipothrixviridae	TTV1	Linear dsDNA (E)	Lytic	Stable at temperature s 85 °C, and pH <3	
Rudiviridae	SIRV1	Linear dsDNA (NE)	Lytic	Can be found in acidic hot spring environment	Prangishvili et al. (2001)
	AFV-1	Linear dsDNA (NE)	Lytic	Stable at hot springs	Prangishvili et al. (2001)
Plasmaviridae	L2	Circular superhelical dsDNA (E)	Lytic	Extremely sensitive to heat, stable at cold	Faquet et al. (2005)
	Kpph1	Circular superhelical dsDNA (E)	Lytic	Stable at temperature up to 55 °C	Huang et al. (2025)

(continued)

Table 1 (continued)

Family	Member	Nucleic acid (structure)	Life cycle (Lytic/ lysogenic)	Sensitivity to external factor (temperature, pH, and salinity)	Reference
Fuselloviridae	SSV1	Circular superhelical dsDNA (NE)	Lytic	Stable at 97 °C and sensitive to pH >11	Faquet et al. (2005)
	His1	Circular super helical dsDNA (NE)	Lytic	Stable at 37 °C and 18% salty concentration	Faquet et al. (2005)
Caudoviridae	St1749	Icosahedral dsDNA (NE)	Lytic	Stable at pH 5–7	Bich van et al. (2025)

Abbreviations: *NE* Non-enveloped, *E* Enveloped, *LPE* Lipoprotein envelope

2 Rationale of the Study

The increasing prevalence of antibiotic-resistance bacteria has become a major global health concern. That is why it is important to choose alternative therapeutic options. Bacteriophages have the ability to target and kill infections, so it is a promising solution to solve this growing problem. Now, phages are rediscovered to treat bacterial infections, especially against antibiotic-resistant bacteria, and in veterinary medicine. So, we use phage therapy because they are self-replicating and eco-friendly. The aim of this review is to study biology, life cycle, and applications; my focus is to study therapeutic potential. It will give a comprehensive review of how phages have a role in medicine and veterinary. These review findings will contribute to the use of phage-based therapies. I can explain to you that with the help of phages, we can fight against antibiotic-resistant bacteria.

3 Mechanisms of Action

Phages needed specific receptors on the surface of the bacteria, such as teichoic acids, lipopolysaccharides, flagella, and proteins, in order to infect them. Because of the specificity of the receptor on the bacterial cell surface and the random nature of the phage–host contact, phages can only infect a certain host. The bacteriophage's ability to cling to the host's surface is also influenced by the virulence, growing conditions, and kind of bacteria (both gram-positive and negative) (Rakhuba et al. 2010). Gram-negative bacteria's outer membrane has an exterior lipopolysaccharide (LPS) embedded with outer proteins (OMPs) for the transportation and diffusion of nutrients. The bacteriophage particle adheres its base plate to the bacterial surface in an irreversible manner after identifying the receptor in a reversible manner. Following that, the tail contraction occurs with ATP assistance, and genetic material is introduced (Fig. 1) (León and Bastías 2015).

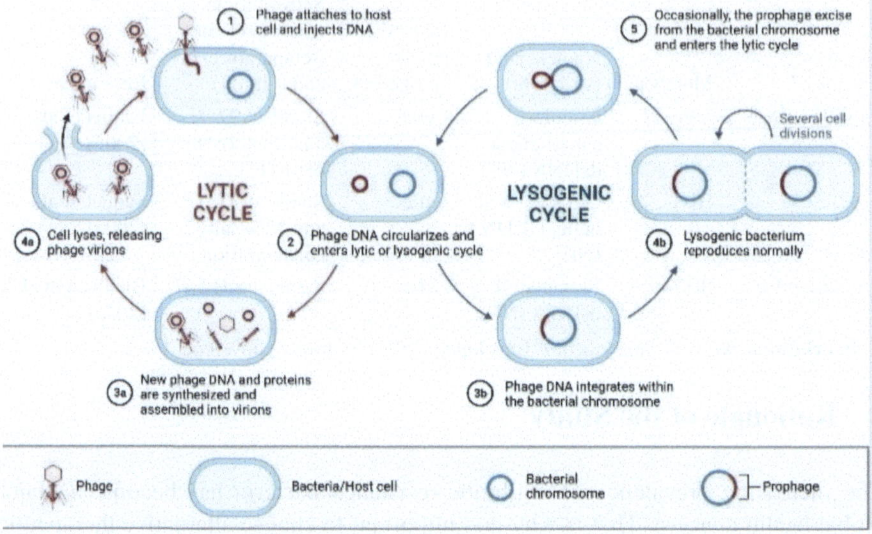

Fig. 1 A figure showing bacteriophages' mechanism of action by lytic and lysogenic cycles

3.1 Applications in Animal Health

Livestock The gastrointestinal tract (GIT) presents itself as the largest immunological organ in animals through its microbial populations, which direct immune system development. Intestine health, alongside shape and structure, undergoes significant changes because of bacteriophage interactions that take place directly or indirectly with gut bacteria. The assessment of intestinal health relies on the measurement of villi height (VH) and crypt depth (CD) in combination with their ratio (VH/CD) as reported by Yan et al. (2023). In a trial using phage supplementation to substitute antibiotics in the diet of weaned piglets, Research from Zeng et al. (2021) showed that adding phages to feed at 40 mg/kg as antibiotic substitutes increased the jejunum and ileum VH while lowering jejunum and ileum CD and enhancing relative mRNA expression.

Aquaculture Aquaculture industries regularly experience economic losses mainly because of unchecked microbial diseases that damage their growth and sustainability (Silva et al. 2014). Fish products are more widely available worldwide since they are an inexpensive source of protein. Aquaculture and related sectors have grown significantly over the past three decades. Nevertheless, harmful bacteria such as *Aeromonas salmonicida, Vibrio vulnificus, Flavobacterium psychrophilum, Vibrio anguillarum photobacterium damselae,* and *Aeromonas hydrophila* thrive and cause illnesses or significant losses in humans due to the use of tainted aquaculture products (Saksida et al. 2006).

Vibriosis is a common disease mainly in marine estuarine fisheries, which causes substantial mortality in fish (upto 100% in larvae). Vibrosis is caused by the bacterial genus *Vibrio*, which includes species such as *V. vulnificus, V. parahaemolyticus, V. anguillarum, V. alginolyticus,* and *V. salmonicida* (Higuera et al. 2013). Treatment with bacteriophages has been found to be effective in treating *V. anguillaraum* infection in fish larvae, while medications have demonstrated decreased efficacy against multidrug-resistant (MDR) bacteria. (Silva et al. 2014).

Companion Animals Wounds are one of the most common sites for infection by different bacteria, including *Staphylococcus aureus*, in impoverished nations (Norbury et al. 2016). Antibiotic resistance in *Staphylococcus aureus* has recently resulted in higher rates of mortality, morbidity, and hospitalization time (Hashemi et al. 2018). It is difficult to manage antibiotic-resistant *S. aureus* in hospitals, particularly in burn units. As a result, researchers are developing novel approaches such as bacteriophage therapy (Fig. 2) (Peng et al. 2019). On a study, Seth et colleagues created a cutaneous lesion on a New Zealand rabbit ear and infected *S. aureus* to build a biofilm at the wound site. The primary treatment for the wounds was debrideme, a phage particular to S. aureus. The research showed debrideme phage used separately or in combination with debridement reduced S. aureus biofilm infections in rabbit wounds, yet the results remained insignificant (Seth et al. 2013).

Environmental Uses *Salmonella Typhimurium* serves as the Gram-negative human pathogen responsible for non-typhoid salmonellosis. The virus spreads through food or water products that become contaminated (Santos et al. 2003). The Centers for Disease Control and Prevention reported non-typhoid salmonellosis afflicted 1.2 million Americans in 2013, where drug-resistant Salmonella caused nearly 100,000 of these cases (CDC 2013). Salmonella ranks as a primary foodborne

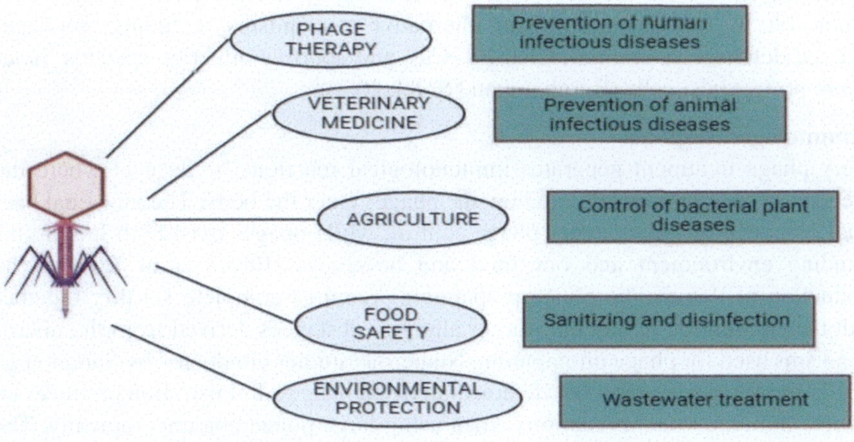

Fig. 2 Applications of phages in animal health

disease cause in Europe because EFSA and ECDC recorded more than 80,000 human cases in 2013 (EFSA and ECDC 2015).

3.2 *Advantages of Phage Therapy: High Specificity for Target Pathogen*

Different broad-spectrum repressive approaches of antibiotics and phages use host specificity because they can recognize certain signal molecules, for example, proteins and polysaccharides on the surface of bacteria (Islam et al. 2024). Phages have the capacity to encode depolymerase, to damage the macromolecules, such as extracellular polysaccharides, capsular polysaccharides, and lipopolysaccharides (LPS) on the exterior barrier of bacteria, which rushes the injection of phage nucleic acid within the bacteria to achieve the invasion process (Pires et al. 2016).

Antibiotic Resistance
Under high stress conditions, phages develop resistance, thus requiring essential methods to counter it. Multiple bacterial strains demonstrate diverse resistance rates together with different resistance effects according to Rohde et al. (2018). Resistances occurred among Acinetobacter baumannii strains during the initial phage therapy run for the described infection. The researchers needed to change to different phage strains because the resistant bacterium responded to those particular strains (Schooley et al. 2017). The recovery of phage-resistant survivors occurs naturally at rates up to 10^{-5} from bacterial cultures raised under laboratory conditions. The development of resistance in vitro occurs less frequently than anticipated, while minimally reaching the detection threshold of 5×10^{-7} for numerous clinical isolates of *Mycobacteroides abscessus* according to Glonti and Pirnay (2022). The survival rates between phage receptor variation and deletion show partial dissimilarities depending on the acceptance criteria. Reception variation is commonly responsible for resistance, but many alternative mechanisms, including prophage-encoded defenses, restriction, CRISPR–Cas, and abortive infection systems, function to protect host cells (Bernheim and Sorek 2020).

Immunologic Response
Every phage treatment generates immunological reactions because of where the infection occurs, combined with how the phages enter the body. The intestinal bacterial composition exists under phage control, while phages exist both in our surrounding environment and our food and beverages (Reyes et al. 2012). The production of therapeutic phage preparations requires complete sterility together with the elimination of any harmful or allergic substances derived from the microorganisms used for phage propagation. Numerous studies conducted by Sarker et al. (2012) and Wright et al. (2009) demonstrated that phage administration produces no adverse immune system reactions when using large phage amounts topically. The human body cannot recognize the peptides that cover phage surfaces, making them

unique. The titers of phages decline quickly after intravenous administration because both innate immune responses, along with phagocytic activity in blood and liver, take effect before adaptive immunity steps in (Sokoloff et al. 2000). Different phage injections lead to better immune response and increased non-neutralizing antibodies, which eventually develop into IgM and finally IgG (Bachrach et al. 2003).

3.3 *Disadvantages of Phage Therapy*

Not Every Phage Is a Good Therapeutic Candidate
Good therapeutic phages should be able to target and kill pathogens with high efficiency and with little chance of negatively changing the ecosystems in which they are used. These characteristics can be reasonably assured as long as phages are obligatorily lytic, stable at standard storage temperatures and conditions, the focus of appropriate efficacy and safety investigations, and, preferably, thoroughly sequenced to confirm the absence of undesirable genes like toxins (Krylov 2001). Notably, a phage that is "obligately lytic" (i.e., unable to display lysogeny and not released over a long duration) is not temperate and is released from infected cells through lysis. The use of temperate phages as therapeutics is difficult because of the encoding of bacterial virulence factors, such as bacterial toxins, and the development of superinfection immunity, which renders phage-sensitive bacteria insensitive (Gill and Hyman 2010).

The Limited Host Range Issue
Antimicrobials displaying selective toxicity will not affect all possible microbial targets. The limited range of phage hosts—a few strains, a few species, or, even less frequently, a few genera of bacteria—usually limits the use of presumed treatment, or treatment regimens that begin before the pathogen's susceptibility to antibacterials, such as specific phages, is known (Kutter et al. 2010). However, because phages are commonly employed in combination with other antibacterial treatments, including other phages (so-called phage cocktails), the lytic spectrum of phage products can be far broader than the spectrum of activity of individual phage types (Kutateladze and Adamia 2010). Even broadly active phage combinations usually show a more specific spectrum of activity than traditional "narrow-spectrum" antibiotics, which is another advantage of phage therapy (Gill and Hyman 2010).

Phages Are Not Special Pharmaceuticals
Phage medicines are protein-based, live-biological agents that can actively multiply, interact with the body's immune system, and even change during manufacturing or use; however, they are not unique in these aspects. For example, bacterial lysing drugs release bacterial toxins in situ; live attenuated vaccines actively multiply and evolve, especially when they infect body tissues; and other protein-based drugs can strengthen immune systems (Skurnik et al. 2007).

Despite these diverse characteristics, complete vaccinations, chemical antibiotics, and protein-based medications have all been authorized for use in the past.

Therefore, it makes sense that medications based on phages shouldn't be rejected because they share similar characteristics (Merabishvili et al. 2009).

4 Challenges and Regulation in Practical Applications

Research has shown phages can eliminate bacteria since 1924, according to Pirnay et al. (2024). To date, no new invention exists that effectively destroys bacteria of clinical importance. Penicillin received its discovery in 1928, before a mass-production process started in 1943, with a feasible price for medical treatment. The governments of numerous states do not regulate the creation of such inventions (Willy et al. 2023). The reason for this lack of knowledge was that bacteriophage biology remained unknown at the time, together with the features of their replication cycle. Researchers fear for phage therapy because they gained more viral knowledge and developed new technology, while bacteria developed resistance, and scientists disagreed about drug amounts and ignored immune system responses, and the treatment needed more promotion (Willy et al. 2023). The safety measures for phage therapy include identifying phages inside experimental animal tissues found in both the brain and the ovary as well as heart regions. Regular human tissue contains about 1016 (pfu) plate-forming units according to Ling and others (2022). Soutourina et al. (2013) discovered through in vitro and in silico tests mRNA with regulatory capabilities for gene expression in bacteria which requires acknowledgment of phage DNA impacts on protein and genetic metabolic regulation of host cells with eukaryotes in sick patients. Researchers have not determined the specific role of phage mRNAs inside human cells.

4.1 Host Immune Response

The interaction of phages with the body's immune cells determines how effective they are against poultry and cattle. With the aid of particular receptor uptake by specific cytophagy, phages in the intestinal lumen can penetrate the intestinal epithelium into the lamina propria under intestinal homeostasis. From there, they can travel via blood into the circulation to interact with immune cells and start an anti-inflammatory response. (Miernikiewicz and Dąbrowska 2022). The immune system of the intestinal mucosa is initiated when the barrier of intestinal structure is damaged; in that way, the immune response is activated. It is noted that phages normally enhance at the place of bacterial colonization, and numerous bacteria in the gut produce a condition for phages to regulate the response of immune system of organisms (Barr et al. 2013).

4.2 Stability and Delivery

For phages, survivability temperature is an important factor (Olson et al. 2004). Temperature plays a major role in phage attachment, penetration, multiplication, and, in the case of lysogenic phages, the length of the latent period is also controlled by temperature. For the penetration of genetic material into bacterial cells, phages required both higher and lower temperatures (Tey et al. 2009). Additionally, temperature determines the viability, occurrence, and storage of bacteriophages. A classic environment in which bacteriophages can remain alive is hot springs (temperature 40–90°C) (Atamer et al. 2009). In experimental animal studies, the most successful and famous of all delivery methods for phages is parenteral delivery because of its rapid distribution into the system circulation. But for successful phage therapy, certain specific administration sites, i.e., intramuscular (IM) and intraperitoneal (IP), are required. Hyman and Abedon (2010) investigated the use of lytic bacteriophages in calves to control *E. coli* septicemia in hens and to determine the effectiveness of intramuscular (IM) administration of 50 ML dilutions (106 PFU) for phage preparations. In the absence of phages, E. coli causes 100% death in 3-week-old and freshly hatched chickens when administered intramuscularly. Phages were merely suspended in Luria-Bertani broth (LB Broth) (Hyman and Abedon 2010).

4.3 Resistance Development

There is a struggle between bacteriophages and the immunity of bacterial infections. Between these two entities, the discriminating pressure is exclusive. Bacteria improve their antiviral mechanism after becoming resistant to bacteriophages, and vice versa. This relationship between bacteria and bacteriophages is an exciting phenomenon high mutation rates are a typical event for bacteria, such as Pseudomonas fluorescents, when they are exposed to reconcile in vitro systems (Buckling and Rainey 2002). Bacteria's resistance to phages can therefore result from a variety of methods. These techniques include adaptive immunity based on CRISPR (Clustered Regularly Interspaced Short Palindromic Repeats) sequences or altering the attachment or adsorption sites (receptors) of phages (Hyman and Abedon 2010). The modification of phage adsorption sites/receptors is the most frequent way that bacteria evade phage invasion and acquire phage resistance. Even so, phages are also capable of adapting to recognize these novel, changed receptors. The production of masking proteins, such as Staphylococcus aureus protein A, or exopolysaccharide (EPS) to obscure the phage receptor, is another tactic used by bacteria. Once more, phages break through the barrier by using polysaccharide hydrolase or polysaccharide lyase to split the EPS layer (Örmälä and Jalasvuori 2013).

5 Limitations of the Study

There is still much to learn about the connection between phages, animal gut microbiota, and immunology. Furthermore, it is unclear how phages entering the animals will affect them. One problem is that LPSs are released into the environment during normal bacterial growth and division, causing hazardous effects, including endotoxins, hemolysis, and tropical inflammation, in addition to entering the environment through bacterial rupture and disintegration (Shen et al. 2023). Because phage lysis of Gram-positive bacteria (like Salmonella) releases LPSs from the bacterial cell wall into the intestinal lumen, which may trigger an inflammatory response, the use of phages against intestinal pathogens must be carefully considered (Ling et al. 2022). Research is still being done to overcome these obstacles, but bacteriophages are still a viable weapon to fight against infections that are resistant to antibiotics.

6 Conclusions

Bacteriophage therapy holds significant promise as a targeted antibacterial approach offering an alternative to combat multidrug-resistant bacterial infections. When combined with antibiotics, phage therapy has the potential to revolutionize infection management. However, the development of bacterial resistance to phages through mechanisms such as loss of phage receptors or degradation of phage nucleic acid presents a challenge that must be addressed to ensure the long-term success of phage therapy. Continue research into optimizing phage formulation, overcoming resistance, and understanding interaction with the host immune system will be essential in realizing the full potential of phage therapy.

Acknowledgments The authors sincerely acknowledged Associate Professor Dr. Saima Naz, Department of Zoology, The Government Sadiq College Women University, Bahawalpur, for her kind suggestions and help related to this work.

References

Abedon ST, García P, Mullany P, Aminov R (2017) Phage therapy: past, present and future. Front Microbiol 8:981

Abd-El Wahab A, Basiouni S, El-Seedi HR, Ahmed MFE, Bielke LR, Hargis B et al (2023) An overview of the use of bacteriophages in the poultry industry: Successes challenges and possibilities for overcoming breakdowns Front Microbiol 14:1136638

Ackermann HW (2001) Frequency of morphological phage descriptions in the year 2000. Arch Virol 146:843–857

Ackermann HW, Tremblay D, Moineau S (2004). Long-term bacteriophage preservation. WFCC Newslett 38:35–40

Atamer Z, Dietrich J, Müller-Merbach M, Neve H, Heller KJ, Hinrichs J (2009) Screening for and characterization of Lactococcus lactis bacteriophages with high thermal resistance. Int Dairy J 19(4):228–235

Bachrach G, Leizerovici-Zigmond M, Zlotkin A, Naor R, Steinberg D (2003) Bacteriophage isolation from human saliva. Lett Appl Microbiol 36(1):50–53. https://doi.org/10.1046/j.1472-765X.2003.01262.x

Barr JJ, Auro R, Furlan M, Whiteson KL, Erb ML, Pogliano J et al (2013) Bacteriophage adhering to mucus provide a non–host-derived immunity. Proc Natl Acad Sci 110(26):10771–10776

Benrabia I, Hamdi TM, Shehata AA, Neubauer H, Wareth G (2020) Methicilli n- resistant Staphylococcus aureus (MRSA) in poultry species in Algeria: Long-term study on prevalence and antimicrobial resistance. Vet Sci 7(2):54. https://doi.org/10.3390/vetsci7020054

Benyamini P (2024) Beyond antibiotics: what the future holds. Antibiotics 13(10):919

Bernheim A, Sorek R (2020) The pan-immune system of bacteria: antiviral defence as a community resource. Nat Rev Microbiol 18(2):113–119

Bich van TT, Anh NTL, Vo VT, Thi NPA (2025) Isolation and characterization of a novel multi-host bacteriophage St1749 and its effectiveness against vibrio Parahaemolyticus and Salmonella Spp. Available at SSRN 5088152

Buckling A, Rainey PB (2002) Antagonistic coevolution between a bacterium and a bacteriophage. Proc R Soc Lond Ser B Biol Sci 269(1494):931–936

Centers for Disease Control and Prevention (U.S.) (2013) Antibiotic resistance threats in the United States. https://stacks.cdc.gov/view/cdc/20705

Chen M, Yu T, Cao X, Pu J, Wang D, Deng H (2025) Isolation and characteriza t ion of Salmonella enteritidis bacteriophage Salmp-p7 isolated from slaughterhouse effluent and its application in food. Arch Microbiol 207(1):7

Clark JR, March JB (2006) Bacteriophages and biotechnology: vaccines, gene therapy and antibacterials. Trends Biotechnol 24(5):212–218

Dadgostar P (2019) Antimicrobial resistance: implications and costs. Infect Drug Resist 12:3903–3910

European Food Safety Authority, & European Centre for Disease Prevention and Control (2015) The European Union summary report on trends and sources of zoonoses, zoonotic agents and food-borne outbreaks in 2013 EFSA J 13(1):3991

Faquet CM, Mayo MA, Maniloff J, Desselberger U, Ball LA (2005) Virus taxonomy: eighth report of the International Committee on Taxonomy of Viruses. London: Elsevier Academic Press.

Favrin SJ, Jassim SA, Griffiths MW (2001) Development and optimization of a novel immunomagnetic separation-bacteriophage assay for detection of Salmonella enterica serovar enteritidis in broth. Appl Environ Microbiol 67(1):217–224

Feng YY, Ong SL, Hu JY, Tan XL, Ng WJ (2003) Effects of pH and temperature on the survival of coliphages MS2 and Qbeta. J Ind Microbiol Biotechnol (30):549–552. https://doi.org/10.1007/s10295-003-0080-y

Fletcher J, Manley R, Fitch C, Bugert C, Moore K, Farbos A et al (2024) The Citizen Phage Library: rapid isolation of phages for the treatment of antibiotic resistant infections in the UK. Microorganisms 12(2):253

Fruciano DE, Bourne S (2007) Phage as an antimicrobial agent: d'Herelle's heretical theories and their role in the decline of phage prophylaxis in the West. Can J Infect Dis Med Microbiol 18(1):19–26

Glonti T, Pirnay JP (2022) In vitro techniques and measurements of phage characteristics that are important for phage therapy success viruses. 14(7):1490. https://doi.org/10.3390/v14071490

Gill JJ, Hyman P (2010) Phage choice, isolation, and preparation for phage therapy. Curr Pharm Biotechnol 11(1):2–14

Hanauer DI, Graham MJ, Sea-Phages, Betancur L, Bobrownicki A, Cresawn SG et al (2017) An inclusive Research Education Community (iREC): impact of the SEA-PHAGES program on research outcomes and student learning. Proc Natl Acad Sci 114(51):13531–13536

Hashemi A, Jaberi S, Shariati A, Ghalavand Z, Bostan Ghadiri N, Doustdar F et al (2018) Identification of inducible clindamycin resistance in staphylococc us aureus strains isolated from hospitalized patients in hospitals of Tehran City (Iran). Qom Univ Med Sci J 11(12):52–60

Higuera G, Bastías R, Tsertsvadze G, Romero J, Espejo RT (2013) Recently discovered Vibrio anguillarum phages can protect against experimentally induced vibriosis in Atlantic salmon, Salmo salar. Aquaculture 392:128–133

Höglund C, Ashbolt N, Stenström TA, Lennart S (2002) Viral persistence in source-separated human urine. Adv Environ Res 6(3):265–275. https://doi.org/10.1016/S1093-0191(01)00057-0

Huang Y, Huang Y, Wu Z, Fan Z, Zheng F, Liu Y, Xu X (2025) Characterization and genomic insights into bacteriophages Kpph1 and Kpph9 against hypervirule nt carbapenem-resistant Klebsiella pneumoniae. Virulence 16(1):2450462. https://doi.org/10.1080/21505594.2025.2450462

Hyman P, Abedon ST (2010) Bacteriophage host range and bacterial resistance. Adv Appl Microbiol 70:217–248

Islam M, Mahbub NU, Shin WS, Oh MH (2024) Phage-encoded depolymerases as a strategy for combating multidrug-resistant Acinetobacter baumannii Front Cell Infect Microbiol 14. https://doi.org/10.3389/fcimb.2024.1462620

Jamal M, Hussain T, Das CR, Andleeb S (2015) Isolation and characterization of a Myoviridae MJ1 bacteriophage against multi-drug resistant Escherichia coli 3. Jundishapur J Microbiol 8(11)

Jepson CD, March JB (2004) Bacteriophage lambda is a highly stable DNA vaccine delivery vehicle. Vaccine 22(19):2413–2419

Kaonga N, Hang'ombe BM, Lupindu AM, Hoza AS (2021) Detection of CTX-M- type extended-spectrum beta-lactamase producing Salmonella Typhimurium in commercial poultry farms in Copperbelt Province, Zambia. Ger J Vet Res 1(2):27–34. https://doi.org/10.51585/gjvr.2021.2.0011

Karn SL, Bhartiya SK, Pratap A, Saroj SK, Kumar R, Sahu M et al (2024) A randomized, placebo-controlled, double-blind clinical trial of bacteriophage cocktails in chronic wound infections. Int J Lower Extremity Wounds:15347346231226342

Kering KK, Zhang X, Nyaruaba R, Yu J, Wei H (2020) Application of adaptive evolution to improve the stability of bacteriophages during storage. Viruses 12(4):423. https://doi.org/10.3390/v12040423

Kirbis A, Krizman M (2015) Spread of antibiotic resistant bacteria from food of animal origin to humans and vice versa. Procedia Food Sc 5:148–151. https://doi.org/10.1016/j.profoo.2015.09.043

Kłak M, Międzybrodzki R, Bubak B, Jończyk E, Weber-Dąbrowska B, Górski A (2010) Studies on the gastrointestinal transit and blood penetration of a therapeutic staphylococcal bacteriophage. Abstract no. 209, First International Congr. Viruses of Microbes, Paris

Krylov VN (2001) Phagotherapy in terms of bacteriophage genetics: hopes, perspectives, safety, limitations. Genetika 37(7):869–887

Kutateladze M, Adamia R (2010) Bacteriophages as potential new therapeutics to replace or supplement antibiotics. Trends in Biotechnol 28(12):591–595

Kutter E, Sulakvelidze A (2004) Bacteriophages: biology and applications. CRC Press

Kutter E, De Vos D, Gvasalia G, Alavidze Z, Gogokhia L, Kuhl S, Abedon ST (2010) Phage therapy in clinical practice: treatment of human infections. Curr Pharm Biotechnol 11(1):69–86

León M, Bastías R (2015) Virulence reduction in bacteriophage resistant bacteria. Front Microbiol 6:343. https://doi.org/10.3389/fmicb.2015.00343

Ling H, Lou X, Luo Q, He Z, Sun M, Sun J (2022) Recent advances in bacteriophage-based therapeutics: Insight into the post-antibiotic era. Acta Pharmaceutica Sinica B 12(12):4348–4364

Lin L, Hong W, Ji X, Han J, Huang L, Wei Y (2010) Isolation and characterization of an extremely long tail Thermus bacteriophage from Tengchong hot springs in China Abstract J Basic Microbiol 50(5):452–456. https://doi.org/10.1002/jobm.201000116

Matsuzaki S, Rashel M, Uchiyama J, Sakurai S, Ujihara T, Kuroda M et al (2005) Bacteriophage therapy: a revitalized therapy against bacterial infectious diseases. J Infect Chemother 11:211–219

Merabishvili M, Pirnay JP, Verbeken G, Chanishvili N, Tediashvili M, Lashkhi N et al (2009) Quality-controlled small-scale production of a well- defined bacteriophage cocktail for use in human clinical trials. PLoS One 4(3):e4944

Miernikiewicz P, Dąbrowska K (2022) Endocytosis of bacteriophages. Curr Opin Virol 52:229–235

Moye ZD, Woolston J, Sulakvelidze A (2018) Bacteriophage applications for food production and processing. Viruses 10(4):205. https://doi.org/10.3390/v10040205

Norbury W, Herndon DN, Tanksley J, Jeschke MG, Finnerty CC, Scientific Study Committee of the Surgical Infection Society (2016) Infection in burns. Surg Infect 17(2):250–255. https://doi.org/10.1089/sur.2013.134

Olson MR, Axler RP, Hicks RE (2004) Effects of freezing and storage temperature on MS2 viability. J Virol Methods 122(2):147–152. https://doi.org/10.1016/j.jviromet.2004.08.010

Örmälä AM, Jalasvuori M (2013) Phage therapy: should bacterial resistance to phages be a concern, even in the long run? Bacteriophage 3(1):e24219. https://doi.org/10.4161/bact.24219

Peng C, Hanawa T, Azam AH, LeBlanc C, Ung P, Matsuda T et al (2019) Silviavirus phage ϕMR003 displays a broad host range against methicillin-resist ant Staphylococcus aureus of human origin. Appl Microbiol Biotechnol 103:7751–7765. https://doi.org/10.1007/s00253-019-10039-2

Pires DP, Oliveira H, Melo LD, Sillankorva S, Azeredo J (2016) Bacteriophage – encoded depolymerases: their diversity and biotechnological applications. Appl Microbiol Biotechnol 100:2141–2151

Pirnay JP, Djebara S, Steurs G, Griselain J, Cochez C, De Soir S et al (2024) Personalized bacteriophage therapy outcomes for 100 consecutive cases: a multicentre, multinational, retrospective observational study. Nat Microbiol:1–20

Prangishvili D, Stedman K, Zillig W (2001) Viruses of the extremely thermophilic archaeon Sulfolobus Trends Microbiol 9(1):39–43. https://doi.org/10.1016/S0966-842X(00)01910-7

Prigent M, Leroy M, Confalonieri F, Dutertre M, DuBow MS (2005) A diversity of bacteriophage forms and genomes can be isolated from the surface sands of the Sahara Desert Extremophiles 9(4):289–296. https://doi.org/10.1007/s00792-005-0444-5

Rakhuba DV, Kolomiets EI, Dey ES, Novik GI (2010) Bacteriophage receptors, mechanisms of phage adsorption and penetration into host cell. Polish J Microbiol 59(3):145

Ramsey J, Rasche H, Maughmer C, Criscione A, Mijalis E, Liu M et al (2020) Galaxy and Apollo as a biologist-friendly interface for high-quality cooperative phage genome annotation. PLoS Comput Biol 16(11):e1008214

Reyes A, Semenkovich NP, Whiteson K, Rohwer F, Gordon JI (2012) Going viral: next-generation sequencing applied to phage populations in the human gut. Nat Rev Microbiol 10(9):607–617

Rohde C, Resch G, Pirnay JP, Blasdel BG, Debarbieux L, Gelman D et al (2018) Expert opinion on three phage therapy related topics: bacterial phage resistance, phage training and prophages in bacterial production strains. Viruses 10(4):178

Sabah AA, Jassim Richard G, Limoges Sabah AA, Jassim Richard G (2017) Limoges bacteriophages: practical applications for nature's biocontrol discussion and general conclusions Springer International Publishing Cham 223–239

Saksida S, Constantine J, Karreman GA, Neville C, Sweeting R, Beamish R (2006) Evaluation of sea lice, Lepeophtheirus salmonis, abundance levels on farmed salmon in British Columbia, Canada. In: The proceedings from the international symposium on veterinary epidemiology and Economics XI. Cairns, pp 6–11

Santos LRD, Nascimento VPD, Oliveira SDD, Rodrigues DP, Reis EMFD, Seki LM et al (2003) Phage types of Salmonella enteritidis isolated from clinical and food samples, and from broiler carcasses in southern Brazil. Revista do Instituto de Medicina Tropical de São Paulo 45:01–04

Sarker SA, McCallin S, Barretto C, Berger B, Pittet AC, Sultana S et al (2012) Oral T4-like phage cocktail application to healthy adult volunteers from Bangladesh. Virology 434(2):222–232

Schooley RT, Biswas B, Gill JJ, Hernandez-Morales A, Lancaster J, Lessor L et al (2017) Development and use of personalized bacteriophage-based therapeutic cocktails to treat a patient with a disseminated resistant Acinetobacter baumannii infection. Antimicrob Agents Chemother 61(10):10–1128

Seth AK, Geringer MR., Nguyen KT, Agnew SP, Dumanian Z, Galiano RD,... & Hong SJ (2013) Bacteriophage therapy for staphylococcus aureus biofilm–infected wounds plastic and reconstructive surgery 131(2):225–234. https://doi.org/10.1097/PRS.0b013e31827e47cd

Shen J, Zhang J, Mo L, Li Y, Li Y, Li C et al (2023) Large-scale phage cultivation for commensal human gut bacteria. Cell Host Microbe 31(4):665–677

Silva YJ, Costa L, Pereira C, Mateus C, Cunha Â, Calado R et al (2014) Phage therapy as an approach to prevent Vibrio anguillarum infections in fish larvae production. PLoS One 9(12):e114197. https://doi.org/10.1371/journal.pone.0114197

Skurnik M, Pajunen M, Kiljunen S (2007) Biotechnological challenges of phage therapy. Biotechnol Lett 29:995–1003. https://doi.org/10.1007/s10529-007-9346-1

Sokoloff AV, Zhang G, Sebestyén MG, Wolff JA, Bock I (2000) The interact ions of peptides with the innate immune system studied with use of T7 phage peptide display. Mol Ther 2(2):131–139

Soutourina OA, Monot M, Boudry P, Saujet L, Pichon C, Sismeiro O et al (2013) Genome-wide identification of regulatory RNAs in the human pathogen Clostridium difficile. PLoS Genetics 9(5):e1003493

Tey BT, Ooi ST, Yong KC, Ng MYT, Ling TC, Tan WS (2009) Production of fusion m13 phage bearing the di-sulphide constrained peptide sequence (C- WSFFSNI-C) that interacts with hepatitis B core antigen. Afr J Biotechnol 8(2):268

Verheust C, Pauwels K, Mahillon J, Helinski DR, Herman P (2010) Contained use of bacteriophages: risk assessment and biosafety recommendations. Appl Biosaf 15(1):32–44

Walker K (2006) Use of bacteriophages as novel food additives. FS06 ANR 490:811

Wick CH, Elashvili I, Mccubbin PE, Birenzvige A (2005) Determination of MS2 bacteriophage stability at high temperatures using the integrated virus detection system (IVDS) 17

Wilson C, Caton TM, Buchheim JA, Buchheim MA, Schneegurt MA, Miller RV (2004) DNA-Repair potential of halomonas spp. from the salt plains microbial observatory of oklahoma. Microb Ecol 48(4):541–549. https://doi.org/10.1007/s00248-004-0243-z

Willy C, Bugert JJ, Classen AY, Deng L, Düchting A, Gross J et al (2023) Phage therapy in Germany—update 2023. Viruses 15(2):588

Wright A, Hawkins CH, Änggård EE, Harper DR (2009) A controlled clinical trial of a therapeutic bacteriophage preparation in chronic otitis due to antibiotic -resistant Pseudomonas aeruginosa; a preliminary report of efficacy. Clin Otolaryngol 34(4):349–357

Yan A, Butcher J, Schramm L, Mack DR, Stintzi A (2023) Multiomic spatial analysis reveals a distinct mucosa-associated virome. Gut Microbes 15(1):2177488

Zeng Y, Wang Z, Zou T, Chen J, Li G, Zheng L et al (2021) Bacteriophage as an alternative to antibiotics promotes growth performance by regulating intestina l inflammation, intestinal barrier function and gut microbiota in weaned piglets. Front Vet Sci 8:623899

Zhang S, Ye Q, Wang M, Zhu D, Jia R, Chen S et al (2025) Isolation and characterization of a broad-spectrum bacteriophage against multi-drug resistant Escherichia coli from waterfowl field. Poult Sci 104:104787

Bacteriophages and Vaccines: A Dual Approach to Immune Enhancement

Faria Javaid, Nimra Riaz, Aneeza Shafqat, Fatima Sarwar, Nayab Batool, Muhammad Umar Zafar Khan, Hafsa Mukhtar, Zil-e-Huma, and Rabia Ghaffar

Abstract Besides the conventional antibiotic therapy, the following options of disease control should be considered as a result of the appearance of the antibiotic resistance phenomenon. The present chapter focuses on new strategies, including bacteriophage therapy, non-conventional antimicrobial drugs, vaccination, and diet as possible alternatives to control resistant bacteria. These strategies offer better, more directed, and longer-lasting solutions to infection control. Probiotics have a positive effect on the immune system, and phage therapy is effective in veterinary medicine. Vaccines proactively boost the immune system and offer long-lasting protection. New oral vaccines, recombinant vectors, adjuvants, and in ovo immunization methods are some technologies in human and veterinary medicine that offer promising approaches to the prevention of diseases. To achieve long-term infection control sustainability and to reduce our dependence on conventional antibiotics, a holistic integrated approach is required. Despite promising results, existing trials are constrained by regulatory barriers, ethical conundrums, and implementation limitations. Future work should focus on enhancing the refinement of precision medicine, coordinating combination therapies, and addressing logistical deployment challenges. This chapter highlights the importance of integrating innovative approaches in the endeavor to achieve lasting disease control and global health security.

Keywords Antibiotic resistance · Bacteriophage therapy · Vaccination · Probiotics · Pathogen elimination

F. Javaid (✉) · N. Riaz · A. Shafqat · F. Sarwar · N. Batool · M. U. Z. Khan · H. Mukhtar · Zil-e-Huma · R. Ghaffar
Institute of Microbiology, University of Agriculture, Faisalabad, Pakistan

R. U. Khan, S. Naz (eds.), *Green Alternatives to Antibiotics in Animal Health*,
https://doi.org/10.1007/978-981-92-0309-3_17

1 Introduction

Pathogens resistant to the most commonly prescribed antibiotics are on the rise, leading to a worldwide health crisis requiring new strategies for pathogen control aside from traditional antibiotics. The rapid adaptation of bacteria to antimicrobial treatments leads to the need for alternative strategies for the management of disease. Developments in targeted pathogen control strategies, as well as innovations in precision medicine and nanotechnology, that together provide new and more specific means of combating infections with less susceptibility to resistance formation (Dai et al. 2025). Furthermore, the potential of biologically driven interventions is also reflected in the resurgence of phage therapy for enhancing immune responses with natural approaches (Allen et al. 2014). The implications of the microbiota for immune modulation, dietary, and probiotic interventions also support the movement toward personalized and ecological protocols of pathogen control (Tillotson and Theriault 2013).

Novel antibiotic strategies rely on tuning into naturally occurring antimicrobial substances, probiotics, and genetically engineered bacteriophages as an alternative infection-fighting strategy, all with reduced risk of forming resistance. Phage therapy, the effort to harness viruses that specifically target bacterial cells, was invented in the early twentieth century but fell out of favor in Western Europe and North America after the introduction of antibiotics in the 1940s, although it was still practiced in Eastern Europe. However, the present world problem of antimicrobial resistance (AMR) has rekindled interest toward this therapeutic approach, and many successful cases are being reported worldwide (Kim et al. 2025).

Similarly, targeted pathogen control strategies leverage the precision of nanotechnology and genetic interventions to disrupt bacterial communication and biofilm formation, enhancing therapeutic outcomes. The recent advances in vaccine research, particularly the exploration of experimental anti-idiotypic vaccine options, are critical to combat infectious diseases that remain a major public health issue (Justiz-Vaillant et al. 2025). By integrating multiple therapeutic approaches, combination treatments enhance both efficacy, and the chance of resistance development is decreased. Another emerging area for pathogen control is augmenting the immune response via natural means. Probiotic bacteria may help to mitigate antibiotic--associated gut microbiota dysbiosis, and these probiotics have the potential to be genetically engineered to produce specific peptides (Saleh et al. 2025). Environmental factors influencing pathogen resistance must also be considered, as ecological changes impact bacterial adaptation and survival.

Nanoparticles exhibit several beneficial attributes for drug delivery, such as increased stability, better solubility, and controlled release profiles. Nanoparticles have their own bactericidal features. However, there are multiple hurdles in the practical application of alternative pathogen control approaches. Regulatory, ethical, and operational barriers need to be identified and solved. These should be of particular interest to animal health professionals as veterinary medicine case studies highlight the practicality of such interventions that can be used more broadly

(Pacyga et al. 2025). These innovative strategies are crucial for the design of sustainable long-term pathogen control approaches within human and animal health.

2 Enhancement of Immune Response Through Bacteriophages and Vaccines

Immune development and homeostasis are both significantly impacted by our microbiome, which is the collective population of different microbes (bacteria, viruses, and fungi) that live inside our bodies. The immune system influences those microbial communities' abundance and makeup. There is still much to learn about our endogenous bacteriophages and the human phage genome despite comprehensive studies of other aspects of the microbiome. The phageome of numerous organs and disease conditions is still unknown. Phage-human immune responses and their direct interactions are only now being thoroughly studied. Research on phage therapy, which uses lytic phages to cure infections caused by harmful bacteria, and engineered phages created for biotechnology applications such as epitope identification, antibody development, and phage vaccines (González-Mora et al. 2020).

Gut microbiota and probiotics play a vital role in immune modulation pathways that either increase or decrease immune responses, which is mediated by T cells, NK cells, dendritic cells, and macrophages. The mucus layer (extracellular component) and the intestinal epithelium (cellular compartment) are physical barriers to gut microbes (Fig. 1). Both contain a variety of specialized cell types. The intestinal epithelium can synthesize and secrete antimicrobial peptides (AMPs) upon contact with enteric bacteria and enterocytes (also known as fluid-transporting cells), as it is made up of entero-endocrine cells, Paneth, and mucus-secreting Goblet cells. The

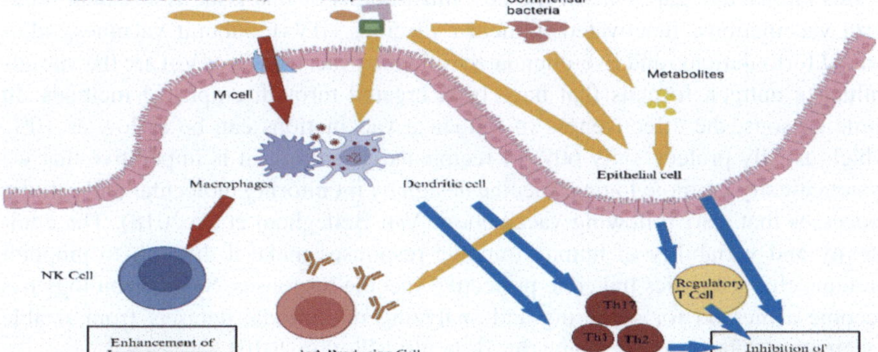

Fig. 1 Schematic representation of relationship between commensal bacteria, food components, probiotics, and the immune system in the gut as well as microorganisms from food

intestinal epithelium in humans regenerates every 3–5 days and eliminates damaged or infected cells (Bischoff et al. 2014).

There are many phages in the human body. Both our bacteria and our own cells are thought to be outnumbered by the estimated 10 phages that reside within each of us. Bacterial colonization sites such as the urogenital tract (UTI), intestine, skin, and upper respiratory tract are home to the majority of these phages. Although our phages exhibit significant genetic diversity and morphological structural heterogeneity, they do share certain characteristics that might be important for immune recognition. They do not have eukaryotic cell membranes and are usually non-enveloped viruses (Tariq et al. 2015). Phages can provide genes to bacteria that are involved in the metabolism of carbohydrates, polysaccharides, and toxins. In rare instances, they can also be a source of antibiotic resistance. *E. coli, Vibrio cholerae, Salmonella, Shigella,* and other microorganisms can produce enzymes that can alter the O-antigen component of LPS, allowing certain phages to modify bacterial antigenicity. Therefore, it's critical to think about how phage interactions with commensal bacteria might change community compositions in ways that affect immune system function and the spread of harmful viruses or even bacteria (Popescu et al. 2021).

Life is a never-ending struggle to live and procreate in a specific ecological niche while competing with other organisms that want to live there and flourish. Such competition frequently occurs between predators and prey rather than between free--dwelling species in search of sufficient resources on their own. If we take out intraspecies conflicts (which would be amazing if that were the case in real life), the true struggle for humans is between the microbial and/or parasitic world as predators and ourselves as prey. Beyond the physical barriers of skin and mucous membranes, the correct operation of our immune system is essential to our success in this fight (Duerkop and Hooper 2013).

Among the best public health interventions, vaccines dramatically lower morbidity and mortality worldwide. However, creating effective vaccines against complex pathogens is still a significant challenge. The remaining dependence on empirical methods and the limited range of correlates of protection (CoPs) are major causes of this knowledge gap. One of the best illustrations of this issue is seasonal influenza vaccinations. Inactivated influenza vaccines (IIVs), subunit vaccines, adjuvanted formulations, and live-attenuated influenza vaccines (LAIVs) are the various influenza antigen formats that have been created through empirical methods. In some seasons, the effectiveness of influenza vaccinations can be as low as 10%, which usually protects only 60% of recipients. Therefore, it is imperative that we systematically promote logical vaccine design by monitoring molecular and cellular processes that start following vaccination (Van Belleghem et al. 2018). The complexity and variability of human immune responses make it difficult to pinpoint immune characteristics linked to protective vaccine responses. Systems biology has become a method for gathering and analyzing multimodal datasets from sizable vaccination cohorts to overcome this difficulty (Plotkin 2010).

Our understanding of the general mechanisms underlying vaccine response heterogeneity has improved thanks to systems immunology, but there are still some problems. Peripheral immune cells have been the primary readout of vaccine response in the majority of previous studies. Although for sampling, blood does not accurately represent the changing immune responses in lymphoid tissues. Furthermore, blood is deficient in many cell types, including those found in germinal centers that support the adaptive immune response. Immune responses within lymphoid tissues can be characterized using immune organoids as an additional method (Forlin et al. 2023). Nonetheless, vaccinations are a vital tool for drastically lowering morbidity and mortality from pathogens. However, our capacity to logically develop vaccines is still restricted (Wagar et al. 2021).

3 Bacteriophages in Animal Health

3.1 Introduction

Bacteriophages are also known as viruses that live as parasites on bacteria. Infections caused by bacteria have been treated with antibiotics long before the nature of antibiotics was understood. Development of treatments as antibiotic substitutes is urgently required to stop clinically significant problems caused by the increase in antibiotic resistance of bacteria. There has been a resurgence of interest in bacteriophages (Bianchessi et al. 2024).

Bacteriophages are categorized as either lysogenic (or temperate) or lytic (or virulent) phages based on how they replicate. After infecting the bacterial cell, lytic phages take over the cellular machinery, break down the DNA of the host cell, and create their own DNA and proteins. This allows the phages to multiply quickly, causing cell lysis and death, which releases the virus to infect other host cells. On the other hand, lysogenic phages insert their genome into the bacterial chromosome after contamination and remain hidden as a prophage, reproducing alongside the cellular chromosome and being passed on to the next offspring for an extended period of time. The lysogenic phase can be converted into the lytic cycle, activated by various stimuli (Hatfull and Hendrix 2011).

Phage binds to a bacterial cell, inserts its DNA, and enters the lytic cycle, which results in the production of new virions and cell lysis, or the lysogenic cycle, where the phage DNA becomes a prophage in the bacterial chromosome. The bacterium can reproduce normally in the lysogenic cycle until specific circumstances cause it to shift to the lytic cycle (Fig. 2). Temperate phages integrate into the bacterial genome, which can shape microbial diversity and functionality in complex ecosystems, in contrast to virulent phages that cause their bacterial hosts to lyse instantly (Mitchell et al. 2007).

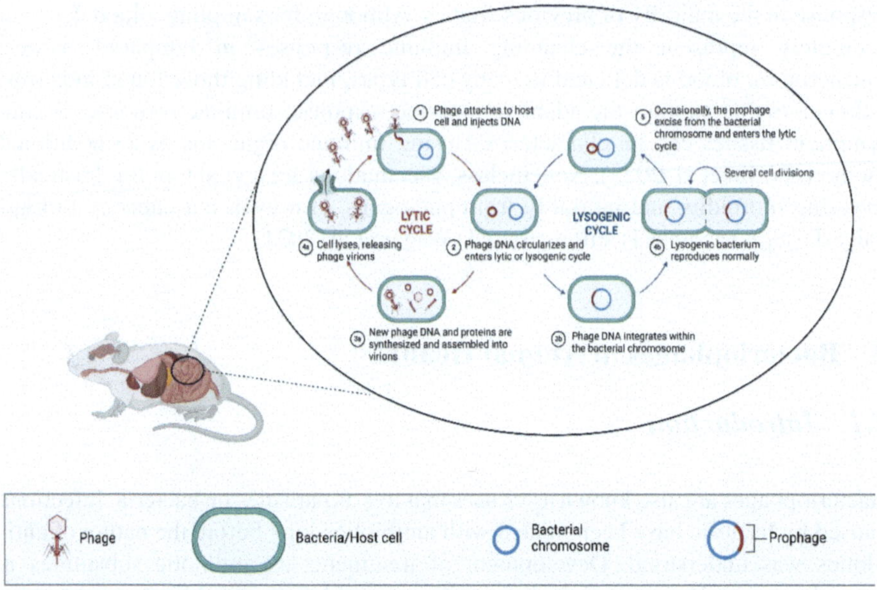

Fig. 2 Schematic representation of bacteriophages through lytic and lysogenic cycles

4 Phage Therapy for Medical and Veterinary Purposes

Bacteriophages have therapeutic applications in veterinary medicine to prevent and treat various bacterial diseases in humans and animals, particularly for cattle and poultry. Despite growing concern about the antibacterial resistance in certain pet--related pathogens and possibly their holder, fewer studies have been conducted in domesticated animals. Although antibiotics are an effective way to treat bacterial infections in farm animals, their use as growth boosters or as treatments for pathogenic diseases in agriculture greatly adds to the problem of antibacterial resistance in global health (Mosimann et al. 2021).

5 Livestock Ruminants

Phage therapy may be a useful treatment for pathogenic infections in ruminants, especially sheep and cattle. Numerous bacterial diseases that affect ruminants can have a detrimental effect on the health of the animals as well as cause farmers to suffer significant financial losses. Furthermore, ruminants might serve as zoonotic pathogen reservoirs. The use of bacteriophages as etiologic agents for specific bacterial infections in ruminants was brought about for *Staphylococcus aureus* and some other bacteria. For both financial (lower milk production quality and quantity) and health (deficient cow health and possible death) objectives, bovine mastitis is the biggest danger to dairy farmers. Commonly isolated bacteria from contagious clinical and subclinical mastitis are *S. aureus* (Bianchessi et al. 2024).

Because of the vast number of microbes that live in the rumen, there isn't much research on adding bacteriophages to ruminant feed. In addition to helping ruminants break down carbohydrates for absorption, microorganisms like bacteria, fungi, and archaea also serve as a "natural medium" for the reproduction and evolution of bacteriophages. Feeding bacteriophages as a supplement has little impact on the microbial structure of the rumen (Jiang et al. 2024).

6 Pigs

As food animals, pigs serve as a harbor for zoonotic pathogens like *Streptococcus suis, S. aureus,* and *Salmonella* spp. Human and pig salmonellosis, which is linked to morbidity and mortality, is caused by the foodborne pathogenic bacterium *Salmonella.* According to the One-Health perspective, weaned pigs were given a cocktail of bacteriophages that were found to be anti-*Salmonella* Serovar *Enterica typhimurium* in the feces of commercial finishing pigs in an attempt to reduce the amount of *Salmonella* that entered the food chain. The results demonstrated that the target bacteria were more successfully infected and killed by a bacteriophage cocktail, which included more than two phages, than by administering a single phage. In fact, cocktails solve the issue of limited host range and reduce the rate at which resistance develops and specific pathogenic strains in the gut. But still more information is required about preparations and dosages of phage administration (Bianchessi et al. 2024).

7 Poultry

Salmonella species, *Campylobacter jejuni, E. coli,* and other zoonotic pathogens that are known to cause illness and financial hardship, as well as poultry-derived food containing antibiotic-resistant bacteria and antibiotic residues, have all been combatted with bacteriophage treatment (Federici et al. 2022).

In addition to preventing surgical site infections, antibiotics are important for treating various bacterial infections in animals, humans, and plants. They are also employed in agriculture to improve fertility, control food-borne bacteria, and preserve food. Bacteria are under tremendous selective pressure worldwide, which has led to the alarming and quick spread of antibiotic-resistant bacterial strains. The use of bacteriophages is one of the most promising methods for controlling bacterial diseases in the current era of antibiotic resistance. Although phage therapy has been largely disregarded in the West, it has recently gained attention again as a substitute for traditional antimicrobials (Mushegian 2020).

Although poultry frequently harbor *Campylobacter*, which typically has no harmful results on them, they serve as reservoirs for human gut pathogens, which

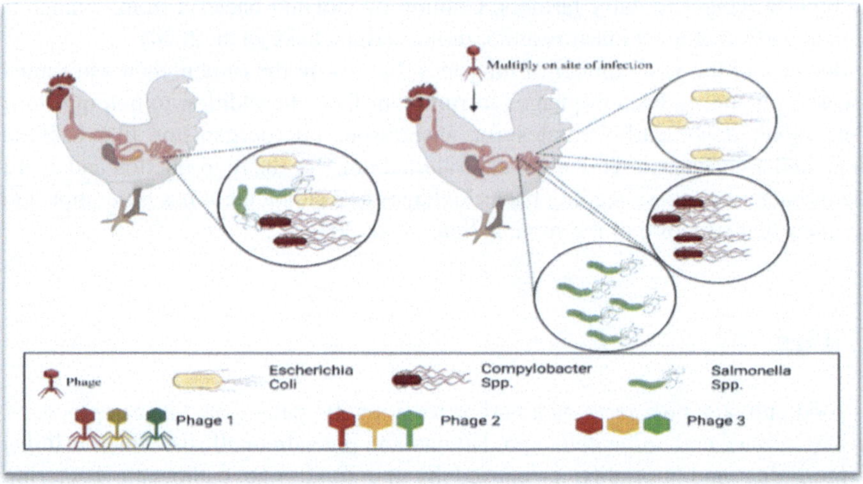

Fig. 3 Schematic representation of the use of particular phages (Phage 1, Phage 2, and Phage 3) to fight off infections in the chicken's digestive tract brought on by *Salmonella* spp., *Campylobacter* spp., and *Escherichia coli*

have recently become more problematic due to its resistance to various antibiotics. Reducing the amount of *Campylobacter, Salmonella,* and *E. coli* in feathered species is necessary to control its entry into the food chain because the majority of its spread takes place at slaughterhouses. According to a recent meta-analysis on the effectiveness of phage treatment in poultry, it can reduce great microbial load (Fig. 3). The efficacy varies depending on the route of administration, but it is generally higher in older chickens in shorter periods of time as compared to younger ones. It can be given as a prophylactic measure, regardless of the quantity of doses (Mosimann et al. 2021).

8 Bovine Species

A very common pathological condition that causes major financial issues for cattle farms is *mastitis,* and financial losses incurred by farms are due to the need to discard the milk produced by cows having mastitis, rather than just the expense of the necessary pharmaceutical treatments. *Acute mastitis* frequently results in the animal's death or, as with recurrent mastitis infections, lower production in terms of both quantity and quality. Because of this, farmers frequently choose to either kill the distressed animals or wait for them to become dry, which results in a loss of milk yield. *Staphylococci, Streptococci, and Enterobacter* are the most frequent pathogens responsible for the development of mastitis. Pathogens that cause mastitis can spread between infected animals or come from the environment. The primary reservoir for the infectious microorganisms *Streptococcus agalactiae* and *S. aureus* is the mammalian gland of the infected cow (Loponte et al. 2021).

Phage K, a bacteriophage, was used to treat *S. aureus*-induced bovine mastitis as early as 1980. The growing drug-resistant *S. aureus*, which was isolated from hospitalized patient samples and other *Staphylococcus* species isolated from bovine infections, was shown to be inhibited by phage K in an in vitro trial in a 2005 study. Inhibition of several *Staphylococcus* species, such as *S. epidermidis, S. aureus, S. saprophyticus, S. chromogenes, S. hyicus, S. hominis, S. hemolyticus, and S. caprae,* is possible with phage K. The results showed that phage preparation reduced *S. aureus*; additionally, it might be applied in teat washes as a defense against mastitis (O'Flaherty et al. 2005).

9 Domesticated Animals

Dogs and cats play different roles in our society. The health and well-being of companion animals are receiving more attention these days. Unfortunately, there may be drawbacks to coexisting with animals. Pets can also serve as reservoirs for the spread of antibiotic resistance in addition to zoonoses. Pets may harbor strains of bacteria that are resistant to common antibiotic treatments, such as *Enterococci* that exhibit vancomycin resistance (VRE) or (MRSA) methicillin-resistant *Staphylococcus pseudointermedius* strains (Loponte et al. 2021).

10 Future Directions for Bacteriophage Therapy in Animal Health

The *immune system* reaction to them poses a serious obstacle and may lessen the effectiveness of treatment. Phage activity within the body is being prolonged due to activation of the immune system. To avoid immune detection, such as immune-suppressive agents or encapsulation techniques are used. Recent research findings showed that bacteriophages cause disruptions in the innate immune response in both birds and mammals. This suggests that, in contrast to other viruses, phages can be used in therapeutic settings without causing severe anaphylactic reactions (Podlacha et al. 2024). Many conventional antibiotics are ineffective against *intracellular bacteria,* which live inside host cells. Phages that can target and lyse intracellular pathogens have been discovered due to recent developments in phage research. Researchers hope to improve treatment capacity to enter and function in intracellular environments by altering phage delivery mechanisms. This calls for creative strategies like designing phages having increased replication and specificity so they can attach to receptors easily that allow them to enter a host cell (Goicoechea Serrano et al. 2024).

Phages are being investigated as platforms for *vaccine development* in addition to infection control. Because target antigens are efficiently identified and displayed by phage display technology, it has the potential to expedite the development of vaccines against emerging pathogens. This strategy has demonstrated potential for producing strong immune responses against a wide range of pathogens. Phage-based vaccines are especially beneficial and useful tools in the fight against infectious diseases (Bianchessi et al. 2024).

11 Vaccination as an Alternative to Antibiotics

The overuse and misuse of antibiotics has become the cause of the emergence of new antibiotic-resistant pathogens. At this time, we see vaccination as a promising alternative to antibiotics so we can prevent and control infectious diseases.

12 How the Vaccine Works

A small and harmless piece of a pathogen (bacteria or virus) is introduced into the body to trigger the immune system so it can produce antibodies and immune cells. When the same pathogen attacks the body, the body is already prepared for it (Senne et al. 2004).

13 Benefits of Vaccination Over Antibiotics

Prevention

In the first place, vaccination prevents infections from occurring and reduces the need for antibiotics.

Targeted Approach

In contrast to antibiotics, vaccines only target specific pathogens and reduce the risk of antibiotic resistance.

Long-Term Protection

Vaccination provides us with long-term protection that lasts for years or even for a lifetime.

Reduced Antibiotic Use

A vaccine provides us with immunity and prevents infection, which leads toward the less or no need for antibiotics and slows down the process of antibiotic resistance.

Herd Immunity

Vaccination provides us with herd immunity and protects vulnerable individuals who are unable to get vaccinated.

Before infection	*After infection*
When a weakened or killed pathogen is inserted into the body through injection, immune cells start recognizing it through dendritic cells, cytotoxic T lymphocytes, and plasma cells. On the other hand, memory cells like T cells and B cells generate long-term memory against that specific pathogen (Fig. 4).	Antibiotics are always used after we get infected. The action mechanism of antibiotics, as compared to a vaccine, is slow. At first, we feel symptoms, then we consult the respective doctor. After several tests, he prescribes us a specific medication. During this process, the infection already spreads throughout the Body (Fig. 4).
If vaccinated.	*If resistant*
If the person is already vaccinated against that specific pathogen, then already present antibodies attach to the virus and inactivate or kill it before the infection spreads (Fig. 4).	After all that procedure of consultation, testing, and medication, in the end, we find out that the respective pathogen is resistant, then the drug will not bind to that targeted site and its efficacy will be reduced or some Time it will be reduced to zero (Fig. 4).
If not vaccinated	*If not resistant*
If the respective person is not vaccinated, then there are no antibodies present to stop the virus, so the infection spreads (Fig. 4).	If the respective pathogen is not a resistant strain, then the drug will bind to the targeted site, and proper binding helps the drug to treat the infection (Fig. 4).

14 Classification of Vaccines

Vaccines can be classified on the basis of various factors such as their composition, mode of preparation, and mechanism of action. Some types of vaccines are given below.

15 Live-Attenuated Vaccines

It has a specific pathogen that is weakened or attenuated but not completely dead. There are several mechanisms to weaken a pathogen; one of them is providing extreme conditions for that pathogen. The benefit of these types of vaccines is that they provide us with a strong immune response; the pathogen is not completely

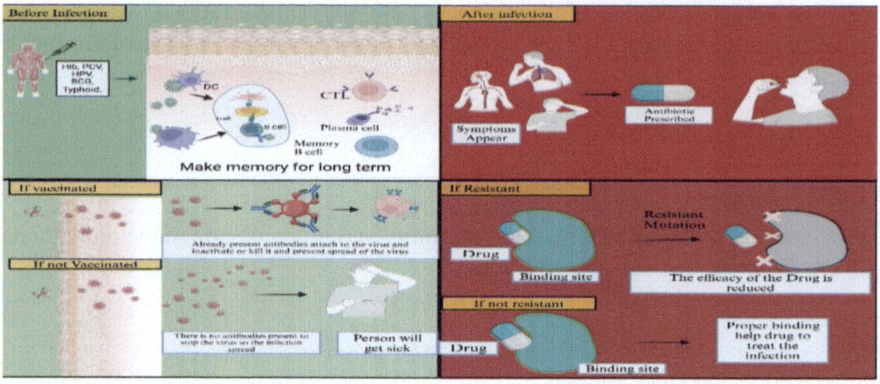

Fig. 4 Schematic representation where the first box represents vaccination before getting infection, the second box the antibody intake after infection happens, the third box the difference between vaccinated and not vaccinated, and the last box the difference between drug resistance and non-resistance

dead, so it goes to its specific targeted site but is unable to multiply there. The body takes it as a real pathogen and shows a response. There are some examples of live attenuated vaccines which are in use, such as the chicken pox vaccine and influenza vaccine (Wareing and Tannock 2001).

16 Inactivated Vaccines

In this type of vaccine, the respective pathogen is killed or inactivated. These kinds of vaccines are mostly used for highly pathogenic strains, or sometimes used for children and the elderly, because in children, we can't use live attenuated vaccines due to their weak immune system. The pathogen is already dead, so infection will not occur. There are some examples of inactivated vaccines which are in use currently: IPV, Hib vaccine, and Hepatitis A vaccines (Sanders et al. 2014).

17 Subunit Vaccines

In this type of vaccine, there is no whole pathogen required, neither live nor dead. In these vaccines, only a component of the respective pathogen, such as a specific protein or subunit, is used. For example, in the case of the ty21a vaccine, there are only O and H factors present in the vaccine that help to produce antibodies but are unable to cause disease itself. There are some examples of subunit vaccines that are in use currently: Hepatitis B vaccine, Hib vaccine, and HPV vaccine (Vartak and Sucheck 2016).

18 Conjugate Vaccines

In some cases, a weakened pathogen is unable to travel to the targeted site, so there is a type of vaccine in which we add a carrier protein with that weakened pathogen so it can travel to that specific site, and the body considers it as a real pathogen, and the immune system becomes activated. Conjugate vaccines provide us with increased immune response when the wild type of that specific pathogen enters our body. There are some examples of conjugate vaccines which are in use currently: the Hib vaccine, pneumococcal conjugate vaccine, and meningococcal conjugate vaccines (Rappuoli et al. 2019).

19 mRNA Vaccines

An advanced type of vaccine is mRNA vaccines. In these vaccines, the only thing that is present is mRNA. mRNA is actually genetic material called messenger ribonucleic acid. It is used to produce specific proteins with the help of ribosomes by instructing the cell. These types of vaccines are mostly used in cases of highly contagious pathogens. The only required part of the pathogen is used, which can stimulate an immune response. The best example of mRNA vaccines is the COVID-19 vaccines, the Pfizer-BioNTech vaccine and the Spikevax vaccine (Barbier et al. 2022).

20 Viral Vector Vaccines

The genetic material of the required pathogen is delivered to the targeted cell with the help of another virus. The virus that is only used for delivery purposes is mostly a nonpathogenic virus. We insert the genetic material of a highly pathogenic virus into a non-pathogenic virus and then insert it into the body. This process helps to get immunity without causing disease because if we use the exact pathogen without making any changes, the person will die because the pathogenicity of these microorganisms is higher than expected. There are some examples of viral vector vaccines, such as the Zabdeno/Myabea Ebola vaccine and the COVID-19 vaccines (Travieso et al. 2022).

21 Toxoid Vaccines

Some pathogens produce toxins; they are not pathogenic themselves, but their toxins cause illness in the host body. So instead of using the whole pathogen or its subunit, we use the toxin of the required pathogen if it produces any. These kind of vaccines helps to elicit the immune response in the host body. The major example of a toxoid vaccine is the diphtheria toxoid vaccine and the tetanus toxoid vaccine (Hu and Zhang 2014).

22 Combination Vaccines

A combination vaccine has a combination of different vaccines in a single injection. The type of vaccines that have the same storage temperature, the same mode of preparation, and the same mechanism of action are usually combined in a single vial or in a single injection. This helps to control and reduce the number of injections and contribute to the control of plastic pollution. The example of a combination vaccine is the MMRV and DTaP vaccine (Decker and Edwards 2000).

23 Different Vaccines for Enhancing Immune Response

23.1 Haemophilus Influenzae Type b (Hib) Vaccine

It provides us with significant advantages over antibiotics. The vaccine prevents *Haemophilus influenzae* infection before the actual pathogen is encountered. But in the case of antibiotic treatment, the infection has already occurred and has shown its potential complications, which lead to meningitis and pneumonia. Some vaccines, like Tetramune, Wyeth Lederle Vaccines, and Pediatrics, were introduced in South Africa in March 1998. Their mode of action is to be analyzed to understand the trend of reported invasive *Haemophilus influenzae* disease in South Africa (Von Gottberg et al. 2006). It is made up from the extract of *Haemophilus influenzae type b* (Hib) bacteria's capsule, i.e., polysaccharides. These polysaccharides are connected to each other through harmless proteins called conjugation. This vaccine, with a conjugation protein, is more effective in young children to stimulate their immune system. As we know, children have a weak immune system, so we can't introduce the exact pathogen because this will cause more loss than benefit (Nix et al. 2012).

23.2 Pneumococcal *Conjugate Vaccine (PCV)*

More use of PCV leads to a decrease in the use of antibiotics and antibiotic-resistant pneumococcal conjugate strains. No more need for antibiotics in treating the Pneumococcal infections, such as pneumonia and meningitis (Chen et al. 2019). The working procedure of this vaccine is that it combines a small portion of inactive pneumococcal bacteria with proteins. The body's immune system recognizes that part of the bacteria and starts making antibodies against that pathogen. Immune response is stronger toward conjugated vaccines as compared to the immune response against polysaccharide vaccines (Shiri et al. 2017).

23.3 Human Papillomavirus (HPV) *Vaccine*

This vaccine prevents infections of HPV, reduces the need for antibiotics, which leads to a significant decrease in the resistant strain of papillomavirus, and prevents cervical cancer and other diseases. The HPV vaccine produces high levels of antibodies, and it is highly effective (Pomfret et al. 2011). The vaccine is made up of virus-like particles (VLPs), not the exact virus, so when a person gets vaccinated, their body's immune system takes it as a real virus and makes antibodies against it. After that, when the real virus enters the body, it will already have antibodies that help to control the infection before the infection spreads further. That is the main reason why we should prefer vaccination over antibiotics. Although vaccinations have some side effects, like syncope (fainting), these side effects are mild and go on their own (Braaten and Laufer 2008).

23.4 Tuberculosis *(BCG) Vaccine*

The Bacillus Calmette-Guérin (BCG) vaccine completely replaces the antibiotics, but it also reduces the antibiotic use and controls the resistant strains. It proves so effective in the case of miliary tuberculosis, which is very severe, particularly in children. This vaccine significantly decreases the overall disease burden, although research is ongoing to develop an improved tuberculosis vaccine (Andersen and Doherty 2005). The procedure of giving the BCG vaccine is quite different. The vaccine is injected into the upper part of the left arm, and after injection, a white lump will appear immediately, and within 2–4 weeks, a red spot or small swelling will appear. Those people who are not exposed to TB yet but are at high risk of exposure should get vaccinated first. The exact mechanism of how the BCG vaccine works on the cellular level starts between 4 and 8 weeks after vaccination, when B cells start showing a response and increase the production of IgG (67), which helps to induce long-lived memory cells so the body will remember the pathogen (Fatima et al. 2020).

24 Typhoid Vaccine (Ty21a, vi Polysaccharide, and Conjugate Vaccine)

Typhoid is a severe bacterial infection, spread through contaminated food and water in regions of poor sanitation. The vaccine helps prevent the *Salmonella typhi* infection and decreases the need for antibiotics (Syed et al. 2020). Three types of vaccines are available Vi polysaccharide vaccine is approved for children 2 and above, and only a single injection is given. Vi polysaccharide vaccine contains a piece of bacterial cell wall, which triggers the immune response but is unable to produce disease. Ty21a is a live attenuated vaccine and is given as 3–4 capsules on alternative days. The Ty21a vaccine has surface antigens of that specific pathogen in it, which help the immune system to generate specific antibodies against the O and H factors. The typhoid conjugate vaccine has vi polysaccharides attached to a carrier protein (Kantele et al. 2013).

25 Rotavirus Vaccine

The rotavirus vaccine prevents rotavirus infection by reducing the risk of severe diarrhea and dehydration. Antibiotics are ineffective against viral infections like rotavirus. The rotavirus vaccine is a live, attenuated vaccine and is administered orally. It multiplies in the intestinal lining, stimulating an immune response. It activates some of the immune cells, such as dendritic cells and T cells, which recognize and respond to the virus. IgA is secreted into the gut to neutralize the virus (Ciarlet and Schödel 2009). The vaccine also induced cellular immunity and provides long-term protection against rotavirus infection. If the infection occurs, the vaccine reduces the severity of the disease by minimizing the risk of complications. It provides protection for at least 2 years after vaccination. Some studies show it provides protection for a longer period of time for up to 5 years or more (Glass et al. 2006).

26 Rabies Vaccine

The rabies vaccine prevents rabies infection and reduces the risk of fatal disease. Antibiotics are ineffective against viral infections like rabies. This is a killed virus vaccine administered via intramuscular injection. Various types of rabies vaccine are currently in use, such as the Human Diploid Cell Vaccine (HDCV), purified Vero cell vaccine, and purified chick embryo cell vaccine (Hicks et al. 2012). The inactivated virus is first detected by antigen-presenting cells (dendritic cells and B cells), which take up the antigen and process it to present it to T cells. T cells, such as CD4+ and CD8+, recognize and respond to antigens. IgG (provides long-term immunity) and IgM (provides initial immunity) are produced as a result of all these procedures. CD8+ cells provide cell-mediated immunity by killing infected cells by releasing cytokines. This results in a rapid and effective immune response upon future exposure to the virus and long-term protection against rabies disease (Mähl et al. 2014).

27 Cholera Vaccine

The cholera vaccine is an oral, inactivated whole-cell vaccine. The inactivated bacteria in the vaccine stimulate an immune response in the gut, where *vibrio cholerae* typically infects. The immune system produces antibodies, specifically immunoglobulin A (IgA), which are secreted into the gut to neutralize the bacteria. The vaccine induces mucosal immunity, which provides protection against *Vibrio cholerae* at the site of infection (the gut). IgA antibodies play a crucial role in neutralizing *Vibrio* cholerae. T cell (CD4+) helps to activate B cells and induce antibody production. This process provides long-term protection against cholerae (Sit et al. 2022). The cholera vaccine provides protection for at least 3–5 years after vaccination, while some studies suggest that the vaccine may provide protection for a longer period, potentially up to 10 years or more. There are some types of oral cholera vaccines: Dukoral (an oral, inactivated whole cell vaccine), Shanchol, and Euvichol. The target population of the cholera vaccine is travelers and residents of areas where cholera is common, high-risk groups such as healthcare workers and emergency responders (Levine 2010).

28 Conclusion

The use of probiotics and bacteriophages has shown promising results in reducing antibiotic resistance. Probiotics have been found to improve gut health, boost immunity, and prevent infections. Bacteriophage, on the other hand, has been used as antibacterial in animals and humans and have shown potential in reducing antibiotic resistance. These viruses, which specifically target bacteria, can be engineered to target antibiotic-resistant strains, making them a valuable tool in the fight against antimicrobial resistance. Additionally, bacteriophages have been found to stimulate the immune system, increasing its ability to recognize and attack bacterial pathogens. Vaccination is another crucial tool for preventing infectious diseases and reducing antibiotic use. Different types of vaccination have been developed to target various pathogens. A vaccine stimulates the immune response to produce antibodies and immune cells, providing long-term protection against specific diseases. Overall, a multi-faceted approach that incorporates probiotics, bacteriophages, and vaccination can help mitigate the rise of antibiotic resistance and promote public health.

References

Allen HK, Trachsel J, Looft T, Casey TA (2014) Finding alternatives to antibiotics. Ann NY Acad Sci 1323(1):91–100

Andersen P, Doherty TM (2005) The success and failure of BCG—implications for a novel tuberculosis vaccine. Nat Rev Microbiol 3(8):656–662

Barbier AJ, Jiang AY, Zhang P, Wooster R, Anderson DG (2022) The clinical progress of mRNA vaccines and immunotherapies. Nat Biotechnol 40(6):840–854

Bianchessi L, De Bernardi G, Vigorelli M, Dall'Ara P, Turin L (2024) Bacteriophage therapy in companion and farm animals. Antibiotics 13(4):1–22

Bischoff SC, Barbara G, Buurman W, Ockhuizen T, Schulzke JD, Serino M, Tilg H, Watson A, Wells JM (2014) Intestinal permeability—a new target for disease prevention and therapy. BMC Gastroenterol 14:189

Braaten KP, Laufer MR (2008) Human papillomavirus (HPV), HPV-related disease, and the HPV vaccine. Rev Obstet Gynecol 1(1):1–2

Chen C, Liceras FC, Flasche S, Sidharta S, Yoong J, Sundaram N, Jit M (2019) Effect and cost-effectiveness of pneumococcal conjugate vaccination: a global modelling analysis. Lancet Glob Health 7(1):e58–e67

Ciarlet M, Schödel F (2009) Development of a rotavirus vaccine: clinical safety, immunogenicity, and efficacy of the pentavalent rotavirus vaccine, RotaTeq®. Vaccine 27:72–81

Dai C, Liu Y, Lv F, Cheng P, Qu S (2025) An alternative approach to combat multidrug-resistant bacteria: new insights into traditional Chinese medicine monomers combined with antibiotics. Adv Biotechnol 3(1):1–6

Decker MD, Edwards KM (2000) Combination vaccines: problems and promise. J Pediatr 137(3):291–295

Duerkop BA, Hooper LV (2013) Resident viruses and their interactions with the immune system. Nat Immunol 14:654–659

Fatima S, Kumari A, Das G, Dwivedi VP (2020) Tuberculosis vaccine: a journey from BCG to present. Life Sci 252:1–10

Federici S, Kredo-Russo S, Valdés-Mas R, Kviatcovsky D, Weinstock E, Matiuhin Y, Silberberg Y, Atarashi K, Furuichi M, Oka A et al (2022) Targeted suppression of human IBD-associated gut microbiota commensals by phage consortia for treatment of intestinal inflammation. Cell 185:2879–2898

Forlin R, James A, Brodin P (2023) Making human immune systems more interpretable through systems immunology. Trends Immunol 44(8):577–584

Glass RI, Parashar UD, Bresee JS, Turcios R, Fischer TK, Widdowson MA et al (2006) Rotavirus vaccines: current prospects and future challenges. Lancet 368(9532):323–332

Goicoechea Serrano E, Blázquez-Bondia C, Jaramillo A (2024) T7 phage-assisted evolution of riboswitches using error-prone replication and dual selection. Sci Rep 14:2377–2377

González-Mora A, Hernández-Pérez J, Iqbal HM, Rito-Palomares M, Benavides J (2020) Bacteriophage-based vaccines: a potent approach for antigen delivery. Vaccine 8(3):1–24s

Hatfull GF, Hendrix RW (2011) Bacteriophages and their genomes. Curr Opin Virol 1(4):298–303

Hicks DJ, Fooks AR, Johnson N (2012) Developments in rabies vaccines. Clin Exp Immunol 169(3):199–204

Hu CMJ, Zhang L (2014) Nanotoxoid vaccines. Nano Today 9(4):401–404

Jiang A, Liu Z, Lv X, Zhou C, Ran T, Tan Z (2024) Prospects and challenges of bacteriophage substitution for antibiotics in livestock and poultry production. Biology 13(1):1–28

Justiz-Vaillant A, Soodeen S, Gopaul D, Arozarena-Fundora R, Thompson R, Unakal C, Akpaka PE (2025) Tackling infectious diseases in the Caribbean and South America: epidemiological insights, antibiotic resistance, associated infectious diseases in immunological disorders, global infection response, and experimental anti-Idiotypic vaccine candidates against microorganisms of public health importance. Microorganisms 13(2):1–37

Kantele A, Pakkanen SH, Karttunen R, Kantele JM (2013) Head-to-head comparison of humoral immune responses to vi capsular polysaccharide and Salmonella Typhi Ty21a typhoid vaccines–a randomized trial. PLoS One 8(4):1–9

Kim MK, Suh GA, Cullen GD, Rodriguez SP, Dharmaraj T, Chang THW et al (2025) Bacteriophage therapy for multidrug-resistant infections: current technologies and therapeutic approaches. J Clin Invest 135(5):1–13

Levine MM (2010) Immunogenicity and efficacy of oral vaccines in developing countries: lessons from a live cholera vaccine. BMC Biol 8:1–10

Loponte R, Pagnini U, Iovane G, Pisanelli G (2021) Phage therapy in veterinary medicine. Antibiotics 10(4):1–21

Mähl P, Cliquet F, Guiot AL, Niin E, Fournials E, Saint-Jean N et al (2014) Twenty year experience of the oral rabies vaccine SAG2 in wildlife: a global review. Vet Res 45:1–17

Mitchell J, Siboo IR, Takamatsu D, Chambers HF, Sullam PM (2007) Mechanism of cell surface expression of the *Streptococcus mitis* platelet binding proteins PblA and PblB. Mol Microbiol 64(3):844–857

Mosimann S, Desiree K, Ebner P (2021) Efficacy of phage therapy in poultry: a systematic review and meta-analysis. Poult Sci 100:1–15

Mushegian AR (2020) Are there 10^{31} virus particles on earth, or more, or fewer? J Bacteriol 202:10–1128

Nix EB, Hawdon N, Gravelle S, Biman B, Brigden M, Malik S et al (2012) Risk of invasive Haemophilus influenzae type b (Hib) disease in adults with secondary immunodeficiency in the post-Hib vaccine era. Clin Vaccine Immunol 19(5):766–771

O'Flaherty S, Ross RP, Flynn J, Meaney WJ, Fitzgerald GF, Coffey A (2005) Isolation and characterization of two anti-staphylococcal bacteriophages specific for pathogenic *Staphylococcus aureus* associated with bovine infections. Lett Appl Microbiol 41(6):482–486

Pacyga K, Pacyga P, Szuba E, Viscardi S, Topola E, Duda-Madej A (2025) Nanotechnology meets phytotherapy: a cutting-edge approach to treat bacterial infections. Int J Mol Sci 26(3):1–26

Plotkin SA (2010) Correlates of protection induced by vaccination. Clin Vaccine Immunol 17(7):1055–1065

Podlacha M et al (2024) Bacteriophage DNA induces an interrupted immune response during phage therapy in a chicken model. Nat Commun 15:2274–2274

Pomfret TC, Gagnon JM Jr, Gilchrist AT (2011) Quadrivalent human papillomavirus (HPV) vaccine: a review of safety, efficacy, and pharmacoeconomics. J Clin Pharm Ther 36(1):1–9

Popescu M, Van Belleghem JD, Khosravi A, Bollyky PL (2021) Bacteriophages and the immune system. Ann Rev Virol 8(1):415–435

Rappuoli R, De Gregorio E, Costantino P (2019) On the mechanisms of conjugate vaccines. Proc Natl Acad Sci 116(1):14–16

Saleh M, Heydari R, Boroujeni MRG, Abiri R (2025) Engineered probiotics that produce antibiotic binding sites: a potential strategy to protect gut microbiome and prevent antibiotic resistance. Med Hypotheses 195:1–6

Sanders B, Koldijk M, Schuitemaker H (2014) Inactivated viral vaccines. In: Vaccine analysis: strategies, principles, and control. Springer Berlin Heidelberg, Berlin/Heidelberg, pp 45–80

Senne D, King D, Kapczynski D (2004) By vaccination. Dev Biol Basel 119:165–170

Shiri T, Datta S, Madan J, Tsertsvadze A, Royle P, Keeling MJ et al (2017) Indirect effects of childhood pneumococcal conjugate vaccination on invasive pneumococcal disease: a systematic review and meta-analysis. Lancet Glob Health 5(1):51–59

Sit B, Fakoya B, Waldor MK (2022) Emerging concepts in cholera vaccine design. Ann Rev Microbiol 76(1):681–702

Syed KA, Saluja T, Cho H, Hsiao A, Shaikh H, Wartel TA et al (2020) Review on the recent advances on typhoid vaccine development and challenges ahead. Clin Infect Dis 71(Supplement_2):S141–S150

Tariq MA, Everest FLC, Cowley LA, De Soyza AD, Holt GS et al (2015) A metagenomic approach to characterize temperate bacteriophage populations from cystic fibrosis and non-cystic fibrosis bronchiectasis patients. Front Microbiol 18:1–12

Tillotson GS, Theriault N (2013) New and alternative approaches to tackling antibiotic resistance. F1000prime Rep 5:1–51

Travieso T, Li J, Mahesh S, Mello JDFRE, Blasi M (2022) The use of viral vectors in vaccine development. NPJ Vaccines 7(1):1–75

Van Belleghem JD, Dąbrowska K, Vaneechoutte M, Barr JJ, Bollyky PL (2018) Interactions between bacteriophage, bacteria, and the mammalian immune system. Viruses 11(1):1–10

Vartak A, Sucheck SJ (2016) Recent advances in subunit vaccine carriers. Vaccine 4(2):1–12

Von Gottberg A, De Gouveia L, Madhi SA, Du Plessis M, Quan V, Soma K et al (2006) Impact of conjugate Haemophilus influenzae type b (Hib) vaccine introduction in South Africa. Bull World Health Organ 84:811–818

Wagar LE, Salahudeen A, Constantz CM, Wendel BS, Lyons MM, Mallajosyula V et al (2021) Modeling human adaptive immune responses with tonsil organoids. Nat Med 27(1):125–135

Wareing MD, Tannock GA (2001) Live attenuated vaccines against influenza; a historical review. Vaccine 19(25–26):3320–3330

Eco-Informed Phage Strategies: Integrative Approaches to Animal Health and Antimicrobial Stewardship

Saima Naz, Ahmad Manan Mustafa Chatha, Sobia Malik, Saba Iqbal, Urwah Ishaque, and Qudrat Ullah

Abstract Integrating bacteriophage therapy into animal health appears to be a promising answer as environmental concerns over antibiotic use increase and antimicrobial resistance (AMR) worsens. By providing tailored bacterial control, phages lessen the need for broad-spectrum antibiotics and delay the emergence of resistant types of bacteria. An eco-informed strategy takes into account the ecological consequences of using phages, such as the dangers of horizontal gene transfer, environmental persistence, and effects on non-target bacteria. This approach uses a multidisciplinary framework that integrates ecology, policy, veterinary science, and microbiology to guarantee the sustainability and efficacy of phage therapy. The creation of phage cocktails to increase host range, the improvement of delivery systems for increased stability and effectiveness, and the establishment of precise regulatory parameters are core techniques. Eco-informed phage therapy is a forward-thinking approach to antimicrobial stewardship and sustainable disease management in animal agriculture, with its roots in the One Health concept, which connects human, animal, and environmental well-being.

Keywords Veterinary medicine · One Health · Eco-informed therapeutics · Sustainable livestock systems · Zoonotic pathogens · Phage therapy

S. Naz (✉) · S. Malik · S. Iqbal · U. Ishaque
Department of Zoology, Government Sadiq College Women University, Bahawalpur, Punjab, Pakistan
e-mail: saima.naz@gscwu.edu.pk

A. M. M. Chatha
Department of Entomology, Faculty of Agriculture and Environment, The Islamia University of Bahawalpur, Bahawalpur, Punjab, Pakistan

Q. Ullah
Department of Theriogenology, Faculty of Veterinary and Animal Sciences, Cholistan University of Veterinary and Animal Sciences, Bahawalpur, Punjab, Pakistan

R. U. Khan, S. Naz (eds.), *Green Alternatives to Antibiotics in Animal Health*,
https://doi.org/10.1007/978-981-92-0309-3_18

1 Introduction

One of the most important issues facing world health in the twenty-first century is antimicrobial resistance (AMR). Food security, environmental stability, animal health, and the efficacy of medical therapies are all at risk. Multidrug-resistant (MDR) bacteria, which can spread resistance genes among animals, humans, and the environment through direct contact or food intake, have become more prevalent as a result of the overuse and abuse of antibiotics in livestock and aquaculture (Abdalhamed et al. 2021; Neogi 2025). Global stakeholders, like the European Union and the United Nations Quadripartite, have responded by adopting a One Health approach that emphasizes the interdependence of environmental, animal, and human health (Larsson et al. 2023).

In this context, bacteriophage therapy has resurfaced as a viable, environmentally friendly substitute for traditional antibiotics. Certain bacterial types can be infected and lysed by viruses called phages. Because of their inherent specificity, they reduce ecological disruption and preserve beneficial bacteria by minimizing off-target impacts (Yang et al. 2025; Chowdhury et al. 2025). Even more therapeutic flexibility is provided by engineered phages. Numerous drawbacks of conventional antibiotics can be addressed by modifying phages using synthetic biology to increase lytic activity, broaden host range, integrate antimicrobial genes, or break up biofilms (Dedrick et al. 2019; Guo et al. 2021).

Enzymes generated from phages, including depolymerases, endolysins, holins, and virion-associated peptidoglycan hydrolases (VAPGHs), have also demonstrated strong antibacterial action against pathogens that form biofilms and are Gram-positive (Sehrawat et al. 2024; Fernandes and São-José 2018). To cure infections that are resistant to drugs and improve the effectiveness of veterinary treatments, recombinant versions of these proteins are being created (Nazir et al. 2023).

Phages are transforming diagnostic technology in addition to medicines. Reporter phages, which are designed to carry genes that indicate the presence of bacteria, allow for the quick and precise identification of living pathogens, frequently surpassing the capabilities of traditional molecular techniques (Meile et al. 2020). Precision veterinary treatment is further supported by integration with IoT-based monitoring and smart farming technology.

Notwithstanding these benefits, there are obstacles to phage therapy's broad use. These consist of phage genetic stability during production and storage, formulation stability, regulatory approval, and manufacturing scalability (Jault et al. 2019; Botka et al. 2019). Adhering to Good Manufacturing Practice (GMP) guidelines is essential for veterinary and clinical applications.

With an emphasis on mechanisms of action, practical applications, environmental effects, regulatory considerations, and alignment with One Health concepts, this chapter examines the potential of phage therapy as a strategy to decrease AMR in animal health (Fig. 1). Phage-based approaches can be combined with probiotics, vaccinations, and surveillance to provide a comprehensive, long-lasting solution to the AMR epidemic.

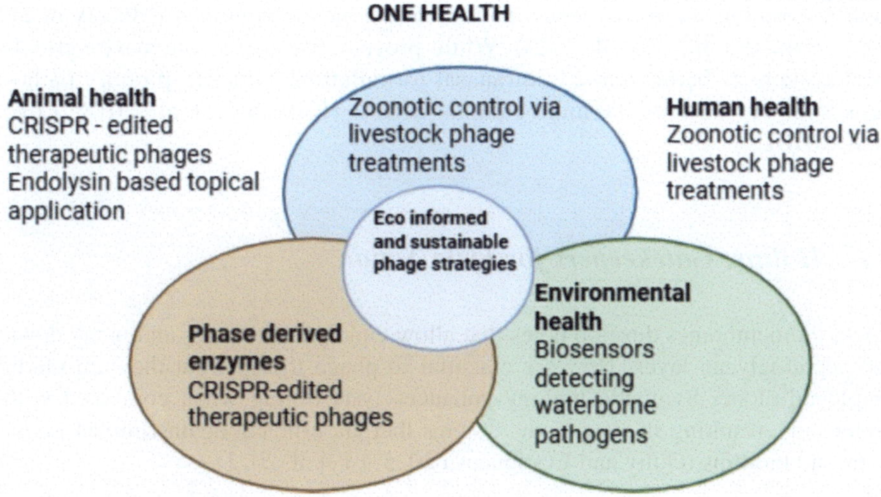

Fig. 1 Eco-informed phage innovations in the One Health triad

2 Rationale of the Study

Antimicrobial resistance (AMR) is a global concern, intensified by antibiotic overuse in veterinary settings, threatening animal health, food safety, and ecosystems. Sustainable alternatives are urgently needed. Bacteriophages offer a targeted, eco-friendly solution, preserving beneficial microbiota and avoiding chemical residues. Advances in bioengineering have enhanced their veterinary applications, yet regulatory gaps, low awareness, and ecological concerns limit their adoption. This study explores integrative, eco-informed phage strategies to reduce antibiotic reliance and support sustainable livestock practices.

3 Phage-Derived Enzymes: Endolysins and Beyond

Gram-positive infections such as *Staphylococcus aureus*, particularly MRSA, increase morbidity, mortality, and healthcare expenses, making antimicrobial resistance (AMR) a major worldwide concern (Golban et al. 2025; Nazir et al. 2023). Enzymes produced from phages, especially endolysins, present a viable substitute. Because the cell walls of Gram-positive bacteria are exposed, these peptidoglycan-degrading enzymes, which are generated by dsDNA bacteriophages, have considerable lytic activity against them (Guliy and Evstigneeva 2025; Oliveira et al. 2019). They are perfect candidates for innovative antimicrobial treatments due to their specificity, ease of engineering, and absence of resistance mechanisms.

Endolysins are used to treat canine dermatitis, horse wounds, and bovine mastitis because they have demonstrated effectiveness against MRSA, Listeria, biofilm-forming *Staphylococcus*, and *Pseudomonas aeruginosa* (Murray et al. 2021; Šumonja and Kotnik 2024). While preventative use in intensive agricultural systems is being studied, intranasal formulations target respiratory pathogens such as Pasteurella multocida in BRDC (Kurćubić et al. 2018; Love et al. 2018).

3.1 Holins: Gatekeepers for Lytic Action

Bacterial membranes develop holes that allow endolysins to enter and break down the peptidoglycan layer. They are essential in phage therapy, but they are rarely employed alone. Synthetic biology enhances lysis timing when combined with endolysins, resulting in synergistic systems that are efficient against mixed infections and biofilms (Guliy and Evstigneeva 2025; Li et al. 2021).

3.2 Depolymerases: Biofilm-Busting Agents

In veterinary contexts, biofilms shield microorganisms and obstruct antibiotic therapies. Depolymerases improve antibiotic access by breaking down biofilm polysaccharides (Araújo et al. 2024). By improving penetration and facilitating phage access, they improve efficacy when used in conjunction with endolysins, antibiotics, or phages (Pal et al. 2024; Shemyakin et al. 2020; Danis-Wlodarczyk et al. 2021). In poultry systems, these enzymes have demonstrated efficacy against Salmonella, *S. aureus*, and *P. aeruginosa* (canine otitis, mastitis, respectively) (Robinson et al. 2019). Additionally, they are included in environmentally friendly disinfectants for agricultural hygiene (Araújo et al. 2024).

3.3 Advantages Over Whole Phage Therapy

Since they don't multiply, phage-derived enzymes have a number of advantages over complete phages, including a decreased risk of environmental disruption and horizontal gene transfer (Munshi et al. 2025). Standardized manufacture, increased storage stability, and a variety of delivery methods (topical, injectable, or nebulized) are made possible by their protein nature (Skurnik et al. 2025; Mujawar and Sankpal 2025). Additionally, they are less prone to cause immunological reactions, which makes them appropriate for animals that are sensitive or immunocompromised.

4 Synthetic Biology and Phage Engineering

To improve phage safety, host range, and antibacterial efficacy, precise genetic alterations are made possible by synthetic biology (Lobocka et al. 2021). Although early methods like as homologous recombination (HR), enabled gene insertion and deletion, they had host restrictions and low effectiveness (~0.1%) (Sun et al. 2023; Kashyap et al. 2025). Recombinant identification is aided by the insertion of selection markers such as trxA (Lv et al. 2023). By enabling the development of "designer phages" unique to particular veterinary diseases, these technologies expand the range of therapeutic applications (Brown et al. 2017).

4.1 Expanding Host Range and Overcoming Resistance

The range of natural phages is limited since they frequently exclusively target particular bacterial strains. By allowing phages to infect new strains or carry antimicrobial payloads, genetic engineering—particularly CRISPR-Cas9—can increase host range and aid in the fight against resistance (Baltrus and Kvitko 2025; Chaudhary et al. 2025). Enzymes, antivirulence factors, or immune modulators that interfere with resistance mechanisms and enhance treatment results can be carried by modified phages (Wu et al. 2025).

4.2 Targeting Biofilms and Anti-virulence Strategies

By interfering with the synthesis of toxins and quorum sensing, engineered phages can inhibit bacterial pathogenicity or break down biofilms using enzymes such as depolymerases (Li et al. 2025; Zhou et al. 2025). For chronic infections such as bovine mastitis, horse wounds, and canine otitis externa, these methods provide focused, resistance-sparing therapies (Touati et al. 2025).

4.3 Phage Engineering for Environmental Stability

Phages are susceptible to stress from their surroundings. Phage capsids can be engineered or encapsulated in nanoparticles to increase their resilience to UV light, pH changes, and temperature changes (Chemla et al. 2025; Lin et al. 2025). According to Hernández-Arteaga et al. (2025), these improvements facilitate prolonged release and boost performance in a variety of veterinary contexts, including topical and spray-based treatments.

4.4 *Phage Delivery Systems: From Passive to Active*

Targeted delivery is still difficult, particularly for diseases that are deep or biofilm-based. While altered phages can carry payloads that boost the immune system or include antimicrobials, phage display methods allow ligand-guided targeting to specific bacteria (Jault et al. 2019). By enhancing accessibility for usage in pigs, poultry, and other livestock, formulations for oral, injectable, or aerosol delivery complement the host immune system (Botka et al. 2019).

5 Phage Therapy Within the One Health Framework

The connection between environmental, animal, and human health is emphasized by the One Health approach. Phage therapy is a perfect fit for the integrated, cross-sector solutions needed to address global problems like antimicrobial resistance (AMR) (Larsson et al. 2023; Neogi 2025).

Animal and public health are at risk due to the growing resistance of zoonotic infections, including Salmonella, Campylobacter, and E. coli, to treatment. Using phages in livestock, particularly before harvest, can significantly lower pathogen burdens, lowering the risk of transmission and improving food safety (Sari 2025; Webb et al. 2025).

Phages do not leave any toxic residues and are biodegradable, unlike antibiotics. By lowering chemical contamination, their focused action promotes ecologically conscious farming. In order to help with early detection and control of resistance strains, engineered phages are also being created for environmental surveillance (Amillano-Cisneros et al. 2025; Abraham et al. 2025).

Phages provide a scalable, affordable solution to cut down on the use of antibiotics worldwide. They are especially useful in rural or low-resource environments because of their simplicity of manufacture and administration, which supports sustainable animal health and the reduction of AMR (Cartledge et al. 2025; Obidi and Ekpunobi 2025).

6 Environmental Persistence and Phage Ecology

Compared to antibiotics, phages have a different ecological footprint because they are biologically generated antimicrobials. Concerns regarding ecological effect, microbial interactions, and environmental persistence are brought up by their growing use in veterinary medicine and agriculture (Liu et al. 2025).

Environmental factors affect phage survival; stability is supported by cool temperatures and neutral pH, but degradation can occur from heat, UV light, and high pH. Although organic materials may provide protection, microbial competition might hasten deterioration. Formulations tailored to various high-organic conditions are necessary for effective field use (Zhao et al. 2025; Wang et al. 2025; Olopade et al. 2025).

Phage populations in soil and water are regulated by ecosystems, facilitating their use in aquaculture, irrigation, and waste management. They are beneficial for biocontrol and exhibit long-term persistence in the right circumstances (Sosnowska et al. 2025; Liu et al. 2025). Lysogenic phages, on the other hand, might promote antibiotic resistance by horizontal gene transfer (HGT). Strategies should prioritize lytic phages, use cocktails sparingly, and adhere to sound agricultural practices in order to reduce hazards (Liang et al. 2025; Zhong et al. 2025). Phages are naturally self-limiting, but if they are abused, they may have an effect on microbial biodiversity. To guarantee safe and responsible deployment, containment mechanisms and regulatory control are crucial, particularly for genetically engineered phages (Gillett et al. 2025).

7 Phage-Based Biosurveillance in Smart Farming

Early identification of infectious organisms is essential since antimicrobial resistance (AMR) still poses a hazard to public health and livestock (Natrah et al. 2025). Phage-based biosurveillance provides a quick and inexpensive substitute for traditional diagnostic techniques, which are frequently slow and expensive. This is especially useful in farm environments with limited resources.

By releasing detectable signals upon infection, genetically modified reporter phages—carrying genes like luciferase or GFP—allow for the real-time detection of certain infections. In a variety of settings, including water systems and animal enclosures, these instruments can detect outbreaks in less than an hour, enabling prompt response and ongoing monitoring (Hao et al. 2025; Guo et al. 2025).

Phage-coated surfaces and diagnostic strips are examples of other developments. These inexpensive instruments provide on-site, user-friendly monitoring by binding to pathogens and indicating their presence through a color shift, similar to lateral flow tests (Jahedi et al. 2025). For automated disease detection and response, these biosurveillance systems are becoming more and more linked with smart agricultural technologies, such as sensor networks, Internet of Things platforms, and data analytics. Through educated, real-time decision-making, this integration lowers the use of antibiotics, enhances animal health and wellbeing, and permits precision disease control (Viet et al. 2025; Nastasijevic et al. 2025; Niyigaba et al. 2025).

8 Phage Applications in Veterinary Medicine

Sustainable agriculture, public health, and animal welfare all heavily rely on veterinary treatment. Veterinarians assist control zoonotic infections and guarantee the safety of products generated from animals by providing preventive care, diagnostics, and treatment. The field integrates environmental, animal, and human health by adopting a One Health approach. This concept is well suited for phage therapy, which provides tailored, environmentally friendly antibiotic substitutes while advancing the more general objective of antimicrobial stewardship (Table 1).

9 Limitations of Phage-Based Therapies in Veterinary Medicine

While promising, phage therapies face several barriers to adoption. Regulatory ambiguity, environmental sensitivity (e.g., to heat, pH, and organic matter), and high production costs limit their scalability. Additional challenges include inconsistent quality, limited veterinary training, and low awareness among farmers. Conventional antibiotics remain favored due to familiarity and ease of use. Progress requires policy reform, targeted education, and investment in research on ecological safety and delivery systems. Despite these hurdles, phages remain a valuable eco-friendly tool in the fight against AMR.

10 Conclusion

By utilizing phages' specificity to lower the risks of AMR, eco-informed phage techniques provide a sustainable substitute for traditional antibiotics. Overcoming obstacles in formulation, delivery, and regulation is essential to successful implementation. Minimal impact on non-target microorganisms and habitats is ensured by incorporating ecological awareness. To mainstream phage use in animal health and support the larger objectives of One Health and antimicrobial stewardship, cross-sector cooperation will be essential.

Table 1 Comparative overview of eco-informed phage applications in veterinary medicine

Sr. no	Technology type	Application domain	Targeted pathogen	Species (Animal)	Method of application	Sustainability advantage	References
1	Encapsulated phages	Health management of swine	*Escherichia coli*	Pig	Microencapsulatio n by oral	Decreased diarrhea in weaned piglets caused by E. coli	Tian et al. (2023)
2	Whole phages	Management for dairy mastitis	*Staphylococcus aureus* (MRSA)	Cattle	Infusion by Intramammary	Effective against mastitis strains that produce biofilms, MRSA, and MDR	Zubareva et al. (2024)
3	Targeted phage therapy	Managements for equine dermatology	*Staphylococcus aureus*	Horses	Application by topical	Lesions from superficial pyoderma improved	Marshall and Marsella (2023)
4	Whole phages	Health management of poultry	*Salmonella enterica*	Poultry	Administration by Oral	Better health and lower infection rates	Wigley (2024)
5	Whole phages	Health management of poultry	*Campylobacter jejuni*	Poultry	Administration by Oral	Broiler colonization has decreased.	Abd-El Wahab et al. (2023)
6	Phage baths	Management b for companion animal health	*Pseudomonas aeruginosa*	Companio n animals	Application by topical	Successfully controlled persistent illnesses	Squires (2021)
7	Immeraion phage therapy	Management for aquaculture health	*Vibrio anguillarum*	Aquacultur e species	Immersion by water	Infections avoided in aquaculture environments	Rai et al. (2024)
8	Immersion phage therapy	Management for aquaculture health	*Aeromonas hydrophila*	Aquacultur e species	Immersion by water	Avoided and treated bacterial infections	Kumari et al. (2023)
9	Targeted phage therapy	Health management for calf	*Escherichia coli*	Calves	Administration by Oral	Decreased incidence of diarrhea	Whon et al. (2021)
10	Phage cocktails	Management for dairy mastitis	*Staphylococcus aureus*	Cattle	Infusion by intramammary	Controlled mastitis in cows	Nale and McEwan (2023)

(continued)

Table 1 (continued)

Sr. no	Technology type	Application domain	Targeted pathogen	Species (Animal)	Method of application	Sustainability advantage	References
11	Whole phages	Management for food safety	*Escherichia coli* O157:H7	Cattle	Administration by oral	Biocontrol of E. coli O157:H7	Holzer et al. (2025)
12	Phage cocktail	Management for food safety	*Escherichia coli* O157:H7	Cattle	Cocktails	Control E. Coli O157:H7 rationally	Śliwka et al. (2025)
13	Phage sprays	Management for dermatology health in companion animals	*Staphylococcus aureus*	Rabbits	Application by topically	Skin infections have decreased.	Plumet et al. (2022)
14	Phage sprays	Management for companion animal otology	*Staphylococcus aureus*	Dogs	Application by topical	Better circumstances for otitis externa	Stefanetti et al. (2024)
15	Whole phages	Health management for domestic animals	*Escherichia coli*	Cattles	Administration by oral	Better gut health and growth performance	Guo et al. (2024)
16	Phage baths	Management for dermatology of companion animals	*Staphylococcus aureus*	Dogs	Application by topical	Controlled persistent skin infections	Vallenas-Sánchez et al. (2022)
17	Phage sprays	Management for dermatology of companion animals	*Staphylococcus pseudintermedi us*	Canine	Application by topical	Cured persistent illnesses	Lynch and Helbig (2021)
18	Phage baths	Management for dermatology of companion animals	*Staphylococcus aureus*	Dogs	Application by topical	Dealt with persistent illnesses	Pyzik et al. (2021)
19	Phage sprays	Management for dermatology of companion animals	*Staphylococcus aeruginosa*	Cats	Application by topical	Chronic infection control	Braunstein et al. (2024)

20	Targeted phage therapy	Health management of poultry	*Salmonella enterditis*	Poultry	Application by topical	Controlled persistent infections	Huang et al. (2022)
21	Targeted phage therapy	Health management of poultry	*Clostridium perfringens*	Poultry	Administration by oral	Decreased incidence of necrotic enteritis	Keerqin et al. (2022)
22	Encapsulated phages	Health management of poultry	*Listeria monocytogenes*	Chickens	As a feed additive	Decreased shedding of pathogens	Grigore-Gurgu et al. (2024)
23	Whole phages	Health management of ruminants	*Mycobacterium avium*	Ruminents	Administration by Oral	Possibility of managing Johne's condition	O'Connell et al. (2023)
24	Targeted phage therapy	Health management for Swine	*Salmonella Typhimurium*	Pigs	Administration by Oral	Decreased shedding and gut colonization	Won et al. (2021)
25	Phage therapy (aerosolized)	Disease management for bovine Respiration	*Listeria innoca*	Cattle	Application by aerosol	Decreased incidence of respiratory infections	Tumuluri (2024)
26	Phage therapy (aerosolized)	Health management for aquaculture	*Vibrio anguillarium*	Shrimp	Immersion in water	Management of Vibriosis outbreaks in shrimp farming	Rai et al. (2024)
27	Phage therapy (immersion)	Health management for aquaculture	*Vibrio harveyi*	Fish	Immersion in water	Decreased incidence of vibriosis in fish farm	Lan (2023)
28	Targeted phage therapy	Health management for aquaculture	*Edwardsiella tarda*	Fish	Immersion in water	Controlled outbreaks of edwardsiellosis	Jun (2024)
29	Whole phages	Health management for cattle reproduction	*Brucella abortus*	Cattle	Administration by oral	Possible decrease in the spread of brucellosis	Pascual et al. (2022)
30	Whole phages	Health management of swine	*Streptococcus suis*	Swine	Administration by Oral	Pigs with controlled streptococcal infections	Lv et al. (2025)

(continued)

References

Abdalhamed AM, Ghazy AA, Zeedan GSG (2021) Studies on multidrug-resistance bacteria in ruminants with special interest on antimicrobial resistances genes. Adv Anim Vet Sci 9(6):835–844

Abd-El Wahab A, Basiouni S, El-Seedi HR, Ahmed MF, Bielke LR, Hargis B et al (2023) An overview of the use of bacteriophages in the poultry industry: successes, challenges, and possibilities for overcoming breakdowns. Front Microbiol 14:1136638

Abraham TJ, Bardhan A, Hoque F, Sarker S (2025) Biosecurity and disease management in aquaculture systems. In: Management of fish diseases. Springer Nature Singapore, Singapore, pp 409–436

Amillano-Cisneros JM, Fuentes-Valencia MA, Leyva-Morales JB, Savín-Amador M, Márquez-Pacheco H, de Jesús Bastidas-Bastidas P et al (2025) Effects of microorganisms in fish aquaculture from a sustainable approach: a review. Microorganisms 13(3):485

Araújo D, Silva AR, Fernandes R, Serra P, Barros MM, Campos AM et al (2024) Emerging approaches for mitigating biofilm-formation-associated infections in farm, wild, and companion animals. Pathogens 13(4):320. https://doi.org/10.3390/pathogens13040320

Baltrus DA, Kvitko BH (2025) Exploring the open (host) range. Cell Host Microbe 33(1):10–12

Botka T, Pantůček R, Mašlaňová I et al (2019) Lytic and genomic properties of spontaneous host-range Kayvirus mutants prove their suitability for upgrading phage therapeutics against staphylococci. Sci Rep 9. https://doi.org/10.1038/s41598-019-41868-w

Braunstein R, Hubanic G, Yerushalmy O, Oren-Alkalay S, Rimon A, Coppenhagen-Glazer S et al (2024) Successful phage-antibiotic therapy of P. aeruginosa implant-associated infection in a Siamese cat. Vet Q 44(1):1–9

Brown R, Lengeling A, Wang B (2017) Phage engineering: how advances in molecular biology and synthetic biology are being utilized to enhance the therapeutic potential of bacteriophages. Q Biol 5:42–54

Cartledge K, Short FL, Hall A, Lambert K, McDonald MJ, Lithgow T (2025) Ethical bioprospecting and microbial assessments for sustainable solutions to the AMR crisis. IUBMB Life 77(1):e2931

Chaudhary N, Sharma K, Kaur H, Prajapati S, Mohan B, Taneja N (2025) CRISPR-Cas-assisted phage engineering for personalized antibacterial treatments. Indian J Med Microbiol 53:100771

Chemla Y, Sweeney CJ, Wozniak CA, Voigt CA (2025) Design and regulation of engineered bacteria for environmental release. Nat Microbiol 10(2):281–300

Chowdhury MAH, Reem CSA, Ashrafudoulla M, Rahman MA, Shaila S, Jie-won Ha A, Ha SD (2025) Role of advanced cleaning and sanitation techniques in biofilm prevention on dairy equipment. Compr Rev Food Sci Food Saf 24(3):e70176

Danis-Wlodarczyk KM, Wozniak DJ, Abedon ST (2021) Treating bacterial infections with bacteriophage-based enzybiotics: in vitro, in vivo and clinical application. Antibiotics 10(12):1497

Dedrick RM, Guerrero-Bustamante CA, Garlena RA, Russell DA, Ford K, Harris K, Gilmour KC, Soothill J, Jacobs-Sera D, Schooley RT (2019) Engineered bacteriophages for treatment of a patient with a disseminated drug-resistant Mycobacterium abscessus. Nat Med 25:730–733

Fernandes S, São-José C (2018) Enzymes and mechanisms employed by tailed bacteriophages to breach the bacterial cell barriers. Viruses 10(8):396

Gillett DL, Selinidis M, Seamons T, George D, Igwe AN, Del Valle I et al (2025) A roadmap to understanding and anticipating microbial gene transfer in soil communities. Microbiol Mol Biol Rev 89:e00225–24

Golban M, Charostad J, Kazemian H, Heidari H (2025) Phage-derived Endolysins against resistant staphylococcus spp.: a review of features, antibacterial activities, and recent applications. Infect Dis Ther 14(1):13–57

Grigore-Gurgu L, Bucur FI, Mihalache OA, Nicolau AI (2024) Comprehensive review on the biocontrol of listeria monocytogenes in food products. Foods 13(5):734

Guliy OI, Evstigneeva SS (2025) Bacteria-and phage-derived proteins in phage infection. Front Biosci Landmark 30(2):24478

Guo D, Chen J, Zhao X, Luo Y, Jin M, Fan F, Park C, Yang X, Sun C, Yan J (2021) Genetic and chemical engineering of phages for controlling multidrug-resistant bacteria. Antibiotics 10:202

Guo X, Luo G, Hou F, Zhou C, Liu X, Lei Z et al (2024) A review of bacteriophage and their application in domestic animals in a post-antibiotic era. Sci Total Environ 949:174931

Guo M, Wang Y, Sun J, Qian C, Lowrie DB, Niu L et al (2025) Rapid detection of drug resistance in mycobacterium tuberculosis clinical isolates for first-line antitubercular drugs by using a novel reporter mycobacteriophage. Front Cell Infect Microbiol 15:1589236

Hao Z, Li M, Zhao Q, Wang L, Liu T, Wang C, Wang C (2025) Construction of Reporter Phage T4:: Nluc and Its Application in the Detection of Escherichia coli in Urinary Tract Infections. iLABMED 3:158

Hernández-Arteaga AM, Mendoza-Corvis F, Salgado-Behaine J, Gontijo M, Batalha LS, Perez G et al (2025) Application of conservation strategies for bacteriophages used in the biocontrol of pathogenic microorganisms in food. J Food Saf 45(1):e70014

Holzer K, Marongiu L, Detert K, Venturelli S, Schmidt H, Hoelzle LE (2025) Phage applications for biocontrol of enterohemorrhagic E. coli O157: H7 and other Shiga toxin-producing Escherichia coli. Int J Food Microbiol 439:111267

Huang Y, Wang W, Zhang Z, Gu Y, Huang A, Wang J, Hao H (2022) Phage products for fighting antimicrobial resistance. Microorganisms 10(7):1324

Jahedi S, Tohidkia MR, Esmaeili M, Bani F, Kaboli S (2025) Phage-displayed antibody fragments in microfluidic paper-based devices: a novel approach for sensitive detection of glycine-extended gastrin 17 biomarker using gold nanoparticles. Microfluid Nanofluid 29(3):16

Jault P, Leclerc T, Jennes S et al (2019) Efficacy and tolerability of a cocktail of bacteriophages to treat burn wounds infected by Pseudomonas aeruginosa (PhagoBurn): a randomised, controlled, double-blind phase 1/2 trial. Lancet Infect Dis 19:35–45

Jun JW (2024) A concise overview of studies on successful real-world applications of bacteriophages in aquaculture. Viruses 16(12):1843

Kashyap R, Tiwari S, Bhattacharya S (2025) Innovative strategies in the fight against bacterial infections: phage therapy, nanotechnology, and new antimicrobial agents for multidrug-resistant pathogens. Results Chem:102304

Keerqin C, McGlashan K, Van TTH, Chinivasagam HN, Moore RJ, Choct M, Wu SB (2022) A lytic bacteriophage isolate reduced Clostridium perfringens induced lesions in necrotic enteritis challenged broilers. Front Vet Sci 9:1058115

Kumari R, Yadav R, Kumar D, Chaube R, Nath G (2023) Evaluation of bacteriophage therapy of Aeromonas hydrophila infection in a freshwater fish, Pangasius buchanani. Front Aquacult 2:1201466

Kurćubić V, Ilić Z, Đoković R, Petrović MP, Caro-Petrović V (2018) Vaccination and medication against bovine respiratory disease complex (BRDC). Biotechnol Anim Husb 34(3):279–287

Lan NGT (2023) Novel vaccination strategy against vibrio harveyi infection in asian seabass (lates calcarifer) using a chitosan–oxygen nanobubble-vaccine complex

Larsson DGJ, Gaze WH, Laxminarayan R, Topp E (2023) AMR, one health and the environment. Nat Microbiol 8(5):754–755

Li X, Zhang C, Wei F, Yu F, Zhao Z (2021) Bactericidal activity of a holin-endolysin system derived from vibrio alginolyticus phage HH109. Microb Pathog 159:105135

Li J, Jia T, Yang L (2025) Targeting anti-virulence factor strategies of bacterial pathogens. Biosafety and Health 7:1

Liang P, Wang H, Chen Q, Luo Q, Cheng J, Tian G et al (2025) Antibiotic resistance gene transfer characteristic and risk evaluation: the entire process from swine manure to biogas slurry and then to soil. J Environ Chem Eng 13(3):116659

Lin Z, Ruan C, Xia R, Liao J, Zhu L, Wang D et al (2025) Bacterium–Phage Interactions Enhance Biofilm Resilience during Membrane Filtration Biofouling under Oxidative and Hydraulic Stresses. Environ Sci Technol 59(17):8614–8628

Liu C, Chen Z, Wang X, Deng Y, Tao L, Zhou X, Deng J (2025) Response of soil phage communities and prokaryote–phage interactions to long-term drought. Environ Sci Technol 59:3054
Lobocka M, Dabrowska K, Gorski A (2021) Engineered bacteriophage therapeutics: rationale, challenges and future. BioDrugs 35(3):255–280
Love MJ, Bhandari D, Dobson RC, Billington C (2018) Potential for bacteriophage endolysins to supplement or replace antibiotics in food production and clinical care. Antibiotics 7(1):17
Lv S, Wang Y, Jiang K, Guo X, Zhang J, Zhou F et al (2023) Genetic engineering and biosynthesis technology: keys to unlocking the chains of phage therapy. Viruses 15(8):1736
Lv R, Zhang W, Sun Z, Si X, Dong H, Liu X (2025) Current prevalence and therapeutic strategies for porcine Streptococcus suis in China. Appl Environ Microbiol 91(3):e02160–e02124
Lynch SA, Helbig KJ (2021) The complex diseases of staphylococcus pseudintermedius in canines: where to next? Vet Sci 8(1):11
Marshall K, Marsella R (2023) Topical bacteriophage therapy for staphylococcal superficial pyoderma in horses: a double-blind, placebo-controlled pilot study. Pathogens 12(6):828
Meile S, Kilcher S, Loessner MJ, Dunne M (2020) Reporter phage-based detection of bacterial pathogens: design guidelines and recent developments. Viruses 12(9):944
Mujawar N, Sankpal PS (2025) Pulse, ultrasound, phage therapy. In: Biofilm associated livestock diseases and their management. Springer, p 171
Munshi M, Jahangir S, Sumaiya S (2025) Phage therapy: navigating the mechanisms, benefits, and challenges in the fight against multidrug-resistant infections. medRxiv, 2025-04
Murray E, Draper LA, Ross RP, Hill C (2021) The advantages and challenges of using endolysins in a clinical setting. Viruses 13(4):680. https://doi.org/10.3390/v13040680
Nale JY, McEwan NR (2023) Bacteriophage therapy to control bovine mastitis: a review. Antibiotics 12(8):1307. https://doi.org/10.3390/antibiotics12081307
Nastasijevic I, Kundacina I, Jaric S, Pavlovic Z, Radovic M, Radonic V (2025) Recent advances in biosensor technologies for meat production chain. Foods 14(5):744
Natrah I, Muthukrishnan S, Mohd-Padil H, Mohamad N, Alipiah NM, Shariff M et al (2025) Antimicrobial resistance in Malaysian shrimp aquaculture and strategies to reduce its occurrence. Rev Aquac 17(3):e70020
Nazir A, Xu X, Liu Y, Chen Y (2023) Phage endolysins: advances in the world of food safety. Cells 12(17):2169
Neogi SB (2025) Addressing the climate-food-health nexus: strengthening ecosystem-based adaptation and one health. Asian-Australas J Food Saf Secur 9(1):1–4
Niyigaba T, Küçükgöz K, Kołożyn-Krajewska D, Królikowski T, Trząskowska M (2025) Advances in fermentation technology: a focus on health and safety. Appl Sci 15(6):3001
O'Connell LM, Coffey A, O'Mahony JM (2023) Alternatives to antibiotics in veterinary medicine: considerations for the management of Johne's disease. Anim Health Res Rev 24(1):12–27
Obidi NO, Ekpunobi NF (2025) A narrative review exploring phage therapy as a sustainable alternative solution to combat antimicrobial resistance in Africa: Applications, challenges and future directions: Une revue narrative explorant la phagothérapie comme solution alternative durable pour lutter contre la résistance aux antimicrobiens en Afrique: applications, défis et orientations futures. Afr J Clin Exp Microbiol 26(2):106–113
Oliveira H, Melo LD, Santos SB (2019) Bacteriophage proteins as antimicrobials to combat antibiotic resistance. In: Antibiotic drug resistance. Wiley, Hoboken, pp 343–406
Olopade BK, Rahman MF, Khyum MMO, Ramkumar S (2025) Persistence and control of viruses on fomites and environmental matrices. Future Virol:1–15
Pal N, Sharma P, Kumawat M, Singh S, Verma V, Tiwari RR et al (2024) Phage therapy: an alternative treatment modality for MDR bacterial infections. Infect Dis 56(10):785–817
Pascual DW, Goodwin ZI, Bhagyaraj E, Hoffman C, Yang X (2022) Activation of mucosal immunity as a novel therapeutic strategy for combating brucellosis. Front Microbiol 13:1018165
Plumet L, Ahmad-Mansour N, Dunyach-Remy C, Kissa K, Sotto A, Lavigne JP et al (2022) Bacteriophage therapy for Staphylococcus aureus infections: a review of animal models, treatments, and clinical trials. Front Cell Infect Microbiol 12:907314

Pyzik E, Radzki RP, Urban-Chmiel R (2021) Experimental phage therapies in companion animals with A historical review. Curr Rev Clin Exp Pharm 16(1):17–29

Rai S, Kaur B, Singh P, Singh A, Benjakul S, Vijay Kumar Reddy S et al (2024) Perspectives on phage therapy for health management in aquaculture. Aquac Int 32(2):1349–1393

Robinson VH, Paterson S, Bennett C, Steen SI (2019) Biofilm production of pseudomonas spp. isolates from canine otitis in three different enrichment broths. Vet Dermatol 30(3):218–e67

Sari DAS (2025) Antimicrobial resistance in companion animals: a growing one health concern. Vet Life Sci Innov 2(1):1–19

Sehrawat S, Nishu, Sangwan Babita Renu, Lamba Mayank, Gangwar Gopal, Nath (2024) Emerging Paradigms for Antibiotic-Resistant Infections: Beyond the Pill Advancing Beyond Antibiotics: The Pioneering Use of Bacteriophage Enzymes Springer Nature Singapore, Singapore 743–768

Shemyakin IG, Firstova VV, Fursova NK, Abaev IV, Filippovich SY, Ignatov SG, Dyatlov IA (2020) Next-generation antibiotics, bacteriophage endolysins, and nanomaterials for combating pathogens. Biochem Mosc 85:1374–1388

Skurnik M, Alkalay-Oren S, Boon M, Clokie M, Sicheritz-Pontén T, Dąbrowska K et al (2025) Phage therapy. Nat Rev Methods Primers 5(1):9

Śliwka P, Moreno DS, Korzeniowski P, Milcarz A, Kuczkowski M, Kolenda R et al (2025) Avian pathogenic Escherichia coli-targeting phages for biofilm biocontrol in the poultry industry. Vet Microbiol 301:110363

Sosnowska M, Łęga T, Olszewski M, Gromadzka B (2025) Phage display technology in ecotoxicology: phage display derived unique peptide for copper identification in aquatic samples. Microb Cell Factories 24(1):11

Squires RA (2021) Bacteriophage therapy for challenging bacterial infections: achievements, limitations and prospects for future clinical use by veterinary dermatologists. Vet Dermatol 32(6):587–e158

Stefanetti V, Passamonti F, Rampacci E (2024) Antimicrobial strategies proposed for the treatment of S. Pseudintermedius and other dermato-pathogenic staphylococcus spp. in companion animals: a narrative review. Vet Sci 11(7):311

Šumonja I, Kotnik T (2024) Skin Dysbiosis in atopic dogs: is phage therapy an alternative to antibiotics? Slovenian Vet Res 61(2)

Sun Q, Shen L, Zhang BL, Yu J, Wei F, Sun Y et al (2023) Advance on engineering of bacteriophages by synthetic biology. Infect Drug Resist 16:1941–1953

Tian Z, Chen J, Lin T, Zhu J, Gan H, Chen F, Zhang S, Guan W (2023) Dietary supplementation with Lysozyme–Cinnamaldehyde conjugates enhances feed conversion efficiency by improving intestinal health and modulating the gut microbiota in weaned piglets infected with enterotoxigenic *Escherichia coli*. Animals 13(22):3497. https://doi.org/10.3390/ani13223497

Touati A, Mairi A, Ibrahim NA, Idres T (2025) Essential oils for biofilm control: mechanisms, synergies, and translational challenges in the era of antimicrobial resistance. Antibiotics 14(5):503

Tumuluri T (2024) Effect of plasma activated mist (PAM) on inactivation of listeria innocua and Klebsiella michiganensis on selected surfaces. Master's thesis, Rutgers The State University of New Jersey, School of Graduate Studies

Vallenas-Sánchez YPA, Bautista-Valles MF, Llaque-Chávarri F, Mendoza-Coello ME (2022) Bacteriophage cocktail as a substitute for antimicrobials in companion animal dermatology. J Selva Andina Anim Sci 9(2):97–117

Viet TD, Keiji S, Xuan TD (2025) The modern technology in developing smart agriculture

Wang J, Liang C, Zhang Y, Liu Z, Huo Y, Zhang Z, Huo M (2025) Impacts of reclaimed water recharge on waterborne viruses decay: characteristics and influencing factors. J Water Process Eng 74:107844

Webb H, Tagg K, Peñil-Celis A, Stapleton G, Ellison Z, Leeper M et al (2025) The Salmonella enterica serotype Hadar pangenome: population structure and dynamics of a zoonotic pathogen

Whon TW, Kim HS, Shin NR, Sung H, Kim MS, Kim JY et al (2021) Calf diarrhea caused by prolonged expansion of autochthonous gut Enterobacteriaceae and their lytic bacteriophages. Msystems 6(2):10–1128

Wigley P (2024) Salmonella and the chicken: reflections on salmonellosis and its control in the United Kingdom. Poult Sci Manage 1(1):1

Won YK, Kim SJ, Han JH (2021) The protective effect of dietary supplementation of salmonella-specific bacteriophages in post-weaning piglets challenged with salmonella typhimurium. J Adv Vet Anim Res 8(3):440

Wu S, Zhang W, Wang D, Balcazar JL, Wang G, Ye M et al (2025) Bacteriophage–bacteria interactions promote ecological multifunctionality in compost-applied soils. Environ Microbiol 27(3):e70074

Yang L, Zhong W, Tang T, He M, Zhang T, Zhou B et al (2025) Phage-based biocontrol strategies and application in aquatic animal disease prevention and control. Rev Aquac 17(3):e70024

Zhao X, Liang X, Zhu Z, Yuan Z, Yu S, Liu Y et al (2025) Phages affect soil dissolved organic matter mineralization by shaping bacterial communities. Environ Sci Technol 59:2070

Zhong W, Zhou Y, Che M, Wang L, Tian X, Wang C et al (2025) Extended-spectrum β-lactamase-producing Escherichia coli isolated from captive primates: characteristics and horizontal gene transfer ability analysis. PLoS One 20(4):e0321514

Zhou S, Liu B, Zheng D, Chen L, Yang J (2025) VFDB 2025: an integrated resource for exploring anti-virulence compounds. Nucleic Acids Res 53(D1):D871–D877

Zubareva VD, Sokolova OV, Bytov MV, Krivonogova AS, Volskaya SV (2024) Alternative treatment methods for bovine mastitis: prospects and limitations. Vet Sci Today:203

Organic Acids: Production, Properties, and Applications in Food Safety and Gut Health

Rabia Maham, Ayesha Sarwar, Zarqa Shaukat, Afaq Mahmood, Sara Mahmood, Gul Naz, Faria Javed, Riffat Shamim Aslam, Umar Zafar Khan, Mubarah Saleem, and Fatima Sarwar

Abstract Naturally occurring substances with acidic properties containing carbon are referred to as organic acids. These substances are known for being weakly acidic and are often categorized as alkanoic acids or aliphatic acids. These aliphatic acids possess at least one functional group consisting of acids, including carboxy, sulfhydryl, and sulfone. Different kinds of organic acids include citrate, ethanoic, sour salt, butanoic, etc. The production of most organic acids and their derivatives occurs through three methods, which are microbial fermentation, enzymatic processes, and chemical synthesis. Different microbes produce significant amounts of these compounds from chemical feedstock waste. Microorganisms generate these acids via their metabolic sequences, which include glycolysis and tricarboxylic acid metabolism. Lactic acid is essential in synthesizing polylactic acid, an aliphatic polyester that is both biodegradable and biocompatible. Ethanoic acid is an important product of gut microbial metabolism and helps maintain a healthy intestinal microbiome. Various other acids, like sour salt, butanoic, and propanoic acids, are described below. These acids contribute to food safety and gut health. With advancements in food processing methods and rising public concerns regarding natural additives, the market for bio-based organic acids has expanded and emerged as a significant category of chemicals.

Keywords Organic acid · Antibacterial · Gut health

R. Maham · Z. Shaukat · A. Mahmood · F. Javed · R. S. Aslam · U. Z. Khan · M. Saleem · F. Sarwar (✉)
Institute of Microbiology, University of Agriculture Faisalabad, Faisalabad, Pakistan

A. Sarwar · S. Mahmood · G. Naz
Institute of Microbiology, Government College University, Faisalabad, Pakistan

R. U. Khan, S. Naz (eds.), *Green Alternatives to Antibiotics in Animal Health*,
https://doi.org/10.1007/978-981-92-0309-3_19

1 Introduction

Naturally occurring substances with acidic properties containing carbon are referred to as organic acids. These substances are known for being weakly acidic and are often categorized as alkanoic acids or aliphatic acids. They dissolve in an aqueous solution and liquid state in the environment (Peng and Biswas 2017). These acids play an important component in metabolic pathways. This involves the Krebs cycle, which exists in almost all living organisms. This cycle occurs in the matrix of mitochondria. Their wide range of characteristics makes them crucial base products across numerous industrial sectors. These include food items, chemical reagents, beauty products, medication, and beverages (Bangar et al. 2022).

Microbial metabolic procedures within the intestines generate volatile acids through which immune defense and protection occur. These substances display properties that suppress tumors, modulate immune responses, and combat enteric bacterial pathogens. They are prevalent in living organisms and crucial against numerous biochemical and geochemical processes. They are utilized in medical healthcare, biomass industries, and food preservation (Liu et al. 2017). In the atmosphere, they form an oxidized organic aerosol, cloud seed, and acid deposition.

These aliphatic acids possess at least one functional group consisting of acids, including carboxy sulfhydryl, and sulfone (Yin et al. 2015). Several carboxyl groups exist in the chemical arrangement of acids. The group consists of methanoic, ethanoic, propanoic, butanoic acids, and carboxy and hydroxy groups. These acids show excellent antibacterial properties. For example, lactate is more effective in combating bacterial infections. Whereas sorbic acid can inhibit mold growth. Some acids, including propanoic, methanoic, and HMB, have broader antimicrobial properties. These acids fight against yeasts, bacteria, and fungi. The combination of these acids has been evaluated as a feed additive because they execute multiple effects (Dibner and Buttin 2002).

Research has demonstrated that when specific acids combine, they exhibit antibacterial effects against various pathogens. 2,4-Hexadienoic and trans-butenedioic acids are known for their antifungal properties. They exhibit bactericidal effects within a pH range of 3–5 and are employed in preserving food and animal feed (Long et al. 2018). This investigation has also shown that commensal *Escherichia coli*, along with *LAB* and *bifidobacteria,* generated beneficial organic acids, such as lactate and acetate. It suggests that *Escherichia coli* improves defense mechanisms against *Pseudomonas aeruginosa* pathogens as well as other infection-causing microorganisms. Bacteriocins produce antibacterial amino acid chains through different gastrointestinal microorganisms (Ayed et al. 2024).

Recently, there has been an increasing focus on utilizing organic acids that enhance gut health and fight against pathogens, especially as the global acceptance of antibiotics as growth enhancers in animal agriculture has declined (Reuben and Torres 2024). The bactericidal properties stem from nonionized elements of the compound, which pierce microbial membranes to disrupt cellular activities. Compounds such as propionate, acetate, lactate, malate, and citrate have demonstrated considerable antimicrobial properties against common foodborne pathogens. These pathogens include *E. coli*, *S. Typhimurium*, and *Listeria species* (Teng et al. 2023).

Various methods exist for producing the bulk of organic acids along with their derived compounds. The production of most organic acids and their derivatives occurs through three methods, which are microbial fermentation, enzymatic processes, and chemical synthesis. Volatile fatty acids are significant compounds created through zymolysis. Lignocellulosic waste and lactoserum have been recognized as viable industrial sugars and feedstocks for producing these compounds. For example, acetic acid can be produced through bacterial fermentation, with certain microorganisms able to incorporate CO_2 in the process (Vidra and Németh 2018).

However, many of these processes are not economically viable because of the higher expenses associated with bioprocesses. Additionally, there is often a lack of understanding of biotechnological approaches and downstream processing methods such as solvent removal, surface binding, electrochemical separation, esterification, and reactive distillation. This further hinders their practical implementation (Yen et al. 2002).

2 Synthesis of Organic Acid

Organic acids are produced through biochemical and industrial pathways. Microorganisms generate these acids via their metabolic sequences, which include glycolysis and tricarboxylic acid metabolism. This method yields lactate, citrate, malate, and itaconate. Another category of acids, gluconate, is synthesized through direct glucose oxidation. There has been an increasing focus on producing organic acids by utilizing metabolically engineered genes from different microbial sources. The global organic acid market has expanded to $4 billion owing to increased demand for diverse chemicals (Lee et al. 2011).

China produces high-quality compounds that are impressive over a specific timeframe. Although organic acids can be synthesized chemically, industries favor microbial production through fermentation techniques. Citric acid, mainly generated through submerged fermentation, boasts the highest production volume, with an annual output of around 1.6 million tons achieved via microbial methods, indicating improved productivity and higher yield (Vandenberghe et al. 2018).

2.1 *Different Types of Organic Acids*

Vishnu et al. (2020) describes in his paper about different organic acids and their pKa values.

Acids	pKa value
Acetic acid	3.7
Formic acid	3.8
Butyric acid	3.4
Propanoic acid	4.5
Fumaric acid	3.6
Tartaric acid	2.9
Benzoic acid	3.78

3 Sources of Carbon-Based Compounds

Various sources of organic compounds play an important role in multiple contexts. They may be derived from biological processes such as plant emissions and the combustion of organic matter. It is produced using biochemical and biotechnological methods. Bio-processing is the primary technique used for the industrial manufacturing of different organic acids. Different microbes produce significant amounts of these compounds from chemical feedstock waste. The mass production of raw materials and microbial sources is essential for improving productivity and yield. Solid-state fermentation, which utilizes renewable feedstocks, is often favored owing to its economic benefits, such as lower cost and easily available resources (Vishnu et al. 2020).

Fermented food, meats, vegetables, and crops are the sources of volatile compounds. For example, fruits and vegetables often contain citric acid, whereas meat often contains creatine. Fermented foods have high acetic acid content, and seasonings contain many organic acids. Wild edible mushrooms contain citrate, ketoglutaric, malate, succinic, and fumaric acids. In soil, organic acids play a crucial role in increasing phosphorus availability. Both aliphatic (e.g., oxalic and citric) and aromatic (e.g., benzoic, 4-hydroxybenzoic, and protocatechuic) organic acids are found in the soil. These acids can naturally occur or be added through organic matter sources such as compost and biochar (Shi et al. 2022).

4 Types of Organic Acids

4.1 Lactic Acids

In 1780, the Swedish scientist Scheele isolated lactate from yogurt. It is also called alpha-hydroxy acid. While chemical methods typically produce an inactive blend of acids, biotechnological approaches can create pure Dextro and levo-lactic acids when specific microorganisms are utilized. Various microbes are responsible for the production of lactate. The fermentation process for producing lactic acid provides multiple advantages, such as replenishable resources and eco-friendly practices (Dusselier et al. 2013).

Lactic acid was initially identified for commercial applications in the food industry, where it serves as a flavor enhancer, pH adjuster, microbial quality enhancer, and mineral supplement. Subsequently, it became vital in the cosmetics industry, functioning as a pH stabilizer, antibacterial agent, moisturizer (owing to its water-retaining capability), and skin brightener (by preventing monophenol hydroxylase formation). In pharmaceuticals, sodium lactate is commonly used as an electrolyte in intravenous (IV) solutions. Calcium lactate is used as an anticancer agent to treat calcium deficiency. In the chemical sector, lactate is used to produce valuable compounds such as propenoic and 2-oxopropanoic acid, bio-based solvents, and esters (Rodrigues et al. 2017a).

Lactic acid is essential in synthesizing polylactic acid, an aliphatic polyester that is both biodegradable and biocompatible. The production of PLA generally involves a melt process that does not use solvents, where the materials are heated until they become liquid. In this process, lactide, which is formed from lactic acid, undergoes ring-opening polymerization. It has been extensively utilized in the medical field through the isomerization of lactate. These include prosthetic devices, suturing materials, orthopedic equipment, tissue-engineering scaffolds, and pharmaceutical administration. In addition, polylactic acid has applications beyond the medical field, such as in packaging materials, structural foams, and textile products (Önder et al. 2016).

4.1.1 Synthesis of Lactic Acids

Certain bacteria and fungi can also produce lactate. Fungal families, such as *Rhizopus*, which belong to the *Mucoromycetes* class, are employed in the aerobic fermentation of starchy materials to convert them into lactic acids. Other bacteria, including *E. coli*, *C. glutamicum*, and various *Bacillus species*, may naturally create comparable amounts of lactic acid (Kim et al. 2022).

Carbohydrates, proteins, and other aromatic compounds are produced by the order *Lactobacillales*. They have aldolase enzymes and convert glucose into lactic acid. These bacteria create organic chemicals that enhance the flavor and aroma of fermented foods. These bacteria have been employed for millennia as preservatives, flavorings, and texturing agents in culinary products. They are now incorporated as starter cultures in several fermentation processes (Narbad and Wang 2018).

Based on metabolic pathways, these bacteria are classified into three physiological categories. The first category is an obligative homofermentor, which mainly generates lactic acid through oxidation. These bacteria have the enzyme 1,6-Diphosphofructose aldolase. The second category, obligative heterofermentor, with the help of Phosphogluconate, generates six-carbon compounds. This metabolic process yields lactate, CO_2, acetate, and ethanol. These bacteria possess phosphoketolase enzymes. The third category, facultative hetero-lactic LAB, metabolizes hexoses using glycolysis. Additionally, they can produce inducible phosphoketolase and metabolize pentoses (Rodrigues et al. 2017b).

Lactobacilli are the most prevalent probiotics in humans and other animals. The primary lactobacilli species prevalent in the intestinal tract include *Lactobacillus gasseri*, *Lactobacillus crispatus, Limnosilactobacillus reuteri, Ligilactobacillus salivarius,* and *Ligilactobacillus ruminis*. These bacteria maintain gut health and provide antimicrobial defense through various mechanisms. These bacteria produce metabolites, including Lantibiotics and other antimicrobial compounds, which play a role in their protective effects against different microbes (Tang et al. 2023).

They can kill target cells by forming pores and dissipating their cytosolic content. PH levels influence the degree of ionization; at a low pH, a significant portion of lactate remains in its undissociated form, which can be harmful to various microorganisms. The most significant obstacles in lactic acid fermentation are the intricate nature of their nutrients and mesophilic properties. Lactic acid bacteria use lignocellulosic materials to produce lactic acid and cellulases. Conversely, lantibiotics are peptides involved in host defense that aim to eliminate pathogens by creating pores in their membranes, leading to the dissipation of cytosolic content. Despite their antimicrobial potential, the widespread use of *Lactobacillus* and LAB species in lactic acid fermentation faces challenges owing to their mesophilic nature and complex nutritional requirements (Ren et al. 2020).

However, recent advancements have shown promise for overcoming these limitations. Some LAB species have demonstrated the ability to utilize lignocellulosic materials for lactic acid production coupled with the synthesis of cellulases. Furthermore, genetic engineering techniques offer promising avenues for enhancing the capabilities of these organisms. The relative ease of genetic manipulation of LAB species provides opportunities to address their inherent limitations and expand their applications in various industrial and biotechnological processes (Ren and Li 2022).

4.2 Ethanoic Acid

It is also called vinegar, and it is the second most common organic compound. It exhibits greater solubility in water than others, which is due to the presence of multiple alkyl groups in its molecular composition (Coban 2020). It is an important compound utilized in multiple sectors such as industry, healthcare, and home use. It can be generated from various sources, including the pyrolysis of lignocellulosic biomass. It is crucial for antimicrobial defense and intestinal health. This acid shows the highest bactericidal activity against *Listeria, Salmonella, and Listeria species*. It is an important product of gut microbial metabolism and helps maintain a healthy intestinal microbiome. Its compound serves multiple vital functions, such as antioxidant, antitumor, and anti-inflammatory, that serve an important role in host defense and immune response (Sarchami et al. 2021).

4.2.1 Industrial Production of Ethanoic Acids

Utilization of refined carbohydrates and alcohol is a primary carbon sources that serve as a substrate in the production of ethanoic acids. Glucose undergoes degradation to pyruvate through glycolysis, facilitated by enzymatic activity. In the heterofermentative process, acetyl-phosphate is formed, yielding acetic acid with the help of the enzyme acetate kinase. In addition, acetyl phosphate can produce acetaldehyde in the presence of coenzyme A. These acetaldehyde molecules are subsequently converted into acid via the reductive acetyl-coenzyme A pathway, which is catalyzed by various dehydrogenase enzymes (Straathof 2014).

Acetic acid production involves continuous extraction and refinement of raw materials from fermentation broth without compromising microbe growth. Multiple sectors utilized traditional methods for the production of this compound. This involves calcification, condensation, and fractional distillation. These methods are cost-effective and consume a very small amount of energy. Nowadays, highly selective membranes produce purer products and decrease energy cost (Pal and Nayak 2017).

Certain bacteria produce acetic acid, such as *Acetobacter* and *Gluconobacter,* via aerobic fermentation. A mixture of *saccharomyces* is also used to produce acetic acid and other SCFAs. These acids contribute to antimicrobial protection. The gut also depends on a wide range of protein antibiotics produced by epithelial cell types. This indicates a complex interaction between the various antimicrobial mechanisms in the intestine. These volatile compounds significantly contribute to gut health and host defense systems. Its production by beneficial microbes underscores the importance of maintaining a well-balanced gut microbiome to maintain optimal health (Woo et al. 2014).

4.3 Sour Salt

It was discovered in the eighteenth century by a German chemist. It was isolated from the citrus fruits. It was used in households for cleaning purposes. It is present in pure white crystals with a low molecular mass. This acid has a low toxic effect on humans and is easily soluble in water. This material is employed in different industries, including healthcare, textile production, fodder, pigment applications, and wastewater treatment. It fulfills roles in several sectors, such as medicine, textile manufacturing, animal nutrition, pigment production, and effluent treatment (Papagianni 2007). The pharmaceutical and cosmetic sectors account for 10% of their usage, with the remaining portion being utilized in medicines, household and cleaning, agriculture, and animal feed. It is known for being biodegradable, environmentally friendly, cost-effective, safe, and versatile in its ability to sequester, buffer, wet, clean, and disperse. CA is incorporated into beverages, extracts, confectionery, and marmalades. This acid acts as an antioxidant in processed meat, canned fruits, and vegetables. It also helps to maintain pH levels in effervescent tablets, jams, gelatins, and cleaning products. Its binding properties are useful in biodegradable detergents (Angumeenal and Venkappayya 2013).

4.3.1 Fermentation and Biosynthesis

Commercially, the production process involves *Aspergillus niger*, glycerine, and 1,3-Dichloropropan-2-one. A variety of microbial communities, such as *P. restrictum, A. flavus, A. wentii, A. nidulans, A. luchensis, Cephalosporium species, T. viride, P. janthinellum, and Phellinus noxius,* are capable of generating citric acid in the presence of inorganic salts and sugars. Among these, *Aspergillus niger* yields the highest amount of citric acid, producing around 112 g for every 100 g of sucrose used (Max et al. 2010).

Three techniques that are essential for the synthesis of these compounds are liquid, solid phase, and static methods. These methods necessitate particular reaction conditions, such as the selection of media, preparation of cultures, and the inclusion of specific growth factors that support the development of cultures. Solid-state fermentation is commonly conducted in growth chambers, Erlenmeyer flasks, and fermenters, including single- and multi-layered bioreactors. Submerged fermentation accounts for about 80% of citric acid production, with industries owing to its high productivity and cost-effectiveness (González-Sáiz et al. 1997).

Small- and medium-scale industries often prefer surface fermentation because of its less price-reduced consumption and easy installation. This process involves pouring inoculum into an enclosed container that is regulated by proper air and humidity. This acid is produced by various physical and biochemical parameters. These include microbial sources, nutrient and substrate availability, and suitable light and air (Soccol and Vandenberghe 2003).

4.4 Butanoic Acid

Production of this acid occurs with the help of hydroformylation by oxidizing butanal (Zigová and Šturdík 2000). In human physiology, various intestinal microbes generate this acid through the breakdown of complex carbohydrates. This results in energy generation and signaling. It is involved in various pathophysiological conditions, including neuroimmune disorders like Alzheimer's syndrome, Idiopathic Parkinson's, and ASD (Jiang Ling et al. 2018).

It acts as an antimicrobial agent and protects the intestine against harmful pathogens. This acid protects the body by diffusing across the microbial plasma membrane and reducing cytoplasmic pH. In poultry, dietary supplementation with butyric acid glycerides has shown positive outcomes, enhanced live performance, and modified the structure of the small intestine wall (Zhang et al. 2020).

This acid protects living organisms from various infectious species, including *Pseudomonas and Klebsiella.* This promotes the growth of healthy microbes in the intestines. It helps in providing energy and promotes cell growth and differentiation in enterocytes (El-Saadony et al. 2022). Butanoic acid shows great potential as an alternative bacteriostatic agent in livestock production, improving gut health, nutrient digestibility, growth performance, and immunity. Its beneficial effects demonstrate its anti-inflammatory, immunoregulatory, and protective properties against various diseases (Xiong et al. 2022).

4.4.1 Production of Butanoic Acids

Butyric acid (butanoic acid), produced by microbes, has gained importance because of its diverse commercial applications and health benefits. Several microorganisms have been studied, especially the genus *Clostridium,* because it produces butyric acid through fermentation. Through metabolic engineering, *Cl acetobutylicum* produces a large amount of butyric acid. By inactivating specific genes (PTA-CTFB-ADHE1), researchers have achieved a shift in metabolism from biphasic to acidogenic fermentation. This helps them produce enhanced butyric acid (Zigová et al. 1999).

Another notable microorganism for this production is *Cl tyrobutyricum* ATCC 25755 T. This showed promising results in co-culture with *Bacillus species.* SGP1 generates this acid by fermenting sucrose, a viable source that is easily available everywhere (Yu et al. 2014). This helps to produce large batches of this acid, which are beneficial for mankind. Interestingly, butyric acid production depends on several factors like pH, temperature, and substrate availability (Dwidar et al. 2013). The amount of undissociated butanoic acid has a linear correlation with specific butanol production rates, suggesting its role as a major driving force for butanol production in *Cl acetobutylicum* (Jang et al. 2013).

4.5 *Propanoic Acids*

Propanoic acid is widely utilized in food preservation, especially in the baking industry, against *Bacillus* species and animal fodder. It acts as an anti-fungal and immune-modulatory, and has been used in agricultural products. It contributes to gut health by providing energy to the intestinal cells and regulates the immune response. In food, this acid provides various vitamins (B1, B2, B12, K, and folic acid) (Antone et al. 2022).

This acid protects the intestinal epithelial lining and barrier function. It influences various endocrine functions and regulates appetite. Some research shows that this short-chain fatty acid was linked with histone acetyltransferase and G-protein receptors to regulate Genetic and cellular energy processes. It disturbs the bacterial morphology by directly acting on flagella and fimbriae. Overall, this acid affects a broad spectrum of immune cells in animals and humans (Langfeld et al. 2021).

4.5.1 Cultivation of Propanoic

Different cultivation techniques have been established for the production of microbial PA. Methods include continuous, discontinuous, and extractive conversion. *Propionibacterium species* are Gram-positive, lack motility, do not produce spores, and are facultative anaerobic bacillus. These species encompass *P. thoenii, P. freudenreichii, P. shermanii, P. acidipropionici, and P. beijingense* (Liu et al. 2012).

It is produced by various microbes through biochemical pathways, which are categorized into three main groups: biosynthesis, catabolism, and fermentation. These diverse pathways demonstrate that a wide range of metabolic processes are involved in the production of propionic acid. Among these, fermentation routes are deemed the most energy-efficient methods for generating propionic acid from an energy-optimization standpoint (Ahmadi et al. 2017).

Propionic acid, a crucial chemical component with diverse industrial uses, is traditionally manufactured using petrochemical methods. However, due to environmental concerns and consumer preferences for bio-based products, there has been a growing interest in using *propionibacteria. Propionibacterium species* are the primary fermentative producers, generating propionic acid as their main fermentation output (Ammar and Philippidis 2021).

5 Antimicrobial Properties of Organic Acids

Most organic acids (OAs) can easily move throughout bacterial cells because of their small size and low molar mass (Theron and Lues 2007b). Consequently, organic acids display antibacterial effects based on their physicochemical properties in the surrounding environment (Ricke 2003). The antibacterial effect of weak OA is effective by factors including lower acid-base level dissociation degree and a specific effect related to the acid molecule itself. These weak, hydrophobic, and nonionized OA compounds can penetrate the bacterial plasma membrane and reduce the pathogenicity of bacteria by creating pores in the membrane (Lianou and Koutsoumanis 2013).

Nonionized organic acids (acetic and citric acids, etc.) demonstrate significant antimicrobial properties, substantially more potent than their dissociated ones (Stratford and Eklund 2003). In low pH settings, the non-dissociated forms of these substances are favored, enabling neutral compounds to cross the cell membrane and enter the cell. In bacterial cells, the cytoplasm has a higher acidic or basic level, the acidic molecules degree of ionization, releasing charged ions that accumulate within the cell. This process continues until the pH gradient across the membrane is balanced, ultimately causing a reduction in intracellular pH and disrupting the proton-motive force across the membrane (Birk et al. 2010).

As a result, the compound's inhibitory effect arises from its passage across the cytoplasmic membrane in the undissociated form. Upon entering the cell, the molecule encounters an elevated pH level, which leads to its dissociation. This causes the release of charged ions, allowing them to traverse the plasma membrane. Consequently, it has been suggested that the suppression of growth caused by weak acid preservatives could result from several actions, such as disruption of membranes, hindrance of vital metabolic processes, challenges to the maintenance of intracellular pH balance, and the buildup of harmful anions (Brul 1999).

In microorganisms, the undissociated form of organic acids (HA) can pass through the microbial membrane when the pH level inside the cell is higher than that of the surrounding environment. Active transport is required to remove protons (H+). Removing excess protons from the cytoplasm utilizes adenosine triphosphate, deprives cellular energy and impacting cellular functions. As the acidity within the cell continues to increase, it can damage or alter the functioning of enzymes, structural proteins, and DNA. Some organic acids, such as malic and citric acids, have been shown to effectively compromise the outer membrane through mechanisms such as chelation or intercalation (Hatti-Kaul et al. 2018).

The effectiveness of organic acids in decontaminating surfaces of meat, vegetables, and fruits is influenced by sanitation, the acid concentration, the properties of the food surface, the degree of bacterial binding, and the penetration of cells within the food (Koutsoumanis and Skandamis 2013). Organic acids have a wide range of antibacterial mechanisms that include the acidic nature of the plasma membrane, osmoregulation, and disruption. Organic Acids as Acidulants (OAAs) are known to effectively inhibit the growth of spoilage and harmful microorganisms in food. However, pathogenic bacteria have still been found in acidified products like apple cider, orange juice, and various pickled or fermented foods (Yoon et al. 2024).

6 Preservative Effect of Organic Acids

With increased microbes, suitable preservation techniques are required to combat food spoilage. Numerous methods of preventing food spoilage include bacterial metabolic compounds, plant, and animal products. Preservatives often include organic acids and their derivatives owing to their beneficial properties. These compounds create an unfavorable environment for microbial growth by lowering pH. Additionally, incorporating acids during food processing can help reduce initial contamination. Acidic conditions also enhance the effectiveness of heat treatments, such as sterilization, in eliminating microorganisms. Furthermore, organic acids and their derivatives exhibit direct antimicrobial effects. For instance, benzoates target spore-forming bacteria when the spore breaks during germination, whereas sorbates impede the development of vegetative cells (James M. Jay 2005).

In the food industry, hurdle technology is employed to eliminate microbes by integrating multiple techniques to suppress microbial growth. This ensures that food is fit for consumption and enhances its shelf life. This includes storing food at low temperatures, preparing it at high temperatures, elevating food acidity, and decreasing the redox potential. Unfortunately, these methods cannot preserve all kinds of food items. For example, fresh meat and ready-made foods remain unspoiled, yet these products are more susceptible to microbial contamination. To inhibit the growth of microbes, these organic compounds are used along with other minerals, which are beneficial for human health (Theron and Lues 2007a).

7 Use of Organic Acids in Food Processing

Organic acids are vital in materials processing because they can easily be removed after treatment in the food industry. Organic acids serve as extraction agents for gelatin production. The process of breaking down collagen crosslinks involves acidification of the raw material and subsequent treatment with warm water. Processing aids can be employed in cleaning processing units. Certain mineral salts dissolve readily in acidic environments. Mild volatile compounds such as citrate can help reduce the buildup of mineral salts on equipment without damaging other construction materials. This approach minimizes fouling while preserving the integrity of the processing units (Giménez et al. 2005).

8 Future Direction

Food safety is a global issue, making quality food items a vital concern in the food industry. Despite the numerous preservation techniques available, food spoilage and poisoning continue to arise. Recently, modern consumer trends have shown a growing preference for foods that are less processed and free from artificial additives. As a result, the importance of ensuring food safety through the use of bio-based compounds and the enhanced production of these substances from various sources has gained considerable attention. Microorganisms can produce organic acids, which are commonly utilized as antimicrobial agents in bio-preservation. With advancements in food processing methods and rising public concerns regarding natural additives, the market for bio-based organic acids has expanded and emerged as a significant category of chemicals. Therefore, more research and experimentation are necessary to boost microbial organic acid production to satisfy global demand.

Moreover, the impact of microbial shape on product synthesis is another factor that must be optimized and regulated to ensure consistent sustainability in the processes. The contamination of food with molds and yeast is still a basic concern, although the food market confirms the best quality food materials. Recent studies have shown that adding microparticles to the fermentation medium for controlling fungal morphology not only boosted overall yield and productivity but also improved system consistency and reproducibility. Hence, additional research should be conducted to explore the influence of microparticle incorporation on organic acid production by fungi (Scallan et al. 2011).

References

Ahmadi N, Khosravi-Darani K, Mortazavian AM (2017) An overview of biotechnological production of propionic acid: from upstream to downstream processes. Electron J Biotechnol 28:67–75

Ammar EM, Philippidis GP (2021) Fermentative production of propionic acid: prospects and limitations of microorganisms and substrates. Appl Microbiol Biotechnol 105:6199–6213

Angumeenal AR, Venkappayya D (2013) An overview of citric acid production. LWT Food Sci Technol 50:367–370

Antone U, Ciprovica I, Zolovs M, Scerbaka R, Liepins J (2022) Propionic acid fermentation—study of substrates, strains, and antimicrobial properties. Fermentation 9:26

Bangar SP, Suri S, Trif M, Ozogul F (2022) Organic acids production from lactic acid bacteria: a preservation approach. Food Biosci 46:101615

Birk T, Grønlund AC, Christensen BB, Knøchel S, Lohse K, Rosenquist H (2010) Effect of organic acids and Marination ingredients on the survival of campylobacter jejuni on meat. J Food Prot 73:258–265

Brul S (1999) Preservative agents in foods mode of action and microbial resistance mechanisms. Int J Food Microbiol 50:1–17

Dibner J, Buttin P (2002) Use of organic acids as a model to study the impact of gut microflora on nutrition and metabolism. J Appl Poult Res 11:453–463

Dusselier M, Van Wouwe P, Dewaele A, Makshina E, Sels BF (2013) Lactic acid as a platform chemical in the biobased economy: the role of chemocatalysis. Energy Environ Sci 6:1415–1442

Dwidar M, Kim S, Jeon BS, Um Y, Mitchell RJ, Sang B-I (2013) Co-culturing a novel bacillus strain with clostridium tyrobutyricum ATCC 25755 to produce butyric acid from sucrose. Biotechnol Biofuels 6:35

El-Saadony MT, Umar M, Hassan F-U, Alagawany M, Arif M, Taha AE, Elnesr SS, El-Tarabily KA, Abd El-Hack ME (2022) Applications of butyric acid in poultry production: the dynamics of gut health, performance, nutrient utilization, egg quality, and osteoporosis. Anim Health Res Rev 23:136–146

Giménez B, Turnay J, Lizarbe MA, Montero P, Gómez-Guillén MC (2005) Use of lactic acid for extraction of fish skin gelatin. Food Hydrocoll 19:941–950

González-Sáiz JM, Fernández-Torroba MA, Pizarro C (1997) Application of weakly basic copolymer polyacrylamide (acrylamide-CO-N,N′-dimethylaminoethyl methacrylate) gels in the recovery of citric acid. Eur Polym J 33:475–485

Hatti-Kaul R, Chen L, Dishisha T, Enshasy HE (2018) Lactic acid bacteria: from starter cultures to producers of chemicals. FEMS Microbiol Lett 365. https://doi.org/10.1093/femsle/fny213

Jang Y-S, Woo HM, Im JA, Kim IH, Lee SY (2013) Metabolic engineering of clostridium acetobutylicum for enhanced production of butyric acid. Appl Microbiol Biotechnol 97:9355–9363

Jay JM (2005) Food protection with chemicals and by biocontrol. In: Modern food microbiology. Springer US, Boston

Jiang Ling JL, Fu Hongxin FH, Yang H, Xu Wei XW, Wang Jufang WJ, Yang Shangtian YS (2018) Butyric acid: applications and recent advances in its bioproduction. Biotechnol Adv 36:2101–2117

Kim J, Kim Y-M, Lebaka VR, Wee Y-J (2022) Lactic acid for green chemical industry: recent advances in and future prospects for production technology, recovery, and applications. Fermentation 8:609

Koutsoumanis K, Skandamis P (2013) New research on organic acids and pathogen behaviour. Elsevier

Langfeld LQ, Du K, Bereswill S, Heimesaat MM (2021) A review of the antimicrobial and immune-modulatory properties of the gut microbiota-derived short chain fatty acid propionate—what is new? Eur J Microbiol Immunol 11:50–56

Lee JW, Kim HU, Choi S, Yi J, Lee SY (2011) Microbial production of building block chemicals and polymers. Curr Opin Biotechnol 22:758–767

Liu L, Zhu Y, Li J, Wang M, Lee P, Du G, Chen J (2012) Microbial production of propionic acid from propionibacteria: current state, challenges and perspectives. Crit Rev Biotechnol 32:374–381

Liu J, Li J, Shin H-D, Liu L, Du G, Chen J (2017) Protein and metabolic engineering for the production of organic acids. Bioresour Technol 239:412–421

Max B, Salgado JM, Rodríguez N, Cortés S, Converti A, Domínguez JM (2010) Biotechnological production of citric acid. Braz J Microbiol 41:862–875

Narbad A, Wang G (2018) Lactic acid bacteria and foodborne pathogens. In: *Lactic acid bacteria in foodborne hazards reduction: physiology to practice*. Springer Singapore, Singapore

Önder ÖC, Yilgör E, Yilgör I (2016) Fabrication of rigid poly (lactic acid) foams via thermally induced phase separation. Polymer 107:240–248

Pal P, Nayak J (2017) Acetic acid production and purification: critical review towards process intensification. Sep Purif Rev 46:44–61

Papagianni M (2007) Advances in citric acid fermentation by aspergillus Niger: biochemical aspects, membrane transport and modeling. Biotechnol Adv 25:244–263

Peng M, Biswas D (2017) Short chain and polyunsaturated fatty acids in host gut health and foodborne bacterial pathogen inhibition. Crit Rev Food Sci Nutr 57:3987–4002

Ren C, Zhang Q, De Haan BJ, Faas MM, Zhang H, De Vos P (2020) Protective effects of lactic acid bacteria on gut epithelial barrier dysfunction are Toll like receptor 2 and protein kinase C dependent. Food Funct 11:1230–1234

Reuben RC, Torres C (2024) Bacteriocins: potentials and prospects in health and agrifood systems. Arch Microbiol 206:233

Rodrigues C, Vandenberghe L, Woiciechowski A, De Oliveira J, Letti L, Soccol C (2017a) Production and application of lactic acid. In: Current developments in biotechnology and bioengineering. Elsevier

Rodrigues C, Vandenberghe LPS, Woiciechowski AL, De Oliveira J, Letti LAJ, Soccol CR (2017b) Production and application of lactic acid. Elsevier

Sarchami T, Batta N, Berruti F (2021) Production and separation of acetic acid from pyrolysis oil of lignocellulosic biomass: a review. Biofuels Bioprod Biorefin 15:1912–1937

Shi Y, Pu D, Zhou X, Zhang Y (2022) Recent Progress in the study of taste characteristics and the nutrition and health properties of organic acids in foods. Foods 11:3408

Soccol CR, Vandenberghe LPS (2003) Overview of applied solid-state fermentation in Brazil. Biochem Eng J 13:205–218

Straathof AJJ (2014) Transformation of biomass into commodity chemicals using enzymes or cells. Chem Rev 114:1871–1908

Tang H, Huang W, Yao YF (2023) The metabolites of lactic acid bacteria: classification, biosynthesis and modulation of gut microbiota. Microb Cell 10:49–62

Teng K, Huang F, Liu Y, Wang Y, Xia T, Yun F, Zhong J (2023) Food and gut originated bacteriocins involved in gut microbe-host interactions. Crit Rev Microbiol 49:515–527

Theron MM, Lues JFR (2007a) Organic acids and meat preservation: a review. Food Rev Intl 23:141–158

Vandenberghe LP, Karp SG, De Oliveira PZ, De Carvalho JC, Rodrigues C, Soccol CR (2018) Solid-state fermentation for the production of organic acids. In: Current developments in biotechnology and bioengineering. Elsevier

Vishnu D, Dhandapani B, Mahadevan S (2020) Recent advances in organic acid production from microbial sources by utilizing agricultural by-products as substrates for industrial applications. In: Jerold M, Arockiasamy S, Sivasubramanian V (eds) Bioprocess engineering for bioremediation: valorization and management techniques. Springer International Publishing, Cham

Woo J-M, Yang K-M, Kim S-U, Blank LM, Park J-B (2014) High temperature stimulates acetic acid accumulation and enhances the growth inhibition and ethanol production by Saccharomyces cerevisiae under fermenting conditions. Appl Microbiol Biotechnol 98:6085–6094

Xiong R-G, Zhou D-D, Wu S-X, Huang S-Y, Saimaiti A, Yang Z-J, Shang A, Zhao C-N, Gan R-Y, Li H-B (2022) Health benefits and side effects of short-chain fatty acids. Foods 11:2863

Yin X, Li J, Shin H-D, Du G, Liu L, Chen J (2015) Metabolic engineering in the biotechnological production of organic acids in the tricarboxylic acid cycle of microorganisms: advances and prospects. Biotechnol Adv 33:830–841

Yu A-Q, Pratomo Juwono NK, Leong SSJ, Chang MW (2014) Production of fatty acid- derived valuable chemicals in synthetic microbes. Front Bioeng Biotechnol 2:78

Zhang S, Dogan B, Guo C, Herlekar D, Stewart K, Scherl EJ, Simpson KW (2020) Short chain fatty acids modulate the growth and virulence of Pathosymbiont Escherichia coli and host response. Antibiotics 9:462

Zigová J, Šturdík E (2000) Advances in biotechnological production of butyric acid. J Ind Microbiol Biotechnol 24:153–160

Zigová J, ŠturdíK E, Vandák D, Schlosser Š (1999) Butyric acid production by clostridium butyricum with integrated extraction and pertraction. Process Biochem 34:835–843

Uncategorized References

Coban HB (2020) Organic acids as antimicrobial food agents: applications and microbial productions. Bioprocess Biosyst Eng 43:569–591

Lianou A, Koutsoumanis KP (2013) Strain variability of the behavior of foodborne bacterial pathogens: a review. Int J Food Microbiol 167:310–321

Ren Y, Li L (2022) The influence of protease hydrolysis of lactic acid bacteria on the fermentation induced soybean protein gel: protein molecule, peptides and amino acids. Food Res Int 156:111284

Ricke S (2003) Perspectives on the use of organic acids and short chain fatty acids as antimicrobials. Poult Sci 82:632–639

Scallan E, Griffin PM, Angulo FJ, Tauxe RV, Hoekstra RM (2011) Foodborne illness acquired in the United States—unspecified agents. Emerg Infect Dis 17:16

Stratford M, Eklund T (2003) Organic acids and esters. In: Food preservatives. Springer

Theron MM, Lues JF (2007b) Organic acids and meat preservation: a review. Food Rev Intl 23:141–158

Vidra A, Németh Á (2018) Bio-produced acetic acid: a review. Period Polytech Chem Eng 62:245–256

Yoon J-H, Oh M-S, Lee S-Y (2024) Effectiveness of organic acids for inactivating pathogenic bacteria inoculated in laboratory media and foods: an updated minireview. Food Sci Biotechnol 33:2715–2728

Enzymes as Antimicrobials: A Look into Their Gut Enhancing Properties and Nutrient Absorption Capacity

Ammara Khaliq, Sundas Mehboob, Arfan Yousaf, Evelyn Saba, and Rabia Ashiq

Abstract Animal health has always been very important as they are critical in the agriculture sector and for maintaining the ecosystem. Their health is closely dependent on a strong gut, as both are strongly correlated. To maintain their gut health, many additives, e.g., exogenous enzymes, are incorporated in the feed because other cost-effective ingredients bring along anti-nutritional factors. When enzymes are added to the feed, they enhance the nutrient absorption and boost the digestibility of the gut by causing the death of harmful bacteria. Cellulases, xylases, and exogenous lysozymes, when added to an animal diet, increase the overall digestion, digestible energy intake, and performance.

Enzymes have their own mechanism of action to kill the pathogenic bacteria by revealing their antimicrobial effects. They are being added to the diet of swine, ruminants, aquatic species, and dogs as they have the ability to boost gut health as well as nutrient absorption.

Keywords Enzymes · Gut digestibility · Nutrient absorption · Antimicrobial effects

---

Ammara Khaliq and Sundas Mehboob contributed equally with all other contributors.

A. Khaliq · S. Mehboob · A. Yousaf
Department of Clinical Studies, Faculty of Veterinary and Animal Sciences, Pir Mehr Ali Shah Arid Agriculture University, Rawalpindi, Pakistan

E. Saba (✉)
Department of Veterinary Biomedical Science, Faculty of Veterinary and Animal Sciences, Pir Mehr Ali Shah Arid Agriculture University, Rawalpindi, Pakistan
e-mail: evelyn.saba@uaar.edu.pk

R. Ashiq
Department of Veterinary Pathology, Faculty of Veterinary and Animal Sciences, Pir Mehr Ali Shah Arid Agriculture University, Rawalpindi, Pakistan

R. U. Khan, S. Naz (eds.), *Green Alternatives to Antibiotics in Animal Health*,
https://doi.org/10.1007/978-981-92-0309-3_20

1 Introduction

Modern agriculture mostly depends on the feed that is of top quality and nutritious as well. However, the cost-friendly ingredients that are added to feed, such as plant material, e.g., badly processed grains and feedstuffs containing high fiber. These ingredients contain the components that are resistant to the endogenous digestive enzymes of the body. Along with this, some ingredients contain anti-nutritional factors (ANFs), resulting in bad digestion and absorption (Pariza and Cook 2010). Research has reported that supplementation of the animal's diet with fiber-digesting exogenous enzymes can increase the animal's utilization of feed as well as animal performance.

Enzyme additions for the ruminant diet, such as cellulases and xylanases, are concentrated extracts from the bacterial or fungal fermentations that have particular enzymatic properties. The increase in the feed digestion is primarily because of the enhanced fiber digestion, resulting in the improved digestible energy intake and increased animal productivity as well as growth (Beauchemin et al. 2003).

Enzyme encapsulation helps in stabilizing the enzymes during the feed processing and then, after digestion release in the gastrointestinal tract GIT of the animal. The activity of these enzymes is affected by factors such as low pH, natural body enzymes, and in ruminants, an additional microbial enzyme that is involved in the fermentation process (Pariza and Cook 2010). These enzymes perform their activity by breaking the larger compounds into smaller ones, and also by another mechanism, which is that they help in the prevention of an increase in digesta viscosity, as increased viscosity can weaken the absorption of beneficial nutrients (Chesson 1993).

Moreover, enzymes bring along their antimicrobial activity against a variety of bacterial species. They cause the death of bacteria through various modes of actions such as hydrolyzing the linkage of different compounds in the bacterial cell wall. Moreover, they also cause the death of bacteria by creating an anaerobic environment and by damaging the quorum-sensing (QS) bacterial network. Enzymes are being added in the diet of swine, ruminants, aquatic species, and poultry in order to enhance the nutrient absorption, improve digestibility, and growth performance.

2 Enzymes

Enzymes are the substances that are involved in accelerating chemical reactions. Without them, life is impossible, as they control every biological process and every reaction in the body, and these enzymes are required for completion. They break down the food, convert their nutrients into simpler forms in order to let them absorb in the blood (Engel 2020). They are extremely effective biological catalysts, having the ability to accelerate chemical reactions millions of times over. Enzymes are proteins in nature with a complex three-dimensional structure. This complex structure impacts their stability during the high-temperature, for instance, during feed manufacture and passing through the gastrointestinal tract.

Enzymes are reported to be used as a catalyst in many industrial processes as well as in products such as juices, wine, leather, baked products, and several others. For enhancing the digestibility of feed ingredients present in monogastric diets, a variety of hydrolase enzymes have shown their effect and are currently being used in the diet. These enzymes are used because of their property to increase the nutritional value of the feed and are involved in the removal of the naturally present anti-nutrients (Choct 2006).

3 Feed Enzymes: Key Players in Digestive and Absorptive Mechanisms

For many years, enzymes as the feed additives have shown their application commercially, and the feed enzyme industry has gone through phases and has evolved up to now. In the first phase, the prime focus was to improve the nutrient digestibility, particularly by decreasing and removing the anti-nutritive effects of the non-starch polysaccharides (NSP). These effects were reported to be more prominent in the broiler diets that are mostly based on viscous grains like wheat, rye, and barley (Choct 2006). For the use in the feed industry commercially, four distinct categories of the enzyme products are currently available, that includes microbial phytases, glycanases that target viscous cereals (e.g., wheat, barley), enzymes that target the nonviscous cereals (e.g., corn, sorghum), and the enzymes having the property of targeting non-cereals (e.g., soybean meal, grain legumes). Many feed enzymes are commercially available, and a few of them are mentioned here: phytases, β-glucanases, xylanases, α-galactosidases, proteases, amylase, lipases, mannanases, cellulases and hemicellulasespectinases (Cowieson et al. 2006).

4 Benefits of Feed Enzymes

The main purpose behind the use of feed enzymes is to improve the performance of animals and to increase the profit. That is done by enhancing the digestibility of indigestible feed products and other components of feed, such as protein, amino acids, starch, and lipids. However, along with these benefits, there are some additional benefits present that contributes in the widespread acceptance of these feed enzymes, such as:

1. Feed enzymes increase the range of feedstuffs for usage and improve the feed formulation. This is done by the reduction and removal of the poorly digested and undigested feed components.

2. These enzymes, as supplementation decreases the variability in ingredient's nutritive value and improve the accuracy of the feed formulation, which is done by uplifting the poor-quality samples.
3. Enzymes help in decreasing the excreta moisture content and reduce the occurrence of wet litter in the birds fed diets with high levels of NSP. Poor litter quality can result in many concerning problems that include foot lesions, hock burns, and the downgrading of carcass.
4. Feed enzymes also enhance the intestinal morphology and integrity, resulting in better feed digestion and improved absorption of the dietary components.
5. The use of enzymes decreases the manure output and nutrient load (nitrogen and phosphorus), which subsequently results in the better use of these components, offering the environmental benefits, particularly in the intensive poultry systems globally (Ravindran 2013).
6. These enzymes improve gut health as they improve the digestibility of feed content, leaving very little indigestible content to reach the lower gut and, along with this, the shift of gut flora toward the favorable gut microflora. Feed enzymes in birds' feed, improve the gut health, and support the immune functions of the birds (Collett 2012).

5 General Mechanism of Action of Feed Enzymes in the Gut

There are two ways by which these feed enzymes enhance the diet digestibility and, in return, animal performance. First, by breaking the larger compounds into the smaller ones that are bound either by the cell wall or within the cell wall. Then the intestinal wall can easily absorb these smaller molecules. These feed enzymes work with the animal body's endogenous enzymes to break these larger compounds for better absorption. For the beneficial results, it is mandatory to select the exogenous enzymes and the body's own exogenous enzymes that complement each other's effect as well as the digestive system of the animal body.

Second, enzymes also help in the prevention of an increase in digesta viscosity, because it can weaken the absorption of beneficial nutrients. Few compounds have this property that, after releasing from the cell walls forms the gel and hence increase the viscosity of digesta. Young animals and birds face more problems because of this increased digesta viscosity. Digesta viscosity can be reduced by the feed enzymes by reducing the chain lengths of the gel-forming NSPs and by not removing their side branches (Chesson 1993).

6 Effect of Feed Enzymes on Gastrointestinal Microbiota

Enzymes perform their function as biological catalysts in chemical reactions by acting on substrates that are specific to them. Feed enzyme characteristics are mostly determined by the organism from which they are being produced, and also the availability of their specific substrates. The effectiveness of the feed enzymes when used in animal feed has different criteria. It is primarily based on their mechanism of action within animal nutrition. There are two perspectives that tell us about the relationship between the gut microflora and feed enzymes. First, the physiology of the GIT and the biochemical properties of digesta are affected by the enzyme substrates. Second, feed enzyme has the ability to amend effects by changing the depletion of the substrates present in the gastrointestinal tract (Adeola and Cowieson 2011).

The relationship between the gut microflora and the feed enzyme is important as the latter is used with the aim to target the substrates like phytate and NSP to eventually degrade these indigestible feed contents. For instance, in broiler chickens, because of soluble NSPs, the intestinal viscosity increases, resulting in poor feed utilization and growth of birds. This increase in viscosity results in triggering many different kinds of physiological reactions in the intestine. That includes the increase in the passage time of the digesta from the small intestine, an increase in the intestinal mass, and more mucosal cell turnover rate.

Along with this, these soluble NSPs are also involved in the stimulation of mucin and carbohydrates countenance in the goblet cells that will lead to the incidence of undigested ingredients. All these physiological responses will lead to an increase in microbial activity, size, and composition of the small intestine. To overcome this issue, supplementation of the diet with NSP-degrading carbohydrases, for instance, xylanase and β-glucanase. They have the ability to break the soluble NSP and hence help in reducing the digesta viscosity, resulting in increased animal performance and good nutrient utilization. Along with these, they can also impact the composition and the metabolic potential of the bacteria present in the intestine (Bedford and Schulze 1998; Hübener et al. 2002; Kiarie et al. 2013).

7 Role of Some Individual Digestive Enzymes

It is being reported that, on average, animals are only able to digest 75–80% of the feed they consume, leaving the rest of the 15–25% feed undigested. The reason is that feed is made up of some indigestible components that pass through the digestive tract without processing and digestion. The animal body lacks specialized enzymes needed for the digestion of such indigestible components present in feed. Feed costs make up to 60–70% of expenses, so for the increase in meat production and decrease in production time and expenses, producers often use enzyme additives in feed. Among these enzymes, the following are some examples of digestive enzymes (Menezes-Blackburn and Greiner 2015).

7.1 *Carbohydrases*

Carbohydrases are a group of enzymes that have the ability to digest or degrade the low-digestible oligosaccharides and polysaccharides into smaller, digestible particles. On the other hand, for the digestion of complex polysaccharides, there is a need for a variety of other complementary enzymes. These will convert the larger molecules of complex polysaccharides into its constitute sugars. Carbohydrases are classified into many enzymes, such as pectinases, α-galactosidases, mannanases, xylanases, and β-glucanases (cellulases) (Barletta 2010).

Carbohydrates from cereals affect the intestinal microflora, nutrient absorption, and the immune response in non-ruminant animals. Cereals contain different kinds of NSPs such as cellulose, resistant starch, arabinoxylans, and β-glucans, which decrease the digestibility of feed and animal health. Some other NSPs, like arabinoxylans and β-glucans, are recognized as they negatively influence animals by increasing the digesta viscosity and intestinal permeability. This influence leads to a decrease in helpful gut microflora, an increase in damaging bacteria, and decreased nutrient digestibility from feed. An increase in the viscosity also disturbs lipid emulsification, short-chain fatty acid formation, and the interaction of nutrients with the intestinal brush border.

On the other hand, nutrient absorption is affected by the insoluble NSP as well because they create a protective barrier around the cellular content. These insoluble NSP also interact with the immune system; they bind with the surface receptors present on the immune cells and trigger the molecular and cellular pathways, and activate the immune system. This activation of the immune system consumes additional energy and nutrients and decreases the resources that can be used for the growth of animals. According to recent research, using enzymes to hydrolyze the polysaccharides into oligosaccharides may change the GIT microbiota and function as a prebiotic. By increasing the production of short-chain fatty acids and reducing unwanted immune activation, this regulation improves gut health (Valente Junior et al. 2024).

7.2 *Proteases*

Proteases are an exceptionally large group of enzymes having many origins and different biochemical properties. These have the ability to hydrolyze the peptide bonds that are present between the amino acids. They break the protein polypeptide chains into the smaller oligopeptides, which are then degraded by carboxypeptidases into amino acids (Menezes-Blackburn et al. 2013). Most of the protease enzymes are very specific and attack the selective amino acid sequence. In poultry and livestock, these enzymes are used for the digestion of storage proteins from plant origin and the degradation of protein with antinutritional factors. A great quantity of storage protein is present in the seeds of the legume family, such as soybeans, which are mostly formed during the seed formation, as a nitrogen source for the new embryo. Proteases also release the starches that are frequently attached to the storage protein and provide the energy source for animals. Trypsin inhibitor, which is a significant anti-nutritional factor, is present in the untreated plant-based protein sources such as soybeans. These inhibitors are involved in decreasing the digestion by blocking the enzyme trypsin, which is released from the pancreas. This enzyme helps in protein degradation in the small intestine. Soybean meal that is processed contains a lesser amount of trypsin and hence has lower digestibility in animals (Imran et al. 2016).

It is reported by research that in the broiler diet, soy protein digestion was improved by the supplementation with proteases. These enzymes work by improving the endogenous peptidases formation, decreasing the need for amino acids and energy. Along with decreasing the trypsin inhibitors, proteases improve the amino acid consumption and decrease the protein turnover (Yu et al. 2007).

Phosphorus is important for the animal's growth because it is involved in the metabolic process and the development of bones. For animal optimal growth, feed is normally supplemented with the calcium phosphate salts that come from phosphate rocks, which are limited resources. On the other hand, plant-based feed, such as grain-based feed, has phosphorus in phytate form, but such a form of phosphorus is not available to non-ruminants. Phosphorus is one of the most important nutrient factors in the ruminant diet that is majorly bound in the mixed salts of the phytic acids in the plant-origin feed. This makes the phosphorus inaccessible to animals without dephosphorylation by the enzymes. Other animals, such as monogastric animals, e.g., pigs and poultry, are not capable of showing enough phytase activity in their digestive tract. That makes the utilization of phosphorus in phytase form difficult for them. For these animals, feed supplemented with phytase increases the phosphorus and calcium utilization, but this degradation of the phytate by phytase enzyme is more shown in pigs than in poultry. However, the performance response by addition of phytase enzyme in feed is reportedly more shown in birds (Raboy 2009; Reddy et al. 1982).

8 Antimicrobial Role of Enzymes

Gut microbiota is imperative for the overall health of animals as it produces necessary vitamins, is involved in immunity, and in the digestion of food. However, due to the use of antibiotics for the growth of animals, they disturb the normal microbial population in the gut. This led to the use of non-antibiotic alternatives to promote the growth of animals. In in vitro studies, one of the enzymes, such as exogenous lysozyme, acted as antimicrobial, particularly against *Clostridium perfringens* in the ileum of broilers (Abdel-Latif et al. 2017). Alpha-amylase inhibits the *Staphylococcus aureus,* and its combination with proteases is effective against *Pseudomonas* sp. (Thallinger et al. 2013).

9 Antimicrobial Mechanism of Action of Enzymes in the Gut

There are various modes of action through which enzymes work as antimicrobials. One of them is the bacteriolytic activity of enzymes by the degradation of the cell wall. They exert their bacteriolytic activity by hydrolyzing the β-1,4-glycosidic linkage between N-acetylmuramic acid and N-acetylglucosamine of the bacterial cell wall, mainly against many Gram-positive bacteria (Abdel-Latif et al. 2017).

Moreover, the other mechanism of action is antimicrobial activity through oxidative stress and depletion of oxygen. In such a mechanism, enzymes accelerate the production of gluconic acid (GA) as well as hydrogen peroxide (H2O2) by the consumption of molecular oxygen. Due to such activity, an anaerobic environment is created which is an unfavorable condition for the bacteria to thrive. Such a mode of action eventually leads to the death of harmful bacteria (Jiang et al. 2024). Besides, enzymes have the ability to damage the microbial network, for instance, QS. Some of the particular enzymes, e.g., lactonases, can damage the QS signal molecules. The QS system is important for the bacteria for its biological activities. In one of the studies, it was demonstrated that the combination of quorum inhibitors and quenching enzymes totally blocked the QS of *Pseudomonas aeruginosa* and made it unable to multiply (Jiang et al. 2024). In addition to this, another mode of action is not specifically or directly antimicrobial, but the enzymes cause immune modulation. Enzymes do not directly kill the bacteria as described in the aforementioned mechanisms. However, they modulate the gut environment by causing a reduction in the unwanted immune activation. It reduces the immune-mediated gut stress and leads to microbiota balance and better nutrient absorption (Jiang et al. 2024).

10 Utilization of Feed Enzymes in Swine Diet: Boosting Nutrient Absorption

It is beneficial to break down the carbohydrates into the absorbable sugars, but it is reported in literature that all monosaccharides are properly utilized by all the species of animalS or all gastrointestinal tracts parts. There are a few sugars, such as xylose, that are competently used by animal species like pigs and birds, such as poultry. But on the other hand, there are a few sugars, such as arabinose, that are not efficiently absorbed inside the hind gut of pigs. They are only absorbed if they are broken down into smaller pieces in the pig's hindgut. Henceforth, for the maximum utilization of energy from the pentose sugars, it is necessary to avoid glycosidases during the feed enzyme formulations.

Sometimes this interaction of the endogenous enzymes and the feed enzymes from the feed can be complex. For example, in piglets, pentosanase enzyme supplementation in the diet containing rye can result in a negative effect on digesta pH, dry matter digestibility, and digesta viscosity. Contrary to that, the rye diet having β-glucanase-supplement and the barley diet having pentosanase- or β- glucanase-supplement showed no such adverse effects. All these data emphasize the importance of understanding the enzyme interaction with feed ingredients and endogenous enzymes (Inborr et al. 1991).

11 Feed Enzymes in Ruminant's Nutrition: Enhanced Digestibility and Absorption

Ruminants have a complex digestive system, as they have a microbial population present in their digestive system that is involved in the fermentation process and secretes different kinds of enzymes. That is why the use of feed enzymes in ruminants was developed slowly, as their gut microflora creates difficulty in understanding the data obtained upon feed enzyme use. Ruminant feed was mostly supplemented with multienzyme complexes that are made up of amylases, cellulases, pectinases, and xylanases. These are mostly used because of their functions, such as enhancing the forage cell walls' digestibility, increasing dairy cattle performance, and enhancing the starch accessibility from cereals, and they have also shown a positive effect on the milk yield (Velázquez-De Lucio et al. 2021).

Its common practise to use agricultural waste as ruminant feed; however, they are usually low in protein, hold anti-nutritionals factors, and high fiber content is present. To improve the quality of such feed, exogenous, such as cellulases are used, and they have shown their effect in dairy cows and small ruminants. Recently, research was conducted by Golder et al. on dairy cows in the United States at three farms to study the effect of fibrolytic enzymes. For a time period of 5 months, eight pens of cows were not given enzymes, and eight pens received 750 ml of the enzymes per ton of feed. Then both of these pens were compared; as a result, the cows that received enzymes in their diet produced 0.70–0.80 kg more milk daily because of the enhanced digestibility of feed. The dry matter intake of the dairy cows was also increased by 0.20 kg, but their body weight was not changed (Golder et al. 2019; Trejo-López et al. 2017).

12 Feed Enzymes in Aquaculture: Improving Digestibility and Growth Performance in Fish

For the development of the sustainable aquaculture industry, the use of feed enzymes plays a very important role. According to the LP Information, Inc. (2022), the global animal feed enzyme market was prized at USD 1340.6 million in 2021 and is expected to expand at a compound annual growth rate (CAGR) of 5.0% from 2022 to 2028. This increase validates the growing significance of the enzymes in the manufacturing of fish feed.

To get optimum output from the fish feed, such as obtaining the required calories and nutrients, their feed should be efficiently digested by suitable enzymes. Fish feed ingredients, such as non-fish raw materials, contain the organic matter having a high molecular weight, resulting in lesser and slower digestion and decomposition of feed in the gastrointestinal tract of fish. A study has reported that the fish growth and digestibility of feed were improved when the enzymes were used for pre-treatment of the plant-derived raw materials. The enzymes that are used in fish feed mainly belong to the hydrolases, and among them, the most efficient are proteases, glucosidases, and lipases (Liang et al. 2022). These enzymes perform their function of breaking the anti-nutritional factors present in feed, such as phytic acid, indigestible oligosaccharides, such as stachyose and raffinose, and antigen protein in feed. These anti-nutritional factors cause hindrance in digestion, cause malnutrition, and decrease growth. Phytases are enzymes that are important for the digestion of the phytase of plant origin, when supplemented in the diet enhances the bioavailability of phosphorus and other minerals and hence increases the growth (Lemos and Tacon 2017).

13 Feed Enzymes: Supporting Digestive Efficiency in Dogs

The dog food industry provides different kinds of commercial diets, which includes pedigree adult chicken and vegetable, etc. These products show the benefits, such as good skin, strong bones, and muscles. Dog foods often contain both animal and plant-based materials and provide a balanced diet to improve the animal's overall health. To obtain these results, degradation of macromolecules such as proteins, fats, and carbohydrates is important. During the manufacturing of pet foods, the processing methods, such as extrusion cooking, can increase palatability and digestibility, but it also reduces the vitamin availability and the quality of protein. Along with these, the addition of vegetable-based components adds anti-nutritional factors in feed. Henceforth, the supplementation of the pet foods with the digestive enzymes can improve the digestibility of ingredients and can improve the nutrient availability to the animal. It is reported through the research that enzymes such as xylanase, protease, β-mannanase, and phytase have shown their beneficial effects and have improved the digestibility of the nutrients present in feed (Jadhav et al. 2023).

Dog feed commonly contains barley and wheat that can decrease the digestibility and negatively affect fecal consistency because of the presence of the high dietary fiber content in it. To solve this problem, glycanases have been used and reported its beneficial effect, such as digesting the soluble fiber content, increasing fermentation in the large intestine of dog and decreasing the anti-nutritive effects. In result, this enzyme leaves a positive result on the digestion and fecal quality (Velázquez-De Lucio et al. 2021).

14 Conclusion

In a nutshell, animal health is dependent on a healthy and maintained gut. To achieve it, certain feed additives, for instance, enzymes, are incorporated. Enzymes break down the complex nutrients into simpler ones, which leads to enhanced absorption of nutrients and gut health. Certain enzymes, e.g., cellulases, xylases, and proteases, are included in the feed. Moreover, enzymes also maintain gut health by inducing their antimicrobial effects and killing the pathogenic gastrointestinal bacteria. They are being included to fight against the phenomenon of antimicrobial resistance. Enzymes are revealing their beneficial properties as alternative feed additives in animal nutrition and health. Further studies are needed to demonstrate their more advantageous effects on animal health.

References

Abdel-Latif MA, El-Far AH, Elbestawy AR, Ghanem R, Mousa SA, Abd El-Hamid HS (2017) Exogenous dietary lysozyme improves the growth performance and gut microbiota in broiler chickens targeting the antioxidant and non-specific immunity mRNA expression. PLoS One 12(10):e0185153

Adeola O, Cowieson AJ (2011) Board-invited review: opportunities and challenges in using exogenous enzymes to improve nonruminant animal production. J Anim Sci 89(10):3189–3218

Barletta A (2010) Introduction: current market and expected developments. In: Enzymes in farm animal nutrition. CABI, Wallingford UK, pp 1–11

Beauchemin KA, Colombatto D, Morgavi DP, Yang WZ (2003) Use of exogenous fibrolytic enzymes to improve feed utilization by ruminants. J Anim Sci 81(14_suppl_2):E37–E47

Bedford MA, Schulze H (1998) Exogenous enzymes for pigs and poultry. Nutr Res Rev 11(1):91–114

Chesson A (1993) Feed enzymes. Anim Feed Sci Technol 45(1):65–79

Choct M (2006) Enzymes for the feed industry: past, present and future. World's Poult Sci J 62(1):5–16

Collett SR (2012) Nutrition and wet litter problems in poultry. Anim Feed Sci Technol 173(1–2):65–75

Cowieson AJ, Hruby M, Pierson EM (2006) Evolving enzyme technology: impact on commercial poultry nutrition. Nutr Res Rev 19(1):90–103

Engel PC (2020) Enzymes: a very short introduction, vol 661. Oxford University Press

Golder HM, Rossow HA, Lean IJ (2019) Effects of in-feed enzymes on milk production and components, reproduction, and health in dairy cows. J Dairy Sci 102(9):8011–8026

Hübener K, Vahjen W, Simon O (2002) Bacterial responses to different dietary cereal types and xylanase supplementation in the intestine of broiler chicken. Arch Anim Nutr 56(3):167–187

Imran M, Nazar M, Saif M, Khan MA, Vardan M, Javed O (2016) Role of enzymes in animal nutrition: a review. PSM Vet Res 1(2):38–45

Inborr J, Bedford MR, Patience JF, Classen HL, Verstegen MWA (1991, September) The influence of supplementary feed enzymes on nutrients disappearance and digesta characteristics in the GI-tract of early weaned pigs. In: Digestive Physiology in Pigs. Proceedings of the 5th International Symposium on Digestive Physiology in Pigs. Wageningen (Doorwerth), The Netherlands, pp 405–410

Jadhav S, Gaonkar T, Joshi M, Rathi A (2023) Modulation of digestibility of canine food using enzyme supplement: an in vitro simulated semi-dynamic digestion study. Front Vet Sci 10:1220198

Jiang Z, Mei L, Li Y, Guo Y, Yang B, Huang Z, Li Y (2024) Enzymatic regulation of the gut microbiota: mechanisms and implications for host health. Biomolecules 14(12):1638

Kiarie E, Romero LF, Nyachoti CM (2013) The role of added feed enzymes in promoting gut health in swine and poultry. Nutr Res Rev 26(1):71–88

Lemos D, Tacon AG (2017) Use of phytases in fish and shrimp feeds: a review. Rev Aquac 9(3):266–282

Liang Q, Yuan M, Xu L, Lio E, Zhang F, Mou H, Secundo F (2022) Application of enzymes as a feed additive in aquaculture. Mar Life Sci Tech 4(2):208–221

LP Information Inc (2022) Global animal feed enzymes market growth 2022–2028. Market Research.com. https://www.marketresearch.com/LP-Information-Inc-v4134/Global-Animal-Feed-EnzymesGrowth-30499852/. Accessed Jan 2022

Menezes-Blackburn D, Greiner R (2015) Enzymes used in animal feed: leading technologies and forthcoming developments. In: Functional polymers in food science: from technology to biology, pp 47–73

Menezes-Blackburn D, Jorquera MA, Greiner R, Gianfreda L, de la Luz Mora M (2013) Phytases and phytase-labile organic phosphorus in manures and soils. Crit Rev Environ Sci Technol 43(9):916–954

Pariza MW, Cook M (2010) Determining the safety of enzymes used in animal feed. Regul Toxicol Pharmacol 56(3):332–342

Raboy V (2009) Approaches and challenges to engineering seed phytate and total phosphorus. Plant Sci 177(4):281–296

Ravindran V (2013) Feed enzymes: The science, practice, and metabolic realities. J Appl Poult Res 22(3):628–636

Reddy NR, Sathe SK, Salunkhe DK (1982) Phytates in legumes and cereals. Adv Food Res 28:1–92

Thallinger B, Prasetyo EN, Nyanhongo GS, Guebitz GM (2013) Antimicrobial enzymes: an emerging strategy to fight microbes and microbial biofilms. Biotechnol J 8(1):97–109

Trejo-López T, Zepeda-Bastida A, Franco-Fernández J, Soto-Simental S, Ojeda-Ramírez D, Ayala-Martínez M (2017) Uso de extracto enzimático de Pleurotus ostreatus sobre los parámetros productivos de cabras. Abanico Vet 7(2):14–21

Valente Junior DT, Genova JL, Kim SW, Saraiva A, Rocha GC (2024) Carbohydrases and phytase in poultry and pig nutrition: a review beyond the nutrients and energy matrix. Animals 14(2):226

Velázquez-De Lucio BS, Hernández-Domínguez EM, Villa-Garcia M, Diaz-Godinez G, Mandujano-Gonzalez V, Mendoza-Mendoza B, Álvarez-Cervantes J (2021) Exogenous enzymes as zootechnical additives in animal feed: a review. Catalysts 11(7):851

Yu B, Wu ST, Liu CC, Gauthier R, Chiou PW (2007) Effects of enzyme inclusion in a maize–soybean diet on broiler performance. Anim Feed Sci Technol 134(3–4):283–294

Implementation, Challenges and Future Prospects

Formulation and Delivery Strategies for Alternatives to Antibiotics in Animal Health: Integration, Efficacy, and Feasibility

Saima Naz, Ahmad Manan Mustafa Chatha, Sobia Malik, Bushra Irfan, Urwah Ishaque, and Qudrat Ullah

Abstract The increasing threat of antibiotic resistance in animal health highlights the urgent need for alternative therapeutic strategies. This review explores known non-antibiotic treatments, including probiotics, prebiotics, immune modulators, antimicrobial peptides, and bacteriophage therapy, emphasizing their safety, efficacy, and cost-effectiveness in veterinary practice. Advanced delivery systems such as nanoparticles, encapsulation technologies, and controlled-release formulations are examined for their potential to enhance stability, bioavailability, and target action. The feasibility of adopting these approaches across diverse animal production systems is discussed with attention to regulatory, economic, and logistical challenges. While research outcomes are promising, further studies are needed to evaluate long- term impacts on animal microbiomes, resistance development, and environmental stability. Collaborative efforts across disciplines will be critical in making these options, such as innovative therapies, into practical, sustainable solutions for enhancing animal health and competing antibiotic resistance.

Keywords Antibiotic alternatives · Animal health · Controlled formulations · Drug delivery systems · Sustainability · Resistance mitigation · Veterinary medicine

S. Naz (✉) · S. Malik · B. Irfan · U. Ishaque
Department of Zoology, Government Sadiq College Women University, Bahawalpur, Punjab, Pakistan
e-mail: saima.naz@gscwu.edu.pk

A. M. M. Chatha
Department of Entomology, Faculty of Agriculture and Environment, The Islamia University of Bahawalpur, Bahawalpur, Punjab, Pakistan

Q. Ullah
Department of Theriogenology, Faculty of Veterinary and Animal Sciences, Cholistan University of Veterinary and Animal Sciences, Bahawalpur, Punjab, Pakistan

R. U. Khan, S. Naz (eds.), *Green Alternatives to Antibiotics in Animal Health*,
https://doi.org/10.1007/978-981-92-0309-3_21

1 Introduction

Over the past century, antibiotics have played a transformative role in both human and animal health. Since resistant bacteria are becoming more and more common, the World Health Organization has recently called attention to the growing issue of antibiotic resistance worldwide. Because of this concerning increase in resistance, standard antibiotics are no longer as effective, and either new medications must be discovered or innovative delivery methods are needed to improve the efficacy of current therapies. The process of developing new antibiotics is a good approach, but it has many limitations, such as high costs, long development periods, and the quick emergence of resistance. It takes more than 10 years and more than $500 million to develop an effective antibiotic, but resistance arises in a few years. There is still without public and private investment in the development of antibiotics (Bethke et al. 2020). A number of alternatives for antibiotics have drawn attention from researchers, including phages, probiotics, antimicrobial peptides, and antibacterial vaccinations. Stability, effectiveness, and application problems are common with these alternatives despite their potential. For example, vaccinations have a narrow range of disease-preventive potential, while antimicrobial peptides lead to low stability in vivo. Phage therapy has limitations because phages are viruses by nature, despite their therapeutic ability (Hathaway et al. 2017).

Antimicrobial nanomaterials and nano-carriers are recent approaches to these problems that have been made possible by novel developments in material science, nanotechnology, and biotechnology. Enhancing current antibiotics with nanotechnology is more feasible and economical than creating new antimicrobial medications. Drug delivery systems (DDS), which first appeared in the twentieth century, encapsulate medications in carriers to enhance controlled release, bioavailability, and targeting. These technologies reduce toxicity and adverse effects, which have major benefits for veterinary medicine (Patra et al. 2018). A special advantage of using DDS in veterinary medicine is the low treatment costs, handling requirements, and stress responses in livestock. DDS makes administering medications to animals easier and increases treatment. DDS has higher early development expenses (Song et al. 2023; Zamboni et al. 2018). For veterinary medicine, different nano-carriers have been observed, including micelles, liposomes, polymeric nanoparticles (NPs), dendrimers, metallic NPs, mesoporous silica NPs, and nano-emulsions. By encapsulating medications, these carriers can enhance the efficacy of pharmacokinetics, target-specific release, and solubility while lowering resistance and side effects. The stability of polymeric nano-carriers in the gastrointestinal tract and their modifiable drug release profiles have received special attention in this research (O'Konek et al. 2020). To address these issues, efforts are also in progress related to nano-medicine in animals that provide food. To analyze the safety of nano- medicines used in animals that are meant to be part of the food chain, for instance, the U.S. FDA has set up programs (Lu et al. 2023; Han et al. 2021). This review categorizes current research on nano-carriers, highlights the role of polymeric drug delivery vehicles in improving antimicrobial efficacy, and discusses challenges in translating these technologies into clinical practices. Additionally, the review addresses the safety concerns, particularly in food-producing animals, where nano-medicines may have implications for human health. By advancing our understanding of these technologies, this work aims to support the development of effective strategies to combat drug-resistant bacterial infection in animal health.

2 Rationale of the Research

New approaches are needed to enhance antimicrobial efficacy in veterinary medicine due to the increasing issue of antibiotic resistance. The rapid development of resistant microorganisms is making existing antibiotics less effective, and the process of creating new antibiotics is time-consuming, expensive, and quickly surpassed by resistance mechanisms. Particularly in veterinary medicine, the demands of intensive livestock production and companion animal care are difficult for current treatments to meet (Rahman et al. 2022). The escalating threat of antibiotic resistance poses a critical challenge to both human and veterinary medicine, jeopardizing the effective treatment of bacterial infection. Antibiotic resistance arises from both intrinsic and acquired mechanisms, with gram-negative pathogens exhibiting particularly robust defensive mechanisms (Yang et al. 2024). Polymer-based nano-carriers and drug delivery systems (DDS) appear to be a potential alternative allowing for controlled release, targeted delivery, and increased antimicrobial agent stability. This study explores veterinary medicine by studying the potential of polymeric nano-platforms, offering a sustainable method to battle drug-resistant diseases, and filling gaps in current antimicrobial therapy (AL-Azzawi et al. 2024).

3 Methodologies for Effective Administration

3.1 Dosing and Frequency of Administration

Frequency and dosing of antimicrobial medicines are important for achieving optimal treatment effects while reducing resistance formation. Correct dosage ensures successful bacterial elimination, but sub-therapeutic dosages may promote resistance. Individualized dosing is one strategy that considers parameters such as the animal's age, weight, and health status, as well as the drug's pharmacodynamics and pharmacokinetics. Long-acting formulations, such as polymeric nanoparticles (PNPs), can enable controlled release to lower the frequency of administration and maintain therapeutic levels for extended periods of time (Caneschi et al. 2023). For example, one promising alternative is the use of bacteriophages. Bacteriophages, which are viruses that target specifically and destroy bacteria, can be modified according to the type of bacterial infection. The type of infection, the severity of the illness, and the particular bacteriophage used all affect the dosage and frequency of bacteriophage delivery. Depending on the treatment plan and the animal's reaction to therapy, bacteriophages can be given orally or by injection at intervals of one dosage to several doses (Callaway et al. 2021).

3.2 Mode of Delivery

Delivery modes are crucial to establishing the efficacy and ensuring animal compliance. Advanced delivery technologies, such as nano-carriers, have been developed to enhance bioavailability, antimicrobial stability, and targeted release (Callaway et al. 2021). Oral delivery mode is always a first choice for cattle due to its ease of administration with encapsulated systems such as cyclodextrins or liposomal carriers used to improve absorption (Seal et al. 2013). Injectable formulations are beneficial for regulated release and quick onset, making them appropriate for systemic infections (Ogunshe et al. 2019). Treatments such as patches and hydrogels allow long-term medication release at the infection site. Inhalation therapies like micro robotic carriers or aerosolized nano-carriers provide lung- specific delivery and are useful for treating respiratory infections. These various delivery systems improve therapeutic effects while reducing resistance risks (Callaway et al. 2021).

3.3 Timing of Administration

Studies have indicated that the timing of the administration of different medications can have significant effects on their safety and effectiveness. Melatonin, for example, has been shown to exhibit phase-shifting effects that are extremely dependent on when it is taken. Lewy et al. showed that the effect of melatonin on circadian rhythms changes depending on when it is administered (Lewy et al. 2009). Timing is essential for aligning medicine delivery and bacterial replication cycles. Precautionary medication is indicated to avoid infections in high-risk situations, although early treatment can improve results in diagnosed cases (Wilkinson et al. 2018). Advances in real-time diagnostics may improve timing accuracy by detecting illnesses in their early stages (Narmani et al. 2023).

3.4 Monitoring and Evaluation

Monitoring emphasizes the process of implementation, ensuring that tasks are completed according to schedule and that resources are used effectively. Evaluation provides insightful information that can guide decisions and enhance future tasks. Evaluations, for instance, have been used to assess how well project activities have achieved anticipated goals and the degree to which program modifications may be linked to the program in the assessment of tech initialization projects (Leeuw & Donaldson 2015). Both are necessary to ensure treatment and identify new resistance trends (Teale and Borriello 2021). A vital method for optimizing results is therapeutic drug monitoring (TDM), which allows dose modifications based on the medication's levels in the animal's system (Fig. 1, Ates et al. 2020). Discovering resistance genes using molecular diagnostics is essential for developing specialized treatment plans that address resistance. So assessments provide valuable information that informs the development of next-generation treatments, enhances dosage schedules, and ensures long-term and efficient disease control (Neculai-Valeanu et al. 2024).

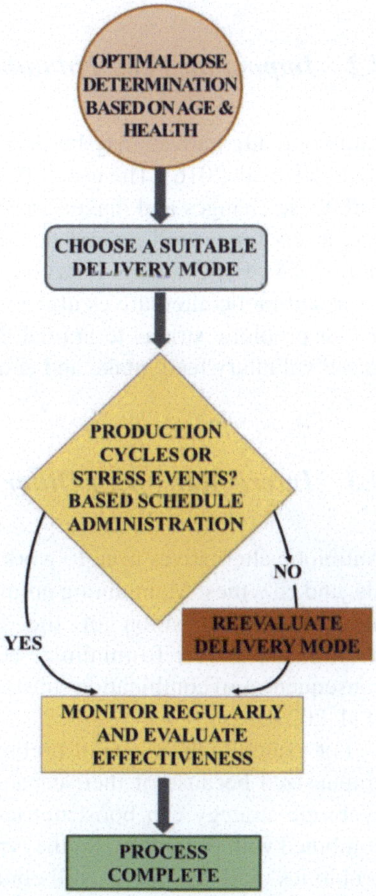

Fig. 1 A flow diagram illustrating the steps involved in effectively administering alternatives

4 Compatibility with Existing Feed Systems

4.1 Integration into Conventional Feed

Integration of antibiotic alternatives into traditional feed systems is possible because they adapt smoothly to different feed forms, including pellets, liquids, and meals (Akintan et al. 2024). These alternatives should be incorporated without interfering with the pelleting and extrusion steps of feed production (Pandey et al. 2019). More effective methods that meet sustainability requirements have resulted from the incorporation of machine learning and computational models in feed design. Through the integration of several data sources, including animal performance indicators, environmental data, and nutritional databases, modern feed formulation has developed into an intricate, data-centric process. Innovative techniques like machine learning and optimization algorithms have produced extremely precise feed formulas suited to particular animal requirements (Berger and Loewy 2024).

4.2 Impact on Feed Palatability and Consumption

Antibiotic alternatives may be impacted by changes to the feed's texture or smell (Jedrejek et al. 2016). The use of flavor and palatability boosters is two ways to deal with these changes and make sure feed stays enticing. For ongoing feed consumption levels for overall animal health, excellent palatability must be ensured (Han et al. 2024). For example, it has been demonstrated that the use of specific probiotics as antibiotic alternatives affects animal feed intake. In some cases, adding particular probiotic strains to animal diets has enhanced feed palatability, which has raised voluntary feed intake and enhanced growth performance (Seal et al. 2013).

4.3 Interactions with Other Feed Additives

Antibiotic alternatives need to work with the current feed, such as vitamins, minerals, and enzymes. Maintaining animal health and production requires balancing the nutritional composition of the feed while implementing these alternatives (Hamasalim 2016). To minimize adverse interactions and optimize the beneficial consequences of combination substances, careful formulation is required (Shahbakht et al. 2024).

For example, the usage of prebiotics such as non-digestible oligosaccharides is emphasized because of their capacity to positively alter the gut microbiota. This synbiotic strategy can boost nutrient absorption and gut health in animals when combined with probiotics. So the type of prebiotics utilized, the particular strains of probiotics used, and the overall composition of the diet can all affect how effective

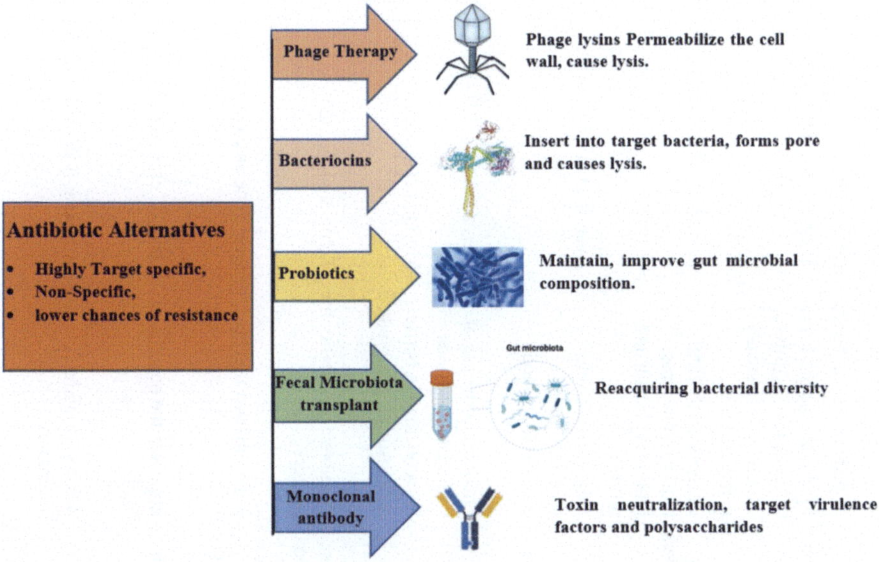

Fig. 2 Different alternatives to antibiotics can be used

such combinations are Fig. 2. To maximize animal performance and health, it is essential that all of these factors be taken seriously into consideration (Markowiak & Śliżewska 2017).

4.4 *Cost-Effectiveness and Feasibility*

Economic feasibility evaluation of using alternatives is significant, balancing the expenses with the long-term benefits of minimizing antibiotics use (Stygar et al. 2022). While these alternatives may have higher production costs, they eventually become more feasible and affordable as they are scaled up. Practical factors like preserving feed quality and economy of scale for large-scale production are crucial for attaining sustained implementation (Rafiq et al. 2022). Bacteriophage therapy, for example, demands the isolation and characterization of particular phages for various bacterial illnesses, which can be expensive and time-consuming. Commercial production of chemicals from plants and antimicrobial peptides can experience difficulties with extraction procedures, yield, and stability (Alaoui Mdarhri et al. 2022). Because of their ability to produce vital biomolecules like vitamins, enzymes, amino acids, and antibiotic compounds, several strains of the *Bacillus subtilis* group are among the most commonly used probiotics in industrial biotechnology (Table 1) (Harwood et al. 2018). Furthermore, a variety of secondary metabolites, such as siderophores, polyketides, terpenes, and different peptides produced via ribosomal and non-ribosomal pathways, are biosynthesized by these bacteria (Caulier et al. 2019).

Table 1 Compatibility of alternatives with existing feed systems: aspect considerations

No.	Aspect	Considerations	Technical challenges	Potential solutions	Impact on animal health	References
1	Feed production	Stability during storage and processing	Degradation of alternatives during pelleting at high temperature s	Develop formulations that tolerate heat or use other further processing techniques	Maintains the medicinal value of feed additives, which leads to steady health benefits	Sang and Blecha (2014)
2	Animal references	Palatability having an impact on feed consumption	Tastes or smells that are unfamiliar cause to consume less	To increase acceptance, use natural taste enhancers or maskers	Ensures a sufficient intake of nutrients, promoting health and growth	Kurt et al. (2019)
3	Cost	Alternatives economic viability	Greater cost of production in comparison with standard antibiotics	Optimize expenses by refining the production process and looking into buying in bulk	Adopt sustainable practices, more accessible methods	Anadón et al. (2019)
4	Technological adoption	Adoption of novel technologies in farming	Resistance to change respective producers	Offer incentives and training courses	Encourages the use of sustainable methods	Mercado et al. (2024)
5	Policies concerning local regulation	Conformity to local feed regulations	Differences in regulations across countries	For localization, collaborate with regulatory experts	Prevents market barriers and ensuring compliance	Mithuna et al. (2024)
6	Universal availability	Access to alternatives in developing countries	Limited supply in environments with few resources	Establish industrial hubs that are limited to a region	Promote accessibility for farmers	Tacon et al. (2022)
7	Consumer priorities	Public approval of antibiotic alternatives	Insecurity toward new technologies	Provide details to consumers about benefits and safety	Refine market uptake	Giacalone and Jaeger (2023)
8	Feed additive interaction	Suitability with the available vitamins and minerals	Negative interactions minimizing efficacy	Test validation under lab conditions	Improves the benefits of feed	Dawood et al. (2018)
9	Production scale	Reliability of large-scale manufacturing	High costs for small- scale production	Collaborate with large manufacturers for economies	Allows for wide acceptance	Laxminarayan et al. (2015)

10	Requirements specific to species	Adjusting possibilities for different animals	Generalized alternatives may not be beneficial for some species	Produce species-specific feed formulation s	Enhances efficiency for target species	Stefańska et al. (2021)
11	Animal welfare	Ensuring integration of alternatives	Observe reactions from untested additives	Investigate behavioral studies before large-scale implementation	Maintains animal welfare standards	Karavolias et al. (2018)
12	Microbial resistance	Reducing antibiotic-resistant bacteria (ARB)	Limited efficiency in reducing ARB	Combine probiotics & prebiotics to ensure antimicrobial effects	Reduces ARB in livestock systems	Ouwehand et al. (2016)
13	Waste management	Effects of animal waste on the environment	Chronic residues in manure	Use alternatives that degrade quickly in the environment	Reduces environmental antibiotic residues	Van Epps and Blaney (2016)
14	Heat stability	Tolerance to high temperatures during processing	Alternatives losing effectiveness under heat	Use encapsulation technologies for heat protection	Sustains active compounds for health benefits	Kolar et al. (2020)
15	Animal health and productivity	Effectiveness in promoting growth and preventing disease	Different responses to different species or breeds	Conduct species- specific trials to new applications	Enhances overall health and productivity, reducing reliance on antibiotics	Jouany and Morgavi (2007)
16	Environmental impact	Effects on waste products & emissions	Potential for increased nitrogen or phosphorus excretion	Select alternatives that improve nutrient utilization efficiency	Decrease environmental pollution from animal agriculture	Low et al. (2021)
17	Nutritional interactions	Compatibility with existing feed components	Potential interactions reducing nutrient availability	Conduct thorough compatibility testing and adjust formulations accordingly	Balanced nutrition, preventing deficiencies	Cheng et al. (2014)

(continued)

Table 1 (continued)

No.	Aspect	Considerations	Technical challenges	Potential solutions	Impact on animal health	References
18	Storage and packaging	Protection against moisture, heat, and contamination	Alternatives losing potency during prolonged storage	Use vacuum-sealed or moisture-resistant packaging	Ensures long- term stability and efficacy	Majid et al. (2018)
19	Supply chain efficiency	Streamlined delivery to farms	Distribution delays affecting feed quality	Establish local distribution hubs to reduce transit times	Improves consistency and availability of alternatives	Founou et al. (2016)
20	Gut microbiota	Supporting a healthy microbiome	Alternatives disrupting microbial balance	Use symbiotic formulation s (probiotics + prebiotics)	Improves gut health and immunity	Thomas et al. (2017)
21	Integration into conventional feed	Feed forms like pellets, liquids, meals	Ensuring a consistent product quality during pelleting, extrusion, etc	Optimization of processing parameters	Potential improvements in digestibility and nutrient absorption	Abdollahi et al. (2013)
22	Mycotoxins	Cereal grains, dried distillers' grains, wheat bran, and corn gluten meal	Dependent on environmental factors (humidity, temperature· O_2 and CO_2)	pH, grain moisture, and water activity play a role in mold growth and mycotoxin production	HACCP food safety aims to control mycotoxin contamination	Gruber-Dorninger et al. (2019)
23	Bacillus subtilis	Lack of pathogenicity, the capacity to secrete proteins, and strong fermentation foundation	Strong resistance to mechanical extrusion, radiation, desiccation, high temperatures, and acids and alkali	Can survive low-pH environments	Good vaccine carrier because of low requirements for nutrition, and large- scale fermentation	Lee et al. (2019)
24	Flexibility among animals	Feasibility of use across different animals	Species- specific differences in metabolism	Conduct species- specific trials before implementation	Expands utility of alternatives across livestock	Kiarie et al. (2013)
25	Bioavailability	Make sure nutrients are effectively absorbed by animals	Limited absorption of certain additives	Usage of bio enhancers to enhance uptake	Increase health benefits of feed alternatives	Sadgrove and Jones (2019)
26	Interaction with medications	Compatibility with other veterinary drugs	Risk of reduced efficacy and adverse interactions	Conduct compatibility tests during development	Focus on animal health is not compromised	Boonyarattanasoonthorn et al. (2021)

27	Suitability in severe circumstance s	Feed appealing in both hot and cold regions	Additives losing flavor or texture in a harsh environments	Develop heat- or cold-resistant formulations	Ensures consistent consumption regardless of weather	Coussement (1999)
28	Age-specific requirements of animals	Different nutritional needs for young and adult animals	Lack of advanced formulations for all life stages	Create life-stage-specific additive formulation s	Maintain growth and health outcomes	Pluske et al. (2018)
29	Compatibility of digestion	Suitability of alternatives for animal digestion	Chances gastrointestinal side effects	Initial screening additives for compatibility	Increase nutrient absorption and gut health	Rinttilä and Apajalahti (2013)
30	Regulatory compliance	Following to food safety and environmental standards	Regional variations in regulations	Standardize compliance checks and certifications	Enhance trust and safety in feed systems	Li and Liu (2023)

5 Limitations of the Study

The study offers insightful information on possible antibiotic substitutes in animal husbandry, but it has a number of drawbacks. It mostly relies on existing literature without presenting new experimental data, which may limit the ability to restrict reliable conclusions (Stanton 2013; Cheng et al. 2014). The effectiveness of alternatives like probiotics and botanical extracts varies depending on aspects such as species and environmental conditions, but the study does not provide precise solutions to manage this variability (Arsene et al. 2021). Furthermore, while alternatives try to prevent antibiotic resistance, such as bacteriophages, may still contribute to resistance development, but this is not studied in depth (Ghosh et al. 2019). This study does not fully evaluate the costs and viability of large-scale execution, economic feasibility is another issue. Overall, even though the study identifies possible substitutes for antibiotics, these deficiencies require additional analysis and comprehensive investigation (Ghosh et al. 2019).

6 Conclusion

Antibiotic resistance in animal health needs alternative therapeutic strategies. It emphasizes the need for effective alternatives due to concerns over antibiotic resistance stemming from the overuse of antibiotics in animal agriculture. This study discusses various alternatives, including probiotics, prebiotics, bacteriophages, and plant extracts, highlighting their potential benefits and limitations. It concludes that further research is necessary to fully understand their efficacy, mechanisms of action, and practical applications in animal husbandry. The importance of developing and implementing effective antibiotic alternatives to ensure sustainable livestock production and mitigate the risk of antibiotic resistance.

References

Akintan O, Gebremedhin KG, Uyeh DD (2024) Animal feed formulation—connecting technologies to build a resilient and sustainable system. Animals 14(10):1497

Alaoui Mdarhri H et al (2022) Alternatives therapeutic approaches to conventional antibiotics: advantages, limitations and potential application in medicine. Antibiotics 11(12):1826

AL-Azzawi MK, Hasan NA, Barrak MM (2024) A review of the development of an understanding of antibiotic interactions, from mechanisms of action to novel resistance and the search for natural alternatives. J Med Genet Clin Biol 1(6):78–102

Anadón A, Ares I, Martínez-Larrañaga MR, Martínez MA (2019) Nutraceuticals used as antibacterial alternatives in animal health and disease. In: Gupta R, Srivastava A, Lall R (eds) Nutraceuticals in veterinary medicine. Springer, Cham, pp 315–343

Abdollahi MR, Ravindran V, Svihus B (2013) Pelleting of broiler diets: An overview with emphasis on pellet quality and nutritional value. Anim Feed Sci Technol 179(1-4):1–23. https://doi.org/10.1016/j.anifeedsci.2012.10.011

Arsene MMJ et al (2021) Short review on the potential alternatives to antibiotics in the era of antibiotic resistance. J Appl Pharm Sci 12(1):029–040

Ates HC, Roberts JA, Lipman J, Cass AE, Urban GA, Dincer C (2020) On-site therapeutic drug monitoring. Trends Biotechnol 38(11):1262–1277

Boonyarattanasoonthorn T, Khemawoot P, Kijtawornrat A (2021) comparing potential drug–drug interactions in companion animal medications using two electronic databases. Vety Sci 8(4):60. https://doi.org/10.3390/vetsci8040060

Berger I, Loewy ZG (2024) Antimicrobial resistance and novel alternative approaches to conventional antibiotics. Bacteria 3(3):171–182

Bethke JH et al (2020) Environmental and genetic determinants of plasmid mobility in pathogenic Escherichia coli. Sci Adv 6(4):eaax3173

Coussement PA (1999) Inulin and Oligofructose: Safe Intakes and Legal Status. The J Nutr 129(7):1412S–1417S. https://doi.org/10.1093/jn/129.7.1412S

Callaway T, Lillehoj H, Chuanchuen R, Gay C (2021) Alternatives to antibiotics: a symposium on the challenges and solutions for animal health and production. Antibiotics 10(5):471

Caneschi A, Bardhi A, Barbarossa A, Zaghini A (2023) The use of antibiotics and antimicrobial resistance in veterinary medicine, a complex phenomenon: a narrative review. Antibiotics 12(3):487

Caulier S, Nannan C, Gillis A, Licciardi F, Bragard C, Mahillon J (2019) Overview of the antimicrobial compounds produced by members of the bacillus subtilis group. Front Microbiol 10. https://doi.org/10.3389/fmicb.2019.00302

Cheng G et al (2014) Antibiotic alternatives: the substitution of antibiotics in animal husbandry? Front Microbiol 5:217

Dawood MA, Koshio S, Esteban MÁ (2018) Beneficial roles of feed additives as immunostimulants in aquaculture: a review. Rev Aquac 10(4):950–974

Founou LL, Founou RC, Essack SY (2016) Antibiotic resistance in the food chain: a developing country-perspective. Front Microbiol 7:1881

Gruber-Dorninger C, Jenkins T, Schatzmayr G (2019) Global mycotoxin occurrence in feed: A ten-year survey toxins 11(7):375. https://doi.org/10.3390/toxins11070375

Ghosh C, Sarkar P, Issa R, Haldar J (2019) Alternatives to conventional antibiotics in the era of antimicrobial resistance. Trends Microbiol 27(4):323–338

Giacalone D, Jaeger SR (2023) Consumer acceptance of novel sustainable food technologies: a multi-country survey. J Cleaner Prod 408:137119

Harwood CR, Mouillon JM, Pohl S, Arnau J (2018) Secondary metabolite production and the safety of industrially important members of the Bacillus subtilis group. FEMS Microbiol Rev 42(6):721–738. https://doi.org/10.1093/femsre/fuy028

Hamasalim HJ (2016) Synbiotic as feed additives relating to animal health and performance. Adv Microbiol 6(4):288–302

Han H-W et al (2021) Selenium nanoparticles as candidates for antibacterial substitutes and supplements against multidrug-resistant bacteria. Biomolecules 11(7):1028

Han D et al (2024) Application and substitution of antibiotics in animal feeding. Medycyna Weterynarjyna-Vet Med Sci Pract 80(1):5–11

Hathaway H et al (2017) Thermally triggered release of the bacteriophage endolysin CHAPK and the bacteriocin lysostaphin for the control of Methicillin Resistant Staphylococcus Aureus (MRSA). J Control Release 245:108–115

Jedrejek D, Lević J, Wallace J, Oleszek W (2016) Animal by-products for feed: characteristics, European regulatory framework, and potential impacts on human and animal health and the environment. J Anim Feed Sci 25(3):189–202

Jouany J-P, Morgavi D (2007) Use of ‘natural’products as alternatives to antibiotic feed additives in ruminant production. Animal 1(10):1443–1466

Kiarie E, Romero LF, Nyachoti CM (2013) The role of added feed enzymes in promoting gut health in swine and poultry. Nutr Res Rev 26(1):71–88. https://doi.org/10.1017/S0954422413000048

Karavolias J, Salois MJ, Baker KT, Watkins K (2018) Raised without antibiotics: impact on animal welfare and implications for food policy. Transl Anim Sci 2(4):337–348

Kolar S, Jurić S, Vlahoviček-Kahlina K, Vinceković M (2020) Potential for application of feed additives encapsulation technology in animal nutrition. Krmiva: Časopis o hranidbi životinja, proizvodnji i tehnologiji krme 62(1):57–66

Kurt T et al (2019) Strategic priorities for research on antibiotic alternatives in animal agriculture—results from an expert workshop. Front Vet Sci 6:429

Lewy S, Alfred J, Ahmed Jeanne M. Latham, Jackson Robert L, Sack (2009) melatonin shifts human orcadian rhythms According to a phase-response curve chronobiology international 9(5) 380–392. https://doi.org/10.3109/07420529209064550

Lee NK, Kim WS, Paik HD (2019) Bacillus strains as human probiotics: characterization safety microbiome and probiotic carrier. Food Sci Biotechnol 28(5):1297–1305. https://doi.org/10.1007/s10068-019-00691-9

Leeuw SI, Donaldson FL (2015) Theory in evaluation: Reducing confusion and encouraging debate Evaluation 21(4):467–480. https://doi.org/10.1177/1356389015607712

Laxminarayan R, Van Boeckel T, Teillant A (2015) The economic costs of withdrawing antimicrobial growth promoters from the livestock sector. OECD Food, Agriculture and Fisheries Papers, OECD Publishing, Paris, No. 78

Low CX et al (2021) Unveiling the impact of antibiotics and alternative methods for animal husbandry: a review. Antibiotics 10(5):578

Lu Y et al (2023) Intranasal administration of edaravone nanoparticles improves its stability and brain bioavailability. J Control Release 359:257–267

Li Q, Liu Q (2023) Impact of digital financial inclusion on residents' income and income structure sustainability 15(3):2196. https://doi.org/10.3390/su15032196

Majid I, Nayik GA, Dar SM, Nanda V (2018) Novel food packaging technologies: innovations and future prospective. J Saudi Soc Agric Sci 17(4):454–462

Mercado SM, Querijero N, Lapitan AV (2024) Contingent effectiveness in technology transfer of two animal feed innovations in a Philippine State University's Public Research Institute. J Int Soc SE Asia Agric Sci 30(2):195–209

Mithuna R et al (2024) Emergence of antibiotic resistance due to the excessive use of antibiotics in medicines and feed additives: a global scenario with emphasis on the Indian perspective. Emerging Contam 10(4):100389

Markowiak P, Śliżewska K, (2017) Effects of probiotics prebiotics and synbiotics on human health nutrients 9(9):102. https://doi.org/10.3390/nu9091021

Narmani A et al (2023) Biomedical applications of PLGA nanoparticles in nanomedicine: advances in drug delivery systems and cancer therapy. Expert Opin Drug Deliv 20(7):937–954

Neculai-Valeanu A-S, Ariton A-M, Radu C, Porosnicu I, Sandulcanu C, Amariții G (2024) From herd health to public health: digital tools for combating antibiotic resistance in dairy farms. Antibiotics 13(7):634

Ogunshe A, Adeola A, Adedokun R, Jagun-Jubril A (2019) Indigenous and cross- border eco-health implications of unethical injectable antimicrobial drugs administration under bovine livestock farming and commercial conditions. J Zoonotic Dis Public Health 3(1):3

O'Konek JJ et al (2020) Intranasal nanoemulsion vaccine confers long-lasting immunomodulation and sustained unresponsiveness in a murine model of milk allergy. Allergy 75(4):872–881

Ouwehand AC, Forssten S, Hibberd AA, Lyra A, Stahl B (2016) Probiotic approach to prevent antibiotic resistance. Ann Med 48(4):246–255

Pluske JR, Turpin DL, Kim JC (2018) Gastrointestinal tract (gut) health in the young pig. Anim Nutr 4(2):187–196. https://doi.org/10.1016/j.aninu.2017.12.004

Pandey AK, Kumar P, Saxena M (2019) Feed additives in animal health. In: Nutraceuticals in veterinary medicine. Springer, pp 345–362

Patra JK et al (2018) Nano based drug delivery systems: recent developments and future prospects. J Nanobiotechnol 16:1–33
Rafiq K et al (2022) Role of different growth enhancers as alternative to in-feed antibiotics in poultry industry. Front Vet Sci 8:794588
Rahman MRT, Fliss I, Biron E (2022) Insights in the development and uses of alternatives to antibiotic growth promoters in poultry and swine production. Antibiotics 11(6):766
Rinttilä T, Apajalahti J (2013) Intestinal microbiota and metabolites—Implications for broiler chicken health and performance. J Appl Poult Res 22(3):647–658. https://doi.org/10.3382/japr.2013-00742
Sang Y, Blecha F (2014) Alternatives to antibiotics in animal agriculture: an ecoimmunological view. Pathogens 4(1):1–19
Seal BS, Lillehoj HS, Donovan DM, Gay CG (2013) Alternatives to antibiotics: a symposium on the challenges and solutions for animal production. Anim Health Res Rev 14(1):78–87
Shahbakht RM et al (2024) The use of phytogenic feed additives in animal nutrition Pakistan scenario. In: Complementary and alternative medicine: feed additives. Unique Scientific Publishers, pp 326–334
Song Y, Day CM, Afinjuomo F, Tan J-QE, Page SW, Garg S (2023) Advanced strategies of drug delivery via oral, topical, and parenteral administration routes: where do equine medications stand? Pharmaceutics 15(1):186
Stanton TB (2013) A call for antibiotic alternatives research. Trends Microbiol 21(3):111–113
Stefańska I, Kwiecień E, Jóźwiak-Piasecka K, Garbowska M, Binek M, Rzewuska M (2021) Antimicrobial susceptibility of lactic acid bacteria strains of potential use as feed additives-the basic safety and usefulness criterion. Front Vet Sci 8:687071
Sadgrove NJ, Jones GL (2019) From petri dish to patient: bioavailability estimation and mechanism of action for antimicrobial and immunomodulatory natural products. Front Microbiol https://doi.org/1010.3389/fmicb.2019.02470
Stygar AH et al (2022) Economic feasibility of interventions targeted at decreasing piglet perinatal and pre-weaning mortality across European countries. Porcine Health Manage 8(1):22
Tacon AG, Metian M, McNevin AA (2022) Future feeds: suggested guidelines for sustainable development. Rev Fish Sci Aquacult 30(2):135–142
Teale C, Borriello P (2021) A proposed scheme for the monitoring of antibiotic resistance in veterinary pathogens of food animals in the UK. Vet Rec 189(3):e201
Thomas M et al (2017) Metagenomic characterization of the effect of feed additives on the gut microbiome and antibiotic resistome of feedlot cattle. Sci Rep 7(1):12257
Van Epps A, Blaney L (2016) Antibiotic residues in animal waste: occurrence and degradation in conventional agricultural waste management practices. Curr Pollut Rep 2:135–155
Wilkinson A, Ebata A, MacGregor H (2018) Interventions to reduce antibiotic prescribing in LMICs: a scoping review of evidence from human and animal health systems. Antibiotics 8(1):2
Yang W, Li J, Yao Z, Li M (2024) A review on the alternatives to antibiotics and the treatment of antibiotic pollution: current development and future prospects. Sci Total Environ 926:171757
Zamboni WC, Szebeni J, Kozlov SV, Lucas AT, Piscitelli JA, Dobrovolskaia MA (2018) Animal models for analysis of immunological responses to nanomaterials: challenges and considerations. Adv Drug Deliv Rev 136:82–96

The Economic and Environmental Impact of Antibiotic Use in Livestock Farming

Muhammad Waqar, Iqra Rafiq, Qudrat Ullah, FNU Nahiduzzaman, Muhammad Waqas, Noor Syaheera Ibrahim, Djilali Benabdelmoumene, Said Dahmouni, Worawan Panpipat, and Manat Chaijan

Abstract The availability and distribution of antibiotics to livestock farming increases the economic value and production, but also causes the problem of AMR. Based on the knowledge of antibiotic use in animal agriculture, the economic impact, together with the environmental impact of AMR, restrictive measures, and chemical contamination, is discussed in this article. Hence, the long-term effects of antibiotics overpower the short-term economic benefits that are accrued from their usage. The study identifies options for aversion to antibiotics, such as probiotics, phytogenics, biosecurity, and precision livestock farming, are the probable strategies. Trends in regulation, as well as changes in consumer preferences for meat, add to the pressure for the industry to switch to non-antibiotic production systems. The results show that while the implementation of divergent and sustainable practices may be financially demanding initially, adaptive access to the market and possible

M. Waqar (✉) · Q. Ullah · W. Panpipat · M. Chaijan
Food Technology and Innovation Research Center of Excellence, School of Agricultural Technology and Food Industry, Walailak University, Tha Sala, Nakhon Si Thammarat, Thailand

I. Rafiq
Department of Microbiology, The Islamia University of Bahawalpur, Bahawalpur, Punjab, Pakistan, Bahawalpur, Pakistan

F. Nahiduzzaman
Department of Microbiology and Hygiene, Bangladesh Agricultural University, Mymensingh, Bangladesh

M. Waqas
Department of Animal Nutrition and Nutritional Diseases, Faculty of Veterinary Medicine, Ondokuz Mayıs University, Atakum, Samsun, Türkiye

N. S. Ibrahim
Faculty of Veterinary Medicine, Universiti Sultan Zainal Abidin, Besut Campus, Besut, Terengganu, Malaysia

D. Benabdelmoumene · S. Dahmouni
Laboratory of Applied Animal Physiology, University Abdelhamid Ibn Badis, Mostaganem, Algeria

R. U. Khan, S. Naz (eds.), *Green Alternatives to Antibiotics in Animal Health*,
https://doi.org/10.1007/978-981-92-0309-3_22

hesitation on the decrease of risks about the compound use of antibiotics outweigh the disadvantages. This work suggests the development of special preventive measures and further research on this phenomenon to create effective policies through the collaboration of the governments to transform the livestock farming industry with less impact on the health of the public, the environment included.

Keywords Green alternative · Sustainable livestock production · Food safety · Antibiotic impact · Precision livestock farming · Precision livestock farming · The economic effect of antibiotics

1 Introduction

The growing consumption of meat and other livestock products around the world has influenced growth and development in the management of animal health, with antibiotics as one of the chief weapons for combating diseases and a growth promoter (Callaway et al. 2021). The use of antibiotics has been greatly practiced in livestock farming, particularly in ensuring food security, increased production, and low rates of mortality. Nonetheless, their utilization has been anathema through excessive prescription and improper consumption, which has raised concerns about AMR, environmental pollution, and sustainability in livestock farming (Ahmed et al. 2024). It is a crucial time to stop relying on antibiotics and find other non-harmful methods of disease prevention and treatment that would not adversely affect the economy and the environment.

Antibiotics have been associated with the growth rate of feed efficiency in modern animal production because of their effectiveness in eradicating bacterial infection (Evangelista et al. 2022). Earlier, the use of antibiotic growth promoters was followed by important cuts in production costs due to a faster rate of weight addition and lesser intake of food (McEwen and Fedorka-Cray 2002). Antibiotic consumption has been most beneficial from the economic perspective in LMICs since veterinary services and biosecurity practices could be restricted (Lhermie et al. 2022). Aside from these economic aspects, antibiotics are vital in managing diseases in the herd so as not to bring about a high mortality rate in the animals that could be a threat to the food production chain. Some diseases like Bovine respiratory disease, mastitis in cattle, and Swine dysentery may cause major loss in productive output, additional costs in terms of veterinary and medical charges, and deterioration in the quality of food products (Neculai-Valeanu et al. 2024). As such, antibiotics have been extremely useful for providing steady and dependable meat, milk, and egg production to feed the world population.

However, the author criticized the use of antibiotics in enhancing livestock production because it has significantly raised the issue of antimicrobial resistance (AMR) within the shortest time possible. Approximately 1.2 million people died as a direct consequence of AMR in 2019, and the numbers suggest that annual fatalities will increase by up to ten million in the absence of measures to address the issue

(Ahmed et al. 2024). The financial implication of AMR is projected to cost the world economy $159 billion per year, and by 2050, it will exert significant pressure on the healthcare and agricultural industries (Tkaczyk et al. 2022). In addition, antibiotic residues in the soil, water, and animal products, which are used as food, are of great concern to the environment. Consequently, the unaltered regulation and elimination of antibiotics through feces and human sewage pose adverse impacts on the environment, such as soil pollution, water seepage, and diminution of the ecosystem, as well as disturbances in the microbes within the soil in question (Talukder et al. 2024). This has seen the detection of tetracycline, macrolide, and sulfonamide in the surface water and agricultural fields, affecting the microbial ecosystem and occasioning gene transfer among bacteria (Serwecińska 2020).

Due to the various economic, environmental, and public health impacts that come with the overuse of antibiotics, the livestock industry is raising new challenges that seek to depart from the traditional use of antibiotics by creating sustainable measures to achieve productivity. Governments and regulatory agencies around the world have placed severe restrictions on the use of AGPs; while the EU banned the products in 2006, the FDA approved the use of the antibiotics for food-producing animals but has imposed strict measures of using them for important medical purposes in humane and veterinary practices (Baynes et al. 2016). Such politics have stimulated a search for possible solutions that enhance animal well-being and production on farms in the same manner that does not jeopardize the future of farming.

Realizing the challenges of using antibiotics to enhance the production of livestock, scientists and actors in the livestock industry are looking for cheaper and environmentally friendly solutions. One of the promising interventions is the introduction of probiotics and prebiotics that help strengthen the gut and immunity, and thus people do not need to use antibiotics as often (Ullah et al. 2024). Lastly, true, the use of phytogenic feed additives such as essential oils and plant extracts containing antimicrobial properties may enhance physical digestion, the reduction of inflammation, and improve growth patterns in the animals (Zhang et al. 2024). Another approach involves precision livestock farming technologies, which can detect diseases on the affected animals only and control them, hence avoiding the spreading of antibiotics to every animal (Ahmed et al. 2024). Enhancing livestock immunity is another measure that also results in low disease incidences, hence reducing the use of antibiotics (Lhermie et al. 2022).

The financial impact of going back to raising animals without antibiotics is a very important factor to consider among the farmers and other stakeholders in the production of livestock animals (Singer et al. 2019). On the other hand, there are some short-term operational costs involved in implementing these strategies for managing disease, which include higher initial costs; however, longer-term benefits resulting from its positive impacts are lower disease treatment costs, lower risk of AMR, and increased competitiveness in the market (Mudenda et al. 2023). Many consumers in developed countries have developed a demand to buy animal products produced without the use of antibiotics; thus, the consumption of sustainable animal products requires a higher price for livestock products (Neculai-Valeanu et al. 2024). Therefore, integrating the antibiotic-free method of production provides farmers with better markets, improved returns, and higher consumer confidence (Haque et al. 2020). Nevertheless, there is still some incongruity posted: Smallholder farmers, especially, may experience major limitations, which may be in terms of finance plus access to technical expertise to afford and adopt these transformations.

The practices toward sustainable livestock production are a complex process that involves many sectors, inclusive of policymakers, researchers, veterinarians, and farmers (Yulianti et al. 2024). Governments must enact favorable laws that allow strict compliance with antibiotic-free measures and offer incentives for their adoption. Technological advancement and investments in research are crucial for the provision of sustainable and affordable technologies that sustain healthy livestock production and reduce effects on the environment (Agregán et al. 2021). Also, effective surveillance in checking on the regularity or increase in antibiotic usage in the food chain and consumer education on products that are free from antibiotics can help create demand for sustainable foods (Haque et al. 2020). To overcome these challenges, concerted efforts from all stakeholders will be useful in making the livestock industry shockproof and economically sustainable.

The authorities of the economic and environmental effects of the use of antibiotics in livestock are discussed in this chapter, including the advantages and disadvantages. It also speaks on the position of antibiotics in AMR and environmental degradation and contamination of foods. It also presents information about legal requirements and international standards concerning the minimization of antibiotic use.

2 The Rationale of the Study

Antibiotics have been widely used as feed in modern livestock farming and have significantly contributed to increasing productivity, controlling diseases, and enhancing production profits. Nevertheless, the increasing rates of AMR threaten the long-term advantages of using drugs and chemicals for disease control. Such bacterial strains are not only dangerous to animal health but also human health, putting in place the need for transitions to a sustainable method. Further, the economic implications of AMR on livestock and the ever-present problem of water and soil pollution from antibiotic residues further stress the need for the review of current management practices. To meet these challenges, a consideration of the costs and benefits of antibiotic use in the economies of the affected countries is essential. This paper aims to present a more subtle analysis of those consequences and investigate possible solutions that may be pro-business while at the same time being environmentally friendly. By combining scientific literature and stakeholder input, this chapter seeks to provide policymakers, farmers, and other relevant stakeholders with practical strategies for reduced antibiotic use and Mexico's movement to a post-antibiotic era without a considerable loss of profit.

3 Economic Implications of Antibiotics Use in the Livestock Industry

Antibiotic use in livestock farming has been widely applied in modern industrial and intensive farming systems as providing certain economic benefits, such as boosting productivity, lowering death rates, and increasing food production (Van et al. 2020). However, the rising problem of AMR, new policies and regulations, and changes in trends in consumers' behaviors are changing the utilization of antibiotics in livestock farming (Van Boeckel et al. 2017). The following section provides more information concerning the possible gains accrued from stock raising, its risks, and likely efficient financial returns that encourage more sustainable production.

3.1 Economic Benefits of Antibiotic Use in Livestock Production

Antibiotics have been applied in animal agriculture practices as feed additives to combat existing bacterial infections, besides acting as feed additives that enhance the growth of animals (Low et al. 2021). These applications have a severe effect on the level of economic efficiency of livestock production. It demonstrates that antibiotics play two roles in disease prevention and treatment, which result in increased productivity, reduced costs, and improvement of economic viability for farmers (Kolomytseva and Eremeenko 2024). They also help to balance some cycles of production to provide foods such as meat, milk, and eggs to consumers consistently.

AGPs make the animals grow to the correct market weight within the shortest time, and less feed is consumed, thus the monetary implications. Research shows that AGPs enhance feed conversion ratio, growth rates in weight, and feed costs, which the poultry and swine industry evaluates to about 5–15% (McEwen and Fedorka-Cray 2002). This impact is most keenly felt in operations in L&M I nations that rely on livestock farming as their biggest economic activity, as well as a source of food.

Moreover, antibiotics improve the salability of livestock products through the elimination of diseases in animals before selling products like meat, milk, and eggs. This makes it easier to access the foreign market, especially as most countries have very tight health standards for animals such as cattle that must be free from communicable diseases. For instance, young people choose to use antibiotics in order to prevent diseases such as mastitis and promote good milk production and quality, hence boosting exportation (Dankar et al. 2022).

3.2 Economic Risks and Challenges of Antibiotic Dependence

Although antibiotics have many advantages in livestock farming, this disadvantage brings about various risks that are dangerous to the sustainable economic status. Closely related to this is the issue of AMR or antimicrobial resistance, by which costs and overall efficacy of treatment have taken a hit, and there are corresponding economic implications here as well. The cost implication for AMR is high, and approximate figures estimated for the global costs of treating AMR ranged between $44 billion in the year 2015 and a projected $159 billion in the year 2050 (Adamie et al. 2023). In the same way, AMR is projected to cause ten million deaths annually and wipe off $100 trillion from the global economy by 2050, with major impacts on health, agriculture, and the economy (Bhatia 2018).

However, the economic impact of AMR extends beyond rising healthcare costs, potentially leading to long-term consequences across various sectors (Ait Ouakrim et al. 2020). Consequently, the productivity of the animal, mortality rate, and veterinary bills are some of the issues that livestock producers experience when antibiotics are rendered ineffective in treating bacterial infections (Bava et al. 2024). If the diseases start spreading, then the entire group of animals may be eliminated, thus incurring serious losses to the farmers, consequently straining the supply chain of the livestock business.

In addition, emerging customers' preference for antibiotic-free products is also evident, especially in HICs, because of growing awareness regarding antibiotic residues in foods and AMR threats. Therefore, large-scale food retailers and restaurants are moving to source antibiotic-free meat, which is likely to force farmers to change their ways (Patel et al. 2020).

This transition may be economically advantageous in the long run since it reduces the cost of drug inputs, but it entails major costs initially, such as biosecurity, improved animal management, and the costs of implementing disease control measures (Lyons et al. 2021). Failure to do so might affect the competitiveness of goods in the market, and the consumer's trust is affected.

3.3 Regulatory Changes and Market Implications

Authorities and the international community are increasingly implementing policies that control the usage of antibiotics in animal farming, which also changes economic characteristics (Lhermie et al. 2022). Most countries have developed policies within which subtherapeutic antibiotic use in feed is prohibited or restricted. This has made farmers search for alternatives. Antibiotic growth promoters were banned in the EU in 2006 and further tightened in the United States and other parts of the world (Andrew Selaledi et al. 2020). This results in further costs that need to be incurred in the other disease prevention strategies, like the use of probiotics, prebiotic food, and strict standards of hygiene and sanitation, among others. These changes create regulations that affect world trade and the openness of markets.

This is because some countries set severe policies for the use of antibiotics and may even be inclined to ban products from abroad that are obtained from animals through restricted policies. This leads to an economic imbalance where the farmers in the regions where antibiotics are administered in large quantities may not be able to market their products in the developed countries. Further, many costs are associated with products that have to meet the criteria of an Antibiotic Free Certification, such as the inspection by third parties and paperwork.

3.4 Cost–Benefit Analysis of Green Alternatives to Antibiotics

Due to the possible negative economic impacts of using antibiotics, there has been an increased focus on economically viable solutions that ensure the productivity of livestock without contributing to AMR. Several solutions are possible:

Probiotics and Prebiotics The bacteria that are considered to be harmless to the human body but are useful for a healthy gut and good immunity, and which minimize the use of antibiotics. Based on the literature, it was alleged that probiotics in feeding livestock enhance feed conversion and reduce disease incidences as much as AGPs, hence enhancing economic value (Zhang et al. 2024).

Organic Acid Substances used in reducing hazardous amines by interacting with ammonia; these antipyretic compounds exist in graded forms, including sodium diphenyl, propylthiouracil, pyridoxine, and trimethadione. From the research carried out, phytogenics have some positive effects on the digestion process and health of animals, resulting in a positive impact on the economy despite the use of antibiotics (Haque et al. 2024).

Relative to the Current System Optimized hygiene, vaccination measures, and proper management lessen the prevalence of diseases, which in turn reduces the use of antibiotics and yet does not affect the profitability of the farms (Lhermie et al. 2022).

Feeding and Nutrition Modifying the feed formulation to improve the immune strength and growth rate of animals, naturally, and reducing the incidence of use of antibiotics (Tona 2018). Comparing these options, it can be stated that although initial investments could be higher, the direct costs of treatment, as well as the risks that are associated with AMR, could be reduced in the long run, while the competitiveness of the market could be higher (Tarín-Pelló et al. 2022). The following are some examples that show increased profitability when farming businesses go for a system of production that is antibiotic-free. For example, free-range and antibiotic-free poultry and beef products can sell at higher prices of 15–35% in retail outlets due to organic prices, which acts as a motivation for sustainable practices (Van et al. 2020).

3.5 Long-Term Economic Sustainability and Future Outlook

The livelihood of livestock farming has changed during the last few years due to the changing focus of client requirements, considering the economic feasibility of the business without compromising on organic results (Fesseha and Aliye 2020). Antibiotics have a long history of increasing the efficiency of agriculture; however, long-term effects of AMR, new rules and regulations, and altering consumer demands call for alternatives (Callaway et al. 2021).

To continue and further develop, the livestock industry needs to implement science-based approaches to policy-making and invest in research to innovate disease control methods in policy-making, farmers, and other researchers (Serra et al. 2020). One positive strategy that governments can pursue to encourage farmers to go for antibiotic-free products is the provision of financial incentives and subsidies, which would help support farmers in their rugged journey. Another way in which governments can contribute to this noble cause is through offering technical assistance to farmers who wish to embrace antibiotic-free products. Bilateral cooperation between the private and public sectors is also important in promoting sustainable and financially sustainable methods in stock raising and production without compromising the environment and the health of human beings.

4 Impact of Antibiotics on Sustainability in Livestock Production

There is a high correlation between sustainable livestock production and the responsible use of antibiotics. On the positive side, antibiotics have been used to increase productivity, eradicate diseases, and ensure food security, but the negative impacts of these antibiotics on the environment and people's health, as shown in Fig. 1, cannot be ignored (Moghadam et al. 2023). The antibiotic administration for food-producing animals has led to AMR, pollution, and an imbalance of ecological niches (Jadeja and Worrich 2022). Therefore, establishing respective Positive or negative influences remain crucial when evaluating the significance of sustainability and searching for ways to minimize adverse effects without compromising the economic feasibility (Chahuán-Jiménez et al. 2023).

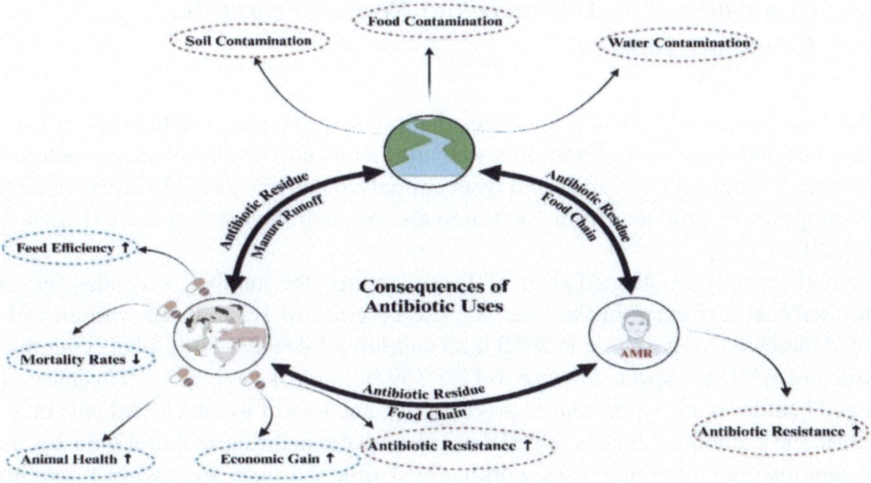

Fig. 1 Showing the effects of antibiotic use on livestock, the environment, and public health

4.1 Environmental Consequences of Antibiotic Use in Livestock Production

The continuous administration of sub-therapeutic doses of antibiotics in livestock farming practices has resulted in the release of antibiotic residues in the environment, such as in the soil and water systems (Talukder et al. 2024). When animals are given antibiotics, a considerable amount is eliminated via urine and feces and is incorporated into manure, which is used in the fields. This leads to pollution of the soil and groundwater with antibiotic residues, which results in a shift in the community of microorganisms and encouragement of antibiotic-resistant bacteria (Serwecińska 2020).

Many researchers have established that tetracyclines, macrolides, and sulfonamides have been analyzed in agricultural soils, river sediments, and even on surface water, hence posing a threat to the fertility of the soils and water bodies (Baynes et al. 2016). By having antibiotic residues on and in the plants, constant use depletes the fertility of the soil as bacteria used for nitrogen fixation, and decomposition of organic matter, among other services, are negatively affected. Besides, it has been evidenced that the presence of antibiotic residues in freshwater sources affects the decrease in the richness and the enhanced transfer of antibiotic-resistance genes among the microorganisms in water (Haque et al. 2024).

4.2 Contribution to Antimicrobial Resistance and Its Economic Burden

AMR remains among the several issues that relate to the use of antibiotics in livestock farming. Ingestion of antibiotics at sub-therapeutic levels enhances bacterial resistance, which can be transferred from animals to humans through direct contact, consumption of products from such animals, or environmental contact (Lhermie et al. 2022).

Another study by Ahmed et al. (2024) identified the number of deaths due to antimicrobial resistance in the year 2019 to be close to 1.2 million, with an estimated increase to ten million in 2050 if no measures are taken. The global economic burden of AMR is expected to rise to US$ 159 billion per year by 2050 because of surged healthcare costs, decreased productivity, and loss of livestock (Adamie et al. 2023). The economic impacts of AMR are also severe in the agricultural industry, as frequent use of antibiotics stops working over time. Farmers are always forced to use more costly substitutes or enhance biosecurity, and therefore, the costs of production rise. Likewise, there might be high mortality in the animals that endangers the productivity and profitability of farming, as shown by Zhang et al. (2024).

4.3 Antibiotic Residues in Animal-Derived Food Products

Antibiotic residues in foods like meat, milk, and eggs are becoming a concern for food handlers. Antibiotic residues are a consequence of an inadequate withdrawal time or overuse of antibiotics in animal production. In vivo, up to 60% of meat samples tested from intensively farmed livestock were positive for antibiotics, which have implications for human health as well as acceptability in the market (Van et al. 2020). In high-income countries, the finances of consumers have improved, and they only want antibiotic-free animal products.

Therefore, there is a high chance that trade barriers will be enhanced. The EU and the United States, for instance, have set MRLs on antibiotics in foods for human consumption and punished food products containing prohibited residues; accordingly, the affected shipments are rejected by customers in the international market (Zay Ya et al. 2024). This added pressure on the farmers as they are now required to implement residue monitoring programs and switch their farming from using regulations as a way of controlling diseases in crops. Yet, the long-term effects are: an increase in consumer confidence, higher prices for products that are free from antibiotics, and increased competitiveness in the market (Neculai-Valeanu et al. 2024).

4.4 Alternatives to Antibiotics for Sustainable Livestock Production

Due to these drawbacks related to the use of antibiotics, scholars and policymakers are now looking forward to the development of other interventions that improve animal health without contributing to the development of AMR and polluting the environment. Some possible and effective sustainable solutions that have been realized are:

Probiotics and Prebiotics Microorganisms that are beneficial for the overall good health of an individual and boost immunity, so that there will be less reliance on antibiotics. According to researchers, the efficacy of the consumed feed can be improved through supplementary intake of probiotic feed, while the bacteria-caused diseases in the animals may also be prevented (Cunningham et al. 2021).

Phytogenic Feed Additives Plant-derived compounds such as essential oils, tannins, and flavonoids with antimicrobial properties. According to the study, phytogenic extracts have the potential to enhance the aspect of digestion, act as a natural anti-inflammatory compound, and optimize the growth of the animal (Zhang et al. 2024).

Better Animal Health Enhancing the implementation of biosecurity measures on farms, the vaccination of animals, and disease surveillance to reduce the incidence of infections and consequently reduce the usage of antibiotics (Baraitareanu and Vidu 2021).

Precision Livestock Farming (PLF) The use of machinery that incorporates various gadgets that help monitor the cattle for disease indicators; then it uses artificial intelligence to treat these diseases to reduce the use of antibiotics (Ahmed et al. 2024).

4.5 Regulatory Initiatives and Policy Recommendations

Most national governments have come up with policies that check the use of antibiotics in animal farming. Thus, antibiotics are either eliminated from Europe as growth promoters or limited in feeding in the USA (FDA) for growth promotion (Baynes et al. 2016). These regulations are pressurizing the industry toward the system of sustainable and antibiotic-free undertakings of livestock.

It is noteworthy that the continued concerns of policymakers and stakeholders can be addressed and managed effectively through the following ways:

- Sharing of financial rewards forms of incentives that could be used to encourage farmers to adopt free-from-antibiotic practices.

- Finance research and exploring more affordable solutions that can surpass the existing ones, for example, vaccines and immunomodulators.
- Upgrading of the surveillance systems that track down the patterns of antibiotic resistance in the animals that are reared for meat production.
- To foster consumer demand for antibiotic-free products, more information needs to be provided to the consumer.

4.6 Future Perspectives on Sustainable Livestock Farming

Sustainability is being integrated with economic viability in feeding Livestock, which will define the future of the livestock business. Although antibiotics have been at the center of livestock health management in the past, the long-term effects imply the requirement for change and embracing long-term and environmentally friendly approaches to managing the health of the animals. The adoption of proper disease control methods, better management practices, and proper standards and policies shall be vital in the sustainable development of livestock production systems in the world.

5 Limitations of the Study

The study's limitation is as follows, although it covers almost all the important aspects of wireless network security as discussed in the previous sections: A weakness is the lack of and uncertainty of regional information about the use of antibiotics, since regulations regarding use and farming might not be the same in all countries or areas of distinct economic status. Such an observation raises questions concerning the generalization of lessons and conclusions. Furthermore, though this paper examines the effective and sustainable substitutes to antibiotics, it lacks information concerning the impact that industrialization of livestock production and smallholder farming have on the matter. Consequently, there exists a need to generate empirical evidence on the consequences of long-term changes in antibiotic consumption on animal yields and disease susceptibility. In addition, the use of literature sources means that there is a risk of methodological bias in the choice of sources, and the same authors may have opposite conclusions. Furthermore, this study does not fully examine the social and cultural factors that may exist in various regions regarding the use of antibiotics and their applicability to transition to other models. However, it establishes the necessary scaffold to explain the main factors under consideration and points out directions for additional research.

6 Conclusion

The dilemma of enhancing the productivity of livestock on the one hand and the appropriate use of Antibiotics on the other is one of the most debated issues in current livestock farming. Despite this, it has been seen that the use of antibiotics has various negative impacts, such as the emergence of AMR and pollution of the environment. Focusing on the impact AMR is expected to have on both the healthcare and livestock industries, where costs are predicted to run into billions, it is very important to adopt sustainable practices. Current actions, such as regulations and active consumer pressure for non-antibiotic integrated food products, indicate that there will be a progressive movement to healthier animal disease prevention mechanisms.

There are more sustainable methodologies, including the use of probiotics, phytogenic feed supplements, and enhanced bio-security systems. However, there is a need for investment in research and farmer education, as well as supportive policies to enable the increased adoption of this practice. They need to combine their efforts to propose actions to create organizational frameworks for both profitability and the optimization of the positive spin-off on the health of the public and the environment. To develop future livestock systems, it is necessary to consider the approaches focused on their effectiveness, along with the development of scientific technologies meeting the goal of high production and sustainability of the industry. It is only going to be through the continued changes in the legal requirements and stakeholder expectations that livestock farming will transform into one with lower impact and higher productivity.

References

Adamie BA, Owusu-Sekyere E, Lindberg M et al (2023) Dairy cow longevity and farm economic performance: evidence from Swedish dairy farms. J Dairy Sci 106:8926–8941. https://doi.org/10.3168/jds.2023-23436

Agregán R, Munekata PES, Feng X et al (2021) Technological advances for sustainable livestock production. In: Sustainable production technology in food. Elsevier, pp 37–47

Ahmed SK, Hussein S, Qurbani K et al (2024) Antimicrobial resistance: impacts, challenges, and future prospects. J Med Surg Public Health 2:100081. https://doi.org/10.1016/j.glmedi.2024.100081

Ait Ouakrim D, Cassini A, Cecchini M, Plachouras D (2020) The health and economic burden of antimicrobial resistance. In: Challenges to tackling antimicrobial resistance. Cambridge University Press, pp 23–44

Andrew Selaledi L, Mohammed Hassan Z, Manyelo TG, Mabelebele M (2020) The current status of the alternative use to antibiotics in poultry production: an African perspective. Antibiotics 9:594. https://doi.org/10.3390/antibiotics9090594

Baraitareanu S, Vidu L (2021) Dairy farms biosecurity to protect against infectious diseases and antibiotics overuse. In: Antimicrobial resistance – a one health perspective. IntechOpen

Bava R, Castagna F, Lupia C et al (2024) Antimicrobial resistance in livestock: a serious threat to public health. Antibiotics 13:551. https://doi.org/10.3390/antibiotics13060551

Baynes RE, Dedonder K, Kissell L et al (2016) Health concerns and management of select veterinary drug residues. Food Chem Toxicol 88:112–122. https://doi.org/10.1016/j.fct.2015.12.020
Bhatia R (2018) Universal health coverage framework to combat antimicrobial resistance. Indian J Med Res 147:228–232. https://doi.org/10.4103/ijmr.IJMR_1462_17
Callaway TR, Lillehoj H, Chuanchuen R, Gay CG (2021) Alternatives to antibiotics: a symposium on the challenges and solutions for animal health and production. Antibiotics 10:471. https://doi.org/10.3390/antibiotics10050471
Chahuán-Jiménez K, Rubilar-Torrealba R, de la Fuente-Mella H, Elórtegui-Gómez C (2023) Unraveling the complex interplay of sustainability, investments, and economic indicators. Sustainability 16:3. https://doi.org/10.3390/su16010003
Cunningham M, Azcarate-Peril MA, Barnard A et al (2021) Shaping the future of probiotics and prebiotics. Trends Microbiol 29:667–685. https://doi.org/10.1016/j.tim.2021.01.003
Dankar I, Hassan H, Serhan M (2022) Knowledge, attitudes, and perceptions of dairy farmers regarding antibiotic use: lessons from a developing country. J Dairy Sci 105:1519–1532. https://doi.org/10.3168/jds.2021-20951
Evangelista AG, Corrêa JAF, Pinto ACSM, Luciano FB (2022) The impact of essential oils on antibiotic use in animal production regarding antimicrobial resistance—a review. Crit Rev Food Sci Nutr 62:5267–5283. https://doi.org/10.1080/10408398.2021.1883548
Fesseha H, Aliye S (2020) Organic foods and public health importance: a review. Vet Med—Open J 5:1–8. https://doi.org/10.17140/VMOJ-5-140
Haque MH, Sarker S, Islam MS et al (2020) Sustainable antibiotic-free broiler meat production: current trends, challenges, and possibilities in a developing country perspective. Biology (Basel) 9:411. https://doi.org/10.3390/biology9110411
Haque MA, Rogerson L, Deb Nath N et al (2024) Sustainable management and valorization of antibiotic waste. Chem Eng J 498:155372. https://doi.org/10.1016/j.cej.2024.155372
Jadeja NB, Worrich A (2022) From gut to mud: dissemination of antimicrobial resistance between animal and agricultural niches. Environ Microbiol 24:3290–3306. https://doi.org/10.1111/1462-2920.15927
Kolomytseva ES, Eremeenko AV (2024) The use of antibiotics in the industrial production of meat and meat products. In: Collection of scientific papers, ООО «ДГТУ-Принт» Адрес полиграфического предприятия: 344003, г. Ростов-на-Дону, пл. Гагарина,1, pp 395–398
Lhermie G, Ndiaye Y, Rushton J, Raboisson D (2022) Economic evaluation of antimicrobial use practices in animal agriculture: a case of poultry farming. JAC Antimicrob Resist 4. https://doi.org/10.1093/jacamr/dlac119
Low CX, Tan LT-H, Ab Mutalib N-S et al (2021) Unveiling the impact of antibiotics and alternative methods for animal husbandry: a review. Antibiotics 10:578. https://doi.org/10.3390/antibiotics10050578
Lyons NA, Afzal M, Toirov F et al (2021) Economic considerations for advancement through the progressive control pathway: cost–benefit analysis of an FMD disease- free zone in Punjab Province, Pakistan. Front Vet Sci 8. https://doi.org/10.3389/fvets.2021.703473
McEwen SA, Fedorka-Cray PJ (2002) Antimicrobial use and resistance in animals. Clin Infect Dis 34:S93–S106. https://doi.org/10.1086/340246
Moghadam AA, Shuai W, Hartmann EM (2023) Anthropogenic antimicrobial micropollutants and their implications for agriculture. Curr Opin Biotechnol 80:102902. https://doi.org/10.1016/j.copbio.2023.102902
Mudenda S, Chabalenge B, Daka V et al (2023) Global strategies to combat antimicrobial resistance: a one health perspective. Pharmacol Pharm 14:271–328. https://doi.org/10.4236/pp.2023.148020
Neculai-Valeanu A-S, Ariton A-M, Radu C et al (2024) From herd health to public health: digital tools for combating antibiotic resistance in dairy farms. Antibiotics 13:634. https://doi.org/10.3390/antibiotics13070634

Patel SJ, Wellington M, Shah RM, Ferreira MJ (2020) Antibiotic stewardship in food- producing animals: challenges, progress, and opportunities. Clin Ther 42:1649–1658. https://doi.org/10.1016/j.clinthera.2020.07.004

Serra R, Kiker GA, Minten B et al (2020) Filling knowledge gaps to strengthen livestock policies in low-income countries. Glob Food Sec 26:100428. https://doi.org/10.1016/j.gfs.2020.100428

Serwecińska L (2020) Antimicrobials and antibiotic-resistant bacteria: a risk to the environment and to public health. Water (Basel) 12:3313. https://doi.org/10.3390/w12123313

Singer RS, Porter LJ, Thomson DU et al (2019) Raising animals without antibiotics: U.S. producer and veterinarian experiences and opinions. Front Vet Sci 6. https://doi.org/10.3389/fvets.2019.00452

Talukder P, Chanda S, Mishra M et al (2024) A study on the impacts of antibiotic pollution and its remediation. Environ Qual Manag 33:783–794. https://doi.org/10.1002/tqem.22152

Tarín-Pelló A, Suay-García B, Pérez-Gracia M-T (2022) Antibiotic resistant bacteria: current situation and treatment options to accelerate the development of a new antimicrobial arsenal. Expert Rev Anti-Infect Ther 20:1095–1108. https://doi.org/10.1080/14787210.2022.2078308

Tkaczyk C, Jones-Nelson O, Shi YY et al (2022) Neutralizing Staphylococcus aureus virulence with AZD6389, a three mAb combination, accelerates closure of a diabetic polymicrobial wound. mSphere 7:e0013022. https://doi.org/10.1128/msphere.00130-22

Tona GO (2018) Current and future improvements in livestock nutrition and feed resources. In: Animal husbandry and nutrition. InTech

Ullah M, Khan FA, Khan MU et al (2024) Revolutionizing livestock management: probiotics as immunomodulators for health and productivity. J Surv Fish Sci. https://doi.org/10.53555/sfs.v11i3.2406

Van Boeckel TP, Glennon EE, Chen D et al (2017) Reducing antimicrobial use in food animals. Science 357:1350–1352. https://doi.org/10.1126/science.aao1495

Van TTH, Yidana Z, Smooker PM, Coloe PJ (2020) Antibiotic use in food animals worldwide, with a focus on Africa: pluses and minuses. J Glob Antimicrob Resist 20:170–177. https://doi.org/10.1016/j.jgar.2019.07.031

Yulianti UA, Muchtar A, Amruddin A (2024) Sustainable livestock management practices: balancing production and environmental conservation. Riwayat Educ J Hist Humanit 7:2923–2930. https://doi.org/10.24815/jr.v7i4.42615

Zay Ya K, Lambiris MJ, Levine GA et al (2024) Coverage of policies to improve antimicrobial stewardship in human medicine in low and middle income countries: results from the Global Survey of Experts on antimicrobial resistance. BMC Public Health 24:2297. https://doi.org/10.1186/s12889-024-19542-2

Zhang T, Nickerson R, Zhang W et al (2024) The impacts of animal agriculture on one health—bacterial zoonosis, antimicrobial resistance, and beyond. One Health 18:100748. https://doi.org/10.1016/j.onehlt.2024.100748

Innovative Approaches to Enhance Economic and Environmental Sustainability in Livestock Production

Sobia Malik, Saima Naz, and Sadia Malik

Abstract Livestock systems' sustainability takes into account social, economic, and environmental factors. Land usage, water consumption, greenhouse gas emissions, and water quality are all aspects of environmental sustainability that are related to cattle production. The industry's long-term survival depends on reducing adverse environmental effects and increasing beneficial effects from animal production. The environment can have an influence on livestock output, but it can also have an impact on animals. Climate change provides new issues for livestock production in the future via effects on feed availability, quality, and potential for increasing heat stress on cattle themselves. All things considered, cattle production will need to adjust to future climatic circumstances and human expectations, which will call for both technical solutions and strong socioeconomic incentives to support their adoption. Future research on environmental sustainability in cattle production may be grounded on the theories of agroecology, socioecological systems, and sustainable living. With a focus on methods like rotational grazing and integrated pest control to reduce environmental degradation, the guidelines support the advancement of sustainable livestock production practices. Research and innovation expenditures are essential for creating new technologies that improve sustainability, and educational programs provide farmers with the skills they need. In order to successfully handle environmental concerns, policymakers are advised to create legislation and offer incentives for sustainable behaviors. Collaboration among stakeholders is emphasized as being crucial. To achieve environmental sustainability in cattle production and strike a balance between conservation and development, a multimodal strategy including innovation, policy, education, and cooperation is necessary.

S. Malik (✉) · S. Naz
Department of Zoology, The Government Sadiq College Women University, Bahawalpur, Pakistan
e-mail: sobia.malik@gscwu.edu.pk

S. Malik
Department of Pharmaceutics, The Islamia University of Bahwalpur, Bahwalpur, Pakistan

R. U. Khan, S. Naz (eds.), *Green Alternatives to Antibiotics in Animal Health*,
https://doi.org/10.1007/978-981-92-0309-3_23

Keywords Environmental regulations · Climate change · Livestock production · Sustainable methods & awareness

1 Rationale of the Study

1.1 Introduction

The production of livestock is essential to both economic growth and global food security, but because of its large environmental impact, its sustainability is coming under increased scrutiny. The cattle industry accounts for around 14.5% of worldwide greenhouse gas emissions, according to the Food and Agriculture Organization, and emissions are predicted to increase as the demand for animal products rises (FAO 2019). Even while worries about how livestock production affects the environment are rising, there is still a significant knowledge gap about the intricate relationships between environmental sustainability and production methods. While many studies have looked at particular aspects of the environmental impact of livestock, like waste management or feed efficiency, there isn't much thorough research that takes into account a number of variables to offer a comprehensive understanding of the sustainability of livestock farming methods.

The term "environmental sustainability" describes the prudent use and management of natural resources to satisfy current demands without endangering the capacity of future generations to satisfy their own. It aims to preserve a balance between human activity and the natural environment by including a number of factors, including ecological integrity, social equality, and economic viability. Addressing urgent problems, including pollution, resource depletion, biodiversity loss, and climate change, is necessary to achieve environmental sustainability. The significance of environmental sustainability in reducing environmental degradation and advancing global well-being is emphasized by sustainable development objectives, according to Miao et al. (2019).

Globally, environmental rules are crucial in determining sustainable cattle agricultural methods. Sustainable agricultural methods have been increasingly popular in industrialized nations like the US in recent years. For example, farms are increasingly using waste management methods like anaerobic digestion and composting (Aziz 2024). The breeding, rearing, and management of domesticated animals for a variety of uses, such as food, fiber, and labor, are all included in livestock production methods. Given their dependence on natural resources and their propensity to exacerbate problems like greenhouse gas emissions, land degradation, and water pollution, these activities have the potential to have a substantial negative influence on environmental sustainability (Herrero et al. 2015). Feed management, which includes the source, formulation, and distribution of animal feed, is an essential component of livestock production. Environmental sustainability may be impacted by the feed's nutritional value, kind, and conversion efficiency. For instance, the ecological footprint of animal production can be decreased by using more sustainable feed components, such as locally obtained fodder or food processing wastes (FAO 2013). In the production of livestock, grazing systems are crucial, particularly for ruminant animals like sheep and cattle. Carbon sequestration, biodiversity, and soil health can all be impacted by various grazing techniques, including rotational and continuous grazing. By improving soil fertility, lowering erosion, and protecting animal habitats, well-managed grazing systems that replicate natural ecosystems can support environmental sustainability. Significant volumes of waste, such as manure, urine, and bedding materials, are produced throughout the livestock production process. Minimizing nutrient runoff and environmental contamination requires effective waste management techniques. To reduce the demand for synthetic fertilizers and mitigate water contamination, technologies like anaerobic digesters, composting, and nutrient management plans can assist in capturing the nutrients from animal waste and recycling them back into agricultural systems (Dinh et al. 2020).

Meeting current requirements without compromising the capacity of future generations to satisfy their own needs is known as sustainability. Although sustainability takes into account social, economic, and environmental factors, discussions concerning the sustainability of meals derived from animals are frequently dominated by environmental concerns, such as climate change. Changes in regulations and investor demands demonstrate the growing attention and pressure on livestock farming to reduce its environmental implications. (Ridoutt 2024). Sustainable livestock farming minimizes adverse effects on ecosystems, conserves natural resources, and improves animal welfare by incorporating environmental, social, and economic considerations into animal production systems, in contrast to today's traditional livestock farming practices. The following strategies that minimize the use of chemical agricultural inputs, save water, and lessen pollution are frequently at the heart of sustainable livestock production. A key topic of discussion in the current agricultural debate is how environmental rules affect sustainable cattle production methods. To lessen the negative environmental consequences of cattle farming, especially with regard to greenhouse gas emissions, water pollution, and land degradation, strict environmental rules and regulations have been implemented worldwide (Jones and Brown 2021; Aziz 2024). In order to encourage sustainability in the livestock sector, these rules frequently include activities such as waste management, pasture management, and animal welfare requirements. However, in order to ensure sustainable livestock farming methods while striking a balance between concerns about food security and economic viability, it is necessary to critically evaluate the efficacy and difficulties related to the execution of these policies (Smith 2018; Aziz 2024).

In the production of livestock, water is an essential resource for both agricultural use and animal hydration. But wasteful water usage and runoff pollution may stress freshwater ecosystems and make water shortages worse. The environmental effect of livestock operations may be decreased, and sustainable water management can be ensured by putting water-efficient techniques like drip irrigation, rainwater collection, and water recycling into practice (FAO 2016). Livestock traits, including growth rate, disease resistance, and feed efficiency, are greatly influenced by genetic selection. Producers can increase livestock systems' production and resource efficiency by breeding animals with desired qualities. By lowering the environmental impact of cattle production, selective breeding for characteristics like resistance to heat or methane reduction can support environmental sustainability (García-Ruiz et al. 2019).

To maintain animal welfare and productivity, disease prevention and management are crucial components of livestock production. The usage of medicines, higher death rates, and environmental pollution are just a few of the serious economic and environmental effects that disease epidemics may have. Environmental sustainability may be promoted by reducing the use of antibiotics and chemical inputs through the use of integrated pest control, immunization programs, and biosecurity measures (Grace et al. 2018). Designs for livestock shelters and housing can affect the comfort, behavior, and environmental effects of the animals. While well-designed housing systems with adequate ventilation and waste management can reduce environmental concerns, confinement systems, such as concentrated animal feeding operations, can pollute the air and water. Alternative living arrangements, including free-range or pasture-based systems, can help the environment by encouraging natural behaviors and lowering the demand for energy-intensive infrastructure (Friggens et al. 2017).

Resource depletion and greenhouse gas emissions are caused by the energy used in livestock production, which includes feed manufacturing, heating, ventilation, and transportation. The environmental impact of livestock operations may be decreased by increasing energy efficiency through the use of sources of renewable energy, energy- efficient machinery, and process optimization. Environmental sustainability may be further improved by integrated energy management techniques that make use of waste heat and byproducts (Cederberg et al. 2017). For grazing, feed manufacturing, and infrastructural expansion, livestock production needs substantial land resources. Deforestation, overgrazing, and soil degradation are examples of unsustainable land use practices that can harm ecosystems and lead to habitat loss and a fall in biodiversity. In livestock production systems, the application of land use planning techniques like rotational grazing, agroforestry, and land zoning may support biodiversity conservation and sustainable land management (Havlík et al. 2014). Regulations and policy frameworks are essential in determining how livestock production methods are carried out and how they affect the environment. Government regulations pertaining to emissions management, land use, water quality, and animal welfare can affect producer behavior and encourage sustainable practices. A worldwide framework for tackling climate change and advancing sustainable development within livestock systems is provided by multilateral accords like the Goals for Sustainable Development and the Paris Agreement (FAO 2015). Ecological, social, and economic effects are greatly impacted by the management and application of livestock production systems, which are closely related to environmental sustainability. Stakeholders may lessen the environmental effects of livestock farming and help create a food system that is more resilient by implementing integrated techniques that put an emphasis on ecosystem health, animal welfare, and resource efficiency (Smith 2023).

1.2 Economic Impacts

Through the process of the Millennium Ecosystem Assessment, or MEA (2005), the idea of ecosystem service (ES) has been widely accepted and developed. Four categories of ecosystem services were identified by the MEA (Cerrato et al. 2023) i.e., (1) Provisioning services, which refer to goods derived from ecosystems (such as food, fiber, fertilizer, fuel, etc.); (2) Regulating services, which refer to the advantages of controlling ecosystem processes (such as recycling waste, preventing land degradation and erosion, etc.); (3) Supporting services, which are essential to the creation of all other ecosystem services (such as preserving soil fertility and structure, preserving genetic diversity, etc.); and (4) Cultural services, which refer to the intangible benefits that humans derive from ecosystems through spiritual development, cognitive growth, introspection, leisure, and visual pleasure (Hoffmann et al. 2014; Cerrato et al. 2023).

The Economic Perspectives of Ecosystems and Biodiversity (Kumar 2010) is a second important initiative on ES. It defined ES as "the both the direct and indirect advantages of ecosystems to human well-being," distinguishing services from benefits in order to clearly identify the services that offer numerous indirect benefits. Because they are important parts of agro-ecosystems and have close interactions with natural ecosystems, livestock and breeders are crucial to the supply of ES (Hoffmann et al. 2014). Because of their higher trophic level, which causes conversion losses and related environmental externalities, cattle interact with other ecosystem elements and processes in a more complicated way than plants do. Specifically, cattle have three traits that influence their distinct functions in ecosystems:

1. The special capacity of livestock to use their digestive tracts to transform indigestible feed and organic waste into useful products;
2. The direct interactions of the trophic network with ecosystems (such as land, vegetation, and soil) through browsing, grazing, and trampling, as well as the production of urine and dung;
3. Their mobility and consequent capacity to adapt to changes in the temporal and spatial fluctuations of ecosystems in terms of resource availability. Furthermore, cattle breeds and species' contributions to ES are closely linked to the production systems with which they are involved and, consequently, to the many human management systems that have an impact on them (Hoffmann et al. 2014; Cerrato et al. 2023).

   In addition, animal production accounts for over 70% of the total agricultural land in the EU and more than 80% of greenhouse gas emissions from agriculture (Leip et al. 2015). Due to their energy consumption in barns, fermented food production, manure management, and deforestation, ruminants are one of the livestock industry's main causes of global warming (Von Greyerz et al. 2023). Life cycle assessment (LCA), a cradle-to-grave methodology to evaluate goods, processes, services, activities, and systems based on the life cycle thinking approach, is one of the most widely used approaches to estimate environmental consequences (Pergola et al. 2021). Based on comparing products and processes with similar functions as well as identifying the subsystems that contribute most to the overall environmental impact, life cycle assessment (LCA) has been shown to be a useful tool for answering questions regarding the environmental impact of different agricultural production systems (Canals et al. 2006; Cerrato et al. 2023).

1.3 Environmental Impacts

The production of livestock interacts with the natural environment and can affect environmental quality in both good and negative ways. Future livestock output may be impacted both directly and indirectly by climate change, which might put pressure on the livestock system's capacity to satisfy public expectations for more productive and ecologically friendly production. In the end, sustainability is complicated and involves balancing a number of factors (social, economic, and environmental) as well as making judgments about what is ideal in each of those areas. Sustainability requires trade-offs, but advancement is also feasible (Place 2024).

1.3.1 Water Management

Life needs water. Agriculture's usage of water is a major environmental sustainability issue that is more regional in scope than emissions of greenhouse gases. The production of feed accounts for the majority of water usage in livestock systems, with agriculture being the main user of water. For instance, irrigation of the crops that feed cattle accounts for 87.5% of the water consumed in the United States' beef production from birth to death (Putman et al. 2023).

Similar to greenhouse gas emissions, there are many classifications for water consumption. The term "blue water use" describes the use of groundwater and surface water for things like animal drinking water and irrigation. Rainfall and evapotranspiration water, the greatest category of water usage, are more difficult to affect through human activity. This is known as "green water use." According to Mekonnen and Hoekstra (2011), gray water is the amount of water needed to compensate for pollution. Reducing water usage, preserving water quality, and increasing agricultural water efficiency are the goals of sustainable water management techniques. Important methods for conserving water include:

- Effective irrigation systems: Conventional flood irrigation systems frequently waste water and cause large amounts of water to evaporate and run off. More effective irrigation techniques, including drip or sprinkler irrigation systems, which supply water straight to plant roots, are being embraced by sustainable farms more and more. These systems guarantee that livestock and crops get the exact quantity of water they require while minimizing water waste (Putman et al. 2023).

- Rainwater harvesting: Reliance on outside water sources can be lessened by gathering and preserving rainwater to utilize during dry spells. For subsequent use in irrigating pastures or watering cattle, rainwater can be taken from building roofs or specifically constructed collection systems and stored in tanks (Putman et al. 2023).
- Riparian buffer zones: Arranging plants beside rivers and streams may help filter water from pastures and keep contaminants like pesticides, fertilizers, and manure out of waterways. Additionally, these buffer zones support biodiversity and offer wildlife habitats (Putman et al. 2023).
- Water recycling: You may drastically cut down on total water use by putting in place technologies that recycle the water used to clean and cool animal buildings. Water used for cleaning milking equipment, for instance, may be filtered and utilized again for cattle drinking or irrigation (Putman et al. 2023).

1.3.2 Land Management

Land usage is another important environmental factor for the sustainability of cattle. In general, meat requires more land per kilogram of food than plant-based alternatives, since land usage is frequently a major point of criticism for diets derived from animals as opposed to plants. This is especially true for foods from ruminants, such as goat, lamb, and cattle. However, for three main reasons—fungibility, quality of use, and multifunctional usage—land use is difficult to evaluate.

- The term "multifunctional use" describes the use of land for various food or commodity reasons as well as for various ecosystem services, or advantages that humans can obtain from the natural world. Multiple revenue streams and human-useful product streams are made possible by these kinds of integrated crop-livestock systems, which can occasionally help with nutrient cycling (Sulc and Franzluebbers 2014).
- The phrase "quality of use" describes the longer-term sustainability factors that are taken into account while using land for food production. For instance, a corn-soy rotation under conventional tillage is probably going to produce a lot more energy per acre than a field with grazing (Cassidy et al. 2013).
- The ability of lands to be utilized for many purposes with equal value is known as fungibility. For instance, a lot of grazing areas are situated on rangelands that are either too steep, rocky, or dry for farming, or that may erode more quickly. Therefore, part of this land is unavailable for the production of plant-based meals, even if the land consumption per kilogram for products derived from animals may be higher (Maczko et al. 2011).

1.3.3 Reduce Methane Emission

Methane emissions from livestock farms are particularly high when ruminants like sheep and cows are raised. Methane is a crucial goal for climate adaptation efforts since it is a greenhouse effect that is significantly more powerful than carbon dioxide. However, there are other ways to lower methane emissions, such as:

- Better diets: The kind of feed that animals are fed may significantly affect how much methane they emit during digestion. Methane emissions can be decreased by using high-quality grains, forages, and feed additives like oils or seaweed. For instance, it has been demonstrated that feeding ruminants red seaweed (asparagopsis) can cut methane generation by as much as 80%. You may learn more about the Alliance's broad global efforts to promote and enhance the adoption of optimum fodder types here (Place 2024).
- Genetic selection: To create cattle with reduced emissions of methane per unit of output, selective breeding procedures are being created. Farmers may lessen their herds' total environmental effect by selecting animals with lower methane emissions and higher feed efficiency (Place 2024).
- Management of manure: When kept in anaerobic conditions, manure is a major source of methane emissions. Methane emissions can be considerably decreased by putting into practice improved manure management techniques, such as composting or the use of anaerobic digesters to generate biogas. Methane from manure is captured by biogas systems, which then transform it into energy from renewable sources that may be sold to the grid or utilized to power agricultural activities (Place 2024).

1.4 Effects of the Climate

Events with excessive precipitation and drought will occur more frequently as a result of climate change. Water supply has declined as a result of, for instance, rising temperatures in eastern Utah and western Colorado, which have reduced snowpacks and accelerated springtime melting of snow and evaporation rates (White et al. 2023). According to the Paris Climate Agreement, the worldwide threshold for temperature change is 2 °C, which has already been experienced by several areas of the southwestern United States (Vose et al. 2017). More frequent intense precipitation events are anticipated, which might lead to an increase in soil erosion and nutrient discharge, in addition to climate change-driven restrictions on water supply (Gowda et al. 2018). Due in part to human management and climate change, the area burnt by wildfires has also increased in both the United States and Canada over the past few decades (Ostoja et al. 2023).

Although burning of grassland and livestock operations can have a direct impact on livestock operations, fires can have an impact that extends beyond the local burn region. Research has shown, for instance, that wildfire particles can negatively affect cattle health in addition to human health. Higher temperatures and levels of CO_2 may boost agricultural yields in some US locations (Kukal and Irmak 2018); nevertheless, quality may suffer for particular fodder crops (Augustine et al. 2018). Indirect consequences on animal output, such as reduced water availability and elevated disease and parasite risks, might be attributed to these effects on pasture quality. Heat stress, which can have detrimental effects affecting animal productivity and raise mortality, is one of the direct effects of changing climates on livestock output (Cheng et al. 2022). More flooding might result from an increase in severe precipitation events, which could have a detrimental effect on livestock output and possibly introduce heavy metals and infections into livestock operations, posing health hazards to both cattle and consumers (Crist et al. 2020). Extreme precipitation and heat stress can also put a strain on individual farm workers and the social structures that sustain agriculture. Long-term livestock systems' sustainability will depend on their capacity to adapt to climate change and build resilience against extreme weather occurrences (Place 2024).

2 Theoretical Framework Environmental Policy Theory

This theory, which dates back to academics like Thomas D. Lynch & Elinor Ostrom, focuses on how governments and regulatory agencies create and carry out environmental policy. It explores the methods used to create rules, how they are implemented, and how they affect different industries, including agriculture. This theory aids in comprehending how policies are created to address environmental issues in livestock farming, such as waste management, emissions reduction, and land use practices, in the context of how environmental regulations affect viable livestock farming methods (Lynch 2020; Aziz 2024).

3 Institutional Theory

This theory, which was created by academics such as Scott, DiMaggio, & Powell, examines how institutions influence behavior and practices in businesses and organizations. Institutional theory aids in comprehending how environmental laws are incorporated into the customs, practices, and organizational frameworks of the agriculture industry in the larger context of environmentally friendly livestock production. In reaction to legal constraints, it emphasizes how institutional forces, norms, and values impact the adoption of sustainable practices (Scott 2019; Aziz 2024).

4 Theory of Socio-Technical Transitions

This theory, which dates back to the work of academics like Geels and Schot, aims to comprehend significant changes in socio-technical systems, such as agriculture. It looks at how social, technical, economic, and regulatory elements interact to propel sustainable shifts. This theory aids in the analysis of how regulatory interventions interact with market dynamics, technological advancements, and social factors to either help or impede the transition to more sustainable farming practices in the context of environmental regulations and sustainable livestock production (Geels and Schot 2021; Aziz 2024).

5 Principles of Socio-Ecological Systems

The theory of socio-ecological systems is concerned with the intricate relationships that exist between ecological and social elements in interconnected human-environment systems. According to Ostrom (2009), this theory acknowledges that human endeavors, such as raising cattle, are influenced by institutions, governmental frameworks, and cultural norms and are situated within larger social and environmental settings. Socioecological systems theory emphasizes how human decision-making, livestock management techniques, and environmental results are all intertwined in the context of cattle production. It is essential to comprehend the institutional and social aspects of livestock production in order to create interventions and policies that support environmental sustainability and cater to the interests and goals of many stakeholders (Smith 2023).

6 Theory of Agroecology

The significance of ecological values in agricultural systems is emphasized by agroecology theory, which has its roots in the research of agronomic and scientist Miguel A. Altieri. Fundamentally, agroecology aims to comprehend and maximize the relationships that exist between humans, animals, plants, and the environment in agricultural settings. According to this hypothesis, sustainable farming methods minimize the need for external inputs like fertilizers and pesticides while enhancing biodiversity, soil health, and ecosystem resilience. According to "Environmental Stability in Livestock" the agroecology concept offers a paradigm for increasing environmental sustainability through the integration of livestock production into agroecosystems. Agroecology theory provides guidance on how to create livestock production systems that are both environmentally sustainable and productive by taking into account the ecological dynamics of livestock systems, such as nutrient cycling, pasture management, and agroforestry integration (Wezel et al. 2009; Smith 2023).

7 An Approach to Sustainable Livelihoods

Robert Chambers and associates developed the sustainable livelihoods approach, which highlights the significance of livelihood methods that improve human well-being while preserving the resilience of ecosystems and natural resources. This strategy aims to enable marginalized people to adopt sustainable livelihood alternatives while acknowledging that livelihoods are impacted by a confluence of social, economic, ecological, and institutional issues. The strategy of sustainable livelihoods offers a perspective through which to view the environmental effects of livestock-based livelihoods while also comprehending how they support rural economies or food security. This method can guide actions that promote the objectives of environmental conservation and human development by finding strategies to improve the sustainability and resilience of livelihoods that rely on cattle (Scoones 2009; Smith 2023).

7.1 The Importance of Livestock Well-Being to Sustainability

It is essential to raise animals in a sustainable manner. In addition to being morally right, ensuring that animals are kept humanely increases farm resilience and output (Aziz 2024). The following are important methods for enhancing animal wellbeing in sustainable livestock farming:

- Systems based on grazing and free-range: Animals' physical and mental health is enhanced when they are allowed to wander freely in their natural habitats. Additionally, it improves the flavor and nutritional value of milk, eggs, and meat.
- Better health via reduced stress: Reducing animal stress through appropriate feeding, housing, and handling strengthens the immune system and lowers the need for antibiotics.
- Natural behaviors: Healthy and happier cattle are produced by designing farming systems that allow animals to exhibit their natural activities, such as rooting, pecking, or grazing.

- Because more people are concerned about the morality of food production, investing in animal welfare increases long-term productivity and customer trust. Enhancing animal well-being offers livestock farmers a number of useful advantages in addition to ethical ones: Farmers who put animal welfare first can also benefit financially from growing customer interest in more moral business methods (Aziz 2024).

7.2 *Obstacles and Difficulties in Implementing Sustainable Livestock Farming Methods*

Despite the many advantages of sustainable livestock farming methods, a number of obstacles may prevent their widespread use (Smith 2023). These consist of:

- *Initial costs:* A large upfront expenditure is necessary for many sustainable initiatives, such as using precision agricultural technologies or setting up biogas plants. These upfront expenses can deter many farmers from participating in the sustainable livestock raising movement because the long-term benefits are yet uncertain.
- *Knowledge and training:* Not all livestock producers are aware of the abovementioned measures since they may not have the technical know-how or access to training required to use sustainable practices successfully.
- *Market access:* Premium markets are frequently needed for sustainable products, and numerous farmers may have trouble finding consumers prepared to pay more for meat, milk, or eggs that are produced responsibly.
- *Policy assistance:* In many areas, farmers are not given enough incentives or assistance by agricultural policy to make the switch to sustainable methods.

8 Future Recommendations

Promoting the industry-wide adoption of sustainable animal production methods is one important suggestion. This entails putting strategies like integrated pest control, varied cropping systems, and rotational grazing into practice to reduce environmental damage while preserving productivity. Policymakers and other stakeholders may promote a more resilient and ecologically friendly cattle industry by pushing farmers to implement sustainable practices. Investing in research and innovation to create new technology and management techniques that improve environmental sustainability in animal production is another important suggestion. This might entail providing financing for studies aimed at increasing feed efficiency, lowering methane emissions, and refining waste management strategies. The business may find more economical and environmentally responsible methods of producing cattle products by utilizing technological breakthroughs.

To provide farmers with the information and abilities required to implement sustainable practices, education and capacity development programs are crucial. Farmers can get helpful advice on adopting eco-friendly practices and overcoming change-related obstacles via extension programs, training sessions, and farmer field schools. Policymakers may aid in the shift to more sustainable animal production systems by funding farmer education. The regulatory landscape and incentives for sustainable farming of livestock are greatly influenced by policymakers. Policies that encourage environmental stewardship, such those governing land conservation, water use, and manure management, should be created and implemented by governments. Furthermore, farmers may be persuaded to embrace sustainable methods by financial incentives, including tax credits, subsidies, and compensation for ecosystem services.

To promote environmental sustainability in cattle production, cooperation between stakeholders—including governmental organizations, trade groups, academic institutions, and non-governmental organizations—is crucial. Policymakers can better solve complex environmental concerns by utilizing the combined resources and skills of multi-stakeholder partnerships. Knowledge exchange, technology transfer, etc., coordinated action toward common sustainability objectives can all be facilitated by collaborative projects. Lastly, encouraging sustainable consumption habits and increasing consumer knowledge of the environmental effects of livestock production can spur improvements across the supply chain. Certification programs, eco-labeling initiatives, and education campaigns can assist customers in making knowledgeable decisions and incentivize manufacturers who use sustainable methods. Customers may encourage manufacturers to give environmental sustainability top priority in their operations by generating consumer demand for sustainable products.

9 Conclusion

To sum up, tackling environmental sustainability in cattle production necessitates a multipronged strategy that incorporates stakeholder engagement, legislative support, education, and technical innovation. Policymakers, stakeholder groups, and consumers may collaborate to support a more resilient and sustainable livestock sector that strikes a balance between environmental preservation, food security, and economic growth by putting the above ideas into practice. First and foremost, legislators and regulatory bodies will acquire important knowledge on the environmental effects of various livestock production methods, enabling them to create evidence-based laws and policies that support sustainable livestock production. A deeper comprehension of the environmental hazards and possibilities related to their operations would also help livestock farmers and industry stakeholders adopt environmentally friendly methods and enhance their environmental performance. Furthermore, customers who are becoming more concerned about how their food choices affect the environment would gain from clear information regarding the environmental sustainability of animal products, which will enable them to make better judgments.

References

Augustine DJ, Blumenthal DM, Springer TL, LeCain DR, Gunter SA, Derner JD (2018) Elevated CO2 induces substantial and persistent declines in forage quality irrespective of warming in mixed grass prairie. Ecol Appl 28(721):735

Aziz W (2024) Impact of environmental regulations on sustainable livestock farming practices in Pakistan. Am J Livest Policy 4(2):36–47

Canals LM, Burnip GM, Cowell SJ (2006) Evaluation of the environmental impacts of apple production using life cycle assessment (LCA): case study in New Zealand. Agric Ecosyst Environ 114:226–238

Cassidy ES, West PC, Gerber JS, Foley JA (2013) Redefining agricultural yields: from tonnes to people nourished per hectare. Environ Res Lett 8:034015

Cederberg C, Flysjö A, Henriksson M (2017) The implications of using life cycle assessment for energy and carbon footprinting in agri-food systems. Curr Opin Environ Sci Health 5:14–21

Cerrato M, Iasi A, Di Bennardo F, Pergola M (2023) Evaluation of the economic and environmental sustainability of livestock farms in Inland Areas. Agriculture 13:1708

Cheng M, McCarl B, Fei C (2022) Climate change and livestock production: a literature review. Atmosphere- Basel 13:140

Crist S, Mori J, Smith RL (2020) Flooding on beef and swine farms: a scoping review of effects in the midwestern United States. Prev Vet Med 184:105158

Dinh QT, Tu VT, Smits K (2020) Nutrient management practices for sustainable vegetable production in the Mekong Delta, Vietnam. J Soil Sci Plant Nutr 20(2):299–317

FAO (2013) Sustainable livestock production: delivering on the potential, Food and Agriculture Organization of the United Nations. Retrieved from http://www.fao.org/3/i3241e/i3241e.pdf

FAO (2015) Global agenda for sustainable livestock: Food and Agriculture Organization of the United Nations. Retrieved from http://www.fao.org/3/a-i4787e.pdf

FAO (2016) Water for sustainable food and agriculture: Food and Agriculture Organization of the United Nations. Retrieved from http://www.fao.org/3/a-i6325e.pdf

FAO (2019) World Livestock: transforming the livestock sector through the Sustainable Development Goals. Food and Agriculture Organization of the United Nations. Retrieved from http://www.fao.org/3/ca0912en/CA0912EN.pdf

Friggens NC, Blanc F, Berry DP, Puillet L, Bigeon J (2017) Review: housing and management of dairy cows: integrating principles and practice from a whole farm perspective. Animal 11(9):1442–1454

García-Ruiz A, Cole JB, VanRaden PM, Wiggans GR, Ruiz-López FJ, Van Tassell CP (2019) Changes in genetic selection differentials and generation intervals in US Holstein dairy cattle as a result of genomic selection. Proc Natl Acad Sci 116(17):8284–8289

Geels FW, Schot JW (2021) Socio-technical transitions to sustainability: a review of concepts, theories, and research approaches. J Econ Geogr 15(4):589–605

Gowda P, Steiner JL, Olson C, Boggess M, Farrigan T, Grusak MA (2018) Agriculture and rural communities [Chapter 10]. In: Reidmiller DR, Avery CW, Easterling DR, Kunkel KE, Lewis KLM, Maycock TK, Stewart BC (eds) Impacts, risks, and adaptation in the United States: fourth national climate assessment. Volume II. U.S. Global Change Research Program, Washington, DC, pp 391–437

Grace D, Dominguez-Salas P, Alonso S, Fahrion AS, Haesler B, Heilmann M, Zinsstag J (2018) Food safety metrics relevant to low and middle income countries. Food Control 87:92–98

Havlík P, Valin H, Herrero M, Obersteiner M, Schmid E, Rufino MC, Mosnier A (2014) Climate change mitigation through livestock system transitions. Proc Natl Acad Sci 111(10):3709–3714

Herrero M, Wirsenius S, Henderson B, Rigolot C, Thornton P, Havlík P, Gerber P (2015) Livestock and the environment: what have we learned in the past decade? Annu Rev Environ Resour 40:177–202

Hoffmann I, From T, Boerma D (2014) Ecosystem services provided by livestock species and breeds, with special consideration to the contributions of small-scale livestock keepers and pastoralists. Commission on Genetic Resources for Food and Agriculture Background Study Paper No. 66 Rev.1

Jones AB, Brown C (2021) Environmental regulations and sustainable livestock farming: a review of current practices. J Sustain Agric 28(2):112–125

Kukal MS, Irmak S (2018) Climate-driven crop yield and yield variability and climate change impacts on the U.S. Great Plains agricultural production. Sci Rep-UK 8:3450

Kumar P (2010) TEEB; The Economics of Ecosystems and Biodiversity: Ecological and Economic Foundations. Earthscan, London, UK/Washington, DC

Leip A, Billen G, Garnier J, Grizzetti B, Lassaletta L, Reis S, Simpson D, Sutton MA, de Vries W, Weiss F (2015) Impacts of European livestock production: nitrogen, sulphur, phosphorus and greenhouse gas emissions, land-use, water eutrophication and biodiversity. Environ Res Lett 10:115004

Lynch TD (2020) Environmental policy theory: foundations and applications. Cambridge University Press

Maczko K, Tanaka JA, Breckenridge R, Hidinger L, Heintz HT, Fox WE, Kreuter UP, Duke CS, Mitchell JE, McCollum DW (2011) Range land ecosystem goods and services: values and evaluation of opportunities for ranchers and land managers. Rangelands 33:30–36

Mekonnen MM, Hoekstra AY (2011) The green, blue and grey water footprint of crops and derived crop products. Hydrol Earth Syst Sci 15:1577–1600

Miao L, Liu X, Wang J (2019) Theoretical research and practical enlightenment of ecological civilization construction. Adv Soc Sci Educ Humanit Res 392:32–35

Millennium Ecosystem Assessment (MEA) (2005) Ecosystems and human well-being: biodiversity synthesis. Island Press, Washington, DC

Ostoja SM, Crimmins AR, Byron RG, East AE, Méndez M, O'Neill SM, Peterson DL, Pierce JR, Raymond C, Tripati A, Vaidyanathan A (2023) Focus on western wild fires. In: Crimmins AR, Avery CW, Easterling DR, Kunkel KE, Stewart BC, Maycock TK (eds) Fifth national climate assessment. U.S. Global Change Research Program, Washington, DC

Ostrom E (2009) A general framework for analyzing sustainability of social-ecological systems. Science 325(5939):419–422

Pergola M, Palese AM, Persiani A, De Francesco P, Cirillo C, Giacca GM, Maffia A, Celano G (2021) Mediterranean cropping systems: the importance of their economic and environmental sustainability. Adv Environ Eng Res 2(4):1–35

Place SE (2024) Environmental sustainability of livestock systems. Meat Muscle Biol 8(1):1–5

Putman B, Rotz CA, Thoma G (2023) A comprehensive environmental assessment of beef production and consump tion in the United States. J Clean Prod 402:136766

Ridoutt B (2024) Drivers of sustainability credentialling in the red meat value chain—A mixed methods study. Agriculture (London) 14:697

Scoones I (2009) Livelihoods perspectives and rural development. J Peasant Stud 36(1):171–196

Scott WR (2019) Institutions and organizations: ideas and interests. SAGE Publications

Smith D (2018) Waste management regulations on dairy farms: greenhouse gas emissions reduction. J Sustain Dev 25(3):245–259

Smith N (2023) Environmental Sustainability in Livestock Production. Int J Livest Policy 2(1):26–38

Sulc RM, Franzluebbers AJ (2014) Exploring integrated crop–livestock systems in different Eco regions of the United States. Eur J Agron 57:21–30

Von Greyerz K, Tidåker P, Karlsson JO, Röös E (2023) A large share of climate impacts of beef and dairy can be attributed to ecosystem services other than food production. J Environ Manag 325:116400

Vose RS, Easterling DR, Kunkel KE, LeGrande AN, Wehner MF (2017) Temperature changes in the United States [Chapter 6]. In: Wuebbles DJ, Fahey DW, Hibbard KA, Dokken DJ, Stewart BC, Maycock TK (eds) Climate science special report: fourth national climate assessment. Volume I. U.S. Global Change Research Program, Washington, DC, pp 185–206

Wezel A, Bellon S, Doré T, Francis C, Vallod D, David C (2009) Agroecology as a science, a movement and a practice. A review. Agron Sustain Dev 29(4):503 515

White DD, Elias EH, Thomas KA, Bradatan CE, Brunson MW, Chischilly AM, Enquist CAF, Fisher LR, Froehlich HE, Koebele EA, Méndez M, Ostoja SM, Steele C, Vanos JK (2023) Southwest [Chapter 28]. In: Crimmins R, Avery CW, Easterling DR, Kunkel KE, Stewart BC, Maycock TK (eds) Fifth national climate assessment. U.S. Global Change Research Program, Washington, DC

Reimagining Veterinary Medicine: The Role of Green Alternatives in Combating Antibiotic Resistance

Saima Naz, Ahmad Manan Mustafa Chatha, Sobia Malik, Muniba Maryam, Urwah Ishaque, and Qudrat Ullah

Abstract With greater emphasis on sustainability and environmental stewardship in this world, it is increasingly important to look for green options. In animal health management. This involves the use of synthetic medicines and chemical treatment. Such use has raised several critical issues, like leaving behind a footprint in environmental degradation, the emergence of resistance against antimicrobials, or the eventual presence of residue in the food chain. All these do affect the ecosystem as well as public health and animal welfare. In this regard, green options such as Phytotherapeutics, probiotics, essential oils, and biopesticides are recommended in the prevention of diseases, stimulation of growth, and control of parasitism, barring damage to the environment. These sustainable approaches not only align with global environmental goals but also support the health of animals, humans, and plants. Advancing research, fostering innovation, and implementing green practices in the animal health system are critical steps toward achieving sustainable agriculture and addressing the challenges of modern veterinary medicine. This chapter discusses the prevalence of antibiotic use in animal health and the mechanism of antibiotic resistance and their impacts, and also involves regulatory responses to curb antibiotic use and studies about challenges in the animal production system.

Keywords Green Options · Animal Health · Medicines · Chemical Treatments · Probiotics · Essential Oils

S. Naz (✉) · S. Malik · M. Maryam · U. Ishaque
Department of Zoology, Government Sadiq College Women University, Bahawalpur, Punjab, Pakistan
e-mail: saima.naz@gscwu.edu.pk

A. M. M. Chatha
Department of Entomology, Faculty of Agriculture and Environment, The Islamia University of Bahawalpur, Bahawalpur, Punjab, Pakistan

Q. Ullah
Department of Theriogenology, Faculty of Veterinary and Animal Sciences, Cholistan University of Veterinary and Animal Sciences, Bahawalpur, Punjab, Pakistan

R. U. Khan, S. Naz (eds.), *Green Alternatives to Antibiotics in Animal Health*,
https://doi.org/10.1007/978-981-92-0309-3_24

1 Introduction

Animal health is the pillar of world agriculture, playing a key role in determining food security, economic well-being, and environmental integrity. Healthy animals are needed to generate safe, healthy, and plentiful food sources for the expanding global population (Ogwu et al. 2024). Livestock diseases can result in major losses in productivity, having a direct effect on farmers' livelihood and countries' economies that rely on agriculture. Animal diseases cost the world economy billions of dollars every year, and this explains why having strong animal health management systems is critical (Kappes et al. 2023).

Animal health management entails prevention, control, and eradication of disease in animals through the application of biosecurity practices, vaccination protocols, and ongoing health checks (Modisane 2009). Good animal health management guarantees the well-being of animals, lessens the threat of zoonotic diseases (transmissible from animals to humans), and promotes the sustainability of agricultural production systems in general. Where livestock production is a main source of livelihood, healthy animals are essential for community health and progress (Hernandez-Patlan et al. 2023). Traditionally, livestock have used antibiotics not just to treat disease but also as growth promoters and preventive drugs against future diseases (Table 1). This broad use has unintentionally hastened the emergence of antibiotic-resistant bacteria (Jamal 2023). The excessive use of antibiotics in food animals has caused the spread of antibiotic-resistant bacteria (ARB) and antibiotic-resistant genes (ARGs) in many environments, such as farms, manure, and wastewater. These resistant strains are a serious public health threat because they can be passed to humans either by direct contact or by the ingestion of contaminated food items (Zalewska et al. 2021). Antimicrobial resistance (AMR) has mounted into a major worldwide health issue, seriously impairing the efficiency of treatment of different infections (Alara and Alara 2024). The World Health Organization (WHO) has identified antimicrobial resistance (AMR) as one of the top ten global public health threats we face today. In 2019 alone, infections resistant to antibiotics were responsible for an estimated 1.27 million deaths worldwide, with around five million additional deaths linked to drug-resistant infections (Salam et al. 2023).

To tackle this issue, researchers are looking into new ways to cut down on antibiotic use in livestock (Tang et al. 2018). For example, a recent study found that treating common infections in dairy cows with a sugar concentrate solution could be an effective alternative to traditional antibiotics. This approach not only benefits animal health but also plays a crucial role in the fight against AMR (Lection et al. 2024). Efforts to combat AMR include developing new antibiotics, encouraging responsible use of antimicrobials, and implementing infection prevention and control measures (Oliveira et al. 2024). However, the challenges remain significant due to the complex factors that contribute to resistance, highlighting the need for a united global response to effectively address this growing threat.

The main goal of the current research is to explore how green alternatives can help reduce reliance on antibiotics in veterinary medicine. This study aims to review

Table 1 Common antibiotics used in livestock, aquaculture, and poultry. categorized by purpose (growth promotion, disease prevention, treatment)

Sr#	Antibiotic	Class	Primary Purpose	Common applications in animals	Dosage Form	Global Usage Trends	References
1	Bacitracin	Polypeptide	Growth Promotion	Used in poultry and swine for enhanced growth rates, feed efficiency, and egg production	Oral (feed additive), Injectable	Widely used in animal feed	Landers et al. (2012)
2	Tylosin	Macrolide	Disease Prevention	Prevents chronic respiratory diseases and bacterial infections in poultry and swine	Oral, Injectable	Increasing in poultry and swine industries	Casewell et al. (2003)
3	Oxytetracycline	Tetracycline	Treatment	Used in livestock and aquaculture to treat bacterial diseases, including respiratory infections in poultry and fish.	Injectable, Oral (feed)	Common in aquaculture and livestock	Van Boeckel et al. (2015)
4	Penicillin	Beta-lactam	Treatment	Treats bacterial infections such as mastitis in dairy cows.	Injectable	Moderate usage in dairy and livestock	Hardefeldt and Prescott (2024)
5	Chlortetracycline	Tetracycline	Growth Promotion	Used in poultry and swine for growth promotion and disease prevention.	Oral (feed additive)	Declining due to regulations	Marshall and Levy (2011)
6	Enrofloxacin	Fluoroquinolone	Treatment	Treats respiratory and digestive infections in poultry and livestock.	Oral, Injectable	Restricted in some countries	Dowling (2024)
7	Neomycin	Aminoglycoside	Disease Prevention	Used to prevent enteric infections in poultry and swine	Oral (water additive)	Widespread in poultry and pig farming	Laurent (2018)
8	Tiamulin	Pleuromutilin	Disease Prevention	Controls swine dysentery and respiratory diseases in pigs	Oral, Injectable	Increasing due to effectiveness	Sperling and Burch (2025)
9	Florfenicol	Amphenicol	Treatment	Used in aquaculture and livestock for respiratory and systemic infections	Injectable, Oral (feed/water)	Common in aquaculture and cattle	See (2022)
10	Colistin	Polymyxin	Treatment	Treats gastrointestinal infections in pigs and poultry	Oral (feed additive)	Declining due to resistance concerns	Rhouma et al. (2016)

(continued)

Table 1 (continued)

Sr#	Antibiotic	Class	Primary Purpose	Common applications in animals	Dosage Form	Global Usage Trends	References
11	Lincomycin	Lincosamide	Growth Promotion	Enhances growth in poultry and swine; also treats respiratory and intestinal infections	Oral, Injectable	Moderate usage in swine	Giguère (2013)
12	Sulfamethazine	Sulfonamide	Disease Prevention	Controls coccidiosis and other bacterial infections in poultry and livestock	Oral (feed/ water additive)	Reduced use due to resistance issues	Das et al. (2025)
13	Spectinomycin	Aminocyclitol	Treatment	Treats respiratory infections in poultry and swine	Injectable, Oral	Moderate usage in poultry	Dowling (2013)
14	Amoxicillin	Beta-lactam	Treatment	Broad-spectrum antibiotic for livestock respiratory and urinary infections.	Injectable, Oral	Common in livestock and poultry	Hardefeldt and Prescott (2024)
15	Doxycycline	Tetracycline	Disease Prevention	Used for respiratory diseases and secondary infections in livestock and poultry	Oral (feed/ water additive)	Increasing due to broad-spectrum activity	Mathers et al. (2011)
16	Erythromycin	Macrolide	Treatment	Treats respiratory and intestinal infections in poultry and livestock.	Oral, Injectable	Restricted in some countries	Trott et al. (2021)
17	Clindamycin	Lincosamide	Disease Prevention	Controls anaerobic infections in poultry and livestock	Oral, Injectable	Limited use due to resistance concerns	Giguère (2013)
18	Kanamycin	Aminoglycoside	Treatment	Used for bacterial infections in livestock	Injectable	Declining due to toxicity issues	Van Duijkeren et al. (2019)
19	Ceftiofur	Cephalosporin	Treatment	Treats respiratory and systemic infections in livestock	Injectable	Increasing in cattle and swine	Hornish and Katarski (2002)

the use of antibiotics in animal health, understand the mechanisms behind antibiotic resistance, and assess their impact on public health and the environment. Additionally, it will analyze existing regulatory policies and pinpoint the challenges in adopting environmentally friendly veterinary practices. Finally, the research will propose strategies to bridge gaps in adoption and strengthen global policies that support sustainable animal health management.

2 Rationale of Study

The growing dependence on antibiotics in veterinary medicine has sparked a concerning trend in antimicrobial resistance (AMR), which poses serious threats to both animal and human health. The overuse of antibiotics in cattle has led to the emergence of resistant bacteria that can spread through food chains, the environment, and even through close human interactions. This crisis has ignited a worldwide search for greener and more sustainable solutions to replace traditional antibiotic use.

Alternatives such as probiotics, essential oils, and biopesticides could be crucial in preventing and managing diseases without contributing to AMR. However, despite their promise, these alternatives aren't widely adopted due to challenges related to regulations, costs, and a lack of awareness. To promote sustainable veterinary practices and lessen our dependence on antibiotics, we need more information on the effectiveness and practicality of these green alternatives. This study aims to delve into how these options can help combat AMR, their benefits and drawbacks, and their potential for widespread application in contemporary veterinary medicine.

3 Antibiotic Use and Resistance

Antibiotic use in animal agriculture has changed quite a bit over the years, reflecting shifts in farming practices, regulations, and increasing worries about antimicrobial resistance. In this chapter, we'll take a look back at the historical trends in antibiotic use across poultry, livestock, and aquaculture, while also highlighting the most frequently used antibiotics for boosting growth, preventing diseases, and treating infections.

3.1 The Prevalence of Antibiotic Use in Animal Health

Antibiotics first made their way into livestock farming back in the 1940s (Mohr 2016). They were primarily used to treat diseases, boost growth, and improve feed efficiency, which led to their widespread inclusion in animal feed (Cheng et al.

2014). In 1948, Merck introduced sulfa quinoxaline, the first antibiotic approved for daily use in poultry feed to combat coccidiosis (Campbell 2008). The use of antibiotics in aquaculture ramped up significantly due to intensive production practices (Lulijwa et al. 2020). By 2017, the number of antimicrobials used in aquaculture reached 164.8 mg/kg, which was notably higher than the amount used in humans per biomass unit. This surge is largely due to the need for disease management in densely stocked aquatic environments (Schar et al. 2020). Globally, antibiotic use in food-producing animals has been substantial. In 2010, livestock consumed around 63,151 tons of antibiotics, and this figure rose to 93,309 tons by 2017 (Robles-Jimenez et al. 2021).

In animal agriculture, antibiotics are typically used for growth promotion, disease prevention, and treatment. Growth promoters help enhance feed efficiency and animal growth, leading to increased production (Upadhayay and Vishwa 2014). Antibiotics aimed at disease prevention help reduce the risk of bacterial infections in intensive farming systems, ensuring the health of the herd (Davies and Wales 2019). Therapeutic antibiotics are used to treat bacterial infections in animals, poultry, and aquaculture, which helps lower mortality rates and boost productivity. Depending on their purpose, these antibiotics can be administered through feed, water, or injections. The application of antibiotics in animal farming is common, but some areas have much higher rates of consumption owing to intensive production, economic growth, and differences in regulation (Kirchhelle 2018). These hotspots across the world are contributing to the increasing threat of antimicrobial resistance, which is a risk to both animal and human health. Asia is the predominant region of antibiotic usage in animal husbandry, with China, India, and Vietnam being the most dependent (Quaik et al. 2020). The demand for animal protein has increased dramatically in these countries, and livestock farming practices have intensified where antibiotics are widely applied for growth promotion, disease prevention, and treatment (Ronquillo and Hernandez 2017). China is the world's largest consumer of veterinary antibiotics, fueled by its enormous livestock sector (Shao et al. 2021). China's pork industry, in especially high demand for antibiotics, has traditionally used them to keep production high (Yang et al. 2019). China's use of antibiotics in its animal agriculture has been associated with antibiotic residues in meat and antibiotic-resistant bacteria (Shao et al. 2021). Brazil is among the largest meat producers and exporters globally, exporting chicken and beef around the world. Brazil's meat sector has significantly depended on the use of antibiotics to maintain high levels of production and avoid outbreaks of disease among intensive farms (Baraibar Norberg 2020).

3.2 Mechanisms of Antibiotic Resistance

Antimicrobial resistance (AMR) is when microorganisms like bacteria, viruses, fungi, and parasites develop resistance to the action of antimicrobial medicines, making infections more difficult to treat and more likely to spread disease, cause severe illness, and death (Tang et al. 2023).

Antibiotics are now a leading medical treatment required for creating sophisticated medical treatments. However, the significant upsurge of antimicrobial resistance in prevalent bacterial pathogens is presently jeopardizing this therapeutic achievement. The World Health Organization has officially listed antibiotic resistance as one of the three major public health risks of the twenty-first century (Munita and Arias 2016). The development of antibiotic resistance in microbial populations is a multifaceted process driven by genetic mutations and horizontal gene transfer (HGT), allowing bacteria to survive and proliferate despite exposure to antimicrobial agents (Muteeb et al. 2023).

Resistance to antibiotics also emerges due to spontaneous chromosomal mutations (Martinez and Baquero 2000) that enhance resistance by modulating antibiotic targets, altering the uptake of the drug, or augmenting the efflux of drugs. Such mutations in ribosomal protein-encoding genes, DNA gyrase genes, or RNA polymerase genes cause a lowering of antibiotic affinity to the molecule, thus the potency of the antimicrobial compounds (Lambert 2005). Further alterations to the cell membrane permeability or efflux pump upregulation enhance efflux of antibiotics from the bacterial cells, further reducing intracellular concentrations of drugs (Kumar and Schweizer 2005).

Horizontal gene transfer is the main mechanism for the efficient spread of ARGs in bacterial populations (Tao et al. 2022). HGT is achieved through three major mechanisms: conjugation, transduction, and transformation. Conjugation is the direct transfer of genetic components by plasmids, specifically resistance plasmids (R-plasmids), which commonly contain multiple resistance determinants (Helinski 2022). Transduction is mediated by bacteriophages that enable the transfer of ARGs between bacterial cells, while transformation is mediated by the uptake and integration of extracellular DNA from lysed bacteria into the recipient genome (Lu and Guo 2024). Mobile genetic elements such as transposons and integrons enhance the mobility of ARGs further, enabling them to persist and spread in microbial populations (Dai et al. 2022).

Environmental reservoirs are important in the dissemination of antibiotic resistance. Agricultural environments, water bodies, and soil communities are important niches for ARG maintenance and transmission (Larsson and Flach 2022).

3.3 The Impact of Antibiotic Resistance

Antibiotic resistance has serious impacts on animal, human, and global economy health. Antimicrobial resistance lowers the quality of treatment in veterinary medicine, which leads to higher illness and mortality in animals for companions and farm animals (Caneschi et al. 2023). Transmission of resistant pathogens from animals to humans is a significant public health issue. Bacteria such as methicillin-resistant Staphylococcus aureus (MRSA) and extended-spectrum beta-lactamase (ESBL)-producing Escherichia coli are present both in veterinary and human medicine (Kristianingtyas et al. 2021). These bacteria transmit through direct contact with infected animals, consumption of contaminated animal products, and environmental exposure. This zoonotic transmission aggravates the global antibiotic resistance pandemic by reducing human infection treatment alternatives and raising the risk of epidemic outbreaks (Endale et al. 2023).

The economic burden of antibiotic resistance goes beyond healthcare expenses; it also reaches into agricultural output and global trade. Farmers incur financial losses with increased veterinary costs, reduced livestock productivity, and trade barriers to antibiotic-contaminated animal products. Healthcare systems incur the economic burden of longer hospitalization, higher diagnostic costs, and more intensive treatment of resistant infections (Gomes 2024).

3.4 Regulatory Responses to Curb Antibiotic Use

In an effort to address the increasing threat of antibiotic resistance, global and regional regulatory frameworks have been established to control the use of antibiotics in medicine for humans and animals. Global organizations such as the World Health Organization (WHO), the World Organization for Animal Health (OIE), and the Food and Agriculture Organization (FAO) have issued guidelines to facilitate the responsible use of antibiotics. The WHO Global Action Plan for Antimicrobial Resistance emphasizes surveillance, antimicrobial stewardship, and reducing the use of non-essential antibiotics in agriculture. The OIE provides guidance for prudent antibiotic use in animal medicine to prevent the spread of resistance in food-producing animals, and the FAO promotes monitoring and control of residues of antibiotics in food products (European Public Health Alliance 2022. Regional activities complement regulatory efforts. In the EU, strict policy prohibits the usage of antibiotics as growth promoters for animal farming and has intensive monitoring programs that observe antimicrobial patterns of resistance in food products and animals. In the United States, such similar restrictions have also been imposed via the U.S. Food and Drug Administration (FDA), wherein veterinary control must be implemented in the usage of antibiotics in food animals under the Veterinary Feed Directive. In developing nations, attempts at regulation are being thwarted by poor enforcement capacity, inadequate infrastructure, and financial constraints. There are, however, countries making progress in the execution of national action plans according to.

international AMR strategies.

Despite such regulatory measures, there are challenges in enforcing antibiotic restrictions. Inadequate compliance, unauthorized sale of antibiotics, and inadequate awareness among stakeholders hinder effective implementation. Additionally, varying regulations across countries offer loopholes for continued misuse of antibiotics in some industries. Enhancing enforcement mechanisms, enhancing public awareness, and increasing international collaboration are required to remove these challenges and attain sustainable antibiotic stewardship practice worldwide (Bhandari et al. 2024). To address the increasing threat of antibiotic resistance, global and regional regulatory frameworks were established to control the application of antibiotics in human and veterinary medicine. Global agencies such as the World Health Organization (WHO), the World Organization for Animal Health (OIE), and the Food and Agriculture Organization (FAO) have issued guidelines to warrant responsible use of antibiotics. The WHO Global Action Plan on Antimicrobial Resistance first focuses on surveillance, antimicrobial stewardship, and reducing the use of non-essential antibiotics in agriculture. The OIE provides guidelines for antibiotic usage in veterinary medicine to prevent resistance transmission in food-producing animals, with the FAO promoting surveillance and control of antibiotic residues in food products (European Public Health Alliance (2022).

Regional activities supplement regulatory measures. Within the European Union (EU), the policies are very strict in prohibiting the use of antibiotics as growth promoters in farm animals, and there are intensive monitoring programs to track antimicrobial resistance patterns in food products and animals. In the United States, such measures have also been implemented by the U.S. Food and Drug Administration (FDA) through a requirement of veterinary control for the use of antibiotics in food animals under the Veterinary Feed Directive. In developing nations, attempts at regulation are thwarted by limited enforcement capacity, inadequate infrastructure, and economic constraints.

But there are countries making progress in rolling out national action plans consistent with international AMR action plans.

Despite such regulatory measures, there are challenges in enforcing antibiotic limitations. Inadequate compliance, unlawful selling of antibiotics, and inadequate awareness among stakeholders hinder effective implementation. Further, varying regulations among countries leave loopholes for continued misuse of antibiotics in certain industries. Enhancing enforcement mechanisms, enhancing public awareness, and increasing global cooperation are needed to nullify these challenges and attain sustainable antibiotic stewardship practice worldwide (Bhandari et al. 2024).

4 Challenges in Animal Production Systems

4.1 *Intensification of Animal Farming*

Intensification is the intensification of the use of external inputs and services to increase the quantity and/or value of output per unit input (Bebe 2003). Intensification of livestock production is greatly encouraged in an effort to meet the increasing demands for livestock products and to enhance the livelihoods of rural households (Udo et al. 2011). The growth of livestock production has led to the spread of industrial farms, where large numbers of animals are housed. The animals are then placed in small enclosures next to each other to increase the productivity efficiency. The system is susceptible to resulting in overpopulation, which causes a lot of stress to be inflicted on the animals and leaves them vulnerable to disease. To counteract the increased risk of disease presented by such high-density settings, there is further dependence on antibiotics. Significantly, the World Health Organization has estimated that in certain nations, the animal industry consumes approximately 80% of medically significant antibiotics, mainly for promoting growth in healthy animals. Among the causes of antibiotic-resistant bacteria, which present a far-reaching sinister threat to global public health, is this widespread usage of antibiotics. Consequently, the intimate and unsanitary environments of intensive farm systems are perfect breeding grounds for disease-promoting bacteria, which further enhance health hazards (Kumar and Patyal 2020).

Therefore, animal farming intensification poses serious challenges to animal welfare and human health and should be carefully analyzed and reviewed for current practices.

4.2 *Economic Pressures on Farmers*

Producers are heavily limited financially between maintaining reasonable costs of animal health and elevating productivity. Animal husbandry intensification is inclined to demand greater investment in feeds, veterinary services, and housing for the maintenance of high production levels (Ma et al. 2020). Economic constraints also vary between industrial-scale farms and small farms.

Small-scale producers generally have few resources and high per-unit prices, and it becomes harder for them to invest in other measures of health (Ewbank 2019). Large industrial farms benefit from economies of scale, making resource allocation for animal health more efficient (De Roest et al. 2018). The adoption of the alternatives to antibiotics, such as vaccines, immune modulators, and probiotics, holds both opportunities and challenges. Such alternatives can enhance animal health alongside the reduction in antibiotic reliance, but are adopted contingent on their price efficiency and strength (Callaway et al. 2021).

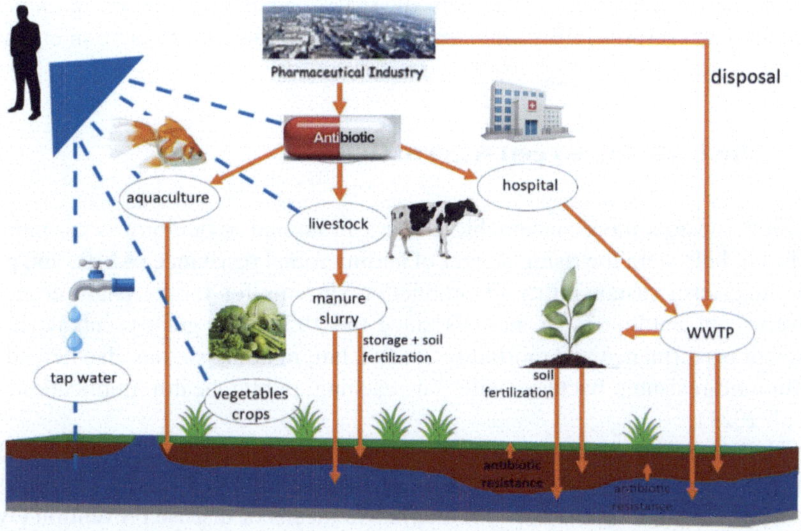

549

Fig. 1 Illustration of antibiotic contamination pathways in the environment

4.3 Environmental Impacts of Current Practices

The shared use of antibiotics in agriculture has resulted in extensive environmental contamination, particularly by runoff from agriculture (Manyi-Loh et al. 2018). Excreted drugs are not fully metabolized by animals and hence end up in manure. When used as fertilizer or stored in anaerobic lagoons, antibiotics and residues have the tendency to leach into the neighboring soil and water systems (Fig. 1). This activity introduces such chemicals into the environment, and it has environmental implications (Zhang et al. 2019).

5 Knowledge Gaps and Awareness

Effective antibiotic stewardship relies on training and making farmers aware of the use of antibiotics. A lack of knowledge among farm owners and workers regarding correct antibiotic practices and resistance consequences has been reported by studies. Additionally, farmers themselves feel that growing resistance to antibiotics can potentially ruin the overall reputation of farming, possibly leading to accusations regarding resistance in the human sector. These results highlight the importance of open communication and education in reconciling public expectations and sustainable farming practices. The fight against AMR demands a multidisciplinary approach that supplements knowledge from other disciplines, other than the conventional biological and medical sciences. According to scholars, social sciences are needed to offset the complexity of AMR, demanding a poor common understanding and responsibility among stakeholders to impede effective intervention.

6 The Case for Green Alternatives

A shift towards more sustainable choices in animal agriculture is essential in an effort to help stem the rising danger of antimicrobial resistance (AMR) and promote the long-term sustainability of production systems for food (Aslam et al. 2024). Overuse of antibiotics in animals since the 1950s has greatly enhanced animal growth performance and morbidity control, but, in doing so, has also helped spread antibiotic-resistant bacteria with far-reaching public health implications (Shao et al. 2021).

Understanding these risks, the World Health Organization has called for the cessation of antibiotics in healthy animals in hopes of decreasing the transmission of resistance, assigning the need for alternative means of disease prevention (WHO).

One of these green options is probiotics. Probiotics are defined by the World Health Organization (WHO) and the Food and Agriculture Organization (FAO) as "live microorganisms which when administered in adequate amounts confer a health benefit on the host" (Wang et al. 2019). By improving the internal microbial balance, probiotics—live beneficial bacteria—are ingested through food or water and help to promote good health. Bacteriocins, siderophores, lysozymes, proteases, and hydrogen peroxides are produced by probiotic bacteria, which prevent the growth of dangerous pathogens (El-Saadony et al. 2021). Lactobacillus spp. was used in various fish species (e.g., Salmon, Tilapia, Trout) in the 1990s to prevent gastrointestinal diseases, improve gut health, and enhance resistance to bacterial infections (e.g., Vibrio spp., Aeromonas) (Ntakirutimana et al. 2023).

7 Limitations of the Study

This study gives valuable information regarding green alternatives to the management of antibiotic resistance in veterinary medicine. There are some limitations. One of them is the lack of long-term data and large-scale studies of green alternatives because it is a fairly new field of study. Economic constraints can also slow down the process, with some farmers and veterinary clinics perhaps considering it too costly to embrace green alternatives. Furthermore, the process of regulatory approval for new green solutions can be lengthy and cumbersome, delaying their market availability. Finally, inadequate knowledge among farmers and veterinarians regarding the benefits and proper use of green alternatives would hinder their complete acceptance. Such limitations underscore the need for extensive research and sound strategies to encourage the practical application of green solutions in veterinary practice.

8 Recommendations

In order to fill the gaps revealed by this study, a number of suggestions can be made. Firstly, subsequent research needs to expand its scope sufficiently to encompass a greater number of green options, including bacteriophages, antimicrobial peptides, and genetically modified probiotics, which have emerged with promise in preventing antibiotic dependence. In addition to that, long-term field trials have to be established in order to establish the long-term effects of such alternatives on animal health, productivity, and resistance. Farmers also need to be educated and sensitized in order to adequately inform livestock farmers on proper antibiotic use and the advantages of alternative methods. Finally, promoting international cooperation among researchers, policymakers, and industry stakeholders can serve to establish homogenized guidelines and best practices for the incorporation of green alternatives in veterinary medicine globally.

9 Conclusion

This study is on the necessity to transition to natural alternatives in veterinary practice to address the growing issue of antibiotic resistance. The improper use of antibiotics in agriculture has resulted in antimicrobial resistance, which is a threat to humans and animals alike. Natural alternatives such as probiotics, essential oils, and biopesticides provide an improved method of supporting animal health with reduced use of antibiotics. But their mass implementation is plagued by a chain of problems such as high expenses, stringent legislation, and poor awareness. To reverse these problems, more research should be conducted, policies should be enforced more firmly, subsidies should be offered to farmers, and individuals should be made more aware of such options. Through the application of sustainable animal health practice, agriculture can be a key force in the fight against antibiotic resistance and the healthier veterinary medicine future.

References

Alara JA, Alara OR (2024) An overview of the global alarming increase of multiple drug resistance: a major challenge in clinical diagnosis. Infect Disord Drug Targets 24(3):26–42

Aslam B, Asghar R, Muzammil S, Shafique M, Siddique AB, Khurshid M et al (2024) AMR and Sustainable Development Goals: at a crossroads. Glob Health 20(1):73

Baraibar Norberg M (2020) Agrofood globalization: the global soybean and beef commodity chains. In: The political economy of agrarian change in Latin America: Argentina, Paraguay and Uruguay. Governance, Development, and Social Inclusion in Latin America. Palgrave Macmillan, Cham, pp 117–163. https://doi.org/10.1007/978-3-030-24586-3_3

Bebe BO (2003) Herd dynamics of smallholder dairy in the Kenya highlands. Wageningen University and Research

Bhandari RK, Pandey AK, Malhotra S, Kakkar AK, Singh S, Cohn J, Shafiq N (2024) Addressing challenges in antibiotic access: barriers, implications and strategies for solution. Pharm Med:1–11

Callaway TR, Lillehoj H, Chuanchuen R, Gay CG (2021) Alternatives to antibiotics: a symposium on the challenges and solutions for animal health and production. Antibiotics 10(5):471

Campbell WC (2008) History of the discovery of sulfaquinoxaline as a coccidiostat. J Parasitol 94(4):934–945

Caneschi A, Bardhi A, Barbarossa A, Zaghini A (2023) The use of antibiotics and antimicrobial resistance in veterinary medicine, a complex phenomenon: a narrative review. Antibiotics 12(3):487

Casewell M, Friis C, Marco E, McMullin P, Phillips I (2003) The European ban on growth-promoting antibiotics and emerging consequences for human and animal health. J Antimicrob Chemother 52(2):159–161. https://doi.org/10.1093/jac/dkg313

Cheng G, Hao H, Xie S, Wang X, Dai M, Huang L, Yuan Z (2014) Antibiotic alternatives: the substitution of antibiotics in animal husbandry? Front Microbiol 5:217

Dai D, Brown C, Bürgmann H, Larsson DJ, Nambi I, Zhang T et al (2022) Long-read metagenomic sequencing reveals shifts in associations of antibiotic resistance genes with mobile genetic elements from sewage to activated sludge. Microbiome 10(1):20

Das B, Sivajothi S, Parsani HR, Suthar AN (2025) Coccidiosis. In: Rana T (ed) Elements of reproduction and reproductive diseases of goats. John Wiley & Sons, Hoboken, NJ, pp 501–516. https://doi.org/10.1002/9781394190089.ch45

Davies R, Wales A (2019) Antimicrobial resistance on farms: a review including biosecurity and the potential role of disinfectants in resistance selection. Compr Rev Food Sci Food Saf 18(3):753–774

De Roest K, Ferrari P, Knickel K (2018) Specialisation and economies of scale or diversification and economies of scope? Assessing different agricultural development pathways. J Rural Stud 59:222–231

Dowling PM (2013) Aminoglycosides and aminocyclitols. In: Giguère S, Prescott JF, Dowling PM (eds) Antimicrobial therapy in veterinary medicine, 5th edn. Wiley-Blackwell, Ames, IA, pp 233–255

Dowling PM (2024) Fluoroquinolones. In: Dowling PM, Prescott JF, Baptiste KE (eds) Antimicrobial therapy in veterinary medicine, 6th edn (chapter 17). Wiley-Blackwell, Hoboken, NJ, pp 325–343

El-Saadony MT, Alagawany M, Patra AK et al (2021) The functionality of probiotics in aquaculture: an overview. Fish Shellfish Immunol 117:36–52

Endale H, Mathewos M, Abdeta D (2023) Potential causes of spread of antimicrobial resistance and preventive measures in one health perspective: a review. Infect Drug Resist 16:7515–7545

European Public Health Alliance (2022) Ending routine farm antibiotic use in Europe

Ewbank R (2019) Conflicts and challenges to enhancing the resilience of small-scale farmers in developing economies. In: Gardner SM, Ramsden SJ, Hails RS (eds) Agricultural resilience: perspectives from ecology and economics. Ecological Reviews. Cambridge University Press, Cambridge, Chapter 9, pp 135–158

Giguère S (2013) Lincosamides, pleuromutilins, and streptogramins. In: Giguère S, Prescott JF, Dowling PM (eds) Antimicrobial therapy in veterinary medicine, 5th edn. Wiley-Blackwell, Ames, IA, pp 199–210

Gomes MP (2024) The convergence of antibiotic contamination, resistance, and climate dynamics in freshwater ecosystems. Water 16(18):2606

Hardefeldt LY, Prescott JF (2024) Beta-lactam antibiotics: penam penicillins. In: Dowling PM, Prescott JF, Baptiste KE (eds) Antimicrobial therapy in veterinary medicine, 6th edn. Wiley-Blackwell, Hoboken, NJ, pp 119–142

Helinski DR (2022) A brief history of plasmids. EcoSal Plus 10(1):eESP-0028

Hernandez-Patlan D, Tellez-Isaias G, Hernandez-Velasco X, Solis-Cruz B (2023) Editorial: technological strategies to improve animal health and production. Front Vet Sci 10:1206170. https://doi.org/10.3389/fvets.2023.1206170

Hornish RE, Katarski SF (2002) Cephalosporins in veterinary medicine-ceftiofur use in food animals. Curr Top Med Chem 2(7):717–731

Jamal A (2023) Antibiotics in contemporary medicine: advances, obstacles, and the future. BULLET: J Multidiscip Ilmu 2:548–557

Munita JM, Arias CA (2016) Mechanisms of Antibiotic Resistance ABSTRACT Microbiology Spectrum 4(2). https://doi.org/10.1128/microbiolspec.VMBF-0016-2015

Kappes A, Tozooneyi T, Shakil G, Railey AF, McIntyre KM, Mayberry DE, Rushton J, Pendell DL, Marsh TL (2023) Livestock health and disease economics: a scoping review of selected literature. Front Vet Sci 10:1168649. https://doi.org/10.3389/fvets.2023.1168649

Tang KL, Caffrey NP, Nóbrega DB, Cork SC, Ronksley PE, Barkema HW, ... & Ghali WA (2018) Restricting the use of antibiotics in food-producing animals and its associations with antibiotic resistance in food-producing animals and human beings: a systematic review and meta-analysis The Lancet Planetary Health 1(8):e316–e327 https://doi.org/10.1016/S2542-5196(17)30141-9

Kirchhelle C (2018) Pharming animals: a global history of antibiotics in food production (1935–2017). Palgrave Commun 4(1)

Kristianingtyas L, Effendi MH, Witaningrum AM, Wardhana DK, Ugbo EN (2021) Prevalence of extended-spectrum β-lactamase-producing Escherichia coli in companion dogs in animal clinics, Surabaya, Indonesia. Int J One Health 7(2):232–236

Kumar A, Patyal A (2020) Impacts of intensive poultry farming on 'One Health' in developing countries: challenges and remedies. Explor Anim Med Res 10(2)

Kumar A, Schweizer HP (2005) Bacterial resistance to antibiotics: active efflux and reduced uptake. Adv Drug Deliv Rev 57(10):1486–1513

Lambert PA (2005) Bacterial resistance to antibiotics: modified target sites. Adv Drug Deliv Rev 57(10):1471–1485

Landers TF, Cohen B, Wittum TE, Larson EL (2012) A review of antibiotic use in food animals: perspective, policy, and potential. Public Health Rep 127(1):4–22. https://doi.org/10.1177/003335491212700103

Larsson DJ, Flach CF (2022) Antibiotic resistance in the environment. Nat Rev Microbiol 20(5):257–269

Laurent JW (2018) Alternatives to common preventive uses of antibiotics for cattle, swine, and chickens. NRDC Report

Lection J, Van Syoc E, Miles A, Hamilton J, Martinez M, Bas S, Silverman J, Barragan A, Ganda E (2024) Use of intrauterine dextrose as an alternative to systemic antibiotics for treatment of clinical metritis in dairy cattle: a microbiome perspective. Front Vet Sci 11:1478288. https://doi.org/10.3389/fvets.2024.1478288

Lu J, Guo J (2024) Prophage induction by non-antibiotic compounds promotes transformation of released antibiotic resistance genes from cell lysis. Water Res 263:122200

Lulijwa R, Rupia EJ, Alfaro AC (2020) Antibiotic use in aquaculture, policies and regulation, health and environmental risks: a review of the top 15 major producers. Rev Aquac 12(2):640–663

Ma W, Bicknell K, Renwick A (2020) Production intensification and animal health expenditure on dairy farms in New Zealand. J Dairy Sci 103(2):1598–1607

Zalewska M, Błażejewska A, Czapko A, & Popowska M (2021) Antibiotics and Antibiotic Resistance Genes in Animal Manure – Consequences of Its Application in Agric Front https://doi.org/1210.3389/fmicb.2021.610656

Manyi-Loh C, Mamphweli S, Meyer E, Okoh A (2018) Antibiotic use in agriculture and its consequential resistance in environmental sources: potential public health implications. Molecules 23(4):795

Marshall BM, Levy SB (2011) Food animals and antimicrobials: impacts on human health. Clin Microbiol Rev 24(4):718–733

Martinez JL, Baquero F (2000) Mutation frequencies and antibiotic resistance. Antimicrob Agents Chemother 44(7):1771–1777

Mathers JJ, Flick SC, Cox LA Jr (2011) Longer-duration uses of tetracyclines and penicillins in US food-producing animals: indications and microbiologic effects. Environ Int 37(5):991–1004

Modisane BM (2009) Field services: eradication and control of animal diseases—Animal health management in the 21st century. Onderstepoort J Vet Res 76(1):115–121. https://doi.org/10.4102/ojvr.v76i1.74

Mohr KI (2016) History of antibiotics research. In: Stadler M, Dersch P (eds) How to overcome the antibiotic crisis: facts, challenges, technologies and future perspectives. Current Topics in Microbiology and Immunology, vol 398. Springer, Cham, pp 237–272. https://doi.org/10.1007/82_2016_499

Muteeb G, Rehman MT, Shahwan M, Aatif M (2023) Origin of antibiotics and antibiotic resistance, and their impacts on drug development: a narrative review. Pharmaceuticals 16(11):1615

Ntakirutimana R, Syanya FJ, Mwangi P (2023) Exploring the impact of probiotics on the gut ecosystem and morpho-histology in fish: current knowledge of tilapia. Asian J Fish Aquat Res 25(3):93–112

Ogwu MC, Izah SC, Ntuli NR, Odubo TC (2024) Food security complexities in the Global South. In: Ogwu MC, Izah SC, Ntuli NR (eds) Food safety and quality in the Global South. Springer Nature, pp 3–33. https://doi.org/10.1007/978-981-97-2428-4_1

Oliveira M, Antunes W, Mota S, Madureira-Carvalho Á, Dinis-Oliveira RJ, da Silva DD (2024) An overview of the recent advances in antimicrobial resistance. Microorganisms 12(9):1920

Quaik S, Embrandiri A, Ravindran B, Hossain K, Al-Dhabi NA, Arasu MV, Ismail N (2020) Veterinary antibiotics in animal manure and manure-laden soil: scenario and challenges in Asian countries. J King Saud Univ-Sci 32(2):1300–1305

Rhouma M, Beaudry F, Letellier A (2016) Resistance to colistin: what is the fate for this antibiotic in pig production? Int J Antimicrob Agents 48(2):119–126

Robles-Jimenez LE, Aranda-Aguirre E, Castelan-Ortega OA, Shettino-Bermudez BS, Ortiz-Salinas R, Miranda M, Gonzalez-Ronquillo M (2021) Worldwide traceability of antibiotic residues from livestock in wastewater and soil: a systematic review. Animals 12(1):60

Ronquillo MG, Hernandez JCA (2017) Antibiotic and synthetic growth promoters in animal diets: review of impact and analytical methods. Food Control 72:255–267

Salam MA, Al-Amin MY, Salam MT, Pawar JS, Akhter N, Rabaan AA, Alqumber MA (2023) Antimicrobial resistance: a growing serious threat for global public health. Healthcare 11(13):1946. Multidisciplinary Digital Publishing Institute

Schar D, Klein EY, Laxminarayan R, Gilbert M, Van Boeckel TP (2020) Global trends in antimicrobial use in aquaculture. Sci Rep 10(1):21878

See A (2022) Investigating bacterial community composition and antimicrobial resistance genes in beef cattle. [Institution or Publisher]

Shao Y, Wang Y, Yuan Y, Xie Y (2021) A systematic review on antibiotics misuse in livestock and aquaculture and regulation implications in China. Sci Total Environ 798:149205

Sperling D, Burch DG (2025) Swine. In: Antimicrobials in livestock 2: choices of prevention and treatment: a European perspective. Cham, Springer Nature Switzerland, pp 79–134

Tang KWK, Millar BC, Moore JE (2023) Antimicrobial resistance (AMR). Br J Biomed Sci 80:11387

Tao S, Chen H, Li N, Wang T, Liang W (2022) The spread of antibiotic resistance genes in vivo model. Can J Infect Dis Med Microbiol 2022(1):3348695

Trott DJ, Turnidge J, Kovac JH, Simjee S, Wilson D, Watts J (2021) Comparative macrolide use in humans and animals: should macrolides be moved off the World Health Organisation's critically important antimicrobial list? J Antimicrob Chemother 76(8):1955–1961

Udo HMJ, Aklilu HA, Phong LT, Bosma RH, Budisatria IGS, Patil BR et al (2011) Impact of intensification of different types of livestock production in smallholder crop-livestock systems. Livest Sci 139(1–2):22–29

Upadhayay UPPDD, Vishwa PCV (2014) Growth promoters and novel feed additives improving poultry production and health, bioactive principles and beneficial applications: the trends and advances—A review. Int J Pharmacol 10(3):129–159

Van Boeckel TP, Brower C, Gilbert M, Grenfell BT, Levin SA, Robinson TP, Teillant A, Laxminarayan R (2015) Global trends in antimicrobial use in food animals. Proc Natl Acad Sci USA 112(18):5649–5654. https://doi.org/10.1073/pnas.1503141112

Van Duijkeren E, Schwarz C, Bouchard D, Catry B, Pomba C, Baptiste KE, Jukes H, Others. (2019) The use of aminoglycosides in animals within the EU: development of resistance in animals and possible impact on human and animal health: a review. J Antimicrob Chemother 74(9):2480–2496

Wang A, Ran C, Wang Y et al (2019) Use of probiotics in aquaculture of China—a review of the past decade. Fish Shellfish Immunol 86:734–755

Yang H, Paruch L, Chen X, Van Eerde A, Skomedal H, Wang Y, Liu Clarke J (2019) Antibiotic application and resistance in swine production in China: current situation and future perspectives. Front Vet Sci 6:136

Zhang M, He LY, Liu YS, Zhao JL, Liu WR, Zhang JN et al (2019) Fate of veterinary antibiotics during animal manure composting. Sci Total Environ 650:1363–1370

The manufacturer's authorised representative in the EU is Springer Nature Customer Service Centre GmbH, Europaplatz 3, 69115 Heidelberg, Germany. If you have any concerns regarding our products, please contact ProductSafety@springernature.com

Printed and bound by CPI Group (UK) Ltd, Croydon, CR0 4YY
21/09/2026
02219991-0018